Life Underground

Life Underground

THE BIOLOGY OF SUBTERRANEAN RODENTS

Eileen A. Lacey, James L. Patton, and Guy N. Cameron, editors

THE UNIVERSITY OF CHICAGO PRESS
CHICAGO AND LONDON

EILEEN A. LACEY is assistant professor of integrative biology at the University of California, Berkeley, and assistant curator of mammals at the Museum of Vertebrate Zoology in Berkeley. **JAMES L. PATTON** is professor of integrative biology at the University of California, Berkeley, and is acting director and curator at the Museum of Vertebrate Zoology in Berkeley. **GUY N. CAMERON** is professor of biological science at the University of Cincinnati.

The University of Chicago Press, Chicago 60637
The University of Chicago Press, Ltd., London
© 2000 by The University of Chicago
All rights reserved. Published 2000
Printed in the United States of America
09 08 07 06 05 04 03 02 01 00   5 4 3 2 1

ISBN (cloth): 0-226-46727-9
ISBN (paper): 0-226-46728-7

Library of Congress Cataloging-in-Publication Data

Life underground : the biology of subterranean rodents / Eileen A. Lacey, James L. Patton, and Guy N. Cameron, editors.
    p.   cm.
  Includes bibliographical references and indexes.
  ISBN 0-226-46727-9—ISBN 0-226-46728-7 (pbk.)
    1. Rodents. 2. Burrowing animals. I. Lacey, Eileen A. II. Patton, James L.
III. Cameron, Guy N.
QL737.R6 L48   2000
599.35′15648—dc21

                                                            99-051019

♾ The paper used in this publication meets the minimum requirements of the American National Standard for Information Sciences—Permanence of Paper for Printed Library Materials, ANSI Z39.48-1992.

For John, Carol, and our many subterranean friends . . .

Contents

Preface ix

Introduction *E. A. Lacey, J. L. Patton, and G. N. Cameron* 1

PART ONE: ORGANISMAL BIOLOGY 15

1. Morphology of Subterranean Rodents *Barbara R. Stein* 19

2. Ecophysiological Responses of Subterranean Rodents to Underground Habitats *Rochelle Buffenstein* 62

3. Sensory Capabilities and Communication in Subterranean Rodents *Gabriel Francescoli* 111

4. Reproduction in Subterranean Rodents *Nigel C. Bennett, Chris G. Faulkes, and Andrew J. Molteno* 145

PART TWO: POPULATION AND COMMUNITY ECOLOGY 179

5. Population Ecology of Subterranean Rodents
Cristina Busch, C. Daniel Antinuchi, J. Cristina del Valle, Marcelo J. Kittlein, Ana I. Malizia, Aldo I. Vassallo, and Roxana R. Zenuto 183

6. Community Ecology of Subterranean Rodents *Guy N. Cameron* 227

7. Spatial and Social Systems of Subterranean Rodents *Eileen A. Lacey* 257

PART THREE: EVOLUTIONARY BIOLOGY 297

8. Genetic Structure and the Geography of Speciation in Subterranean Rodents: Opportunities and Constraints for Evolutionary Diversification   *Eleanor K. Steinberg and James L. Patton*   301

9. Paleontology, Phylogenetic Patterns, and Macroevolutionary Processes in Subterranean Rodents   *Joseph A. Cook, Enrique P. Lessa, and Elizabeth A. Hadly*   332

10. Coevolution and Subterranean Rodents   *Mark S. Hafner, James W. Demastes, and Theresa A. Spradling*   370

11. The Evolution of Subterranean Rodents: A Synthesis   *Enrique P. Lessa*   389

List of Contributors   421
Taxonomic Index   425
Subject Index   439

Preface

The idea for a volume on subterranean rodents originated with Guy and Jim. Both have long-standing interests in these animals, as evidenced by their research on pocket gophers. Nearly a decade ago, Guy and Jim began discussing the need for a resource that summarized our knowledge of the rodents that exploit subterranean habitats. Their belief that such a publication was both timely and important was not dampened by initial reactions from editors that, effectively, "a book on gophers would not be very interesting." The project continued to gestate until 1996, when active work on the volume began as part of plans to convene a symposium on subterranean rodents at the Seventh International Theriological Congress, to be held in Acapulco, Mexico, in September of 1997.

During the period between the initial discussions by Guy and Jim and the 1997 symposium, several things transpired that helped to bring the project into focus. First, a symposium on subterranean mammals held at the Fifth International Theriological Congress in 1989 resulted in the publication of an influential volume edited by Eviatar Nevo and Osvaldo Reig (1990) titled *Evolution of Subterranean Mammals at the Organismal and Molecular Levels*. Shortly thereafter, *The Biology of the Naked Mole-rat* (1991) edited by Paul Sherman, Jennifer Jarvis, and Richard Alexander, was published. Both works drew attention to subterranean rodents and helped to stimulate interest in these animals as subjects for research in a variety of biological disciplines.

Second, recent technological advances have greatly improved our ability to study the biology of secretive subterranean animals. For example, the growing accessibility of radiotelemetry and other forms of remote monitoring allows us to collect data on free-living animals that otherwise could not be obtained. The molecular genetic revolution has had an even greater impact on studies of subterranean taxa. The ever-growing list of molecular techniques available to biologists has enabled students of subterranean rodents to gain new insights into a variety of topics, including behavior, physiology, population dynamics, and evolutionary relationships. Although molecular genetic research was a primary focus of the volume edited by Nevo and Reig, the pace at which this technology has advanced has resulted in considerable new research over the past decade.

Finally, there has been an explosion of research on subterranean rodents, fueled by both the new technology and the publication of the 1990 volume. This burst of enthusiasm is clearly evident in the number of references in this volume that date from 1990 or later. In contrast to early studies, which focused almost exclusively on geomyid pocket gophers from North America, current research encompasses members of five rodent families that occur on five different continents. Studies of these animals address almost every aspect of organismal and evolutionary biology, with research topics ranging from the mechanisms of vitamin D metabolism to rates and patterns of evolutionary diversification in different subterranean lineages. Areas of research that have undergone particularly rapid expansion in recent years include reproductive endocrinology, social behavior, and population genetic structure. The ever-growing list of questions being addressed with data from subterranean rodents is testament to their increasing importance as subjects for biological research.

With this wave of interest growing, Guy and Jim enlisted the aid of Eileen to see the project to fruition. As the ITC symposium and resulting volume began to take shape, several themes began to emerge. Foremost among these was the role of the subterranean niche in shaping the biology of the taxa under consideration. In particular, how have the properties of this niche both facilitated and constrained the evolution of the rodents that inhabit it? What challenges do subterranean species face that are not shared by their surface-dwelling counterparts? The second theme that runs throughout the volume concerns the nature and extent of convergence among subterranean species. Over the years, a number of generalizations have arisen regarding phenotypic traits shared by these rodents. How valid are these generalizations? Do subterranean rodents display extensive convergence in response to pressures imposed by the subterranean niche, or do gross similarities in biology mask considerable variability in response to environmental challenges?

In addressing these questions, we agreed that information on subterranean rodents would be most effectively disseminated if data were organized by topic, rather than by taxon. In particular, we believed that this arrangement would encourage the cross-taxon comparisons needed to address issues of convergence and divergence. At the same time, by presenting material topically, we hoped to reach a broader audience of biologists. Specifically, we were interested in making data from subterranean rodents accessible to researchers studying similar questions in other groups of organisms.

Having identified a series of subjects to be covered, we were faced with the difficult task of selecting contributors to the volume. Our choices were dictated by several factors, beginning with research interests and expertise. In addition, we were eager to recruit individuals whose work, collectively, would encompass a variety of geographic regions and rodent taxa. Finally, when possible, we tried to select younger researchers who could offer fresh perspectives on the issues raised in the volume.

As with any endeavor of this sort, we have only imperfectly achieved the goals established at the beginning of the project. No volume that purports to summarize "the biology" of anything can truly be complete, and this one is no exception. In many cases, our ability to review a topic comprehensively is limited by the quantity of data available or the number of species that have been studied. For the same reason, our ability to address the two underlying themes of the book varies among chapters; while the literature on some topics emphasizes variation within subterranean taxa, research on other topics focuses more on comparisons between subterranean and surface-dwelling species. Realizing that such inequities were likely to occur, we have encouraged contributors to be forthright in identifying data limitations in the hope of stimulating the research needed to fill these gaps. If nothing else, the preparation of this volume has dramatically underscored our belief that lifetimes of research remain to be conducted on subterranean rodents.

Finally, we would like to acknowledge the contributions of many others to this project. In particular, we would like to draw attention to the work of the many researchers cited in this volume who did not participate in the ITC symposium. Clearly, we owe a great debt to these individuals, whose research adds immeasurably to our understanding of subterranean species. In addition, we would like to thank Christie Henry of the University of Chicago Press for her patience and encouragement throughout the preparation of this volume. Christie was adept at gently prodding us to move forward with the project when angst threatened to subsume productivity. It has been a challenging but rewarding experience that has led each of us to a greater understanding of our own study animals.

Introduction

E. A. Lacey, J. L. Patton, and G. N. Cameron

This volume examines the biology of subterranean rodents, a fascinating group of animals that exhibits numerous adaptations for life underground. As discussed in detail below, both the number of rodent lineages that have independently adopted a subterranean way of life and the wide geographic distribution of these animals make this collection of taxa ideal for studies of evolutionary convergence and divergence. One of the primary goals of this volume is to compare and contrast how these different taxa have responded to the evolutionary challenges associated with life underground.

Because the group of mammals considered in this volume is defined on the basis of the habitat they use, rather than their taxonomic or phylogenetic affiliations, a short discussion of the characters that make a species "subterranean" seems warranted. In brief, we define subterranean rodents as those species that live in underground burrows and that conduct the vast majority of their life activities below the soil surface. The transition from surface-dwelling to subterranean most likely represents a continuum of specializations for life underground; we have elected to focus on species at one end of this continuum, namely, those exhibiting the greatest restriction to and greatest specialization for life in subterranean habitats. Although this distinction is somewhat arbitrary, we believe that the almost complete restriction of activities to underground burrows creates selective pressures that differ from those experienced by animals that are routinely active on the soil surface. Thus we attempt to distinguish between taxa that rarely, if ever, leave their burrows and those that may use subterranean burrows but that spend considerable portions of their lives above ground.

In our effort to define subterranean rodents, a terminological issue arises that requires clarification. Traditionally, the terms "fossorial" and "subterranean" have been used interchangeably to describe the taxa considered in this volume. The term *fossorial* refers to animals that are specialized for digging. Although this definition clearly includes the rodent lineages listed below, it also includes a number of taxa, such as ground squirrels *(Spermophilus)*, mountain beaver *(Aplodontia)*, and some genera of voles *(Hyperacrius)*, that use subterranean burrows or runways and that are modified for digging in one way or another, but whose activities are not restricted to beneath the soil surface. For this reason, we prefer the term *subterranean* to describe the species considered in this volume.

Which Rodents Are Subterranean?

Because distinctions between surface-dwelling, semi-subterranean, and truly subterranean rodents are continuous and somewhat arbitrary, we have compiled a brief catalog of the taxa that are the focus of this volume. Included are data on geographic distribution, ecology, and taxonomic relationships, as well as comments regarding our general understanding of the biology of each group. This information is intended to introduce readers to these animals, as well as to delineate the taxa covered in the following, topical discussions of subterranean rodents.

Subterranean rodents occur on every major continental land mass except Australia. Most, but not all, of these taxa inhabit open areas such as savannas, grasslands, steppes, or alpine meadows, although a few species occur in densely vegetated shrub or forests. Virtually all are denizens of moderately moist to dry soils; none occur where the soil is consistently saturated with water or permanently frozen. In addition, these animals tend to prefer loamy to sandy soils; they are rarely found in indurate clays. Given these habitat preferences, it is not surprising that populations of subterranean rodents are often patchily distributed, occurring where soil and vegetation conditions are appropriate. As detailed in several of the chapters that follow, this pattern of spatial distribution has fundamental consequences for many aspects of the biology of these animals.

Although we know a great deal about some lineages of subterranean rodents, others remain virtually unknown. For example, the North American pocket gophers (Geomyidae) and Middle Eastern mole-rats (Spalacinae) have been studied extensively. In contrast, the Eurasian zokors (Myospalacinae) and the Asiatic and African bamboo and root rats (Rhizomyinae) have received only cursory attention. These inequities in coverage are evident for every topic considered in this volume, ranging from the summary of biogeographic and phylogenetic data presented here to the more detailed discussions of organismal, population, and evolutionary biology that follow.

For simplicity, we have adopted the classification scheme and taxonomy provided Wilson and Reeder (1993). We recognize, however, that a comprehensive taxonomy for these animals is lacking and that much systematic work remains to be done. An overview of the fossil records for these lineages is provided in chapter 9 of this volume. Information regarding the morphological attributes, ecology, and life history of each group can be found in the generic accounts of Nowack (1991), as well as in the specific references provided in the following chapters.

The rodent taxa considered in this volume (arranged by continent and taxon) are as follows (see also table 1).

Eurasia

1. MOLE-VOLES: FAMILY MURIDAE, SUBFAMILY ARVICOLINAE. There are two genera of arvicoline rodents that are exclusively subterranean *(Ellobius* and *Prometheomys)*, although other genera contain species that also exhibit considerable specialization for life underground. Included among the latter are *Hyperacrius wynnei*, an inhabitant of moist temperate forests and meadows in northern Pakistan and adjacent Kashmir (Roberts 1977), and the subterranean form of the water vole, *Arvicola terrestris,* which occurs in central Europe (Meylan 1977).

Musser and Carleton (1993) list five species of mole-voles *(Ellobius)* and provide a brief synopsis of their taxonomic history. Other authors list as few as two species (Corbet and Hill 1992). The genus occurs in steppe and semidesert habitats ranging from southeastern Europe north of the Black Sea (Ukraine) through the Caucasus Mountains of Georgia into Turkey, Iran, Afghanistan, Pakistan, Turkmenistan, and Kazakhstan (fig. 1). Here, they apparently prefer the deep and moist soils of streamsides and lake margins, where they construct elaborate, branching burrows. The animals feed primarily on underground plant parts. Their current range represents only a subset of a much larger Pleistocene distribution that extended through the Middle East to northern Africa (Jaeger 1988).

The long-clawed mole-vole, *Prometheomys,* consists of a single species that occurs in alpine and subalpine meadows of the montane regions of Georgia and extreme northeastern Turkey (fig. 1). It builds complex burrow systems and feeds on both belowground and aboveground plant parts.

2. ZOKORS: FAMILY MURIDAE, SUBFAMILY MYOSPALACINAE. Seven species of *Myospalax* are currently recognized (Musser and Carleton 1993). Five of these species are limited to parts of China, although the remaining two are geographically more widespread (*M. aspalax,* from the Upper Amur basin of Russia east to Mongolia, and *M. myospalax,* from Russia and Kazakhstan)

TABLE 1 SUBTERRANEAN RODENT TAXA CONSIDERED IN THIS VOLUME

Family	Genera	Number of species	Body size	General habitat range	General geographic range
Geomyidae	*Geomys*	5	Moderate (300–450 g)	Loose sandy soil in open or sparsely wooded areas	Southeastern and central U.S.; northeastern Mexico
	Orthogeomys	11	Moderate to large (500–800 g)	Arid tropical lowlands to montane forests from sea level to 3,000 m	Southern Mexico to northwestern Colombia
	Pappogeomys	9	Small to large (250–900 g)	Wide habitat range from deserts and grassy plains to montane meadows and tropical scrublands; sea level to 3,700 m	South-central U.S. to central Mexico
	Thomomys	9	Very small (50 g) to moderate (550 g)	Wide range of soils and habitats from desert scrub to montane meadows; below sea level to above 4,000 m	Southern Canada, western U.S, to central Mexico
	Zygogeomys	1	Moderate (300–550 g)	Montane meadows above 2,200 m	Michoacan, Mexico
Muridae: Arvicolinae	*Ellobius*	5	Very small (75 g)	Steppe and semidesert	Ukraine and Crimea south to Turkey and east to Iran, Afghanistan, Pakistan, Turkmenistan, and Kazakhstan
	Prometheomys	1	Very small (70 g)	Alpine and subalpine meadows (1,500 to 2,800 m)	Caucasus Mtns., Georgia, and northeastern Turkey
Muridae: Myospalacinae	*Myospalax*	7	Moderate (150–600 g)	Cultivated and wooded habitats in montane valleys (900 and 2,200 m)	Central Russia to northeastern China
Muridae: Rhizomyinae	*Cannomys*	1	Large (500–800 g)	Meadows and bamboo forests in hilly or mountainous areas	Eastern Nepal east to southern China, Thailand, and Cambodia
	Rhizomys	3	Very large (1–4 kg)	Bamboo thickets in upland areas between 1,000 and 4,000 m	Northern India to central China and south through the Malay Peninsula and Sumatra
	Tachyoryctes	11	Small (150–300 g)	Moist open grasslands, moorlands, and sparse savannas up to 4,100 m	Ethiopia south to Tanzania and west to Uganda, Rwanda, Burundi, and eastern Zaire

Family/Subfamily	Genus	No. of species	Body size	Habitat	Distribution
Muridae: Spalacinae	*Nannospalax*	3	Small (100–250 g)	Sandy or loamy soils in a variety of habitats, from dry shrublands and woodlands to desert scrub; sea level to 2,600 m	Ukraine south through Balkan states, east through Turkey to Georgia, hence south through Middle East to coastal Egypt and Libya
	Spalax	5	Moderate (200–600 g)	Wide range of soils in plains below sea level, upland steppes, and hilly regions	Russia and Ukraine east to western shore of Caspian Sea
Ctenomyidae	*Ctenomys*	56	Small (100 g) to large (> 750 g)	Extremely wide range of habitats and soils from sea level to above 4,000 m	Southern Peru south to Tierra del Fuego and east to southeastern Brazil
Octodontidae	*Spalacopus*	1	Small (110 g)	Moist to semiarid open shrublands from coastal areas to montane slopes above 3,000 m	Central Chile
Bathyergidae	*Bathyergus*	2	Large (850–1,500 g)	Coastal sand dunes and sand flats from the coast to 1,500 m	Namibia and South Africa
	Cryptomys	7	Small (200 g)	Compact or sandy soils in woodlands, savannas, and secondary forests at elevations to 2,200 m	Ghana east to Tanzania and south to South Africa
	Georychus	1	Small (180 g)	Wide range of soils	South Africa
	Heliophobius	1	Small (160 g)	Sandy soils in dry, open plains or woodlands	Zaire and Kenya south through Tanzania, Mozambique, Zambia, and Zimbabwe
	Heterocephalus	1	Very small (30–80 g)	Arid open shrub or grasslands in a variety of soils from 400 to 1,500 m	Somalia, Ethiopia, and Kenya

Note: Taxonomy follows references in Wilson and Reeder 1993.

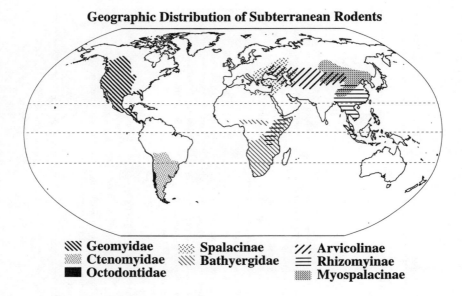

FIGURE 1. Geographic distribution of subterranean rodents by family or subfamily. Taxonomy follows Wilson and Reeder (1993).

(fig. 1). All are moderately large animals with enlarged foreclaws. They frequent montane valleys and are known for their remarkable digging abilities. Little information has been published on the biology of these species, and both geographic limits and species boundaries are poorly understood.

Eurasia and Africa

3. MOLE-RATS: FAMILY MURIDAE, SUBFAMILY SPALACINAE. Two closely related genera are currently recognized: *Nannospalax* and *Spalax*. The former contains three species and the latter contains five (Musser and Carleton 1993). Extreme karyotypic diversity in these animals (Savic and Nevo 1990), however, suggests a greater level of species diversity than is currently recognized, and traditional morphological species such as *N. ehrenbergi* and *N. leucodon* are often referred to as "superspecies" complexes. Species of *Nannospalax* are distributed parapatrically from Yugoslavia east through Hungary, Romania, Bulgaria, Greece, Turkey, Armenia, and Georgia, and south through Lebanon, Syria, Israel, and Jordan in the Middle East to isolated populations on the northern coast of Africa in Egypt and Libya (fig. 1). The *N. ehrenbergi* superspecies has been exceedingly well studied by Eviatar Nevo and his colleagues in Israel (Nevo 1991), and these animals have, to many, come to represent the archetypical subterranean rodent. The *N. leucodon* complex in the Balkan states and Turkey (Nevo et al. 1995; Savic 1973) has also been thoroughly studied, but to a lesser extent than *N. ehrenbergi*.

Species of *Spalax* occur to the north of those of *Nannospalax*, from southeastern Poland, southwestern Russia, the Ukraine, and Romania across the northern margins of the Black Sea and east to the western edge of the Caspian Sea and south to Ciscaucasia (fig. 1). Species ranges are either allopatric or parapatric. Most aspects of the biology of this complex are poorly known, particularly when compared with the detailed information available on *Nannospalax*. Both *Spalax* and *Nannospalax* inhabit open grassland, steppe, and semidesert soils.

4. BAMBOO AND ROOT RATS: FAMILY MURIDAE, SUBFAMILY RHIZOMYINAE. The genera in this subfamily have widely disjunct distributions, with *Rhizomys* and *Cannomys* in southeastern Asia (from northern India and Nepal to central China and south through Indochina to Sumatra: see fig. 1) and *Tachyoryctes* in east-central Africa, from Ethiopia south through Kenya to Tanzania and west to Uganda, Rwanda, Burundi, and eastern Zaire (fig. 1). None of these animals have received much attention in terms of ecology, life history, and genetic variation, although their fossil history has been reviewed and a cladistic hypothesis for relationships among genera has been proposed (Flynn 1990; Cook, Lessa, and Hadly, chap. 9, this volume). Root rats, *Tachyoryctes*, favor the deep and moist soils of open grasslands or upland savannas in mountain valleys. These animals are solitary and feed on both belowground and surface plant materials (Kingdon 1974). Musser and Carleton (1993) recognize nine species, most with allopatric and highly localized montane ranges. However, the geographic distributions and species limits of these animals are poorly known.

The bamboo rats *(Rhizomys* and *Cannomys)* collectively have a broader geographic range, but exhibit less species-level diversity, than the root-rats. The single species of *Cannomys* ranges from eastern Nepal through northern and northeastern India (Bhutan, Sikkim, Assam), southeastern Bangladesh, Burma, southern China, Thailand, northwestern Vietnam, and Cambodia (Corbet and Hill 1992). *Rhizomys,* which consists of three species that largely replace one another from north to south, is distributed from southern and eastern coastal China through Indochina (Laos, Vietnam, Cambodia, Thailand, eastern Burma), the Malay Peninsula, and the island of Sumatra (Corbet and Hill 1992). Both genera are large-bodied (table 1) and occupy bamboo thickets or other forested habitats in hilly or mountainous regions, rather than the open grasslands or meadows characteristic of most subterranean taxa. *Cannomys* and *Rhizomys* are less specialized for subterranean life than is *Tachyoryctes*. While *Rhizomys* appears to specialize on bamboo, *Cannomys* eats a wider array of plant materials.

Africa

5. BLESMOLS OR AFRICAN MOLE-RATS: FAMILY BATHYERGIDAE. The bathyergids have been extensively studied both in the field and in the laboratory, particularly during the last two decades. As a result, they are among the best known of the subterranean rodents. The long-term research programs of Jennifer Jarvis and Nigel Bennett in South Africa, Paul Sherman in the United States, and their many students and collaborators recently culminated in the publication of *The Biology of the Naked Mole-rat* (Sherman, Jarvis, and Alexander 1991). Despite the apparent taxonomic limitation of the volume, it summarizes the evolution, ecology, life history, and social biology of all genera in the family. The distribution, taxonomy, and phyletic history of the five extant genera are reviewed by Honeycutt et al. (1991; see also Honeycutt 1992).

Cryptomys is the most speciose genus in the family, with seven species currently recognized (Woods 1993 follows the taxonomy suggested in Honeycutt et al. 1991). Species are, however, poorly delineated due to extreme variation in morphological features, so species limits, and thus geographic distributions, require further study. Three species have disjunct distributions north of the equator, one in Ghana, one in Nigeria, and the third extending from southeastern Nigeria across Cameroon, the Central African Republic, southern Sudan, and northern Zaire, into Uganda. The other four species are distributed parapatrically from Tanzania west through southern Zaire and Angola, and south to include all, or parts, of Zambia, Malawi, Zimbabwe, Mozambique, Namibia, Botswana, and South Africa (fig. 1). *Bathyergus* currently contains two species, both limited to the coastal margins of South Africa and Namibia (fig. 1). The other three genera are monotypic: *Georychus* is limited to several disjunct populations from both coastal and interior parts of South Africa; *Heliophobius* ranges widely from southern Kenya through Tanzania and southeastern Zaire into eastern Zambia, Malawi, and northern Mozambique; and *Heterocephalus* occurs from the Horn of Africa in Somalia to central and southern Ethiopia and northern Kenya (fig. 1).

Most blesmol species are relatively small-bodied animals, although individuals of *Bathyergus* may exceed 1 kg in weight. General accounts of their biology can be found in Jarvis 1984 and Jarvis and Bennett 1990, in addition to several papers in Sherman, Jarvis, and Alexander 1991. *Cryptomys* inhabits the broadest range of habitats, extending from forested regions through woodlands to savannas. The other genera are more typically found in semiarid open grasslands or shrublands, or coastal dunes and sandy flats. All genera, with one exception, are tooth diggers; *Bathyergus* uses well-developed claws on the forefeet. Perhaps the best-known feature of blesmols is the elaborate social structure found in some taxa, as exemplified by *Heterocephalus* (Jarvis 1981; Lacey, chap. 7, this volume).

North America

6. POCKET GOPHERS: FAMILY GEOMYIDAE. Given the common name "pocket gopher" because of their fur-lined external cheek pouches, these animals have been studied more extensively and over a longer period of time than any other group of subterranean rodents. The family has a long and rich fossil history (Cook, Lessa, and Hadly, chap. 9, this volume) and has been the focus of intensive investigation of nearly all aspects of its ecology, population and social biology, evolutionary genetics, and taxonomy. Chase, Howard, and Roseberry (1982) and Patton (1990) provide excellent summaries of the vast literature on these animals. Indeed, research on these animals has so dominated studies of subterranean taxa that the world of subterranean rodents is often viewed—erroneously, or at least narrowly—through the eyes of the pocket gopher researcher.

Five genera of geomyids are recognized in the Recent record (Patton 1993). As in many other subterranean taxa, however, species limits remain debatable (Steinberg and Patton, chap. 8, this volume), in this case despite extensive studies at both morphological and genetic levels. Wahlert (1985) provided a classification of the superfamily Geomyoidea, including the Geomyidae, and Russell (1968) reviewed the fossil and Recent genera, to the exclusion of *Thomomys*. Additional details of the fossil record and phylogeny are given by Cook, Lessa, and Hadly (chap. 9, this volume).

Geomys comprises at least five species. These are moderate-sized animals that exhibit disjunct distributions in the southeastern United States, the Great Plains of North America from south-central Canada to Texas, and into coastal Tamaulipas in northeastern Mexico. Nine species of *Pappogeomys* are currently recognized. All have localized distributions in the trans-volcanic belt of central Mexico except *P. castanops*, which occurs more widely, ranging from the southern Great Plains of the United States throughout the Central Plateau of Mexico. *Orthogeomys* consists of eleven species of large-bodied animals, most of which are known only from single localities or otherwise restricted areas from southern Mexico through Central America to extreme northwestern Colombia. There are nine currently recognized species of *Thomomys*, some with exceedingly large geographic ranges and others that are highly restricted geographically. The genus itself ranges widely, from south-central and southwestern Canada throughout the mountainous and intermontane regions of the western United States, and most of northern Mexico as far south as Mexico City. Finally, the monotypic *Zygogeomys* has an extremely limited distribution in the state of Michoacán, Mexico.

All pocket gophers exhibit similar lifestyles and ecologies, occupying individual burrows and consuming both aboveground and belowground plant materials. Most of these animals are scratch diggers, although some will use their teeth in hard soils. All inhabit open grasslands, sparse shrublands, and/or montane meadows, although different species have widely separate elevational and, thus, habitat ranges (table 1). Genera, as well as species, are characteristically allopatric or parapatric in their respective distributions. Where they are in contact, a reasonably "strict" set of ecological ranges is found (Miller 1964; Thaeler 1969).

South America

7. TUCO-TUCOS: FAMILY CTENOMYIDAE. The ctenomyids constitute by far the most speciose group of extant subterranean rodents, containing fifty to sixty species in the single genus *Ctenomys*. The genus is widely distributed, ranging from the high bunchgrass of the Andean Altiplano in southern Peru and Bolivia to sea level in Chile, and throughout much of Argentina into southeastern Brazil (fig. 1). Their habitat range is equally broad, but all species occur in relatively open areas of dry to moist grasslands, shrublands, or forest meadows. The animals are typically solitary, with individually occupied burrows, although burrow sharing has been reported for several species (Lacey, chap. 7, this volume). The forefeet are equipped with enlarged claws that, along with the incisors, aid in digging. They feed on both belowground and aboveground plant materials. The common name "tuco-tuco" is onomatopoeic for their characteristic vocalizations. While most species have received only cursory, if any, attention, a few species are the subject of intensive research programs pursuing a wide array of ecological and evolutionary questions.

8. CORURO: FAMILY OCTODONTIDAE. A single species, *Spalacopus cyanus*, occurs in central Chile, from sea level to about 3,000 m on the western Andean slopes. The animals are particularly common in semiarid areas containing scattered clumps of vegetation, as well as in wet streamside flats. This species is colonial and, like the closely related *Ctenomys*, is quite vocal. Also like *Ctenomys*, *Spalacopus* digs with both forefeet and incisors, but it feeds almost exclusively on underground plant tubers and stems. Although these animals remained unstudied for many years following an early report by Reig (1970), they have recently become the focus of studies focusing on community and physiological ecology (Contreras 1986; Contreras et al. 1993).

The Subterranean Niche

Having introduced the taxa that are the focus of this volume, we now ask, what is it about these animals that makes them particularly intriguing as research subjects? Much of our interest in these taxa stems from efforts to understand the role of the subterranean niche in shaping evolutionary convergences and divergences among these animals. The subterranean niche differs from surface habitats in that individuals routinely make use of a third dimension: depth. This third dimension, in conjunction with the soil from which tunnels are excavated, generates a distinct set of selective pressures that both create and constrain opportunities for adaptive responses to environmental conditions.

It is generally assumed that the subterranean niche is relatively invariant (Nevo 1979), despite the diversity of geographic areas and surface habitats in which subterranean rodents occur. This assumption has no doubt been reinforced by the apparent morphological similarities among subterranean taxa, which are thought to reflect constraints imposed by life underground (Stein, chap. 1, this volume; McNab 1979). Because the subterranean milieu is thought to play such a central role in shaping the biology of the species considered in this volume, it seems reasonable to consider the physical characteristics of underground burrows before attempting to relate these conditions to specific aspects of rodent biology.

In general, burrow systems are characterized by the absence of light, an excess of moisture, relatively invariant temperatures, and low ratios of oxygen to carbon dioxide. Because light is absorbed by the burrow walls, even open tunnels become dark within a short distance of the tunnel entrance. All burrows have high relative humidities, and in closed tunnels, humidity may approach 100% (Kennerly 1964). Air flow is generally reduced relative to surface habitats, leading to poor ventilation and elevated levels of CO_2 (Darden 1972; Arieli 1979).

These conditions suggest that subterranean rodents must contend with a variety of physiological and other problems that are not encountered by surface-dwelling taxa (Buffenstein, chap. 2, this volume). For example, activity patterns, blood physiology, and metabolic rates may all be influenced by low levels of gas exchange within subterranean burrows. At the same time, the absence of convective cooling suggests that heat loss may be a problem, perhaps resulting in behavioral, morphological, and physiological adaptations that increase an animal's ability to dump heat into its environment. Even the absence of light may be important, as it may contribute to morphological and other modifications, including the evolution of specialized modes of communication (Francescoli, chap. 3, this volume).

The evolutionary opportunities associated with life underground have received less attention. Use of subterranean burrows may free animals from some of the pressures experienced by surface-dwelling taxa. For example, burrows may decrease exposure to certain predators or create new opportunities for deterring predators, such as blocking tunnels containing snakes or other subterranean threats (Busch et al., chap. 5; Lacey, chap. 7, this volume). At the same time, because environmental conditions within burrows tend to be somewhat buffered against surface conditions (e.g., rapid changes in temperature), subterranean rodents may be able to exploit habitats that would be much less accessible to surface-dwelling species (Busch et al., chap. 5, this volume).

In addition to physical conditions within burrows, the environmental conditions required for excavating tunnels may influence the biology of these animals. Morphological modifications associated with digging are one of the most conspicuous ways in which the challenges of tunneling through soil are manifest (Stein, chap. 1, this volume). Perhaps less obvious are the effects of soil conditions on the distributions of subterranean rodents: only where soil composition is appropriate can these animals excavate tunnels, which may generate the patchy distribution of populations for which subterranean taxa are well known. This distribution, in turn, may constrain dispersal and gene flow, thereby creating new opportunities for evolutionary diversification (Steinberg and Patton, chap. 8, this volume). In short, there are numerous ways in which the subterranean niche may affect its inhabitants. By generating conditions not experienced by surface-dwelling taxa, the subterranean niche has undoubtedly influenced countless aspects of rodent biology.

Why Study Subterranean Rodents?

Biologists have long been intrigued by apparent convergences among subterranean rodents. The geographic variation and diversity of the surface habitats used by these animals is readily apparent, leading to considerable speculation regarding the nature of the selective pressures that have generated morphological, ecological, and other similarities among these taxa. As our understanding of phylogenetic relationships among subterranean lineages has increased, our ability to identify examples of evolutionary convergence and divergence in subterranean taxa has improved, leading to further interest in these animals as subjects for evolutionary research.

Although we are undoubtedly biased, we can think of few other collections of mammal species that provide the same opportunities to explore convergent and divergent responses to environmental conditions. Some of the similarities and differences among subterranean rodents have long been evident to biologists; others are only now being identified. Thus our knowledge of these animals consists of both well-established generalizations that require testing and newly emerging relationships whose generality has yet to be determined. The result is an exciting mix of theoretical and empirical work that serves to generate more questions than it answers. We hope that the contents of this volume will convince readers—as they have convinced us—that there is still much to be learned from studies of subterranean rodents.

Literature Cited

Arieli, R. 1979. The atmospheric environment of a fossorial mole rat *(Spalax ehrenbergi)*: Effect of season, soil texture, rain, temperature and activity. *Comparative Biochemistry and Physiology* 63A: 569–75.
Chase, J. D., W. E. Howard, and J. J. Roseberry. 1982. Pocket gophers (Geomyidae). In *Wild mammals of North America: Biology, management, and economics*, edited by J. A. Chapman and G. A. Feldhammer, 239–55. Baltimore: Johns Hopkins University Press.
Contreras, L. C. 1986. Bioenergetics and distribution of fossorial *Spalacopus cyanus* (Rodentia): Thermal stress, or cost of burrowing? *Physiological Zoology* 59:20–28.
Contreras, L. C., J. R. Gutiérrez, V. Valverde, and G. W. Cox. 1993. Ecological relevance of subterranean herbivorous rodents in semiarid coastal Chile. *Revista Chilena de Historia Natural* 66:357–68.
Corbet, G. B., and J. E. Hill. 1992. Mammals of the Indomalayan region: A systematic review. Oxford: Oxford University Press.
Darden, T. R. 1972. Respiratory adaptations of a fossorial mammal, the pocket gopher, *Thomomys bottae*. *Journal of Comparative Physiology* 78:121–37.
Flynn, L. J. 1990. The natural history of rhizomyid rodents. In *Evolution of subterranean mammals at the organismal and molecular levels*, edited by E. Nevo and O. A. Reig, 155–83. Progress in Clinical and Biological Research, vol. 335. New York: Wiley-Liss.
Honeycutt, R. L. 1992. Naked mole-rats. *Scientific American* 80:43–53.
Honeycutt, R. L., M. W. Allard, S. L. Edwards, and D. A. Schlitter. 1991. Systematics and evolution of the family Bathyergidae. In *The biology of the naked mole-rat*, edited by P. W. Sherman, J. U. M. Jarvis, and R. D. Alexander, 45–65. Princeton, NJ: Princeton University Press.
Jaeger, J.-J. 1988. Origine et evolution du genre *Ellobius* (Mammalia, Rodentia) en Afrique Nord-Occidentale. *Folia Quaternaria* 57:3–50.
Jarvis, J. U. M. 1981. Eusociality in a mammal: Cooperative breeding in naked mole-rat colonies. *Science* 212:571–73.
———. 1984. African mole-rats. In *The encyclopedia of mammals*, edited by D. Macdonald, 708–11. New York: Facts on File Publications.
Jarvis, J. U. M., and N. C. Bennett. 1990. The evolutionary history, population biology, and social structure of African mole-rats: Family Bathyergidae. In *Evolution of subterranean mammals at the organismal and molecular levels*, edited by E. Nevo and O. A. Reig, 97–128. Progress in Clinical and Biological Research, vol. 335. New York: Wiley-Liss.
Kennerly, T. E. 1964. Microenvironmental conditions of the pocket gopher burrow. *Texas Journal of Science* 16:395–441.
Kingdon, J. 1974. *East African mammals: An atlas of evolution in Africa*. Vol. II, part B: *Hares and rodents*. London and New York: Academic Press.

McNab, B. K. 1979. The influence of body size on the energetics and distribution of fossorial and burrowing mammals. *Ecology* 60:1010–21.

Meylan, A. 1977. Fossorial forms of the water vole, *Arvicola terrestris* (L.), in Europe. *EPPO Bulletin* 7:209–21.

Miller, R. S. 1964. Ecology and distribution of pocket gophers (Geomyidae) in Colorado. *Ecology* 45:256–72.

Musser, G. G., and M. D. Carleton. 1993. Family Muridae. In *Mammal species of the world: A taxonomic and geographic reference*, 2d ed., edited by D. E. Wilson and D. M. Reeder, 501–755. Washington, DC: Smithsonian Institution Press.

Nevo, E. 1979. Adaptive convergence and divergence of subterranean mammals. *Annual Review of Ecology and Systematics* 10:269–308.

———. 1991. Evolutionary theory and processes of active speciation and adaptive radiation in subterranean mole rats, *Spalax ehrenbergi* superspecies, in Israel. *Evolutionary Biology* 25:1–125.

Nevo, E., M. G. Filippucci, C. Redi, S. Simson, G. Heth, and A. Beiles. 1995. Karyotype and genetic evolution in speciation of subterranean mole rats of *Spalax* in Turkey. *Biological Journal of the Linnean Society* 54:203–24.

Nowack, R. M. 1991. *Walker's mammals of the world*. 5th ed. Baltimore: Johns Hopkins University Press.

Patton, J. L. 1990. Geomyid evolution: The historical, selective, and random basis for divergence patterns within and among species. In *Evolution of subterranean mammals at the organismal and molecular levels,* edited by E. Nevo and O. A. Reig, 49–69. Progress in Clinical and Biological Research, vol. 335. New York: Wiley-Liss.

———. 1993. Family Geomyidae. In *Mammal species of the world: A taxonomic and geographic reference*, 2d ed., edited by D. E. Wilson and D. M. Reeder, 469–76. Washington, DC: Smithsonian Institution Press.

Reig, O. A. 1970. Ecological notes on the fossorial octodont rodent *Spalacopus cyanus* (Molina). *Journal of Mammalogy* 51:592–601.

Roberts, T. J. 1977. *The mammals of Pakistan*. London: Ernest Benn. xxvi + 361 pp.

Russell, R. J. 1968. Evolution and classification of the pocket gophers of the subfamily Geomyinae. *Miscellaneous Publications, Museum of Natural History, University of Kansas* 16:473–579.

Savic, I. R. 1973. Ecology of the mole rat *Spalax leucodon* Nordm. in Yugoslavia. (In Serbo-Croatian with English summary). *Proceedings in Natural Sciences, Department of Natural Sciences Matica Srpska, Novi Sad* 44:5–70.

Savic, I. R., and E. Nevo. 1990. The Spalacidae: Evolutionary history, speciation, and population biology. In *Evolution of subterranean mammals at the organismal and molecular levels,* edited by E. Nevo and O. A. Reig, 129–53. Progress in Clinical and Biological Research, vol. 335. New York: Wiley-Liss.

Sherman, P. W., J. U. M. Jarvis, and R. D. Alexander, eds. 1991. *The biology of the naked mole-rat*. Princeton, NJ: Princeton University Press.

Thaeler, C. S., Jr. 1969. An analysis of the distribution of pocket gopher species in Northeastern California (genus *Thomomys*). University of California Publications in Zoology 86:1–46.

Wahlert, J. H. 1985. Skull morphology and relationships of geomyoid rodents. *American Museum Novitates* 281:1–20.

Wilson, D. E., and D. M. Reeder. 1993. *Mammal species of the world: A taxonomic and geographic reference*. 2d ed. Washington, DC: Smithsonian Institution Press.

Woods, C. A. 1993. Suborder Hystricognathi. In *Mammal species of the world: A taxonomic and geographic reference*, 2d ed., edited by D. E. Wilson and D. M. Reeder, 771–806. Washington, DC: Smithsonian Institution Press.

Part One: Organismal Biology

The evolutionary challenges and opportunities of living in a subterranean environment are addressed by adaptations at various levels of organization, ranging from individuals to populations and communities of organisms. These adaptations may affect numerous aspects of an individual's phenotype, including its behavior, morphology, and physiology. The chapters in the first part of this volume focus on a fundamental component of the biology of subterranean rodents: the adaptation of the individual organism to life underground. Morphological, physiological, sensory, and reproductive adaptations to subterranean environments are explored in detail. These discussions serve both to characterize the biology of individuals that live underground and to lay the groundwork for subsequent discussions of the population, community, and evolutionary biology of subterranean rodents.

Adaptive modifications of basic rodent biology are crucial to life underground because of the unique biotic and abiotic challenges encountered in the subterranean world. Subterranean rodents expend considerable time and energy digging, and these expenditures are reflected in the morphological adaptations exhibited by these animals. Life in a dense, dark, and semisolid medium with high concentrations of carbon dioxide and low concentrations of oxygen has produced numerous physiological modifications that allow individuals to cope with these conditions. Subterranean environments create special problems for communication among individuals, and these challenges appear to have led subterranean taxa to exploit sensory and communication modalities that are not often used by surface-dwelling taxa. At the same time, some subterranean taxa exhibit extreme reproductive specializations that are unusual, if not unique, among mammals. Understanding how the subterranean ecotope has influenced these aspects of the biology of rodents is a recurrent theme among the first four chapters of this volume.

It is frequently asserted that although subterranean rodents represent diverse taxa from numerous different geographic regions, these animals exhibit convergent adaptations that have led to morphological uniformity. Stein (chap. 1) challenges this dogma by demonstrating marked intergeneric variation in the morphology of subterranean rodents. She describes fundamental differences in the digging apparatus of these animals, including variations in external anatomy, osteology, and myology that appear to reflect the independent evolutionary origins of the different subterranean lineages (see Cook, Lessa, and Hadly, chap. 9, this volume). Stein concludes that further studies of descriptive anatomy are needed to understand completely both the phylogenetic relationships among taxa and the relative contributions of phylogeny and geographic variation to constraints on morphology.

In addition to morphological adaptations, subterranean rodents exhibit physiological adaptations for life underground. Buffenstein (chap. 2) explores physiological differences between surface-dwelling and subterranean rodents, with emphasis on how the latter have solved the physiological challenges imposed by their distinctive environments. Subterranean rodents are characterized by increased length of the hindgut and microbial fermentation, both of which aid in the digestion of poor-quality diets. Other physiological attributes that appear to be shared among these animals include low metabolic rates, high thermoneutral ranges, and low body temperatures. Poor ventilation in subterranean burrows leads to both low rates of evaporative water loss and high levels of hypoxic and hypercapnic stress, particularly when multiple individuals occupy a burrow system, as occurs in some taxa (see Lacey, chap. 7, this volume). Perhaps less conspicuous are the physiological challenges imposed by the low levels of sunlight in subterranean burrows and the resulting deficiencies in vitamin D, which appear to be linked to adaptations that function to increase the efficiency of mineral absorption. Thus the subterranean world seems to have influenced the physiology of rodents in a number of initially subtle, unexpected ways.

The physical medium in which subterranean rodents live makes it difficult for individuals to communicate using the "typical" mammalian sensory modalities of vision and olfaction. Francescoli (chap. 3) assesses the physical requirements of visual, olfactory, auditory, and tactile communication and explores the role of each in the communication systems of subterranean rodents. In particular, Francescoli attempts to discern why some of these taxa have evolved specialized forms of seismic communication, whereas others appear to lack this mode of interaction. The resulting discussion places the available data on communication in subterranean rodents in an adaptive framework that should serve to generate considerable new research on this topic.

Reproduction is one of the most fundamental aspects of mammalian biology, yet surprisingly little is known about reproduction by subterranean taxa. Information on attributes such as ovarian cycle length, mode of ovulation (spontaneous or induced), length of gestation, and litter size is surprisingly scarce, and the majority of data that have been collected apply to only a few groups, notably geomyids and bathyergids. Even within these taxa, however, many aspects of reproduction remain poorly understood. Bennett, Faulkes, and Molteno (chap. 4) have collated data on the reproductive biology of subterranean taxa, including behavioral information regarding patterns of courtship, mating, and paternity. Their discussion of these topics underscores how much remains to be learned, while at the same time highlighting one of the most unusual patterns of reproduction known among mammals, the reproductive division of labor within groups of social bathyergids.

As each of the chapters in part 1 indicates, more research is needed on all aspects of the organismal biology of subterranean rodents. Our knowledge of these animals clearly varies among taxa. The best understood of the subterranean rodents are the Geomyidae, Ctenomyidae, and Bathyergidae, particularly when data for these taxa are compared with the lack of information for other groups such as the Myospalacinae and Rhizomyinae. As our knowledge of the morphology, physiology, communication systems, and reproductive biology of these fascinating animals increases, we will be able to understand more clearly how they have adapted to subterranean environments, and how these adaptations affect other aspects of their biology such as ecology, population genetics, and evolutionary diversification.

CHAPTER ONE

Morphology of Subterranean Rodents

Barbara R. Stein

Descriptive anatomy represents an essential component of phylogenetic and biomechanical analyses. In the recent volume entitled *The Evolution of Subterranean Mammals at the Organismal and Molecular Levels* (Nevo and Reig 1990), however, descriptive information on morphology was not emphasized. Instead, authors discussing morphology chose to focus on more conceptual issues, such as questions concerning the functional morphology of the sensory apparati (Burda, Bruns, and Müller 1990) and morphological constraints on evolution (Lessa 1990). A common justification for this approach was that previous knowledge of morphology in subterranean mammals had already been adequately summarized by authors such as Ellerman (1959), Dubost (1968), Nevo (1979), and Hildebrand (1985).

Recent perusal of the literature on rodent morphology, however, indicates that this is not the case. Information, even in summary articles, is often of a general nature, and taxonomic treatment is uneven. For many subterranean genera, detailed descriptions of the postcranial skeleton, muscle structure, or internal organs are limited (e.g., *Ctenomys*) or are sorely lacking (e.g., *Spalacopus* and *Ellobius*). Because emphasis has been placed on those aspects of gross anatomy that relate most prominently to morphological convergence in these mammals, little literature exists regarding anatomical structures that are not associated with burrowing.

Similarities in external body form among morphologically convergent rodent genera often belie differences in their musculoskeletal systems that reflect their independent phylogenetic histories (Stein 1994). Convergence and parallelism within the order Rodentia, as evidenced by similarities in cranial and dental morphology, have been well documented (Wood 1936, 1955), and this review will not include a repetition of those characters. In contrast, scant attention has been paid to the degree of convergence and parallelism present in postcranial structures. Any discussion of morphology and convergence in subterranean rodents must also contend with uncertain taxonomic affinities that have yet to be resolved, including the intrafamilial affinities of *Myospalax* and *Bathyergus,* familial relationships among New World caviomorphs, and the monophyly of caviomorphs with respect to Old World hystricomorphs (summarized in Carleton and Musser 1985, Wilson and Reeder 1993, and Luckett and Hartenberger 1985).

This chapter attempts to summarize in detail the extent of current knowledge concerning the morphology of subterranean rodents and to highlight topics in need of additional research. It is also intended that this compilation of descriptive information will underscore the importance of phylogenetic and functional data to studies of all morphologically convergent assemblages of mammals. Because this chapter is primarily a review of our current knowledge and understanding of morphology in subterranean rodents, the following discussion will be most comprehensive for those aspects of gross anatomy that relate prominently to burrowing. However, an attempt has been made to survey current knowledge in all areas of gross morphology for these animals. For details of cranial and dental anatomy not mentioned below, readers are directed to Anderson and Jones 1985 for family diagnoses, Ellerman 1940 for genus-level descriptions, Ognev 1950 and Allen 1940 for Asian taxa, Kingdon 1984 for African taxa, Harrison and Bates 1991 for Arabian genera, Hinton 1926 for arvicolids, Wahlert 1985 for geomyoids, and Honeycutt et al. 1991 for bathyergids.

External Morphology

Within mammals, no group exhibits a more intriguing suite of morphological characters than those taxa adapted to a subterranean way of life. The morphology of subterranean rodents reflects their need to move in dark, confined tunnels that are often humid and low in oxygen (Hildebrand 1985; Buffenstein, chap. 2, this volume). Subterranean habitats are also marked by a relatively constant ambient temperature, as well as a paucity of organic nutrients and a relatively large concentration of minerals for a given surface area, factors that are assumed to constrain the form and function of the animals inhabiting this environment. These conditions are thought to have produced a body plan that is exemplified by a fusiform shape, a relative reduction in both limb length and the size of external protuberances, and specialized sensory structures for the subterranean detection of food, predators, and potential mates.

Perhaps more than any other suite of morphological characters, the fusiform shape and reduced limbs of subterranean rodents are acknowledged as adaptations to maneuvering in narrow and confined spaces. Deviations from this generalized form in terms of body size and shape, pelage color and texture, and presence or absence of external protuberances reflect the influences of both local environmental parameters and phylogenetic history on these lineages. Responses to local environmental parameters are of two types: those that involve direct selection on genetic variation (e.g., color crypticity), and those that are ecophenotypic and nongenetic (e.g., body size). Carefully crafted studies (e.g., Patton and Brylski 1987) can be used to differentiate between these types of responses and should contribute significantly to our understanding of the factors responsible for the morphological variation observed among these genera.

Body Form and Size

Although a fusiform body shape is often considered a hallmark of subterranean mammals, other phenotypic traits, such as body size, pelage color and texture, and tail length, are not as uniform. *Ellobius* and *Prometheomys* are the smallest of the subterranean rodents, with head and body lengths (HB) of 100–150 mm and 125–160 mm, respectively (Nowak 1991). At the opposite end of the spectrum are the rhizomyines. *Rhizomys*, the largest of the three genera in this subfamily, has an HB of 230–480 mm (Nowak 1991). Between these extremes lies an almost complete gradation in body size, with some subterranean genera showing spectacular intra- and interspecific differences in this character (e.g., *Cryptomys* and *Thomomys*). Ranges of body lengths and weights for subterranean taxa are not listed here, but are given by Nowak (1991), Harrison and Bates (1991), Allen (1940), and others. Sexual dimorphism in body size has been examined in rhizomyines (Jarvis 1973), *Bathyergus* (Jarvis and Bennett 1990), and *Thomomys* (Daly and Patton 1986), but comparable data for the remaining subterranean taxa are not available.

Among pocket gophers of the genus *Thomomys*, body size is plastic, with variation directly attributable to the nutritional quality of the food available in a given habitat (Patton and Brylski 1987). Male pocket gophers living in nutrient-rich alfalfa fields were, on average, 25% larger in cranial length and weighed nearly twice as much as those inhabiting adjacent or nearby natural desert regions where food quality and quantity were vastly inferior. Comparable studies are lacking for other genera, although Jarvis and Bennett (1991) reported that body size in *Heterocephalus* decreased with increasing soil aridity. Jarvis, O'Riain, and McDaid (1991) showed that in this genus, body size is also strongly influenced by the animals' social environment, a variable that awaits examination in other taxa.

Pelage and Skin

Pelage color in subterranean mammals has been shown to vary regionally with humidity and locally with substrate color and composition, tending toward crypticity in all cases (Nevo 1979; Heth, Beiles, and Nevo 1988). When present, fur is generally short, fluffy, and often upright, brushing easily in both directions in most of these taxa. The fur of *Spalacopus* seems to be the least compliant in this regard, being dark and glossy but not particularly soft and fluffy. *Heterocephalus,* of course, exhibits the most extreme pelage modification, having lost all of its body fur. In *Rhizomys* and large-bodied pocket gophers such as *Orthogeomys,* pelage composition varies latitudinally, from soft and thick in northern areas to harsh and scanty in the tropics (Nowak 1991). In *Cannomys* and *Tachyoryctes,* the pelage is generally thick and soft throughout each species' geographic range.

Little attention has been paid to the thermoregulatory properties of skin or to the mechanical stresses that affect skin as a result of burrowing (McNab 1966). Hildebrand (1974) reported that in taxa such as geomyids, bathyergids, and *Ctenomys,* the skin is only loosely attached to the underlying muscle, enabling these animals to literally turn 180° within their skin, with the skin following once the turn is completed. Such behavior facilitates reversing direction in narrow tunnels where friction between fur and the adjacent dirt walls is greatest.

Recently, Klauer, Burda, and Nevo (1997) conducted histological analyses of macroscopically distinct regions of skin (hairy skin, vibrissae, buccal ridge, and rhinarium) on the head of *Nannospalax,* which frames its face with a row of stiff fringe hairs. The authors related their findings to the unique method of head-lift digging in that genus (see below). Additional data of this sort are needed for the remaining taxa.

Tachyoryctes also has stiff facial hairs, but these are presumed to function as tactile organs in this genus (Nowak 1991). In *Prometheomys,* vibrissae line each side of the face between the eye and nose, but no function has been reported for these structures (Ognev 1950). Vibrissae surrounding the mouth (as opposed to the face) are discussed below.

While digging, subterranean mammals must effectively exclude soil from external orifices such as the eyes, ears, nose, mouth, and anus, where the presence of particulate matter may impede respiration and digestion or damage delicate sensory organs. In bathyergids, slightly vulvar external nares reduce entry of dirt into the nasal cavities (Jarvis and Sale 1971), and well-haired lip folds behind the upper incisors prevent dirt from entering the mouth (Shimer 1903; Macdonald 1984). Furred lips that close behind procumbent incisors also are present in other chisel-tooth diggers such as geomyids and *Ctenomys* (Pearson 1959). In *Prometheomys,* the labial lobes are drawn into the space between the incisors and the molars, where they meet, but do not join. In *Ellobius,* however, the lobes do join (Vinogradov 1926; Ognev 1950). Gromov and Polyakov (1992) reported that inferior labial flaps in *Nannospalax,* like those in *Prometheomys,* do not fuse behind the incisors.

Cheek pouches in rodents may be either internal or external to the oral cavity. The fur-lined external cheek pouches found in pocket gophers are a synapomorphy for the superfamily Geomyoidea, which contains not only pocket gophers but also kangaroo rats and pocket mice. No other rodent genera have external cheek pouches, and no other subterranean genera have cheek pouches of any sort. Hill (1935b) discussed the muscles and nerves associated with the retractor muscle of pocket gopher pouches. Brylski and Hall (1988) addressed the developmental origins of both internal and external cheek pouches, and Ryan (1986) detailed their comparative morphology and evolution.

Pinnae

In general, reduction in size of the pinnae is evident in subterranean rodents. Large pinnae (i.e., external ears) would seemingly impede burrowing and be subject to continual rubbing and friction against burrow walls. Among bathyergids, *Bathyergus* has reduced each of its pinnae to a ring of skin surrounding its auditory aperture. The remaining genera in this family have reduced the pinnae to low ridges of skin that leave the openings of the auditory canals exposed. Despite this reduction, hearing in at least one bathyergid *(Heterocephalus)* remains acute (Jarvis and Bennett 1991). *Nannospalax* has lost its pinnae almost entirely, and the large, tubular opening of the auditory canal is clearly visible externally. Ognev (1950) reported that the ear opening in *Ellobius* is reduced to an oblique slit no more than 2 mm in diameter that is hidden in the body fur. The external ear in this genus is reduced to a slight elevation of the superior margin around the opening. In contrast, the pinnae of *Prometheomys* and *Myospalax* have been described as moderately large (Hinton 1926; Vinogradov 1926). Hinton (1926) reported the presence of an antitragus in *Prometheomys,* and Ognev (1950) observed that the auditory ossicles of this genus differ from those in most other voles. In the remaining subterranean taxa, the pinnae are small (geomyids, *Ctenomys*) to moderate (rhizomyines, *Spalacopus*) in size, and the auditory canal is exposed.

Tails

Tail length varies almost as greatly as body size in subterranean rodents, both absolutely and in terms of length relative to that of the head and body (see Nowak 1991 for measurements). In *Nannospalax,* the tail is virtually nonexistent; in *Ellobius,* it is barely evident. In all bathyergids except *Heterocephalus,* the tail is relatively small. The tail is moderately well developed in geomyids, rhizomyines, *Spalacopus,* and *Myospalax* and is relatively long in *Ctenomys* and *Heterocephalus.* In *Prometheomys,* tail length may be approximately one-half head and body length. In all genera except *Heterocephalus,* the tail is haired, albeit lightly in some taxa.

Multiple functions have been proposed for the tails of subterranean rodents. Bathyergids other than *Heterocephalus* are distinct in having long lateral fringe hairs on each side of the tail that aid in dirt removal when the animals kick posteriorly. These hairs give their short tails a broad, flat, whisk broom-like appearance. The degree to which the tail is tactilely sensitive is unknown for many subterranean genera, although it has been reported to be so in *Heterocephalus* (Jarvis and Bennett 1991) and geomyids (McLaughlin 1984). In the latter group, the tail is highly vascularized and well supplied with nerves. It has also been suggested that vascularization of the tail may be a means of dumping heat (McNab 1966). The tails of pocket gophers are also reported to brace the body and to aid in orientation during burrow excavation (Hickman 1984). Hickman (1985) reported that *Ctenomys* uses its tail as a prop that supports the body in concert with either the forelimbs or hind limbs. In *Heterocephalus,* one mode of orientation involves a side-to-side sweeping motion of the tail (Lacey et al. 1991). This behavior is particularly pronounced when animals move backward through their burrow systems.

Forefeet and Hind Feet

Another general characteristic of subterranean mammals is relatively broad feet. This modification increases the surface area available for moving soil. The ventral surfaces of the manus and pes are generally naked, supporting large palmar, plantar, and often digital pads. The feet and digits are ringed by a fringe of stiff, coarse hairs that increases the ventral surface area and aids in trapping soil and pushing it out of tunnels. These hairs are present even in *Heterocephalus,* and are conspicuous on both the forefeet and hind feet in all bathyergids, particularly *Bathyergus.* These hairs are also well developed on the forefeet of *Thomomys, Geomys, Ctenomys,* and *Spalacopus,* and they are readily observable on the hind feet of *Tachyoryctes, Ctenomys, Nannospalax, Ellobius,* and *Prometheomys.* They are, however, less prominent on the hind feet of other genera.

Sensory Systems

Whereas characteristics of external morphology in subterranean rodents have been viewed primarily as adaptations to enhance digging and the construction of tunnel systems, the sensory systems in these taxa exhibit modifications that are best correlated with the unusual environmental parameters that exist underground (e.g., low levels of light and a dense medium through which sound must be transmitted). Adaptations of the sensory systems in subterranean mammals were summarized by Burda, Bruns, and Müller (1990) and those of rodents are reviewed by Francescoli in chapter 3 of this volume. The following brief discussion is merely a summary of morphological trends in the visual, auditory, chemical, and mechano-sensory systems of these animals.

Reduction and even loss of the visual sense organs in subterranean rodents is countered by hypertrophy of the auditory and mechanical receptors. Enhanced acoustic and tactile sensitivities presumably aid in underground navigation, in detection and avoidance of predators, in finding mates, and in foraging (Eloff 1951). Subterranean rodents also tend to decrease the length of external protuberances (e.g., ear pinnae) associated with their sensory systems to prevent irritation due to friction while burrowing and to decrease or minimize the exposure of delicate sense organs (see above). Given these dramatic anatomical shifts, parallel modifications in the nervous system might be expected. However, other than inferences that can be derived from examination of cranial and cervical foramina (e.g., Wahlert 1978, 1985) and limited data on the innervation of muscles in these taxa (e.g., Woods 1975), no information exists concerning the structure or function of the nervous system in subterranean rodents.

Visual System

A subterranean way of life has not resulted in visual capabilities comparable to those of nocturnal mammals. Instead, there is a tendency toward both reduction in the size of the eye and structural regression of the visual system (Burda, Bruns, and Müller 1990). However, there is no direct correlation between the presence and size of external eyes in subterranean rodents and their degree of photosensitivity (de Jong et al. 1990). Changes in the visual apparatus may involve a decrease in size of the eyeball itself or in the number and size of the ocular muscles that protrude and retract the eye. Similarly, reduction in vision may be due to a thickening of the cornea or to retinal degeneration. Thus visual capacity may be relatively independent of the size of the eye. The extent of regression of the visual system is not correlated with the phylogenetic age of the lineages under consideration, nor with the degree to which a rodent is subterranean (Burda, Bruns, and Müller 1990).

In *Nannospalax,* the eyes are completely degenerate and lie subcutaneously. Nevertheless, the animals are able to detect photoperiod and to differentiate light from dark, although they cannot see images (de Jong et al. 1990). Whether this loss is related to the specialized mode of head-lift digging (see below) in this genus has not been investigated. Geomyids, bathyergids, rhizomyines, *Myospalax,* and *Ellobius* all have small to minute eyes. *Georychus,* which spends some time at the soil surface, has larger eyes than other bathyergid genera (Macdonald 1984). The eyes of bathyergids are partly degenerate and lack photosensitivity (Eloff 1951), although Poduschka (1978) reported that *Cryptomys* can perceive light. Other evidence suggests that the surface of the bathyergid eye may be able to detect air currents that would indicate damage to the burrow system (Macdonald 1984). In rhizomyines, the eyes are partly degenerate but perceive light. In *Tachyoryctes,* they are sufficiently functional to detect predators (Jarvis 1973). Lay (1967) observed that *Ellobius* allowed to wander freely frequently fell into holes, over drop-offs, and into streams, behavior that he felt might be attributed to poor visual acuity. In *Prometheomys,* the eyes are reduced, and this animal is reported to avoid light (Ognev 1950). In *Ctenomys* and *Spalacopus,* the eyes are relatively large and positioned high on the head (Camin, Madoery, and Roig 1995). Since *Spalacopus* rarely emerges from its burrow, such placement may aid it in scanning the landscape for predators prior to making an occasional dash from one hole to the next (Reig 1970). Whereas reduction in size of external eyes is typically viewed as loss due to disuse, diminution of eye size in these rodents may also decrease the risk of injury to this delicate and complex organ.

Auditory System

Although pinnae are known to both amplify sounds and localize the direction from which sounds come, these functions may not be extremely important to animals living underground. To date, however, no one has studied the consequences of this structural loss for auditory function in subterranean mammals. The range of detectable sound frequencies for each of these genera currently is not known. The structure of their auditory systems may be related to their degree of sociality, the amount of time spent underground, soil density and composition, and whether or not vocalizations are detected through tunnels or through solid dirt (Francescoli, chap. 3, this volume).

It is expected that a large external auditory canal might permit dirt and soil to enter the ears, causing disequilibrium and decreasing an animal's ability to detect sounds of a certain frequency. However, Burda, Bruns, and Müller (1990) noted that the outer ear canal, which is a part of the external ear, is always normally developed. A functional explanation for this constancy of size has not been proffered. Agrawal (1967) reported that the external auditory meatus is spout-shaped and projects posteriorly in *Rhizomys* and *Cannomys* owing to strong movements of the mandibular articulation during digging and chewing. In a somewhat related vein, Lindenlaub, Burda, and Nevo (1995) compared the morphology of the vestibular organs of *Cryptomys* and *Nannospalax* with those of *Rattus* and concluded that specialization of this structure in subterranean taxa may compensate for the loss or reduction of other senses in these genera.

Chemoreceptors and Mechanoreceptors

If not actually more "highly developed," the tactile senses are undoubtedly more frequently employed by subterranean rodents than by their surface-dwelling relatives (Burda, Bruns, and Müller 1990; Francescoli, chap. 3, this volume). The nose, vibrissae, and foot pads are all used to detect external stimuli such as food, potential mates, or the presence of predators. In addition, the feet and vocal cords may be used to generate audible or vibratory signals that can be detected by others underground.

Sokolov and Kulikov (1987) reported on the structure and function of rodent vibrissae in a variety of surface-dwelling taxa. Based on their examination of macro- and micromorphology, these authors concluded that vibrissae are used, in combination with other senses, for close-range orientation and control of body position in relation to the substrate. In some taxa, the vibrissae may also aid in controlling the position of food taken into the mouth. Jarvis and Bennett (1991) reported that in *Heterocephalus*, sensory hairs scattered along the relatively long tail guide the animal as it moves posteriorly, a conclusion in keeping with Sokolov and Kulikov's result.

Olfaction is an important sense in subterranean mammals, although there is variation among taxa in its degree of development. Jarvis and Sale (1971) reported that the large, flat snouts of *Heliophobius* and *Heterocephalus* are extremely sensitive. Burda, Bruns, and Müller (1990) argued that the subterranean ecotope does not, in and of itself, represent a selective force that either promotes or depresses the development of chemoreceptive systems. By way of example, they noted that olfactory brain centers are well developed in *Nannospalax*, which lacks external eyes, but that they are quantitatively regressed in *Cryptomys*, a genus in which the eyes are only partly degenerate.

From this brief review it is clear that studies of the sensory systems of subterranean rodents are still in their infancy. Much work remains to be done before a complete understanding of how these mammals perceive and interact with their environments is achieved. Both descriptive and functional studies that consider multiple taxa are needed.

The Nature of Digging

Despite the apparent constancy and predictability of life underground, striking differences in the structural elements of the digging apparatus abound. Within the basic morphological blueprint outlined above, there has been a proliferation of structural modifications to facilitate tunnel excavation in a variety of soils. For example, Lessa and Thaeler (1989) identified two alternative morphological strategies related to digging: an increase in incisor procumbency versus an enlargement of the forearms, as exemplified by thomomyine and geomyine pocket gophers, respectively. These alternative responses are thought to reflect the diversity of soil types in which these rodents are found. The degree to which subterranean mammals locomote above ground to disperse, to forage, or to find mates is also thought to influence the nature of the digging apparatus.

What is immediately apparent upon examination of the digging behavior of subterranean rodents is that morphological variations among genera frequently do not reflect currently accepted phylogenies (table 1.1; Cook, Lessa, and Hadly, chap. 9, this volume). For example, *Ellobius* and *Prometheomys*, both members of the murid subfamily Arvicolinae, do not break soil or remove dirt in a similar manner (Hinton 1926; Gambaryan and Gasc 1993). The same is true of *Bathyergus* relative to the remaining members of its clade, although the phylogenetic relationship of *Bathyergus* to the other four bathyergid genera remains uncertain (Maier and Schrenk 1987; Honeycutt et al. 1991; Faulkes et al. 1997). Given the absence of a clear phylogenetic component, structural variation in the digging apparatus of subterranean rodents must be viewed as the outcome of complex interactions among phylogenetic history, soil type, and the duration and nature of surface activities.

Hildebrand (1985) divided the digging activities of all subterranean mammals into the following functional units: (1) producing and transmitting force, (2) transporting soil, (3) resisting loads, and (4) sustaining high or long-term activity. In contrast, Gambaryan and Gasc (1993) characterized digging as a four-step process comprising the sequence of events that must be performed during tunnel excavation. First, the soil is broken. Second, the loosened dirt is raked up under or next to the body. Third, the accumulated dirt is moved posterior to the animal with five or six kicks of its hind feet. Fourth, the animal deposits the dirt outside of its burrow. The following

TABLE 1.1 COMPARISON OF THE FUNCTIONAL AND STRUCTURAL ELEMENTS OF THE DIGGING APPARATUS IN SUBTERRANEAN RODENTS

	Primary digging mode	Secondary digging mode	Dirt removal method	Present above ground
Bathyergidae				
Bathyergus	Scratch	??	Back kick	Rarely
Cryptomys	Chisel-tooth	Snout	Back kick	Rarely
Georychus	Chisel-tooth	Snout	Back kick	Rarely
Heliophobius	Chisel-tooth	Snout	Back kick	Rarely
Heterocephalus	Chisel-tooth	Snout	Back kick	Rarely
Ctenomyidae				
Ctenomys	Scratch	Incisors	Back kick	Rarely
Octodontidae				
Spalacopus	Chisel-tooth	Claws	Back kick	Diurnally to feed
Muridae				
Myospalax	Scratch	Head	Turn 180°	Rarely
Ellobius	Head-lift	Incisors	Turn 180° + back kick	Frequently
Prometheomys	Scratch	??	Turn 180°	To disperse + nocturnally to feed
Nannospalax	Head-lift	Snout	Turn 180°	Nocturnally to feed
Cannomys	Chisel-tooth	Claws	Turn 180°	Nocturnally to feed
Rhizomys	Chisel-tooth	Claws	Turn 180°	Nocturnally to feed
Tachyoryctes	Chisel-tooth	Claws	Turn 180° + back kick	Nocturnally to feed
Geomyidae				
Geomys	Scratch	Incisors	Turn 180°	Occasionally
Thomomys	Scratch	Incisors	Turn 180°	To disperse
Orthogeomys	Scratch	Incisors	Turn 180°	??
Pappogeomys	Scratch	Incisors	Turn 180°	??
Zygogeomys	Scratch	Incisors	Turn 180°	??

discussion considers both of these conceptual approaches to digging and their implications for morphological specialization.

Breaking Up Soil

Among subterranean rodents, three methods of breaking up soil have been recognized. These techniques are referred to as scratch digging, chisel-tooth digging, and head-lift digging (Hildebrand 1985).

SCRATCH DIGGING. Scratch digging is characterized by the alternate flexion and extension of the forelimbs. Soil is broken and loosened with the claws, after which it is pushed or flung posteriorly with the pads of the forefeet. This method of digging predominates among geomyids and is used almost exclusively by *Geomys* (E. P. Lessa et al., unpublished data). *Ctenomys, Bathyergus, Myospalax,* and *Prometheomys* also employ this technique (Ognev 1950; Landry 1957; Lessa and Thaeler 1989; Gambaryan and Gasc 1993; Camin, Madoery, and Roig 1995). Among scratch diggers, some geomyids and ctenomyids differ from other taxa by using their teeth as well as their claws to break up highly compact soils (Lessa and Thaeler 1989). Although *Myospalax* loosens earth with its claws, it moves soil with its head, in a manner similar to that used by *Nannospalax* to remove dirt (see below; Gambaryan and Gasc 1993). In hard soils, *Myospalax* continuously flexes and extends its claws until the soil is loosened, then sweeps it posteriorly. In soft soils, the act of loosening dirt is combined with raking it under the body in a single process, evidence that digging modes are not rigid even within a taxon and may be influenced by soil type and density.

CHISEL-TOOTH DIGGING. Chisel-tooth digging is characterized by use of the procumbent incisors to break up soil. In conjunction with powerful head and jaw muscles, the incisors loosen soil, which is subsequently removed from the burrow with the head or feet. This mode of digging has been observed in rhizomyines, *Spalacopus,* bathyergids other than *Bathyergus,* and some pocket gophers of the genus *Thomomys* (Holliger 1916; Landry 1957; Genelly 1965; Reig 1970; Jarvis and Sale 1971; Jarvis 1973; Lessa and Thaeler 1989; Flynn 1990). Bathyergids also employ their snouts while digging (Jarvis and Sale 1971), and both rhizomyines and *Thomomys* use their claws (Xu 1984; Lessa and Thaeler 1989). Similarly, Gambaryan and Gasc (1993) observed that in soft soils, *Ellobius* uses its incisors to loosen dirt before raking it under the belly with its forefeet, while Reig (1970) reported that *Spalacopus* uses both its incisors and forefeet to loosen soil. Although *Tachyoryctes* is considered a chisel-tooth digger, it has been reported that these animals use their forefeet to dig and their incisors merely to cut intervening roots (Nowak 1991).

HEAD-LIFT DIGGING. Head-lift digging is characterized by use of the incisors in concert with the skull to form a powerful drill and shovel combination that is capable of loosening and removing soil (Hinton 1926; Nevo 1961; Watson 1961; Formozov 1966; Agrawal 1967; Lay 1967; Laville et al. 1989; Gambaryan and Gasc 1993). This mode of digging is employed by *Ellobius* and *Nannospalax* to excavate shallow tunnels or to compact soil in deeper runways. Although the head and fur of these genera are highly modified, the fore- and hind feet are broad but otherwise relatively unspecialized. *Nannospalax* also has a broad, tough, muscular nose pad and a snout callus, both of which are used to tamp dirt against the burrow walls and onto the surface mounds formed around burrow entrances. The rhinarium in *Ellobius* is naked and somewhat flattened, perhaps reflecting similar uses during tunnel excavation.

Removing Loosened Soil

Once the soil has been broken up and a sufficient quantity of debris has accumulated around or beneath an animal, it must be removed from the tunnel under construction (see table 1.1). Most commonly, mounds of dirt are removed as an animal backs up in the burrow, kicking the dirt behind itself (Gambaryan and Gasc 1993). Bathyergids, *Ctenomys, Spalacopus,* and *Ellobius* have been reported to move backward, throwing dirt posteriorly with a single kick or with simultaneous thrusts of their hind feet (Genelly 1965; Lay 1967; Reig 1970; Hickman 1985; Harrison and Bates 1991; Camin, Madoery, and Roig 1995). In the eusocial *Heterocephalus,* colony members may work individually or in relay to accomplish this task (Jarvis and Sale 1971).

Other subterranean rodents are reported to turn 180° in their tunnels and push the loosened dirt from their burrows face first, like a bulldozer. Gambaryan and Gasc (1993) observed that *Ellobius* employs this technique as well, pushing soil with its breast and upper incisors. Pocket gophers use their forefeet, palms turned outward and juxtaposed in front of their faces, to accomplish this task (Breckenridge 1929). Rhizomyines use their forefeet, chin, and chest. *Myospalax* pushes the dirt with its forefeet and then tamps the dirt with its head (Dubost 1968). *Nannospalax* uses its nose, chest, and the top of its head (Ognev 1950; Nevo 1961). *Prometheomys* is molelike and uses only its head to expel soil (Hinton 1926; Dubost 1968).

In contrast to methods for breaking up soil, the manner in which dirt is removed from a tunnel seems to be phylogenetically conserved, with a major dichotomy between sciurognath and hystricognath taxa. As this discussion indicates, however, the descriptive categories used to characterize the two major phases of tunnel excavation are not rigid, and the same species may employ elements of both methods. For example, it has been reported that *Tachyoryctes* removes most soil using simultaneous kicks of its hind feet (Hickman 1983). Others have observed that once a mass of dirt has accumulated within a tunnel, these animals expel soil with the muzzle (Yalden 1975), use the chest as a scoop (Lehmann 1963), or use one side of the face and one forefoot to expel dirt (Jarvis and Sale 1971). Such contradictory reports call for further examination of these behaviors.

Primary Components of the Digging Apparatus

Given that digging is a major activity in which all subterranean rodents are involved, it is not surprising that the overwhelming majority of morphological studies of these mammals have focused on examining, describing, and measuring the primary structures used in digging. Regardless of digging mode, all subterranean taxa show some degree of modification of the teeth, head, neck, and forelimbs that is thought to be associated with specialization for life underground. As Nevo (1995) has pointed out, the overall body plan in fossorial mammals may be viewed as a series of contrasts, with anatomical reductions in external body elements opposing hypertrophies in the digging apparatus.

Because of their use in phylogenetic analyses, most details of cranial osteology and dental morphology have been fully described for subterranean rodents. Diagnostic characters for each family, subfamily, and genus can be found in the literature (see above). Nevertheless, characteristics of the skull and teeth that are convergent among these taxa warrant summary and some additional comment.

For readers interested in general reviews, Bekele (1983a,b) describes details of cranial osteology and myology in *Tachyoryctes splendens,* Yalden (1985) covers cranial osteology in *T. macrocephalus,* and Flynn (1990) reviews general rhizomyine morphological features. Lehmann (1963) should be consulted for details concerning the forearm myology in *Tachyoryctes,* which she compares with that in *Geomys* and *Ctenomys.* Orcutt (1940) describes the muscles of the head, neck, and pectoral appendages in *Geomys bursarius,* and Holliger (1916) considers the anatomical adaptations in the forelimb of *Thomomys.* Holliger's work was followed two decades later by Hill's (1937) description of all osteology and myology in *Thomomys.* Woods (1975) describes the morphology of the hyoid, laryngeal, and pharyngeal regions in bathyergid rodents and other selected hystricomorph genera. Sterba's (1969) description and comparison of pectoral muscles in several rodent genera includes *Nannospalax.* Parsons (1896) summarizes the myology of several myomorph genera, including *Rhizomys, Georychus,* and *Bathyergus.* Despite these efforts, morphological descriptions and functional analyses of the head, neck, and pectoral regions are lacking for most subterranean rodents.

Teeth

INCISORS. Because incisors act like picks, delivering a great deal of force to a restricted area, they are intimately involved in the digging process, even if they are not the primary digging tool. In general, the upper incisors of sub-

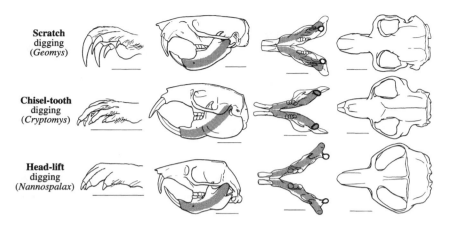

FIGURE 1.1. Primary elements of the digging apparatus in scratch digging, chisel-tooth digging, and head-lift digging subterranean rodents, showing differences in skull shape, degree of incisor procumbency, placement of incisor roots, and size of the claws of the forefoot.

terranean rodents are more chisel-shaped than those of surface-dwelling taxa, thus enhancing the animals' ability not only to loosen soil but also to cut through roots, tubers, and the soft parts of plants. Additionally, the more pressure that is applied to an incisor tip, the greater the functional advantage in having a longer incisor that tapers sharply at the tip, because soil resistance is decreased and applied force increased with a reduction in the cross-sectional area of the incisors (Lessa 1990). An examination of the upper incisors of chisel-tooth diggers and head-lift diggers versus scratch diggers reflects this strategy and reveals a gradation in incisor width and shape that correlates with both the degree of procumbency and the extent to which the teeth are used in digging by each of these taxa.

Incisor procumbency is influenced by both the degree of curvature of the teeth and their position in the rostrum (Landry 1957; Akersten 1981) (fig. 1.1). Characteristically, the large, elongated roots of the upper incisors of subterranean rodents extend to (e.g., in *Prometheomys*) or beyond (e.g., in *Cannomys, Spalacopus, Ellobius*) the first molar (M^1). In all bathyergids except *Bathyergus,* the roots extend above or posterior to the upper molar tooth row, appearing between the tooth rows in a palatal view of the skull. This condition is unique within rodents (Ellerman 1940). In *Nannospalax,* the roots form a slight knob on the palate anterior to the first upper molar. In those thomomyine pocket gophers that exhibit procumbency (subgenus *Megascaphus*), the incisor root is shifted slightly more posteriorly than in other *Thomomys (sensu stricto)* (Thaeler 1968, 1980; Lessa and Patton 1989).

Roots of the lower incisors in subterranean rodents are also shifted posteriorly and insert close to or into the mandibular condyles (fig. 1.1). *Ellobius* and *Nannospalax*, both head-lift diggers, seem to exhibit the most extreme condition in this regard. The roots of their lower incisors extend posterior to the second lower molar and the region of the glenoid process, forming a prominent knob on the inner margin of the mandibular ramus (Gromov and Polyakov 1992).

Incisor curvature is a function of both the radius of curvature of the teeth and tooth length. Curvature is produced by the difference in growth rate between the dorsal and ventral incisor surfaces—that is, between the deposition rates of enamel and dentine. Hard enamel, which exists only on the anterior surface of rodent incisors, ensures that the softer dentine portion of the tooth will wear away more quickly, producing and maintaining a self-sharpening edge that is critical to survival in these mammals. Since all rodent incisors have this asymmetrical pattern of enamel and dentine, subterranean rodents compensate for increased wear on their incisors by dramatically increasing the overall growth rate of these teeth. Although data on incisor growth rates are lacking for most subterranean species, the upper incisors of pocket gophers have been reported to grow an average of 0.35 mm/day in *Geomys* and more than 0.5 mm/day in *Thomomys* (Miller 1958; Manaro 1959), rates that are almost double those recorded for non-subterranean taxa (Howard and Smith 1952).

Although chisel-tooth digging is accomplished primarily with the lower incisors, it is the upper incisors that show greater variability in their degree of procumbency. This apparent contradiction is at least partially explained by the large anterior expansion of the glenoid fossa in most subterranean rodents. In all rodents, some degree of expansion permits the mandible to slide anteriorly, allowing the lower incisors to approximate the uppers. While this expansion appears relatively greater in subterranean taxa, no studies have quantified this structural variation with respect to function among rodent taxa that differ in their primary incisor use (e.g., gnawing wood, biting seeds, breaking soil). In contrast, because the upper incisors are fixed within the cranium, varying their degree of procumbency is the means by which an appropriate angle of action is maintained relative to the lower incisors (Lessa 1990).

Cryptomys and *Heliophobius* have the most procument upper incisors of any rodent, with tips that project far forward of the rostrum (Landry 1957) (see fig. 1.1). *Heterocephalus* and *Georychus* also have procument incisors, but in *Bathyergus*, the scratch-digging member of this clade, the incisors are not procumbent. In *Ellobius*, the incisors are so procumbent that their whole anterior surface is visible dorsally (Harrison and Bates 1991). In *Prometheomys*, the upper incisors are strongly curved but are not procumbent, extending only as far anteriorly as the nasal bones (Hinton 1926). Within the rhizomyines, the incisors of *Rhizomys* are nearly vertical, whereas those of *Cannomys* and *Tachyoryctes* are procumbent. Among pocket gophers, species and subspecies in the genus *Thomomys* tend to show greater incisor procumbency than do most members of the scratch-digging genus *Geomys*. The highly speciose genus *Ctenomys* exhibits a wide range of variation in incisor procumbency that is comparable to the variation reported among species of *Thomomys*. In contrast, its closest subterranean relative, the genus *Spalacopus*, contains only one species, which possess highly procumbent incisors. In *Nannospalax*, the upper incisors are massive but essentially vertical in orientation.

Lower incisors are generally less procumbent than upper incisors because their primary function is to manipulate dirt rather than to break up soil (Hildebrand 1985). The exceptions to this functional rule are the chisel-tooth-digging bathyergid genera and *Nannospalax*. In *Heliophobius*, *Heterocephalus*, *Georychus*, and *Cryptomys*, the two halves of the mandible are not ankylosed; the lower incisors, which are intimately involved in digging, are capable of moving separately (Eloff 1951). In *Nannospalax*, the lower incisors protrude anteriorly to such a degree that they are permanently outside the mouth, permitting this rodent to dig without opening that orifice (Topachevskii 1976). (For a discussion of furred lips that exclude dirt, see above.) Because the degree of incisor procumbency may vary among subterranean members of a single family or clade, comparative functional analyses of dental morphology in closely related (e.g., intrafamilial) taxa may be illuminating.

Subterranean rodents generally have an unusually loose articulation between the mandibular condyles and the glenoid fossae that permits the mandibles to move in an anterior-posterior plane. As a result, the lower jaw (and the teeth it contains) is more easily manipulated than the upper jaw. Consequently, the lower incisors wear more quickly and exhibit a growth rate approximately twice that of the upper incisors (Miller 1958). In *Geomys*, the lower incisors grow an average of 0.67 mm/day, while in *Thomomys* they grow almost 1.0 mm/day. These rates are more than double those recorded for lower incisor growth in non-subterranean taxa (Howard and Smith 1952).

Visual inspection of the teeth of subterranean rodents shows a trend toward decreasing color with increasing procumbency. In *Rhizomys*, *Tachyoryctes*, *Ctenomys*, *Geomys*, and *Myospalax*, the enamel is dark orange. In contrast, in *Thomomys*, *Cannomys*, *Spalacopus*, *Nannospalax*, and *Prometheomys*, it is a pale orange, and in the bathyergids and in *Ellobius* it is white. Within the genus *Thomomys*, the more highly procumbent species exhibit the palest incisor color. As one might expect from its name, *T. bottae leucodon*, from northern California, has the most procumbent incisors of any subspecies in that taxon. Given these observations, it would be interesting to know whether the deposition rates of orange and white enamel differ and, if they do, whether they explain the observed relationship between tooth color and procumbency. The topic of calcium dumping in subterranean rodents and its possible relationship to tooth color is addressed by Buffenstein in chapter 2 of this volume.

Although much has been written about the enamel microstructure of rodent incisors and its relationship to the phylogenetic history of members of that order (Koenigswald 1985; Martin 1992), far less is known or understood about the relationship of microstructure to incisor function. Flynn, Nevo, and Heth (1987) examined incisor enamel in four chromosomal species of *Nannospalax ehrenbergi* and compared its microstructure with that in other subterranean genera. Their results suggested that enamel thickness and the composition of enamel layers are correlated with the extent to which the incisors are used in digging. A study by Justo, Bozzolo, and De Santis (1995) of selected ctenomyid and octodontid rodents reached a similar conclusion, as did Buzas-Stephens and Dalquest (1991), who examined three geomyid genera. Buzas-Stephens and Dalquest (1991) attributed variation in enamel ultrastructure between incisors and premolars/molars to differences in primary tooth function (i.e., use of the incisors for digging versus use of the premolars and molars in mastication).

The presence of grooves on the anterior surface of the upper incisors has been interpreted in several ways. Merriam (1895) hypothesized that the serrate tip formed by the groove is useful in biting food objects. Russell (1968) proffered an almost opposite explanation, suggesting that the grooves might aid in extracting the teeth from a food item. He also proposed that they serve to strengthen the teeth without adding to their overall width. Akersten (1973) analyzed the incisor grooves of pocket gophers within a phylogenetic context and concluded, like Russell, that the primary function of the median groove was to strengthen the incisor. In addition, Akersten hypothesized that the lingual groove served to reinforce the lateral edge of the incisor against wear and chipping during burrowing.

In *Bathyergus, Geomys, Orthogeomys, Pappogeomys,* and *Zygogeomys* the upper incisors are heavily grooved on their anterior surface. In *Heterocephalus* and *Prometheomys* the anterior surface appears to be only lightly etched. In the remaining taxa, no incisor grooves are evident. This taxonomic distribution of grooved incisors does not support Akersten's functional explanation. If scratch diggers had weaker incisors and therefore required grooves to strengthen their teeth (as suggested by the presence of strong grooves in *Bathyergus* and in pocket gophers other than *Thomomys*), then one would expect grooves also to have evolved in *Ctenomys* and *Myospalax*, but not in *Heterocephalus*. This is not the pattern observed. Additional hypotheses and functional tests that might explain the distribution and degree of definition of these structures in rodents are needed.

MOLARS. It has been presumed that the number and configuration of rodent molars primarily reflect the phylogenetic histories of the various lineages within this order. This assumption is exemplified most clearly in the preponderant use of dental characters such as enamel structure, complexity of the molar pattern, and rootedness of the cheek teeth in phylogenetic analyses. The only general modification of the molars apparently associated with a subterranean way of life is the tendency of the molar tooth row to be relatively greater in length than that of non-subterranean taxa. Among subterranean rodents, however, intergeneric differences in molar structure are evident. For example, in *Ellobius,* the enamel pattern of the molars is considerably modified from the usual arvicolid type (Allen 1940), and *Heliophobius* has more molars than any other living rodent (Ellerman 1940). In *Spalacopus,* the cheek teeth are little modified from the typical octodontid pattern, but the diastema is relatively longer than in *Ctenomys,* a modification presumed to permit increased protraction and retraction of the jaw (Bekele 1983a). Only secondarily have these dental differences among genera been correlated with any specified function. In *Nannospalax,* there is evidence that molar pattern varies among populations, responding to local differences in soil type, climate, and vegetation (Butler et al. 1993). In addition, both the overall shape of the molars (which ranges from cuspidate to laminate-prismatic) and their relative lengths often reflect intergeneric differences in diet (Busch et al., chap. 5, this volume).

Head and Neck

Regardless of digging mode, subterranean mammals must exert forces capable of breaking hard soils. That is, they must generate out-forces (F_o) that exceed the resistance of the soil, as described by

$$F_o = F_i l_i / l_o$$

where F_i = in-force generated by the muscles, l_i = in-lever acting to produce the in-force, and l_o = out-lever associated with the out-force (Hildebrand 1974; Lessa 1990) (see fig. 1.2). The design of the bone-muscle systems of the head, neck, and forelimbs reflects the morphological compromise between this need and the constraints imposed by both phylogenetic history and the other functions for which these structures are used (e.g., locomotion, food gathering).

OSTEOLOGY. In general, modifications of the skull and its degree of adaptation to a subterranean way of life appear to be correlated with the role that a given structure plays in the digging process and the types of food that an animal eats (Agrawal 1967). As might be expected, modifications in cranial and cervical osteology are more apparent in chisel-tooth and head-lift diggers than in scratch diggers. In their studies of mastication in pocket gophers, Wilkins and Woods (1983) and Wilkins (1988) supported this notion by concluding that skull shape in pocket gophers was primarily an adaptation to fossoriality and only secondarily reflected masticatory modes.

The skulls of subterranean rodents are generally characterized as being more massive than those of their surface-dwelling relatives. In particular, the crania tend to be dorsally flattened and to be relatively broader and deeper than the skulls of non-subterranean taxa, modifications that increase the surface area available for muscle attachments (see fig. 1.1). In addition to accommodating larger, more powerful muscles, the resultant change in skull shape also causes muscle insertions to shift in ways that allow the jaw muscles to exert increased leverage or force against a resistant surface.

The occiput of the skull is often enlarged and inclined anterodorsally in subterranean rodents, providing increased area for insertion of robust neck muscles. This structural tendency is especially apparent in *Nannospalax*, in *Ellobius*, and, to a lesser extent, in the rhizomyines, and it contributes to the wedgelike or triangular shape of the skull that is characteristic of subterranean rodents when viewed dorsally or in profile. The posteriorly widened zygomatic arches in subterranean rodents serve as origins for the enlarged jaw muscles, principally the *masseter*, a primary jaw closer. These flaring structures also contribute to the wedge-shaped appearance of the head. Topachevskii (1976) suggested that the strongly triangular shape of the skull in the head-lift digger *Nannospalax* corresponded to that of a spade, reflecting its role in moving loosened dirt (see fig. 1.1).

In subterranean rodents, the interorbital and temporal ridges generally fuse over the frontal and parietal bones to form a continuous dorsal crest that protrudes from the skull and often flares posteriorly on the surface of the interparietal. This crest tends to increase in size and proximity to the midline with increasing age. The squamosal bones, which serve as the point of origin for the *temporalis*, also may be relatively larger in these genera, extending anteriorly over the frontal bones and encroaching posteriorly on the parietal and interparietal bones. In *Rhizomys*, *Cannomys*, and *Ellobius*, the parietals are reduced laterally, and adults lack an interparietal.

In head-lift diggers and, to a lesser extent, in chisel-tooth diggers, a massive rostrum and flaring premaxillae support the nose pad. The premaxillae extend anterior to the nasals in taxa such as *Ellobius*, in which additional support of the highly procumbent incisors is needed. Often, prenasal ossicles are present while the nasals themselves are generally reduced anteriorly to prevent their interference with digging. These rodents may also have enlarged and inflated tympanic bullae that measure more than 20% of the occipitonasal length, a modification thought to aid in resonating and amplifying sound vibrations (Agrawal 1967; Francescoli, chap. 3, this volume). In *Spalacopus,* however, the auditory bullae are not enlarged, a condition that differs from that found in other octodontids (Woods and Hermanson 1985).

On the ventral surface of the cranium, there is a tendency toward posterior shortening of the incisive (anterior palatine) foramina as a means to accommodate increasing pressure from the deep-rooted upper incisors that lie dorsal to it. Similarly, the palatal portion of the skull is relatively more robust in subterranean taxa. Although the function of the vomeronasal (Jacobson's) organ in mammals is obscure (Hildebrand 1974; Francescoli, chap. 3, this volume), the size and placement of the incisive foramina may also relate to the degree to which chemoreception is important in subterranean taxa.

The mandible in subterranean rodents tends to be longer and deeper than that in surface-dwelling genera to accommodate the elongated lower incisors and to increase the area available for attachment of the enlarged jaw muscles in most taxa. There also is a prominent masseteric crest over the mandibles for insertion of the *masseter.* Bekele (1983a) reported that the upper and lower diastemas are proportionately longer in *Tachyoryctes* than in *Rattus,* and he suggested that this change allowed for increased protraction and retraction of the jaw. Corti et al. (1996) examined morphometric variations in the size and shape of the mandible in multiple populations of four chromosomal species of *Nannospalax ehrenbergi.* They concluded that size differences between the sexes and across species are related to chromosome number and ecogeography, but that the biomechanical significance of these differences in shape is not easily interpreted and warrants additional study.

Other general osteological modifications of the skull and cervical region include an increase in size of the lambdoidal (nuchal) crest for insertion of the robust neck muscles, a general shortening of the neck, a thickened axial spine, and a reduction of the transverse processes on the cervical vertebrae. This last condition is particularly pronounced in *Nannospalax* and *Cryptomys.* In addition, in *Pappogeomys (Cratogeomys)* (Gupta 1966) and in *Nannospalax* (Laville et al. 1989), several of the cervical vertebrae are closely fitted or fused as a means to stabilize and buttress the neck.

MYOLOGY. The principal myological modification of the head and neck that appears to be associated with burrowing is a dramatic increase in size of the muscles that close the jaw *(masseter* and *temporalis)* and that raise and lower the head and stabilize the neck *(rhomboideus capitis* and *cervicis, splenius, rectus capitis dorsalis major* and *minor,* and *obliquus capitis dorsalis major).* Along these lines, Bekele (1983b) and Hill (1937) have reported a dramatic decease in size and a complete absence of some neck muscles associated with transverse movements of the head (e.g., *sternomastoid, longissimus capitis,* and *semispinalis capitis*) in *Tachyoryctes* and *Thomomys,* respectively. Bekele hypothesized that this condition was due to either the unimportance of transverse movements in *Tachyoryctes* or the usurpation of this function by other muscles. Both suggestions warrant testing.

A decrease in size and importance of the eyes in subterranean mammals (see above; Francescoli, chap. 3, this volume) also has resulted in a migration of the origins of the jaw musculature to include the orbits in many taxa. In *Ellobius,* the enlarged *temporalis* muscle extends just beyond the lateral borders of the parietal bones, creating an indistinct ridge on each side of the head (Allen 1940). Krapp (1965) provides a complete description of the muscles of mastication in *Nannospalax* and relates his findings to the animal's subterranean way of life.

Forelimbs

In addition to the teeth and the head and neck, the forelimbs constitute the third primary component of the digging apparatus in subterranean rodents. Overall, these structures appear relatively short, with massive bones that are dense and able to resist both torsion and bending (see below). Even chisel-tooth diggers profit from the relative increase in force that such modified forelimbs can exert because these taxa use the downward pressure of their forefeet to counterbalance the upward pressure that their lower incisors exert against the soil. Similarly, subterranean rodents that use their heads to move dirt must use their forelimbs for support and force transfer, thereby increasing pressure on those structural elements relative to non-subterranean taxa. Those rodents that remove dirt with simultaneous thrusts of the hind feet often stand on their forefeet while the posterior portion of the body is dorsally inclined (Genelly 1965).

OSTEOLOGY. Because burrowing results in greater force being generated by and placed on the limbs, most digging mammals have relatively short limb bones, exclusive of the claws on the manus (Casinos, Quintana, and Viladiu 1993). Functionally, this change in shape is equivalent to increasing the out-force by decreasing the out-lever of the forearm (fig. 1.2). The limb bones also have pronounced processes and tuberosities for muscle attachment that clearly reflect their participation in digging. Proximally, there is a marked increase in size of the teres major process of the scapula, which can be seen as a ventral prolongation of its vertebral border. The posterior scapular fossa is also enlarged to accommodate the increased size of the origins of the *teres*

MORPHOLOGY OF SUBTERRANEAN RODENTS   41

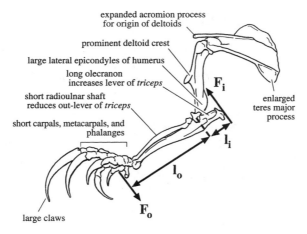

FIGURE 1.2. Lateral view of the forearm in a pocket gopher *(Geomys)*. The in-lever (l_i) and out-lever (l_o) which produce the in-force (F_i) and out-force (F_o) necessary to dig are shown in relation to many of the osteological and myological modifications commonly observed in subterranean rodents.

major muscle and the long head of the *triceps*. The scapular spine, with a long, flaring acromion process, often extends beyond the glenoid fossa and serves as point of origin for an enlarged *deltoideus*. The coracoid process of the scapula is generally stouter and more prominent than in surface-dwelling taxa, reflecting a decrease in lateral movement around the shoulder joint in subterranean species. Similarly, the glenoid fossa, which articulates with the head of the humerus and is typically round in shape, is ovoid or elongated along the anterior-posterior axis in subterranean taxa.

Biknevicius (1992, 1993) has shown that differences in burrowing styles in at least some subterranean rodent genera are reflected in specific differences in the diaphyseal cross-sections of the humerus, where increased buttressing by cortical bone is observed. An increase in the cortical area of the bone shaft would help to both counter the large compressive stresses to which the humerus is routinely subjected and resist bending and torsion of that bone during excavation.

The humerus in subterranean rodents usually possesses a prominent and somewhat distally placed deltoid crest for insertion of an enlarged *deltoideus* muscle. An enlarged medial epicondyle of the humerus serves as the point of attachment for the origin of the *pronator* muscles and the carpal flexors. An enlarged lateral epicondyle is the site of origin for the *supinator* and the forearm extensors. The humeral head is generally ovoid rather than round (see above), and the capsular ligament of the glenohumeral joint may be loose. These last two modifications increase the possible range of motion in an anterior-posterior plane at that joint, while restricting lateral movement.

With respect to the distal long bones of the forearm in subterranean rodents, the ulna is relatively shorter than in surface-dwelling taxa, and its shaft may be noticeably curved. A decrease in length of the ulna is, in part, indicative of a decrease in the area available for insertion of the out-lever muscles, in this case, the medial head of the *triceps*. Correspondingly, the relatively greater length of the olecranon process in subterranean mammals increases the length of the in-lever for insertion of the long and lateral heads of the *triceps*. Hildebrand (1974) noted that in pocket gophers and bathyergid rodents, the olecranon process of the ulna is equal to approximately one-third the length of the ulna distal to the elbow joint. In spalacines, the distance is greater than one-third that length (Topachevskii 1976). Decreasing the relative length of the ulnar shaft (out-lever) while increasing the relative length of the olecranon process (in-lever) is an additional means of increasing the force that a subterranean rodent can exert against the soil during burrow excavation.

Both the radius and the ulna may show more pronounced articulations with the humerus in subterranean rodents than in their non-subterranean relatives. For the radius, this is accomplished via an ovoid (versus round) articular surface; for the ulna, it is achieved via a deep semilunar notch. These modifications are ways in which subterranean rodents can strengthen and maintain the integrity of the elbow joint despite the increasing pressure that their forelimbs experience while burrowing. A prominent radial-collateral ligament in these taxa also helps to prevent dislocation of the elbow while digging.

In *Thomomys*, two grooves on the ulnar shaft permit passage of most of the forearm extensor tendons (Hill 1937). Lehmann (1963) observed a moderately deep groove in the ulnar shaft in *Tachyoryctes*, and she reported a relatively larger coronoid process on the ulna in the subterranean taxa that she examined. Lehmann (1963) also observed elongated acromion and olecranon processes in *Tachyoryctes*, but noted that members of that genus did not show other forearm modifications associated with digging that are found in *Ctenomys* and *Geomys*. Such morphological differences among subterranean genera might be expected given that *Tachyoryctes* is a chisel-tooth digger whereas *Geomys* and *Ctenomys* are scratch diggers.

Subterranean rodents generally show a marked reduction in the length of the carpals, metacarpals, and phalanges when compared with non-subterranean taxa. All genera have five digits on the forefoot, although a reduction of the first and fifth digits occurs in some taxa. Ubilla and Altuna (1990) reported that in *Ctenomys*, the thenar pad on the palmar surface acted as a supernumerary finger during prehension, replacing the atrophied pollex characteristic of that genus. The presence of additional sesamoid bones along the ventral surface of the distal phalanges and within the digital flexor tendons of most subterranean taxa increases the angle at which those tendons insert on the manus, thereby increasing the force that can be generated during muscle contraction. The presence of an elongated pisiform bone at the wrist also increases the in-lever of the flexor muscles inserting on the manus. Lehmann (1963) noted the presence of a pseudostyloid process at the distal end of the ulnar side of the radius, and proposed that this structure represents an additional modification to decrease lateral movement of the bones at the wrist joint.

Claws on the manus of many subterranean rodents are stout and elongate. Claws that are used to break up soil (versus those that move soil) tend, in addition, to be laterally compressed and to have bony phalanges surrounded by an unguis. The foreclaws in *Bathyergus* are perhaps larger and more striking in appearance than those in any other subterranean rodent genus, although those of *Geomys* are also impressive (see fig. 1.1). In *Myospalax*, the front claws are so enlarged that, during locomotion, they are doubled under and the animal walks on its knuckles (Macdonald 1984).

In geomyids and in *Ctenomys*, the middle three or lateral four claws may grow to three-quarters the length of the forearm. Growth of the center foreclaw may reach 90 mm/year in geomyids and 72 mm/year in *Ctenomys* (Hildebrand 1974).

Claw morphology, however, does not readily distinguish scratch diggers from chisel-tooth diggers. *Thomomys*, a scratch digger, and *Cannomys* and *Rhizomys*, both chisel-tooth diggers, also have strong claws, with the third digit bearing the longest nail. In *Tachyoryctes*, the manus is well developed, but the claws are smaller than those of the other rhizomyine genera. The claws of *Spalacopus* are relatively similar in size to those observed in *Ctenomys*, despite the difference in primary digging mode between these two genera. Among the chisel-tooth-digging bathyergid genera, *Cryptomys* and *Georychus* bear strong claws, whereas the claws of *Heterocephalus* and *Heliophobius* are relatively small. *Nannospalax*, primarily a head-lift digger, does not show specially enlarged forefeet or claws, although the claws are robust. *Ellobius* also has small but broad forefeet and small claws. In *Prometheomys*, the middle digit bears an extremely enlarged claw.

MYOLOGY. Another way in which subterranean mammals can increase the out-force generated by the forelimbs is to increase the components of the force equation representing the in-force and the in-lever. Increasing the in-force may be accomplished by increasing the size of the forearm muscles by expanding the area on the skeleton that is available for muscle origin and insertion (see above), by increasing muscle length, or by altering muscle configuration. This last method includes evolving pinnate rather than parallel-fibered muscles, increasing the number of muscles present on the forelimb, or increasing the number of heads of a muscle available to perform a given function (see fig. 1.2). By increasing the number of fibers that insert on a tendon within a given area, pinnate muscles increase the cross-sectional area of a muscle relative to muscles in which the fibers run parallel to one another. This modification, in turn, increases the force that the muscle can exert when contracting (Hildebrand 1974). Increasing the in-lever may be accomplished myologically by moving muscle insertions toward the distal portion of the forearm and away from the joints that they turn, as exemplified by the condition of the *pronator teres* in pocket gophers (Lessa and Stein 1992).

Several myological modifications that alter the in-force and the in-lever are found in the forelimbs of pocket gophers (Lessa and Stein 1992). Examples include relatively larger and more distal insertions of the *deltoideus* and *latissimus dorsi* muscles along the shaft of the humerus. Similarly, a more proximal insertion of the *supinator* along the humeral shaft should cause that muscle to act as a manus flexor as well as a supinator of the forearm. Gambaryan and Gasc (1993) commented on the enormous additional olecranon head of the *flexor digitorum profundus* in *Myospalax*, as well as on increases in size of the *flexor digitorum superficialis*, the muscles of the *pectoralis* complex, and the *serratus ventralis* in this genus. Strong development of the *pectoralis* in *Nannospalax* and *Myospalax* is presumed to relate to use of the forelimbs to remove dirt from the burrow, although Gambaryan and Gasc also noted that these two genera differ in the structure of muscles used to throw dirt with the head. Gambaryan and Gasc further reported that in *Myospalax*, the *serratus ventralis posterior* aids in anterior and dorsal movement of the body relative to the pectoral girdle, the *serratus ventralis anterior* helps to balance the head and neck while dirt is being lifted, and the middle portion of the *serratus* stabilizes the scapula against the body. They also observed that *Myospalax* retains an *omotransversarius superior*, a muscle that is not found in any other myomorphs and one that is only weakly present in some modern hystricomorphs and sciuromorphs.

Both the *subscapularis* and several forearm flexors show increased pinnation in many subterranean rodents relative to their surface-dwelling counterparts. In *Ctenomys*, the *subscapularis minor* inserts by a tendon distinct from that of the *subscapularis* and acts as an additional humeral flexor. Additionally, the number of forearm extensor muscles is generally greater in subterranean mammals than in non-subterranean taxa. This modification aids both in burrowing and in resisting joint flexion, particularly among head-lift diggers. One example is the presence of the *dorsoepitrochlearis* in geomyid rodents, which functions as a fourth head of the *triceps*. An increase in the size of the forearm extensors that insert on the manus (e.g., *extensor digitorum communis*) may also contribute to an increase in the force that the digits can exert when extending.

As an additional means of increasing out-force, subterranean rodents might be expected to exhibit modifications in fiber type composition of muscles associated with the digging apparatus. Little in the literature, however, addresses this question. Goldstein (1971) suggested that proportional differences in red versus white fiber types (i.e., slow-contracting versus fast-contracting fibers) within homologous muscles might be important strength-related modifications among subterranean taxa, although he did not test this hypothesis functionally. Instead, he commented only that the proportions and types of fibers present in a muscle would depend primarily on the range of intensity needed to perform a given function and on the number and array of functions with which that muscle was involved.

Summary of Primary Components of the Digging Apparatus

Regardless of digging mode, all subterranean rodents show general modifications of the head, neck, teeth, and forelimbs that have been interpreted as specializations to enhance digging and the excavation of burrows. However, there is no strict concordance between the observed osteological or myological variation among taxa and either function or phylogeny. Skulls in subterranean rodents tend to be dorsally flattened and posteriorly broadened to increase the area available for the attachment of enlarged muscles that close the jaw, raise and lower the head, and stabilize the neck. Cervical vertebrae tend to be relatively short, more robust, and less flexible than in non-subterranean species to minimize transverse motions. A dramatic increase in incisor procumbency is also characteristic of those taxa that dig primarily with their teeth. In addition, there is a tendency for the bones of the forelimbs in all subterranean rodents to be relatively short and robust, with pronounced processes and tuberosities that increase the area available for attachment of enlarged flexor and extensor muscles. The degree to which each of these traits is expressed in a taxon correlates loosely with the digging mode that has been observed in that genus.

Secondary Components of the Digging Apparatus

As subterranean rodents break up soil with their teeth or forefeet, piles of dirt quickly accumulate underneath or beside their bodies. Periodically, this dirt must be removed from the burrow. Regardless of the technique employed (see above and table 1.1), the hind limbs, pelvis, and associated elements of the axial skeleton constitute important secondary components of the digging apparatus. These structures brace the animal's body, deliver horizontal thrust, move the body forward, throw dirt backward, allow an animal to back up in its burrow, and facilitate aboveground locomotion. Given this multiplicity of functions, it is surprising that so little attention has been paid to the descriptive, comparative, or functional morphology of this portion of the body.

The amount of time that subterranean rodents spend above ground is dependent upon the number and nature of the activities carried out there (see table 1.1; Busch et al., chap. 5, this volume). Juvenile pocket gophers reportedly locomote above ground to disperse, but as adults they rarely, if ever, leave their burrow systems (Bryant 1913; Daly and Patton 1990). Although *Ctenomys* forages within its tunnel system, these animals feed primarily above ground on grass, fresh stems, and other plant parts (Reig 1970). Two species *(C. mendocinus* and *C. opimus)* have been observed on the surface in a bipedal stance, a typical scanning behavior employed by prairie dogs and ground squirrels (Camin, Madoery, and Roig 1995). Spalacine and rhizomyid rodents commonly emerge at night to forage (Reed 1958; Wiles 1981) and, between March and May, juvenile *Nannospalax* disperse during the day (Nevo 1961). Although rhizomyines may completely exit their burrows in search of food, their hind feet are often kept within the burrow and braced against its walls for rapid retreat in the event that a predator is detected (Jarvis 1973). Although *Ellobius* frequently emerges above ground (Harrison and Bates 1991), *Prometheomys* and the bathyergid rodents seldom leave their burrows except when they are flooded or when the animals disperse, although *Heliophobius* and *Cryptomys* may wander more than other genera (Ognev 1950; Scharff and Grütjen 1997). Similarly, *Spalacopus* and *Myospalax* rarely appear above ground (Macdonald 1984).

Morphological studies that have examined the secondary components of the digging apparatus include Vinogradov's (1926) description of the osteology of *Prometheomys,* Hill's (1937) classic monograph on the morphology of *Thomomys,* Parsons's (1896) myological descriptions of *Rhizomys, Georychus,* and *Bathyergus,* and Stein's (1993) comparative study of hind limb musculature in geomyine and thomomyine pocket gophers. Two comparative studies of individual hind limb elements also deserve mention. Hildebrand (1978) detailed the relationship of the medial tarsal bone to the insertion and function of flexor muscles on the inside of the foot in rodents, including several subterranean genera. Stains (1959) discussed variation in the structure and use of the calcaneum in a variety of mammal genera, including pocket gophers.

Pelvis and Axial Skeleton

As secondary components of the digging apparatus, the pelvis and axial skeleton anchor the body and are the point of articulation for the hind limbs. Pressure on the hind limbs of subterranean rodents generated by bracing the body against the tunnel walls has, in most taxa, resulted in an elongation of the sacrum and in the fusion of the sacral vertebrae. Despite this fusion, the spines of subterranean rodents are remarkably flexible, as evidenced by the animals' ability to turn themselves 180° in a burrow barely wider than their bodies. Such flexibility has been observed in *Heliophobius*, which arches its back sharply each time its hind feet approximate its forefeet prior to thrusting backward to throw dirt from its burrow (Jarvis and Sale 1971). Similarly, *Heterocephalus* doubles over ventrally so that its mouth is in contact with its anogenital region each time it practices coprophagy (Lacey et al. 1991).

The pelvis of subterranean rodents is generally reduced relative to that in non-subterranean taxa, presumably in response to their need to turn within the narrow confines of a burrow (fig. 1.3). However, it is also fused to the sacrum. The ischium and ilium are rodlike and lie parallel to the vertebral column. The pubic bones may diverge posteriorly to aid in bracing the body and to accommodate the more robust hind limb muscles (Chapman 1919). In most taxa, the pubic symphysis is weak or short, and in some it is absent. The symphysis of female pocket gophers is resorbed at first pregnancy to enlarge the birth canal, and it remains open for the remainder of their lives (Hisaw 1924). In males, however, the symphysis persists throughout life. The acetabulum of the pelvis, which articulates with the head of the femur, is located dorsally relative to the level of the spine in subterranean rodents, and is often oriented laterally (e.g., in geomyids). This shift in orientation presumably decreases compressive forces at the pubic symphysis and enhances the ability of these animals to walk or brace their bodies with splayed legs.

Flynn (1990) stated that the pelvis in rhizomyines shows "typical burrowing modifications," including a reduced symphysis. Chapman (1919) described the pelvis of bathyergids and *Myospalax* as similar to that of pocket gophers. Examination of skeletal material reveals that in *Nannospalax, Geomys, Thomomys,* and *Tachyoryctes,* the pubis is reduced to a thin sliver of bone and the ischium is expanded anteriorly. This modification of the ischium is presumed to accommodate the origins of the enlarged limb muscles.

Chapman (1919) also reported that when the pelvis is reduced there is a commensurate reduction in the *rectus abdominus* muscles that originate from that bone. He observed that, in this circumstance, the posterior fibers of the muscles cross, with the left muscle arising from the right pubis and the right muscle arising from the left pubis. In instances in which the pubic

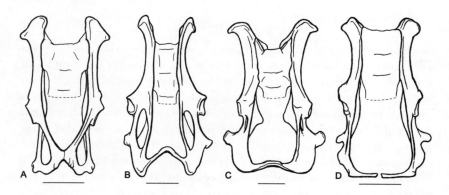

FIGURE 1.3. Differences in shape and degree of fusion in the pelves of (A) *Ctenomys*, a scratch digger, (B) *Thomomys*, a chisel-tooth digger, (C) *Nannospalax*, a head-lift digger, and (D) *Tachyoryctes*, (D) a chisel-tooth digger.

symphysis is lacking altogether, the muscles spread out into a broad aponeurosis, attaching to the pubic ligaments or to the bones lateral to the symphyseal opening.

Hind Limbs

In subterranean rodents, the hind limbs power locomotion, brace the body while the animal digs, and in some cases assist in dirt removal. As a result, hind limbs and forelimbs are nearly equal in length in these genera. The femora of most subterranean rodents support a well-developed trochanter in close proximity to the femoral head. There is also extensive fusion of the tibiae and fibulae in *Nannospalax, Georychus, Cryptomys, Geomys, Thomomys, Tachyoryctes,* and *Cannomys*, although this character alone is not indicative of a subterranean way of life. The tarsals and metatarsals are shortened in subterranean genera, and all taxa bear five digits with well-developed claws on the hind feet.

As is true of the forelimbs, the long bones of the hind limbs are thickened to reduce bending and torsion. Biknevicius (1993) reported that among fossorial and non-fossorial caviomorph rodents, the cortical cross-sectional area of the femora scaled similarly with body weight regardless of bone type. The sole exception to this pattern was *Ctenomys*, whose femora differed from those of non-burrowing caviomorphs of similar size by having significantly smaller moments of area. Biknevicius suggested that this variation reflects different patterns of limb use between the two groups, in particular, the decreased tendency toward cursoriality in *Ctenomys* relative to other caviomorphs. Biknevicius (1992) also observed that pocket gopher femora are relatively stronger than those of selected non-fossorial rodents. She related the increased femoral strength in these taxa to the need to brace the hind limbs against tunnel walls to counteract the propulsive forces that are generated when dirt is pushed from the burrow.

Despite their similar involvement in the digging process and in aboveground locomotion, considerable variation in hind limb musculature was noted between geomyine and thomomyine pocket gophers (Stein 1993). The observed differences were found to reflect the currently accepted phylogeny for genera and subgenera within the Geomyidae. This variation contrasts with that observed in the pectoral regions of these taxa (see above), in which differences in limb musculature are related to differences in the digging modes employed by members of these two clades (Lessa and Stein 1992). Because differences in limb myology among mammals frequently offer characters that are useful in phylogenetic analyses (Rinker 1954; Klingener 1964; Stein 1987, 1990), it is anticipated that examination of this character set in other subterranean rodent taxa may help to resolve the placement of several genera within their respective clades, as well as to illuminate the relationship between muscle structure and function.

Summary of Secondary Components of the Digging Apparatus

Regardless of the technique used to break up soil, all subterranean rodents use their hind limbs to brace their bodies against tunnel walls, deliver horizontal thrust, move their bodies forward, throw dirt backward, back up in their burrows, and locomote above ground. In contrast to the morphological variations observed in the forelimb, limited data suggest that variation in hind limb morphology among these taxa may be explained by examining the phylogenetic histories of these lineages and the nature and extent of aboveground activities in each taxon (e.g., dispersal, foraging, escaping flooded burrows). The major skeletal modifications in the hind limbs of subterranean rodents are an elongation of the sacrum and fusion of this bone to a reduced pelvis. Surprisingly, these changes have occurred with no apparent loss of flexibility. In most taxa the femora are also more robust than in surface-dwelling rodents, and the tibiae and fibulae are fused. These modifications resist the bending and torsional forces to which the hind limbs are continually subjected.

Internal Anatomy

Although less conspicuous than external anatomy, variations in internal morphology among subterranean rodents would be expected to reflect adaptations to the unique physical and physiological requirements of life underground. Food requirements, foraging habits, reproductive biology, and many aspects of behavior can be explored through examination of the morphology of the digestive system, the circulatory system, the urogenital tract, and the glandular systems of mammals. Overall, however, little attention has been paid to the gross morphology of the internal organs and organ systems in rodents, and no studies have compared these structures across subterranean taxa. General comparative works on internal anatomy in rodents include a study of carotid circulation by Guthrie (1963), comparisons of stomach morphology in cricetine and microtine rodents by Carleton (1973, 1981), a discussion of the comparative anatomy of the diastemal plate in microtine genera (Quay 1954a), and several studies detailing the morphology of the glans penis in a variety of rodent taxa (Hooper and Musser 1964; Voss and Linzey 1981; Lidicker and Brylski 1987).

Circulatory and Respiratory Systems

Direct studies of circulation in rodents have been limited to descriptions of the carotid arterial systems in different taxa (Guthrie 1963; Bugge 1970, 1971a,b,c, 1974; Woods 1975). Although these investigations have revealed a pattern of bifurcation among blood vessels that is strongly reflective of ontogenetic patterns and current classifications at the family and subfamily levels, these analyses do not include most subterranean genera. Because carotid circulation is intimately linked to the number and location of cranial foramina, readers interested in this topic also should consult Hill (1935a) and Wahlert (1978, 1985).

In the only study of its kind concerning subterranean genera, Maina, Maloiy, and Makanya (1992) compared lung morphology in *Tachyoryctes* and *Heterocephalus* both microscopically and morphometrically. Morphologically, they found that the cellular composition and arterial arrangement of the lungs of *Tachyoryctes* are similar to those of surface-dwelling mammals, whereas the lungs of *Heterocephalus* resemble an early developmental stage of the mammalian lung. Morphometrically, the lungs of *Tachyoryctes* have notably greater pulmonary volume and diffusing capabilities than those of *Heterocephalus*. The paedomorphic characteristics observed in *Heterocephalus* suggest that additional studies are needed to distinguish phylogenetic effects from the effects of selection for a subterranean way of life. In contrast, the lung structure of *Tachyoryctes* may reflect the greater percentage of time that members of this genus spend above ground.

While not directly addressing differences in circulatory system morphology among subterranean rodents, McNab's (1966) classic paper on metabolism in fossorial rodents provides indirect information in the form of data on oxygen consumption and thermoregulatory physiology in this highly modified assemblage. Information on other structural aspects of the circulatory and respiratory systems in these animals is lacking.

Digestive System

Although most subterranean rodents forage primarily underground, a wide variety of food types is consumed (Busch et al., chap. 5, this volume). Differences in diet should translate into both genetic and nongenetic differences in the structure and tissue composition of the stomach and associated digestive organs, although little research on this topic has been conducted. Details concerning intestinal morphology for *Ellobius* and *Prometheomys* were reported by Carleton (1981). Snipes, Nevo, and Sust (1990) investigated the anatomy of the caecum in *Nannospalax*, and Reig et al. (1990) included a description of the stomach and intestines in their review of the biology and evolution of *Ctenomys*. One example of ecophenotypic variation in gut morphology is an increase in the size of the caecum and large intestine in pocket gophers in response to decreased food quality (Loeb, Schwab, and Demment 1991). With these few exceptions, all descriptive information concerning the digestive systems of subterranean rodents comes from Carleton and Musser (1985).

Urogenital System

The configuration of the urogenital system may reflect differences in diet, habitat, or social structure in rodents. For instance, it is expected that kidney morphology in subterranean rodents will be influenced by habitat, specifically the availability of water (freestanding or from food), although there are almost no data to substantiate or refute this hypothesis. Nevo and his coworkers (e.g., Heth et al. 1996; Heth, Nevo, and Todrank 1996) have examined the production of, and response to, urine odors in *Nannospalax*, but similar data are lacking for other subterranean genera. Comparative data relating to urine composition and concentration, both of which may influence urogenital morphology, are also lacking.

It is generally acknowledged that the testes of subterranean rodents will be abdominal, given the short limbs of these rodents and the narrow confines of their tunnel systems. There are few descriptions, however, of either the gross morphology or microanatomy of the baculum, glans penis, ovaries, testes, or their associated ducts and glands in these species. Williams (1982) described the phalli of pocket gopher genera, and Williams, Schlitter, and Robbins (1983) included a description of the phallus and spermatozoa of *Cryptomys* in their discussion of morphological variation in that genus. Penile and sperm morphology in *Ctenomys* have also been described (Altuna and Lessa 1985; Lessa and Cook 1989; Vitullo, Roldan, and Merani 1988; Feito and Gallardo 1982). Harrison and Bates (1991) described the baculum of *Nannospalax,* and Ognev (1950) noted that that structure is distinctive in *Ellobius* and *Prometheomys.* Although Arata (1964) described the anatomy of the male accessory reproductive glands in twenty-four genera of muroid rodents, no subterranean taxa were included in his review. Carleton and Musser (1985) reported that accessory reproductive glands are present in male *Ellobius.* Morphological descriptions of urogenital structures for other subterranean taxa are lacking.

Glandular Systems

The number and placement of the mammary glands is generally known for subterranean genera (see Nowak 1991 for a summary). This wealth of data, however, contrasts sharply with the paucity of information available on the structure of other glandular systems in these rodents. Several studies, each relatively narrow in scope, have examined a single component of the glandular system in subterranean taxa, but descriptive and functional data are lacking for most genera and most systems. For example, in many subterranean rodents, well-developed lacrimal glands produce a thick liquid that continually cleans the cornea, but no comparative studies detailing the structure or function of these glands have been conducted. Similarly, in his extensive review of mammalian skin, Sokolov (1982) discussed the presence of glands in some subterranean rodents, but he did not elaborate on their morphology. Information on meibomian glands, the modified sebaceous glands that occur in the eyelids, was summarized Carleton and Musser (1985) for *Ellobius* and *Prometheomys,* two genera that were omitted from Quay's (1954b) anatomical study of those structures in voles and lemmings.

Sweat (apocrine sudoriferous) glands are unique to mammals, but in many taxa they are restricted to the muzzle and feet. Such a configuration might be expected in subterranean species given the unique environmental parameters that exist in their burrow systems. Jarvis and Bennett (1991) noted that *Heterocephalus glaber* lacks sweat glands, but that sebaceous glands are present. It is presumed that the oily secretions from these glands prevent excessive drying of the animal's thin, naked skin, even under the conditions of high relative humidity experienced underground. McNab (1966) observed that there was no visible evidence of sweating in the fossorial rodents whose metabolism he studied.

The comprehensive review of the rodent preputial gland done by Brown and Williams (1972) contains information on the subterranean genera *Geomys, Ellobius,* and *Rhizomys.* This gland is a source of olfactory stimuli that have been shown to mediate behavioral responses in social and sexual interactions between conspecifics in selected genera. Additional studies in these taxa, and examination of the structure and function of the preputial gland in other species, may help to explain why some subterranean rodents are social and others solitary.

Summary and Future Directions

The notion of morphological convergence has both fascinated and frustrated biologists since before the start of the twentieth century (Merriam 1895; Shimer 1903). As researchers have attempted to understand function, elucidate phylogenetic histories, and tease apart local factors responsible for differences in size and shape, apparent similarities in form have presented them with a wealth of testable hypotheses and, occasionally, unexpected results (Stein 1993). While life underground is generally assumed to have constrained the morphology of subterranean rodents, a tremendous amount of variation persists. Fundamental differences in the structural elements of the digging apparatus (a fundamental component of life underground) have been observed across taxa. Variation in external anatomy (e.g., pelage color and texture) has also been noted. Similarly, differences in cranial osteology and hind limb musculature have been demonstrated. This intergeneric variation challenges the perception of morphological uniformity that is generally used to characterize subterranean rodents.

Without further descriptive studies of the anatomy of these animals, biomechanical and phylogenetic analyses will continue to suffer from an incomplete understanding of the morphological characters on which they are based. Throughout this chapter, failure to discuss the nature of a specific morphological feature or failure to cite a specific taxonomic group should be interpreted as evidence that the relevant information is lacking. Most previous studies of convergence in subterranean mammals have focused primarily on the morphology of the head and forelimbs, often to the exclusion of other potentially informative data sets. Questions addressing the development of the musculoskeletal system and ontogenetic changes in morphology, as well as questions regarding microanatomy, phenotypic plasticity, and sexual dimorphism, all demand attention (although see Burda 1989 on development in rodents; Bennett et al. 1991 on growth and development in bathyergids; Malizia and Busch 1991 on growth in *Ctenomys;* Hill 1934 and Sudman, Burns, and Choate 1986 on external characters and postnatal development in newborn pocket gophers). Better understanding of these issues may contribute to a clarification of phylogenetic relationships, particularly among murid rodent genera, and may help to resolve questions regarding phenotypic novelties, such as the distinctive (among bathyergids) mode of digging displayed by *Bathyergus*.

In addition, several larger issues remain regarding the morphology of subterranean rodents. On the one hand, there is a need to identify factors that developmentally and allometrically constrain the morphological variation noted above (e.g., Demeter 1992). On the other hand, there is a need to better understand the variable abiotic factors and conflicting functional demands that contribute to the morphological differences observed among these taxa, despite what is perceived to be a relatively constant environment and a strict selective regime. Because the limited vagility and patchy geographic distribution of subterranean rodents have resulted in extensive interspecific morphological variation in some genera, it seems reasonable to ask to what extent morphological variation reflects environmental factors associated with differences in species distributions (Nevo 1979) versus the phylogenetic constraints that have produced these distinct rodent lineages.

Acknowledgments

I wish to thank the editors of this volume for inviting me to participate, and E. P. Lessa for coaxing me until I agreed to examine limb morphology in pocket gophers.

Literature Cited

Agrawal, V. C. 1967. Skull adaptations in fossorial rodents. *Mammalia* 31:300–12.

Akersten, W. A. 1973. Upper incisor grooves in the Geomyinae. *Journal of Mammalogy* 54:349–55.

———. 1981. A graphic method for describing the lateral profile of isolated rodent incisors. *Journal of Vertebrate Paleontology* 1:231–34.

Allen, G. M. 1940. *The mammals of China and Mongolia.* New York: American Museum of Natural History.

Altuna, C. A., and E. P. Lessa. 1985. Penial morphology in Uruguayan species of *Ctenomys* (Rodentia: Octodontidae). *Journal of Mammalogy* 66:483–88.

Anderson, S., and J. K. Jones, Jr., eds. 1985. *Orders and families of Recent mammals of the world.* New York: John Wiley & Sons.

Arata, A. A. 1964. The anatomy and taxonomic significance of the male accessory reproductive glands of muroid rodents. *Bulletin of the Florida State Museum* 9:1–42.

Bekele, A. 1983a. The comparative functional morphology of some head muscles of the rodents *Tachyoryctes splendens* and *Rattus rattus*. I. Jaw muscles. *Mammalia* 47:395–420.

———. 1983b. The comparative functional morphology of some head muscles of the rodents *Tachyoryctes splendens* and *Rattus rattus*. II. Cervical muscles. *Mammalia* 47:549–72.

Bennett, N. C., J. U. M. Jarvis, G. H. Aguilar, and E. J. McDaid. 1991. Growth and development in six species of African mole-rats (Rodentia: Bathyergidae). *Journal of Zoology, London* 225:13–26.

Biknevicius, A. R. 1992. Limb use and skeletal differentiation in burrowing rodents. Abstract no. 142, Annual Meeting of the American Society of Mammalogists, Salt Lake City, UT.
———. 1993. Biomechanical scaling of limb bones and differential limb use in caviomorph rodents. *Journal of Mammalogy* 74:95–107.
Breckenridge, W. J. 1929. Actions of the pocket gopher *(Geomys bursarius)*. *Journal of Mammalogy* 10:336–39.
Brown, J. C., and J. D. Williams. 1972. The rodent preputial gland. *Mammal Review* 2:105–49.
Bryant, H. C. 1913. Nocturnal wanderings of the California pocket gopher. *University of California Publications in Zoology* 12:25–29.
Brylski, P., and B. K. Hall. 1988. Epithelial behaviors and threshold effects in the development and evolution of internal and external cheek pouches in rodents. *Zeitschrift für Zoologische Systematik und Evolutionsforschung* 26:144–54.
Bugge, J. 1970. The contribution of the stapedial artery to the cephalic arterial supply in muroid rodents. *Acta Anatomica* 76:313–36.
———. 1971a. The cephalic arterial system in mole-rats *(Spalacidae)*, bamboo rats *(Rhizomyinae)*, jumping mice and jerboas *(Dipodoidea)* and dormice *(Glioidea)* with special reference to the systematic classification of rodents. *Acta Anatomica* 79:165–80.
———. 1971b. The cephalic arterial system in New and Old World hystricomorphs, and in bathyergoids, with special reference to the systematic classification of rodents. *Acta Anatomica* 80:516–36.
———. 1971c. The cephalic arterial system in sciuromorphs with special reference to the systematic classification of rodents. *Acta Anatomica* 80:336–61.
———. 1974. The cephalic arteries of hystricomorph rodents. In *The biology of hystricomorph rodents*, edited by I. W. Rowlands and B. J. Weir, 61–78. Symposia of the Zoological Society of London, 34. London: Academic Press.
Burda, H. 1989. Relationships among rodent taxa as indicated by reproductive biology. *Zeitschrift für Zoologische Systematik und Evolutionsforschung* 27:49–57.
Burda, H., V. Bruns, and A. M. Müller. 1990. Sensory adaptations in subterranean mammals. In *Evolution of subterranean mammals at the organismal and molecular levels*, edited by E. Nevo and O. A. Reig, 269–93. Progress in Clinical and Biological Research, vol. 335. New York: Wiley-Liss.
Butler, P. M., E. Nevo, A. Beiles, and S. Simson. 1993. Variations of molar morphology in the *Spalax ehrenbergi* superspecies: Adaptive and phylogenetic significance. *Journal of Zoology, London* 229:191–216.
Buzas-Stephens, P., and W. W. Dalquest. 1991. Enamel ultrastructure of incisors, premolars, and molars in *Thomomys, Cratogeomys*, and *Geomys* (Rodentia: Geomyidae). *Texas Journal of Science* 43:65–74.
Camin, S., L. Madoery, and V. Roig. 1995. The burrowing behavior of *Ctenomys mendocinus* (Rodentia). *Mammalia* 59:9–17.
Carleton, M. D. 1973. A survey of gross stomach morphology in New World Cricetinae (Rodentia, Muroidea), with comments on functional interpretations. Miscellaneous Publications, Museum of Zoology, University of Michigan, no. 146:1–43.
———. 1981. A survey of gross stomach morphology in Microtinae (Rodentia: Muroidea). *Zeitschrift für Säugetierkunde* 46:93–108.
Carleton, M. D., and G. G. Musser. 1985. Muroid rodents. In *Orders and families of Recent mammals of the world*, edited by S. Anderson and J. K. Jones, Jr., 289–379. New York: John Wiley & Sons.
Casinos, A., C. Quintana, and C. Viladiu. 1993. Allometry and adaptation in the long bones of a digging group of rodents (Ctenomyinae). *Zoological Journal of the Linnean Society* 107:107–15.

Chapman, R. N. 1919. A study of the correlation of the pelvic structure and the habits of certain burrowing mammals. *American Journal of Anatomy* 25:185–219.

Corti, M., C. Fadda, S. Simson, and E. Nevo. 1996. Size and shape variation in the mandible of the fossorial rodent *Spalax ehrenbergi:* A Procrustes analysis of three dimensions. In *Advances in morphometrics,* edited by L. F. Marcus, M. Corti, A. Loy, C. J. P. Naylor, and D. E. Slice, 303–20. New York: Plenum Press.

Daly, J. C., and J. L. Patton. 1986. Growth, reproduction, and sexual dimorphism in *Thomomys bottae* pocket gophers. *Journal of Mammalogy* 67:256–65.

———. 1990. Dispersal, gene flow, and allelic diversity between local populations of *Thomomys bottae* pocket gophers in the coastal ranges of California. *Evolution* 44:1283–94.

de Jong, W. W., W. Hendriks, S. Sanyal, and E. Nevo. 1990. The eye of the blind mole rat *(Spalax ehrenbergi):* Regressive evolution at the molecular level. In *Evolution of subterranean mammals at the organismal and molecular levels,* edited by E. Nevo and O. A. Reig, 383–95. Progress in Clinical and Biological Research, vol. 335. New York: Wiley-Liss.

Demeter, A. 1992. Allometry and morphological integration in the skull of the lesser mole-rat *(Spalax leucodon)* and the East African mole-rat *(Tachyoryctes splendens). Israel Journal of Zoology* 38:419–20.

Dubost, G. 1968. Les mammifères souterrains. *Revue d'Écologie et de Biologie du Sol* 5:99–197.

Ellerman, J. R. 1940. *The families and genera of living rodents.* London: British Museum of Natural History.

———. 1959. The subterranean mammals of the world. *Transactions of the Royal Society of South Africa* 35:11–20.

Eloff, G. 1951. Adaptation in rodent moles and insectivorous moles, and the theory of convergence. *Nature* 168:1001–2.

Faulkes, C. G., N. C. Bennett, M. W. Bruford, H. P. O'Brien, G. H. Aguilar, and J. U. M. Jarvis. 1997. Ecological constraints drive social evolution in the African mole-rats. *Proceedings of the Royal Society of London* B 264:1619–27.

Feito, R., and M. Gallardo. 1982. Sperm morphology of the Chilean species of *Ctenomys* (Octodontidae). *Journal of Mammalogy* 63:658–61.

Flynn, L. J. 1990. The natural history of rhizomyine rodents. In *Evolution of subterranean mammals at the organismal and molecular levels,* edited by E. Nevo and O. A. Reig, 155–83. Progress in Clinical and Biological Research, vol. 335. New York: Wiley-Liss.

Flynn, L. J., E. Nevo, and G. Heth. 1987. Incisor enamel microstructure in blind mole rats: Adaptive and phylogenetic significance. *Journal of Mammalogy* 68:500–507.

Formozov, A. N. 1966. Adaptive modifications of behavior in mammals of the Eurasian steppes. *Journal of Mammalogy* 47:208–23.

Gambaryan, P. P., and J.-P. Gasc. 1993. Adaptive properties of the musculoskeletal system in the mole-rat *Myospalax* (Mammalia, Rodentia): Cinefluorographical, anatomical, and biomechanical analyses of burrowing. *Zoologische Jahrbücher. Abteilung für Anatomie und Ontogenie der Tiere* 123:363–401.

Genelly, R. E. 1965. Ecology of the common mole-rat *(Cryptomys hottentotus)* in Rhodesia. *Journal of Mammalogy* 46:647–65.

Goldstein, B. 1971. Heterogeneity of muscle fibers in some burrowing mammals. *Journal of Mammalogy* 52:515–27.

Gromov, I. M., and I. Y. Polyakov. 1992. *Voles (Microtinae).* New Delhi: Oxonian Press Pvt. Ltd.

Gupta, B. B. 1966. Fusion of cervical vertebrae in rodents. *Mammalia* 30:25–29.

Guthrie, D. A. 1963. The carotid circulation in the Rodentia. *Bulletin of the Museum of Comparative Zoology, Harvard University* 128:455–81.

Harrison, D. L., and P. J. J. Bates. 1991. *The mammals of Arabia.* Kent: Harrison Zoological Museum.

Heth, G., G. K. Beauchamp, E. Nevo, and K. Yamazaki. 1996. Species, population and individual specific odors in urine of mole rats *(Spalax ehrenbergi)* detected by laboratory rats. *Chemoecology* 7:107–11.

Heth, G., A. Beiles, and E. Nevo. 1988. Adaptive variation of pelage color within and between species of the subterranean mole rat *(Spalax ehrenbergi)* in Israel. *Oecologia* 74:617–22.

Heth, G., E. Nevo, and J. Todrank. 1996. Seasonal changes in urinary odors and in responses to them by blind subterranean mole rats. *Physiology and Behavior* 60:963–68.

Hickman, G. C. 1983. Burrows, surface movement, and swimming of *Tachyoryctes splendens* (Rodentia: Rhizomyidae) during flood conditions in Kenya. *Journal of Zoology, London* 200:71–82.

———. 1984. Behavior of North American geomyids during surface movement and construction of earth mounds. *Special Publications, The Museum, Texas Tech University* 22:165–86.

———. 1985. Surface-mound formation by the Tuco-tuco, *Ctenomys fulvus* (Rodentia: Ctenomyidae), with comments on earth-pushing in other fossorial mammals. *Journal of Zoology, London* A 205:385–90.

Hildebrand, M. 1974. *Analysis of vertebrate structure.* New York: John Wiley & Sons.

———. 1978. Insertions and functions of certain flexor muscles in the hind leg of rodents. *Journal of Morphology* 155:111–22.

———. 1985. Digging of quadrupeds. In *Functional vertebrate morphology,* edited by M. Hildebrand, D. M. Bramble, K. F. Liem, and D. B. Wake, 98–109. Cambridge, MA: Harvard University Press.

Hill, J. E. 1934. External characters of newborn pocket gophers. *Journal of Mammalogy* 15:244–45.

———. 1935a. The cranial foramina in rodents. *Journal of Mammalogy* 16:121–29.

———. 1935b. The retractor muscle of the pouch in the Geomyidae. *Science* 81:160.

———. 1937. Morphology of the pocket gopher, mammalian genus *Thomomys. University of California Publications in Zoology* 42:81–172.

Hinton, M. A. C. 1926. *Monograph of the voles and lemmings (Microtinae).* London: British Museum of Natural History.

Hisaw, F. L. 1924. The absorption of the pubic symphysis of the pocket gopher, *Geomys bursarius* (Shaw). *American Naturalist* 58:93–96.

Holliger, C. D. 1916. Anatomical adaptations in the thoracic limb of the California pocket gopher and other rodents. *University of California Publications in Zoology* 13:447–94.

Honeycutt, R. L., M. W. Allard, S. V. Edwards, and D. A. Schlitter. 1991. Systematics and evolution of the family Bathyergidae. In *The biology of the naked mole-rat,* edited by P. W. Sherman, J. U. M. Jarvis, and R. D. Alexander, 45–65. Princeton, NJ: Princeton University Press.

Hooper, E. T., and G. G. Musser. 1964. The glans penis in Neotropical cricetines (Family Muridae) with comments on the classification of muroid rodents. Miscellaneous Publications, Museum of Zoology, University of Michigan, no. 123:1–57.

Howard, W. E., and M. E. Smith. 1952. Rate of extrusive growth of incisors of pocket gophers. *Journal of Mammalogy* 33:485–87.

Jarvis, J. U. M. 1973. Activity patterns in the mole-rats *Tachyoryctes splendens* and *Heliophobius argenteocinereus. Zoologica Africana* 8:101–19.

Jarvis, J. U. M., and N. C. Bennett. 1990. The evolutionary history, population biology and social structure of African mole-rats: Family Bathyergidae. In *Evolution of subterranean mammals at the organismal and molecular levels,* edited by E. Nevo and O. A. Reig, 97–128. Progress in Clinical and Biological Research, vol. 335. New York: Wiley-Liss.

———. 1991. Ecology and behavior of the family Bathyergidae. In *The biology of the naked mole-rat,* edited by P. W. Sherman, J. U. M. Jarvis, and R. D. Alexander, 66–96. Princeton, NJ: Princeton University Press.

Jarvis, J. U. M., J. J. O'Riain, and E. McDaid. 1991. Growth and factors affecting body size in naked mole-rats. In *The biology of the naked mole-rat*, edited by P. W. Sherman, J. U. M. Jarvis, and R. D. Alexander, 358–83. Princeton, NJ: Princeton University Press.

Jarvis, J. U. M., and J. B. Sale. 1971. Burrowing and burrow patterns of East African mole-rats *Tachyoryctes, Heliophobius* and *Heterocephalus*. *Journal of Zoology, London* 163:451–79.

Justo, E. R., L. E. Bozzolo, and L. J. M. De Santis. 1995. Microstructure of the enamel of the incisors of some ctenomyid and octodontid rodents (Rodentia, Caviomorpha). *Mastozoologia Neotropical* 2:43–51.

Kingdon, J. 1984. *East African mammals: An atlas of evolution in Africa*. Chicago: University of Chicago Press.

Klauer, G., H. Burda, and E. Nevo. 1997. Adaptive differentiations of the skin of the head in a subterranean rodent, *Spalax ehrenbergi*. *Journal of Morphology* 233:53–66.

Klingener, D. 1964. The comparative myology of four dipodoid rodents (Genera *Zapus, Napaeozapus, Sicista*, and *Jaculus*). Miscellaneous Publications, Museum of Zoology, University of Michigan, no. 124:1–100.

Koenigswald, W. V. 1985. Evolutionary trends in the enamel of rodent incisors. In *Evolutionary relationships among rodents: A multidisciplinary analysis*, edited by W. P. Luckett and J.-L. Hartenberger, 403–22. New York: Plenum Press.

Krapp, F. 1965. Schädel und Kaumuskulatur von *Spalax leucodon* (Nordmann, 1840). *Zeitschrift für Wissenschaftliche Zoologie* A 173:1–71.

Lacey, E. A., R. D. Alexander, S. H. Braude, P. W. Sherman, and J. U. M. Jarvis. 1991. An ethogram for the naked mole-rat: Non-vocal behaviors. In *The biology of the naked mole-rat*, edited by P. W. Sherman, J. U. M. Jarvis, and R. D. Alexander, 209–42. Princeton, NJ: Princeton University Press.

Landry, S. O., Jr. 1957. Factors affecting the procumbency of rodent upper incisors. *Journal of Mammalogy* 38:223–34.

Laville, P. E., A. Casinos, J.-P. Gasc, S. Renous, and J. Bou. 1989. Les mécanismes du fouissage chez *Arvicola terrestris* et *Spalax ehrenbergi*: Étude fonctionelle et Évolutive. Digging mechanisms in *Arvicola terrestris* and *Spalax ehrenbergi*. *Anatomischer Anzeiger* 169:131–44.

Lay, D. M. 1967. A study of the mammals of Iran resulting from the Street Expedition of 1962–1963. *Fieldiana: Zoology* 54:1–282.

Lehmann, W. H. 1963. The forelimb architecture of some fossorial rodents. *Journal of Morphology* 113:59–76.

Lessa, E. P. 1990. Morphological evolution of subterranean mammals: Integrating structural, functional, and ecological perspectives. In *Evolution of subterranean mammals at the organismal and molecular levels*, edited by E. Nevo and O. A. Reig, 211–30. Progress in Clinical and Biological Research, vol. 335. New York: Wiley-Liss.

Lessa, E. P., and J. A. Cook. 1989. Interspecific variation in penial characters in the genus *Ctenomys* (Rodentia: Octodontidae). *Journal of Mammalogy* 70:856–60.

Lessa, E. P., and J. L. Patton. 1989. Structural constraints, recurrent shapes, and allometry in pocket gophers (genus *Thomomys*). *Biological Journal of the Linnean Society* 36:349–63.

Lessa, E. P., and B. R. Stein. 1992. Morphological constraints in the digging apparatus of pocket gophers (Mammalia: Geomyidae). *Biological Journal of the Linnean Society* 47: 439–53.

Lessa, E. P., and C. S. Thaeler, Jr. 1989. A reassessment of morphological specializations for digging in pocket gophers. *Journal of Mammalogy* 70:689–700.

Lidicker, W. Z., Jr., and P. V. Brylski. 1987. The canilurine rodent radiation of Australia, analyzed on the basis of phallic morphology. *Journal of Mammalogy* 68:617–41.

Lindenlaub, T., H. Burda, and E. Nevo. 1995. Convergent evolution of the vestibular organ in the subterranean mole-rats, *Cryptomys* and *Spalax*, as compared with the aboveground rat, *Rattus*. *Journal of Morphology* 224:303–11.

Loeb, S. C., R. G. Schwab, and M. W. Demment. 1991. Responses of pocket gophers *(Thomomys bottae)* to changes in diet quality. *Oecologia* 86:542–51.
Luckett, W. P., and J.-L. Hartenberger, eds. 1985. *Evolutionary relationships among rodents: A multidisciplinary analysis.* New York: Plenum Press, in cooperation with NATO Scientific Affairs Division.
Macdonald, D., ed. 1984. *The encyclopedia of mammals.* New York: Facts on File Publications.
Maier, W., and F. Schrenk. 1987. The hystricomorphy of the Bathyergidae, as determined from ontogenetic evidence. *Zeitschrift für Säugetierkunde* 52:156–64.
Maina, J. N., G. M. O. Maloiy, and A. N. Makanya. 1992. Morphology and morphometry of the lungs of two East African mole rats, *Tachyoryctes splendens* and *Heterocephalus glaber* (Mammalia, Rodentia). *Zoomorphology* 112:167–79.
Malizia, A. I., and C. Busch. 1991. Reproductive parameters and growth in the fossorial rodent *Ctenomys talarum* (Rodentia: Octodontidae). *Mammalia* 55:293–305.
Manaro, A. J. 1959. Extrusive incisor growth in the rodent genera *Geomys*, *Peromyscus*, and *Sigmodon*. *Quarterly Journal of the Florida Academy of Sciences* 22:25–31.
Martin, T. 1992. Schmelzmikrostruktur in den inzisiven alt- und neuweltlicher hystricognather nagetiere. *Paleovertebrata, Memoire extra* 1–68.
McLaughlin, C. A. 1984. Protrogomorph, Sciuromorph, Castorimorph, Myomorph (Geomyoid, Anomaluroid, Pedetoid, and Ctenodactyloid) rodents. In *Orders and families of Recent mammals of the world,* edited by S. Anderson and J. K. Jones, Jr., 267–88. New York: John Wiley & Sons.
McNab, B. K. 1966. The metabolism of fossorial rodents: A study of convergence. *Ecology* 47:712–33.
Merriam, C. H. 1895. Monographic revision of the pocket gophers, family Geomyidae (exclusive of the species *Thomomys*). *North American Fauna* 8:1–258.
Miller, R. S. 1958. Rate of incisor growth in the mountain pocket gopher. *Journal of Mammalogy* 39:380–85.
Nevo, E. 1961. Observations on Israeli populations of the mole rat *Spalax ehrenbergi* Nehring 1898. *Mammalia* 25:127–44.
———. 1979. Adaptive convergence and divergence of subterranean mammals. *Annual Review of Ecology and Systematics* 10:269–308.
———. 1995. Mammalian evolution underground: The ecological-genetic-phenetic interfaces. *Acta Theriologica,* supplement 3:9–31.
Nevo, E., and O. A. Reig, eds. 1990. *Evolution of subterranean mammals at the organismal and molecular levels.* Progress in Clinical and Biological Research, vol. 335. New York: Wiley-Liss.
Nowak, R. M. 1991. *Walker's mammals of the world.* Baltimore: Johns Hopkins University Press.
Ognev, S. I. 1950. *Mammals of the U.S.S.R. and adjacent countries.* Washington, DC: Smithsonian Institution and NSF.
Orcutt, E. E. 1940. Studies of the muscles of the head, neck, and pectoral appendages of *Geomys bursarius. Journal of Mammalogy* 21:37–52.
Parsons, F. G. 1896. Myology of rodents. Part II. An account of the myology of the Myomorpha, together with a comparison of the muscles of the various suborders of rodents. *Proceedings of the Zoological Society of London* 1896:159–92.
Patton, J. L., and P. V. Brylski. 1987. Pocket gophers in alfalfa fields: Causes and consequences of habitat-related body size variation. *American Naturalist* 130:493–506.
Pearson, O. P. 1959. Biology of the subterranean rodents, *Ctenomys,* in Peru. *Memorias del Museo de Historia Natural "Javier Prado"* 9:1–56.
Poduschka, W. 1978. Zur Frage der Wahrnehmung von Lichtreizen durch die Mullratte, *Cryptomys hottentotus* (Lesson, 1826). *Säugetierkundliche Mitteilungen* 26:269–74.

Quay, W. B. 1954a. The anatomy of the diastemal palate in microtine rodents. Miscellaneous Publications, Museum of Zoology, University of Michigan, no. 86:1–41.
———. 1954b. The meibomian glands of voles and lemmings. Miscellaneous Publications, Museum of Zoology, University of Michigan, no. 82:1–17.
Reed, C. A. 1958. Observations on the burrowing rodent *Spalax* in Iraq. *Journal of Mammalogy* 39:386–89.
Reig, O. A. 1970. Ecological notes on the fossorial octodont rodent *Spalacopus cyanus* (Molina). *Journal of Mammalogy* 51:592–601.
Reig, O. A., C. Busch, M. O. Ortells, and J. R. Contreras. 1990. An overview of evolution, systematics, population biology, cytogenetics, molecular biology and speciation in *Ctenomys*. In *Evolution of subterranean mammals at the organismal and molecular levels*, edited by E. Nevo and O. A. Reig, 71–96. Progress in Clinical and Biological Research, vol. 335. New York: Wiley-Liss.
Rinker, G. C. 1954. The comparative myology of four mammalian genera, *Sigmodon*, *Oryzomys*, *Neotoma*, and *Peromyscus* (Cricetinae), with remarks on their intergeneric relationships. Miscellaneous Publications, Museum of Zoology, University of Michigan, no. 83:1–124.
Russell, R. J. 1968. Evolution and classification of the pocket gophers of the subfamily Geomyinae. *Miscellaneous Publications, Museum of Natural History, University of Kansas* 16:473–579.
Ryan, J. M. 1986. Comparative morphology and evolution of cheek pouches in rodents. *Journal of Morphology* 190:27–41.
Scharff, A., and O. Grütjen. 1997. Evidence for aboveground activity of Zambian Molerats (*Cryptomys*, Bathyergidae, Rodentia). *Zeitschrift für Säugetierkunde* 62:253–54.
Shimer, H. W. 1903. Adaptations to aquatic, arboreal, fossorial, and cursorial habits in mammals. III. Fossorial adaptations. *American Naturalist* 37:819–25.
Snipes, R. L., E. Nevo, and H. Sust. 1990. Anatomy of the caecum of the Israeli mole rat, *Spalax ehrenbergi* (Mammalia). *Zoologischer Anzeiger* 224:307–20.
Sokolov, V. 1982. *Mammal skin*. Berkeley: University of California Press.
Sokolov, V. E., and V. F. Kulikov. 1987. The structure and function of the vibrissal apparatus in some rodents. *Mammalia* 51:125–38.
Stains, H. J. 1959. Use of the calcanium in studies of taxonomy and food habits. *Journal of Mammalogy* 40:392–401.
Stein, B. R. 1987. Phylogenetic relationships among four arvicolid genera. *Zeitschrift für Säugetierkunde* 52:140–56.
———. 1990. Limb myology and phylogenetic relationships in the superfamily Dipodoidea (birch mice, jumping mice, and jerboas). *Zeitschrift für Zoologische Systematik und Evolutionsforschung* 28:299–314.
———. 1993. Comparative hind limb morphology in geomyine and thomomyine pocket gophers. *Journal of Mammalogy* 74:86–94.
———. 1994. Forces shaping muscle structure in semiaquatic mammals. Abstract, International Conference on Vertebrate Morphology, Chicago, IL.
Sterba, O. 1969. Pectoral muscles in some rodents. *Zoologicke Listy* 18:1–6.
Sudman, P. D., J. C. Burns, and J. R. Choate. 1986. Gestation and postnatal development of the Plains pocket gopher. *Texas Journal of Science* 38:91–94.
Thaeler, C. S., Jr. 1968. An analysis of the distribution of pocket gopher species in northeastern California (genus *Thomomys*). *University of California Publications in Zoology* 86:1–46.
———. 1980. Chromosome numbers and systematic relations in the genus *Thomomys* (Rodentia: Geomyidae). *Journal of Mammalogy* 61:414–22.
Topachevskii, V. A. 1976. *Fauna of the USSR*. Vol. 3. *Mammals*. No. 3. *Mole rats, Spalacidae*. Leningrad: Nauka Publishers, Leningrad Section.

Ubilla, M., and C. A. Altuna. 1990. Analyse de la morphologie de la main chez des espèces de *Ctenomys* de l'Uruguay (Rodentia: Octodontidae). Adaptations au fouissage et implications évolutives. *Mammalia* 54:107–17.
Vinogradov, B. S. 1926. Some external and osteological characters of *Prometheomys schaposchnikovi* Satunin. *Proceedings of the Zoological Society of London* 1926:401–11.
Vitullo, A. D., E. R. S. Roldan, and M. S. Merani. 1988. On the morphology of the spermatozoa of tuco-tucos, *Ctenomys* (Rodentia: Ctenomyidae): New data and its implications for the evolution of the genus. *Journal of Zoology, London* 215:675–83.
Voss, R. S., and A. V. Linzey. 1981. Comparative gross morphology of male accessory glands among Neotropical Muridae (Mammalia: Rodentia) with comments on systematic implications. Miscellaneous Publications, Museum of Zoology, University of Michigan, no. 159:1–41.
Wahlert, J. H. 1978. Cranial foramina and relationships of the Eomyoidea (Rodentia, Geomorpha). Skull and upper teeth of *Kansasimys*. *American Museum Novitates* 2645:1–16.
———. 1985. Skull morphology and relationships of geomyoid rodents. *American Museum Novitates* 2812:1–20.
Watson, G. E. III. 1961. Behavioral and ecological notes on *Spalax leucodon*. *Journal of Mammalogy* 42:359–65.
Wiles, G. J. 1981. Abundance and habitat preferences of small mammals in southwestern Thailand. *Natural History Bulletin of the Siam Society* 29:44–54.
Wilkins, K. T. 1988. Prediction of direction of chewing from cranial and dental characters in *Thomomys* pocket gophers. *Journal of Mammalogy* 69:46–56.
Wilkins, K. T., and C. A. Woods. 1983. Modes of mastication in pocket gophers. *Journal of Mammalogy* 64:636–41.
Williams, S. L. 1982. Phalli of Recent genera and species of the family Geomyidae (Mammalia: Rodentia). *Bulletin of the Carnegie Museum of Natural History* 20:1–62.
Williams, S. L., D. A. Schlitter, and L. W. Robbins. 1983. Morphological variation in a natural population of *Cryptomys* (Rodentia: Bathyergidae) from Cameroon. *Annales Musée Royal de l'Afrique Centrale, Sciences Zoologiques* 237:159–72.
Wilson, D. E., and D. M. Reeder, eds. 1993. *Mammal species of the world: A taxonomic and geographic reference*, 2d ed. Washington, DC: Smithsonian Institution Press.
Wood, A. E. 1936. Parallel radiation among geomyoid rodents. *Journal of Mammalogy* 18:171–76.
———. 1955. A revised classification of rodents. *Journal of Mammalogy* 36:165–87.
Woods, C. A. 1975. The hyoid, laryngeal and pharyngeal regions of bathyergid and other selected rodents. *Journal of Morphology* 147:229–50.
Woods, C. A., and J. W. Hermanson. 1985. Myology of hystricognath rodents: An analysis of form, function and phylogeny. In *Evolutionary relationships among rodents: A multidisciplinary analysis*, edited by W. P. Luckett and J.-L. Hartenberger, 515–48. New York: Plenum Press.
Xu, L.-H. 1984. Studies on the biology of the hoary bamboo rat (*Rhizomys pruinosus* Blyth). *Acta Theriologica Sinica* 4:99–105.
Yalden, D. W. 1975. Some observations on the giant mole-rat *Tachyoryctes macrocephalus* (Ruppell, 1842) (Mammalia: Rhizomyidae) of Ethiopia. *Monitore Zoologico Italiano* 6:275–303.
———. 1985. *Tachyoryctes macrocephalus*. *Mammalian Species* 237:1–3.

CHAPTER TWO

Ecophysiological Responses of Subterranean Rodents to Underground Habitats

Rochelle Buffenstein

A primary factor influencing the distribution, abundance, and evolution of organisms is their ability to tolerate different environmental conditions (Krebs 1978). Ecophysiology explores the internal processes and regulatory mechanisms that permit organisms to survive and reproduce in a multitude of different and often adverse environments, with emphasis on the interaction between organism and environment (Feder et al. 1987; Degen 1997). As a result, this chapter, which addresses ecophysiological responses to a subterranean milieu, is broad-based, and briefly embraces such diverse aspects of biology as behavior, ecology, and morphology in an attempt to answer both proximate and ultimate questions regarding physiological responses to the challenges imposed by life underground. It also compares the physiological attributes of subterranean animals with those of surface-dwelling rodents. Where adequate data sets exist, I also explore apparent convergences and divergences in the physiological responses displayed by the different species of subterranean rodents. I address physiological processes including neuroendocrine responses, energy and water balance, gas diffusion, and mineral absorption and regulation, while posing the generic questions of how these physiological systems differ from those of surface-dwelling animals and whether they may be considered specialized for an underground milieu.

Physiological Constraints of a Subterranean Ecotope

The subterranean "niche" occurs in a wide range of climatic zones, altitudes, and soil types. Despite these pronounced environmental differences, similar microclimatic conditions are thought to prevail below ground (Busch et al., chap. 5, this volume). Subterranean burrows are characterized as dark, dank environments with low primary productivity and limited ventilation. In general, these habitats are considered more stable and buffered than aboveground habitats, with smaller daily and seasonal fluctuations in temperature and humidity (Mayer 1955; Bennett, Jarvis, and Davies 1988). At the same time, subterranean habitats are believed to present animals with a suite of environmental conditions that are physically more stressful than those encountered above ground. As a result of both the similarity and inhospitability of their habitats, subterranean rodents are expected to share a variety of physiological specializations not found in surface-dwelling taxa.

The ability of the subterranean habitat to shape rodent physiology becomes evident if we consider the effects of selected aspects of the burrow environment on its inhabitants. Among the most conspicuous environmental challenges imposed by subterranean habitats are the absence of light, restricted gas exchange, and the energetic demands of burrowing. Each of these parameters is expected to influence the physiology of subterranean rodents in a variety of ways.

Even in porous and sandy soils, little light penetrates into subterranean burrows. This absence of light affects not only the carrying capacity of the habitat, but also vitamin D_3 formation and patterns of thermoregulation in animals that lead a strictly subterranean (chthonic) existence (Nevo 1979; Ar 1986; Buffenstein 1996). In taxa that rarely emerge above ground, the absence of light may also influence the maintenance of circadian and circannual patterns of activity (Rado et al. 1991).

Gas exchange is particularly important in subterranean habitats. Gas exchange between subterranean burrows and the atmospheric air is primarily dependent upon the diffusion properties of the soil, which vary with soil type and moisture content (Arieli 1979). Limited convection occurs in burrows as gases expand and contract due to changes in barometric pressure and temperature (Ar 1986). Ventilation of burrow systems is thought to be achieved primarily by the movement of animals, with the result that areas of low activity (e.g., the nest) are generally considered to be poorly ventilated (Olszewski and Skoczen 1965; Ar 1986). Due to these constraints on gas exchange, the gas composition of the burrow atmosphere differs considerably from the atmospheric conditions found above ground. Specifically, both models of diffusion-mediated gas exchange (Withers 1978) and experimental data (MacLean 1981) indicate that unless the soil is completely devoid of biotic substances, burrow atmospheres will always be hypoxic (low in oxygen) and hypercapnic (high in carbon dioxide) relative to surface atmospheres. These pronounced differences in atmospheric conditions are expected to have marked effects on the metabolic, respiratory, and cardiovascular physiology of subterranean taxa.

The excavation of new burrows requires considerable expenditure of energy. While digging, animals must break up compacted soils and remove the resulting debris from the burrow (Stein, chap. 1, this volume). The energetic costs of these activities vary with soil type, burrow radius, and burrow length (Contreras and McNab 1990), as well as with the moisture content of the soil (Lovegrove 1989). Nevertheless, burrowing is expected to have a substantial influence on patterns of energy flux in these animals, particularly given that most subterranean species forage by excavating new tunnels. Increased energy requirements for burrowing activity should also result in increased respiration, which may be problematic under hypoxic conditions. The increased movement within tunnels generated by digging activity, however, may lead to increased ventilation and gas exchange, thus improving atmospheric conditions for the animals.

Clearly, the subterranean habitat has multiple effects on the physiology of its inhabitants. Like morphological adaptations (Stein, chap. 1, this volume), physiological responses to subterranean life appear to converge, with independent lineages displaying similar solutions to environmental challenges. In particular, convergent patterns of energy flux, heat transfer, water balance, respiration, and neuroendocrinology are evident among these animals. Although some of these responses (e.g., thermoregulation) have been the subject of intensive study across a variety of taxa, others (e.g., mineral absorption) have received comparatively little attention. Thus, in many ways, the extent of physiological convergence among subterranean taxa has yet to be fully explored. Similarly, for many physiological processes, the degree of differentiation between subterranean and surface taxa remains to be determined.

Just as certain physiological processes have received more attention than others, some subterranean taxa have been the focus of more research than others. Most studies of physiological systems in subterranean rodents have focused on the Spalacinae and Bathyergidae (Ar 1986; Jarvis and Bennett 1991; Nevo 1995), including the invaluable contributions from the laboratory of Eviator Nevo. By comparison, the ecophysiology of other subterranean rodents is only poorly known. In addition, with the exception of thermoregulatory physiology, few physiological systems have been fully examined in more than one family of subterranean rodents. Thus considerable work remains to be done to understand the adaptive solutions employed by subterranean rodents in response to commonly encountered physiological stresses.

Neuroendocrine and Sensory Physiology

Sensory Physiology

A critical component of any physiological response to external stimuli is the ability to perceive environmental conditions. Although much remains to be learned about the sensory physiology of subterranean rodents, electrophysiological studies conducted to date (e.g., Heth et al. 1991; Rehkämper, Necker, and Nevo 1994; Mann et al. 1997) tend to confirm morphological suggestions that vision and hearing are not important sensory modalities in these animals.

One of the most conspicuous morphological changes found in subterranean taxa is the reduction in size of the eyes (Stein, chap. 1, this volume). Although this reduction suggests a decreased reliance on visual stimuli, the visual abilities of most species remain poorly known. All subterranean rodents, even those lacking external eyes (e.g., *Nannospalax ehrenbergi*), are thought to be capable of detecting light (Cooper, Herbin, and Nevo 1993a,b; Herbin, Reperant, and Cooper 1994). Their ability to detect visual images, however, is more controversial, and appears to vary among species. At one extreme are species such as *N. ehrenbergi*, which are apparently unable to detect images; at the other extreme are some ctenomyids, which clearly respond to moving images (Francescoli, chap. 3, this volume). The extent to which visual perception and acuity vary predictably with factors such as eye size or extent of aboveground activity remains the subject of some debate (Stein, chap. 1; Francescoli, chap. 3, this volume).

The external ear is also modified in subterranean rodents and is typically reduced in size relative to surface-dwelling species (Nevo 1995; Stein, chap. 1, this volume). This reduction may contribute to a limited ability to localize sounds. Poor hearing ability (as measured by evoked potentials) over the entire auditory range has been reported for geomyids, bathyergids, and spalacines (Müller and Burda 1989; Burda, Bruns, and Müller 1990; Heffner and Heffner 1992, 1993; Heffner et al. 1994; Francescoli, chap. 3, this volume). Among these rodents, the greatest hearing sensitivity is at low frequencies. This observation concurs with what would be expected for animals that receive signals transmitted through soil, rather than through air (Heth, Frankenberg, and Nevo 1986). In semi-subterranean species (e.g., black-tailed prairie dogs, *Cynomys ludovicianus*), both the range of frequencies detected and the sensitivity to auditory signals is greater than in strictly subterranean species, although both abilities remain inferior to those of surface-dwelling taxa (Heffner et al. 1994).

The ability to detect tactile and possibly also olfactory cues may be enhanced in subterranean rodents (Jarvis and Bennett 1991; Mann et al.

1997; Stein, chap. 1; Francescoli, chap. 3, this volume). Potential somatosensory organs include the mystacial vibrissae and well-developed piloRuffini complexes (Klauer, Burda, and Nevo 1997; Stein, chap. 1, this volume). These possible sensory structures are often coupled with a markedly hypertrophied somatosensory cortex. For example, in *N. ehrenbergi*, the somatosensory cortex is 1.7 times larger than it is in the surface-dwelling, more visually oriented laboratory rat (Nevo 1995; Mann et al. 1997). This increase in the size of the brain structure associated with tactile information should facilitate greater reliance upon such cues to assess environmental conditions. An enhanced ability to process tactile information should also tend to promote seismic (rather than visual or auditory) communication in subterranean rodents (Heth et al. 1987; Rado et al. 1987; Nevo, Heth, and Pratt 1991; Mann et al. 1997; Narins et al. 1997; Francescoli, chap. 3, this volume). Although it is often speculated that olfactory abilities are augmented in subterranean animals (Jarvis and Bennett 1991) and may be of particular importance in communication among social species, this component of the sensory physiology of subterranean rodents remains to be explored.

Neuroendocrinology

Potential responses to external stimuli include chemical changes within the body. The neuroendocrine system directs short-term and long-term chemical responses to both external and internal stimuli. Short-term responses are regulated primarily by neurotransmitters acting on the sympathetic nervous system. In contrast, long-term responses are regulated primarily by bloodborne hormones produced by endocrine glands. Neither aspect of chemical regulation has been well studied in subterranean rodents.

Most neuroendocrinological studies of subterranean taxa have focused on brain morphology rather than the role of these structures in neuro- or electrophysiological responses (Necker, Rehkämper, and Nevo 1992; Towe and Mann 1995; Frahm, Rehkämper, and Nevo 1997; Mann et al. 1997). Neuroendocrine studies are limited to spalacines (Rehkämper, Necker, and Nevo 1994; Mann et al. 1997) and bathyergids (Bennett, Faulkes, and Molteno 1996, chap. 4, this volume), and in the latter family, these studies almost exclusively target reproductive suppression in eusocial species.

Circadian Rhythms

Daily (circadian) rhythms exhibit an approximately 24-hour cycle that is maintained by endogenous pacemakers. These pacemakers are synchronized (entrained) by photoreceptive mechanisms that respond to changes in light irradiance at dawn and dusk (Bunning 1973; Argamaso et al. 1995). The functions of circadian rhythms range from optimizing foraging efficiency to synchronizing social interactions among conspecifics. Both the outcomes and the regulation of these activity patterns are intimately linked to the ecophysiology of the organisms that exhibit them.

Like many surface-dwelling rodents, the majority of subterranean species display pronounced bimodal patterns of activity over the course of a 24-hour period, with peak activity occurring at night; examples include *Thomomys bottae* (Reiter et al. 1994), *Cryptomys damarensis* (Lovegrove, Heldmaier, and Ruf 1993), *Georychus capensis* (Lovegrove and Muir 1996), *Heterocephalus glaber* (N. Herhold, personal communication), *Tachyoryctes splendens* (Jarvis 1973), and some subspecies of *N. ehrenbergi* (Sanyal et al. 1990; Rado et al. 1991). Other subterranean species, however, lack this marked activity peak, instead displaying multiple, arrhythmic bouts of activity during a 24-hour period; examples include *Heliophobius argenteocinereus* (Jarvis 1973), *Cryptomys hottentotus* (Hickman 1980) and some subspecies of *N. ehrenbergi* (Ben Shlomo, Ritte, and Nevo 1995). Specialization for life in underground habitats does not appear, therefore, to have had a consistent effect on the temporal patterning of activity in subterranean rodents.

In surface-dwelling animals, circadian rhythms are thought to be entrained by changes in light intensity, rather than by cyclic changes in other environmental parameters such as temperature. Physical parameters within the subterranean burrow system tend to remain unchanged over the course of a 24-hour period, providing few viable alternatives to light as the basis for entraining circadian rhythms. Indeed, even temperature changes are attenuated in all but the most superficial burrows, and the remarkable daily and seasonal constancy of burrow temperatures (Bennett, Jarvis, and Davies 1988) precludes the use of this variable in circadian entrainment.

Because truly subterranean animals live in permanent darkness, it has been questioned whether photoperiodic entrainment occurs in these animals. Despite the frequent regression of both visual pathways and the pineal gland (Quay 1981; Pevet et al. 1984), subterranean rodents do perceive changes in photoperiod (Haim et al. 1983; Pevet et al. 1984; Francescoli, chap. 3, this volume). Both the eye and the infraorbital Harderian glands are likely sources of light detection and activation of the associated neuroendocrine pathways (Rado et al. 1993) that are thought to regulate circadian rhythms. In *N. ehrenbergi*, experimental removal of the Harderian glands has confirmed the role of these structures in detecting light cues (Rado et al. 1993), although animals whose Harderian glands were removed eventually did respond to changes in photoperiod, suggesting that other, nonocular photoreceptors exist. To date, however, the exact location of these alternative photoreceptors remains unknown, even for surface-dwelling taxa (Argamaso et al. 1995).

Secretion of the hormone melatonin, which regulates circadian rhythms (Bunning 1973; Reiter et al. 1983), is generally regulated by photoperiod. In surface-dwelling species, the pineal and the Harderian glands secrete melatonin (Reiter et al. 1983) in response to darkness. In the subterranean mole-rats *C. damarensis* and *H. glaber*, the Harderian gland is disproportionately large, accounting for about 0.2% of the body mass of these animals (R. Buffenstein, personal observation; van Jaarsveld, Mhatve, and Reiter 1989). Similarly, this structure appears to be enlarged in *N. ehrenbergi* (Gilad et al. 1997). These findings suggest a possible connection between persistent darkness, melatonin secretion, and the regulation of circadian activities. Harderian gland function in *N. ehrenbergi*, however, is influenced by sex steroids as well as by photoperiod (Gilad et al. 1997). Furthermore, rhythmicity of activity is not correlated with melatonin secretion (Ben Shlomo et al. 1996), casting doubt on the role of melatonin in regulating the activity patterns of these burrow dwellers.

As these brief discussions of sensory capabilities, hormonal responses, and circadian rhythms clearly indicate, additional studies are needed to characterize the neuroendocrinological physiology of subterranean rodents. In addition to a better understanding of the sensory capabilities of these animals, we need a better understanding of the pathways that connect sensory receptors to the responses that they elicit.

Energy Flux

Energy flux is a critical component of the sustained survival of an animal, and both energy intake and energy utilization are focal points of ecophysiological research. These processes are necessarily related to numerous other aspects of the animal's biology, including its behavior, ecology, and life history, for energy intake must be sufficient to balance all the energetic costs of living in a given environment (Feder et al. 1987).

The energetic costs associated with life in a subterranean habitat are typically assumed to be high, especially in species that seldom, if ever, emerge above ground. Foraging and other underground activities often involve extensive burrowing. Depending on soil type, this may result in energy expenditures more than three hundred times as great as those required to move the same distance across the soil surface (Vleck 1979). The poor quality of the food items available underground further compounds the problem of maintaining energy balance (Jarvis, Bennett, and Spinks 1998; Busch et al., chap. 5, this volume). How do subterranean rodents meet these challenges?

Energy Acquisition

FORAGING. Foraging in subterranean habitats is generally assumed to be a "blind" process (Jarvis, Bennett, and Spinks 1998) in that animals extend their burrows in randomly chosen directions until a food source is encountered (Busch et al., chap. 5, this volume). Regardless of the mode of digging employed (Stein, chap. 1, this volume), tunnel excavation is an energetically demanding process; in geomyids and bathyergids, tunnel excavation can result in a 300–4,000-fold increase in metabolic rate compared with resting values (Vleck 1979; du Toit, Jarvis, and Louw 1985; Lovegrove 1989). The cost of foraging is often exacerbated by the patchy distribution of food supplies, which requires that animals tunnel considerable distances to reach new food resources (Lovegrove and Wissel 1988; Busch et al., chap. 5, this volume). For animals living in arid regions, the cost of foraging may also be influenced by rainfall, being lowest when precipitation renders hard, dry soils more workable (Vleck 1981; Lovegrove 1989; Jarvis, Bennett, and Spinks 1998).

Perhaps because of their random foraging behavior (Jarvis, Bennett, and Spinks 1998) and the fact that food is costly to locate, subterranean species are generally less selective about the food items that they consume than are surface-dwelling rodents (Heth, Goldenberg, and Nevo 1989; de Villiers 1993; Jarvis, Bennett, and Spinks 1998; Busch et al., chap. 5, this volume). Although some subterranean genera feed primarily upon roots, corms, bulbs, and tubers, others feed on the aerial portions of plants (Jarvis and Sale 1971; Reichman and Smith 1985; Stuebe and Anderson 1985; Lovegrove and Jarvis 1986; Comparatore, Cid, and Busch 1995; Jarvis, Bennett, and Spinks 1998; Busch et al., chap. 5, this volume). When the diets of strictly subterranean rodents are compared with those of rodents foraging above ground, it appears that subterranean rodents tend to favor food items of higher nutritional quality with concomitantly lower fiber content, such as geophytes and perennial dicots (Reichman and Jarvis 1989; Jarvis, Bennett, and Spinks 1998). However, most species consume whatever is available. At certain times of the year, even the most favored foods may have a relatively high fiber and low protein content, in addition to containing secondary chemical compounds considered toxic to most domestic livestock (Buffenstein and Yahav 1994; Jarvis, Bennett, and Spinks 1998; Busch et al., chap. 5, this volume). Insensitivity to these poisons or detoxification of the harmful substances must therefore be common, although the mechanisms employed are unknown.

DIGESTION. Subterranean rodents readily extract energy from foods and exhibit high digestive efficiencies on both natural and domestic diets

TABLE 2.1 DIGESTIVE EFFICIENCIES OF SUBTERRANEAN RODENTS

Species	Diet	Digestive efficiency	Source
Bathyergus janetta	Field tuber	83.8	Bennett and Jarvis 1995
Bathyergus suillus	Grass stems and leaves	86.8	Bennett and Jarvis 1995
Cryptomys damarensis	Gemsbok cucumber	65.0	Bennett and Jarvis 1995
	Wild onion	94.8	Bennett and Jarvis 1995
	Sweet potato	97.8	Bennett and Jarvis 1995
	Carrots	87.9	Yahav, Carlston, and Buffenstein 1993
Cryptomys hottentotus	*Romulea* corms	97.1	Bennett and Jarvis 1995
Georychus capensis	Sweet potato	97.4	Du Toit, Jarvis, and Louw 1985
		96.8	Bennett and Jarvis 1995
Heterocephalus glaber	*Macrotyloma* and *Pyrenacantha* species	60–70	Bennett and Jarvis 1995
	Sweet potato	94.6	Buffenstein and Yahav 1991a
	Carrots	89.4	Buffenstein and Yahav 1991a
Nannospalax ehrenbergi			
$n = 52$	Carrots	92.3	Yahav, Simson, and Nevo 1988
$n = 54$		92.4	Yahav, Simson, and Nevo 1988
$n = 58$		93.4	Yahav, Simson, and Nevo 1988
$n = 60$		95.6	Yahav, Simson, and Nevo 1988

(Buffenstein and Yahav 1991a; Bennett and Jarvis 1995) (table 2.1). Indeed, the digestive efficiencies (>70%) of these species far exceed those of large domestic ungulates (e.g., cows: 50–60%: van Soest 1982) and surface-dwelling rodents (Bennett and Jarvis 1995) for diets of similar fiber content. This efficiency increases the energetic return per foraging effort, allowing the animals to maximally exploit the food resources available to them.

Subterranean rodents are largely herbivorous (Miller 1964; Pearson 1959; Busch et al., chap. 5, this volume), and, because they feed primarily on the fibrous portions of plants, they consume large quantities of cellulose, hemicellulose, and lignin. The digestive enzymes present in the mammalian gut are unable to decompose these complex sugars, and thus subterranean rodents, like other mammalian herbivores, must rely on symbiotic microorganisms to digest these compounds by anaerobic fermentation (van Soest

1982). The high digestive efficiencies of subterranean rodents are attributed to comparatively long retention times in the digestive tract (Yahav and Choshniak 1990). Food is held in the considerably enlarged caecum and hindgut (Perrin and Curtis 1980; Loeb, Schwab, and Demment 1991; Buffenstein and Yahav 1994) until fermentation has liberated all available energy.

The abundant microorganisms in the hindguts of subterranean rodents (e.g., Porter 1957; Buffenstein and Yahav 1991a) provide their hosts with high-energy compounds in the form of volatile fatty acids such as acetic, propionic, and butyric acid. These fatty acids are rapidly absorbed into an herbivore's bloodstream and are used in the citric acid cycle to provide the animal with more than 60% of its basal energetic requirements (Parra 1978). Because both the quantity of volatile short-chain fatty acids produced and the rate of their production increase with increasing fiber content of the diet, microbial fermentation allows subterranean rodents to obtain adequate energy even on low-quality/high-fiber diets (Buffenstein and Yahav 1991a, 1994).

To further maximize the energy extracted from their diets, subterranean rodents may also engage in coprophagy. Coprophagy is common among bathyergids (Jarvis and Bennett 1991), and has also been reported in some ctenomyids and geomyids (Kenagy and Hoyt 1980; Altuna, Bacigalupe, and Corte 1998). Functionally, coprophagy provides a good source of dietary protein in the form of digested microbes (Lee and Houston 1993). In addition, in some species, coprophagy may provide an important dietary supplement for weaning pups and lactating females. In *H. glaber*, for instance, young pups and pregnant and lactating females routinely solicit feces from other animals (Jarvis 1991). This rich, high-quality food source may provide individuals with both protein and the microbes needed for anaerobic fermentation.

Microbial fermentation of plant fiber also has important implications for the thermal physiology of subterranean rodents. In most herbivores, microbial fermentation occurs at a comparatively high temperature (41°C: van Soest 1982). Among bathyergids, however, the optimal fermentation temperature is considerably lower (33°C: Yahav and Buffenstein 1991a), in keeping with the lower body temperatures generally reported for these animals. It is not known whether lower optimal fermentation temperatures are a general feature of subterranean or other rodents because fermentation experiments are typically conducted at 41°C (van Soest 1982), with no attempt to determine the species-specific optimum. If, as seems likely, optimal fermentation temperatures correspond to "typical" body temperatures, than lower fermentation temperatures would be expected for subterranean rodents.

In most ruminants, microbial fermentation is a heat-liberating process

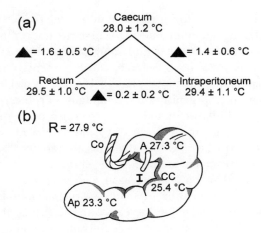

FIGURE 2.1. (a) Differences between rectal, intraperitoneal, and caecal temperatures. In all cases the caecum is more than 1°C cooler than the abdominal surrounds. (b) The caecum of the naked mole-rat, showing the temperatures measured in the three main regions of the caecum (A, *ampulla caeci;* CC, *corpus caeci;* Ap, *apex caeci*) as well as in the rectum (R), proximal colon (Co), and ileum (I).

that causes the rumen to be 2–3°C warmer than the surrounding abdomen (van Soest 1982). In *H. glaber* and *C. damarensis,* however, *in vivo* caecal temperatures are more than 1°C lower than core body temperatures, with a distinct temperature gradient evident within the caecum (fig. 2.1b). For example, in *H. glaber,* the apex caeci is the coolest region of the gut, with greater than a 1°C temperature differential relative to the rectum (fig. 2.1a; Yahav and Buffenstein 1992). These data suggest that although microbial fermentation liberates energy in the form of heat from short-chain fatty acids, in bathyergid mole-rats this heat is absorbed by the animal, rather than being released to the environment (Yahav and Buffenstein 1992). While heat absorption may be an undocumented feature of all hindgut fermentors, it is probably particularly advantageous in the burrow environment because an internal heat sink would decrease the risk of overheating in a confined and humid habitat. Clearly, more research is needed to establish the generality of this phenomenon and to assess its adaptive significance for subterranean species.

Energy Utilization

The rate at which an organism uses energy to fuel physiological processes is typically measured in terms of its basal metabolic rate (BMR), which is defined as the amount of energy or oxygen consumed per unit time by a resting, unstressed, mature animal kept at metabolically neutral environmental temperatures. Typically, BMR is measured under specified laboratory conditions that facilitate comparisons within and among taxa. As a result of these unnatural features, the reliability of BMR as an index of energy consumption by free-living animals is subject to considerable debate (Koteja 1991; Speakman, McDevitt, and Cole 1993).

Marked differences in BMR have been reported between recently field-captured bathyergids and conspecifics that have been held in captivity for extended periods (Bennett, Clarke, and Jarvis 1992; Bennett, Taylor, and Aguilar 1993), sounding a potential note of caution for researchers attempting to measure BMR in laboratory colonies of these animals. In particular, a greater than 40% decrease in BMR was reported for *C. h. hottentotus* that had been held in captivity for more than 2 months (Bennett, Clarke, and Jarvis 1992). Bennett and co-workers attributed these changes in captive individuals to acclimation to the laboratory climate, including habituation to captivity and human contact, acclimation to lower and more variable ambient temperatures, and acclimation to atmospheres with higher partial pressures of oxygen. Alternatively, these changes in BMR may have reflected weight changes among the captive animals due to maturation, ad libitum food supplies, and reduced activity in a captive setting. Although the sample sizes reported in the study are too small to be statistically meaningful, these marked metabolic differences between free-living and captive animals suggest that further research is needed to determine the effects of acclimation on laboratory studies of metabolism in subterranean rodents.

This concern notwithstanding, subterranean rodents typically have lower BMRs than their surface-dwelling counterparts (table 2.2). In geomyids, bathyergids, and spalacines, the observed BMR is lower than that predicted on the basis of body size (McNab 1979; Lovegrove and Wissel 1988; Buffenstein and Yahav 1991b; Bennett et al. 1994; Bennett, Cotterill, and Spinks 1996; Marhold and Nagel 1995). This reduction in BMR relative to body size is shared by desert-adapted rodents occupying ecotopes with poor, patchy, and unpredictable food supplies (Hart 1971; Schmidt-Nielsen 1975) and probably reflects the need to conserve energy in challenging habitats. Most (but not all) of the subterranean rodents that display large deviations from predicted metabolic rates inhabit arid areas; this observation suggests that aridity-related factors may be influential in selecting for low metabolic rates. Energetic constraints related to high foraging costs, especially in arid areas where food is patchily distributed, may provide selection pressure favoring parsimonious energy usage and consequent low metabolic rates (Lovegrove 1986a).

In addition to reducing the costs imposed by life in arid environments, low metabolic rates may be advantageous to subterranean rodents in more

TABLE 2.2 BASAL METABOLIC RATES OF SUBTERRANEAN RODENTS

Species	Habitat	Social status	MR[a] (ml/g/h)	M (%)[b]	Mass (g)	Source
Pitymys pinetorum	Mesic	Solitary	2.29	134	26	McNab 1979
Thomomys talpoides	Mesic	Solitary	1.22	115	106	Bradley, Miller, and Yousef 1974
Nannospalax ehrenbergi (n = 52)	Mesic	Solitary	0.95	92	116	Nevo and Shkolnik 1974
Tachyoryctes splendens	Mesic	Solitary	0.78	90	191	McNab 1979
Geomys pinetis	Mesic	Solitary	0.83	88	152	McNab 1979
Nannospalax ehrenbergi (n = 58)	Mesic/semiarid	Solitary	0.86	84	106	Nevo and Shkolnik 1974
Bathyergus suillus	Mesic	Solitary	0.49	83	102	Lovegrove 1986b
Geomys bursarius	Mesic	Solitary	0.70	81	101	Bradley and Yousef 1975
Bathyergus janetta	Semiarid	Solitary	0.53	78	406	Lovegrove 1986b
Thomomys umbrinus	Semiarid	Solitary	0.85	77	95	Bradley, Miller, and Yousef 1974
Cryptomys hottentotus darlingi	Semiarid	Social	0.98	76	60	Bennett, Taylor, and Aquilar 1993
Heliophobius argenteocinereus	Semiarid	Solitary	0.85	75	88	McNab 1979
Cryptomys mechowi	Mesic	—	0.60	78	272	Bennett et al. 1994
Heterocephalus glaber	Semiarid/arid	Social	1.00	69	42	Buffenstein and Yahav 1991b
Cryptomys damarensis	Semiarid/arid	Social	0.66	66	131	Bennett et al. 1994
Cryptomys damarensis	Semiarid/arid	Social	0.57	57	125	Lovegrove 1986a
Cryptomys hottentotus amatus	Mesic	Social	0.63	53	77	Bennett, Jarvis, and Cotterill 1993b
Heterocephalus glaber	Semiarid/arid	Social	0.66	45	40	McNab 1979

[a]Minimal resting metabolic rate, as measured in the thermoneutral zone.
[b]Percentage of metabolic rate, predicted on the basis of body size, using the allometric equation of Hayssen and Lacy (1985).

mesic habitats. In subterranean burrows, poor gas exchange may lead to hypoxic atmospheres. This problem may be especially pronounced in mesic habitats, in which wetter soils lead to particularly poor gas diffusion. As a result, even in mesic habitats, reduced oxygen consumption should be selected for. Furthermore, by reducing their overall metabolic rate, subterranean rodents may be able to reduce both oxygen consumption and the oxygen gradient across the alveolar capillary network, thereby improving their ability to exploit hypoxic environments.

Yet another factor that may influence the metabolic rates of subterranean rodents is their degree of sociality. Regardless of other environmental parameters, social species, in which multiple animals nest together in underground chambers (Lacey, chap. 7, this volume), are expected to experience greater degrees of hypoxia and hypercapnia than solitary species, in which only a single animal inhabits a burrow system. Selection pressure favoring lower metabolic rates should therefore be greater in social species. Empirical data, however, do not support this prediction. Although the BMRs of social subterranean species are low, they do not depart any more from predicted values than do those for solitary subterranean species. These data suggest that hypoxia and hypercapnia are not the primary determinants of BMR in these animals.

Low BMRs may also be advantageous if they increase the gap between the minimum and maximum metabolic rates that an animal can achieve. This greater metabolic scope may be particularly important during times of sustained high-energy demand, such as during prolonged tunnel excavation (Vleck 1979; du Toit, Jarvis, and Louw 1985; Lovegrove 1989). The metabolic scope of mole-rats appears to be higher than that of their surface-dwelling counterparts, with maximum rates 300–4,000 times BMR reported in both gophers and bathyergids (Vleck 1979; Lovegrove 1989).

Given the variety of selective pressures that appear to favor lowered BMR in subterranean taxa, how do these rodents achieve reduced levels of resting metabolic activity? Metabolic costs can be reduced by decreasing the energetic costs associated with specific biological processes. For example, the energetic requirements of neural tissue are extremely high, and the brain alone consumes more than 20% of the body's total energy intake, despite the fact that it constitutes only a small fraction of total body mass (Cooper, Herbin, and Nevo 1993b). Consequently, reduction in size of the eyes or other neurally expensive structures could have a substantial effect on overall metabolic rate (Cooper, Herbin, and Nevo 1993b). Thus microphthalmia and auditory insensitivity in bathyergids and spalacines (Burda, Bruns, and Müller 1990; Heffner and Heffner 1992, 1993) may be energy-conserving adaptations (Cooper, Herbin, and Nevo 1993b), as may reduced brain sizes in some geomyids (Towe and Mann 1995).

Low rates of metabolism in subterranean rodents may, in turn, have significant implications for other physiological and life history attributes in these animals. For example, low metabolic rates may be related to the slow rates of growth and development observed in bathyergids (Jarvis 1991; Bennett et al. 1991) as well as the unusually long life spans of geomyids (Smolen, Genoways, and Baker 1980) and bathyergids (Brett 1991; Jarvis and Bennett 1990).

Low BMR may conserve oxidative energy and lower both food and water requirements while limiting the amount of heat requiring dissipation. While the advantages of this strategy are obvious, this attenuated metabolism affects many aspects of the biology of subterranean rodents. Considerable work still needs to be done to understand the mechanisms by which metabolic costs are reduced and the effects thereof.

Energy Balance

THERMAL FLUX. Energy in the form of heat is regularly exchanged between an organism and its environment, with the net flow of energy down a thermal gradient. While transient changes in heat flux are common, prolonged survival requires that an individual be at thermal equilibrium, with heat gain and heat loss approximately equal (fig. 2.2). Heat exchange may be accomplished by radiation, conduction, convection, and evaporation (Seagrave 1971; Cossins and Bowler 1987). Both the mechanism and the rate of heat exchange depend upon the physical characteristics of the animal and its environment.

Radiant heat loads (acquired primarily through exposure to solar energy) are typically reduced in subterranean taxa, although the animals may be subjected to infrared thermal radiation from the ground and from conspecifics. Similarly, both convective and evaporative heat exchange are impeded underground by the lack of air movement and the high humidity commonly found in burrows. As a direct consequence of these limited ave-

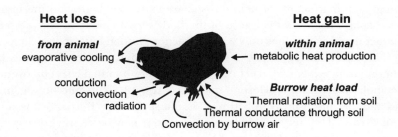

FIGURE 2.2. Heat exchange between a subterranean rodent and its underground habitat.

TABLE 2.3 THERMAL CONDUCTANCE IN SUBTERRANEAN SPECIES

Species	Mass (g)	Thermal conductance observed (ml O_2/g, h. °C)	Predicted (%)[a]	Source
Heterocephalus glaber	39	0.387	254	McNab 1979
Geomys bursarius	197	0.121	184	Bradley and Yousef 1975
Heliophobius argentocinereus	88	0.139	139	McNab 1979
Tachyoryctes splendens	191	0.090	135	McNab 1979
Cryptomys species	82	0.144	134	Marhold and Nagel 1995
Thomomys umbrinus	85	0.121	119	Bradley, Miller, and Yousef 1974
Thomomys talpoides	106	0.099	109	Bradley, Miller, and Yousef 1974

[a]Expressed as the percentage predicted by body mass using the allometric equation $C(\%) = 100 \times TC/(1.02)\text{mass}(g)^{-0.519}$ (Aschoff 1981).

nues of heat exchange, heat exchange via conduction is of particular importance in an underground environment.

Conduction requires direct physical contact between the animal and the object with which it is exchanging heat (Cossins and Bowler 1987), and is dependent upon the conductive properties of the surfaces in question as well as the area of the surfaces that are in direct contact. Maximal conduction typically occurs when animals adpress the ventral surface of the body against the substrate (thigmothermic regulation), as *H. glaber* commonly does when attempting to gain heat in warm, shallow tunnels (Brett 1991; Buffenstein and Yahav 1991b).

Laboratory studies indicate that rates of conductive heat loss in subterranean rodents often far exceed those predicted on the basis of body size (table 2.3) (McNab 1980; Aschoff 1981). These high rates of conduction are considered advantageous for animals living in enclosed burrows (McNab 1966, 1979), as conduction is the only effective means of heat loss available to subterranean species. Limited air circulation combined with high humidity precludes the use of evaporative cooling, and the insulative properties of soil restrict radiant heat loss. As a result, high rates of conductance between the animal and the soil provide the primary means of dissipating metabolic heat, particularly during periods of high activity. Heat loss may be further facilitated by physiological changes such as vasodilation of the feet, rhinarium, or other furless portions of the body (R. Buffenstein, unpublished data).

BODY TEMPERATURE. Body temperature represents the balance between metabolic heat production and heat loss to the environment. Among subterranean rodents, low metabolic rates coupled with high rates of conductive heat loss and variable mechanisms of temperature regulation contribute to the extreme variation in body temperature observed (table 2.4) (Contreras and McNab 1990).

The extent to which body temperature is regulated varies among the taxa considered in this volume. At one extreme are species such as *N. ehrenbergi* and *T. splendens,* in which heat exchange is tightly controlled and body temperature fluctuates only slightly (Nevo and Shkolnik 1974). At the other extreme is *H. glaber,* which lacks pelage or any other form of insulation and is thus effectively poikilothermic (Buffenstein and Yahav 1991b). In this species, body temperature is directly dependent on the physical rates of heat transfer between the animal and the environment (Buffenstein and Yahav 1991b). Body temperature may drop by more than 5°C if the animals are housed in ambient temperatures below 25°C. Numerous other species exhibit poikilothermic traits and stenothermy even though they possess fur (Lovegrove 1986b; Bennett, Jarvis, and Cotterill 1993; Bennett et al. 1994). At present, the reasons for these interspecific differences in thermolability are unknown. Further research is needed to assess both the phylogenetic and ecophysiological bases for the different patterns of thermoregulation in subterranean rodents.

THERMONEUTRAL ZONES. Subterranean rodents tend to have broad thermoneutral zones, with temperature-independent minimal metabolic effort occurring at ambient temperatures of 28–33°C (see table 2.4). Perhaps not surprisingly, the thermoneutral zones of these animals tend to match ambient temperatures within burrow systems (e.g., Bennett, Jarvis, and Davies 1988).

In surface-dwelling rodents, body temperature tends to be closely regulated and varies little, if at all, in response to environmental conditions (Hart 1971; French 1993). In contrast, subterranean rodents display a wide variety of responses to temperatures that fall outside their thermoneutral zones. In some species (e.g., *C. damarensis* and *N. ehrenbergi*), individuals respond to decreases in ambient temperature with marked increases in metabolic rate (Nevo and Shkolnik 1974; Lovegrove 1986a). In others (e.g., some *Cryptomys*), a more modest metabolic response is associated with a decrease in body temperature of 2–5°C (Lovegrove 1986b; Marhold and Nagel 1995). Still other species (e.g., *C. h. darlingi* and *H. glaber*) display little metabolic response, instead allowing body temperature to vary markedly with changes in ambient temperature (Buffenstein and Yahav 1991b; Bennett, Jarvis, and Cotterill 1993). Again, however, the evolutionary and ecological reasons for these differences in metabolic response remain to be determined.

One response to extreme ambient temperatures that has not been re-

TABLE 2.4 TEMPERATURE REGULATION IN SUBTERRANEAN RODENTS

Species	Mean body mass (g)	Body temperature	Thermoneutral zone	Precision of thermoregulation[a]	Source
Heterocephalus glaber	42	33	31–34	L	Buffenstein and Yahav 1991b
Cryptomys hottentotus darlingi	60	33	28–31.5	L	Bennett, Jarvis, and Cotterill 1993b
Cryptomys hottentotus amatus	50–93	34	28–32	L	Bennett et al. 1994
Cryptomys hottentotus n = 68	82	36	32.5	L	Marhold and Nagel 1995
Cryptomys hottentotus nimrodi	81	34	31–32	H	Bennett, Cotterill, and Spinks 1996a
Heliophobius kapeti	99	34	28–30	H	McNab 1966
	88	35	28–33	H	
Cryptomys bocagei	66–122	34	31.5–32.5	L	Bennett et al. 1994
Cryptomys hottentotus hottentotus	60–142	34	30–32	H	Haim and Fairall 1986
	67–104	34	27–30	H	Bennett, Clarke, and Jarvis 1992
Cryptomys hottentotus natalensis	60–182	33	30–31.5	H	Bennett, Taylor, and Aquilar 1993a
Cryptomys damarensis	124	35	27–31	H	Lovegrove 1986a
	130–192	35	28–31	H	Bennett, Clarke, and Jarvis 1992
Nannospalax ehrenbergi n = 52	128	36	28.3–N	H	Nevo and Shkolnik 1974
n = 58	124	36	28.4–N	H	
n = 60	121	36	30–	H	
Tachyoryctes splendens	191	36	27–34	H	McNab 1966
Geomys bursarius	197	36		H	Bradley and Yousef 1975
Cryptomys mechowi	267	34	29–30	H	Bennett et al. 1994
Bathyergus janetta	406	35	27–34	L	Lovegrove 1986b
Bathyergus suillus	620	35	25–31	H	Lovegrove 1986b

[a] L, thermolabile; H, homeothermic

ported for subterranean rodents is torpor. In a number of surface-dwelling rodents (e.g., some heteromyids and cricetines), daily or seasonal torpor is commonly employed as a response to low environmental temperatures or periods of reduced food availability (French 1993). Subterranean species, however, do not appear to employ this means of reducing metabolic costs during periods of environmental stress. It has been argued that because the subterranean environment is relatively invariant compared with surface habitats (Nevo 1979; Bennett, Jarvis, and Davies 1988), the inhabitants of subterranean burrows are rarely faced with the types of environmental extremes that favor torpor. As studies of subterranean habitats increase (Busch et al., chap. 5, this volume), the quantitative data needed to address this possibility should become available.

NON-SHIVERING THERMOGENESIS. Non-shivering thermogenesis (NST) is thought to be the primary means of endogenous heat production by small mammals housed at cold temperatures (Carneheim, Nedergaard, and Cannon 1984; Heldmaier, Klaus, and Wiesinger 1990). Among subterranean rodents, NST has been studied primarily in bathyergids and spalacines, in which it is associated with marked increases in brown adipose tissue, metabolic rate, and body temperature (table 2.5). Because NST does not interfere with muscular function, it may enhance the ability of these small mammals to thermoregulate while remaining physically active, as when foraging in superficial tunnels in which temperatures are likely to be cooler (Bennett 1990). Seasonal changes in burrow temperature have been reported for *C. damarensis* (Bennett, Jarvis, and Davies 1988; Bennett 1990), suggesting that the efficacy of NST may vary seasonally, as has been reported for non-subterranean rodents (Foster 1984). Although *H. glaber* also employs NST, the absence of insulation (e.g., fur, subcutaneous fat) in this species renders this mode of thermoregulation ineffective, as heat is rapidly lost to the environment (Hislop and Buffenstein 1994). Instead, *H. glaber* must rely on behavioral mechanisms of thermoregulation to maintain body temperature (Brett 1991; Yahav and Buffenstein 1991b).

TABLE 2.5 NON-SHIVERING THERMOGENIC CAPACITY IN SUBTERRANEAN RODENTS

Species	NST[a] capacity	Tb (°C)	Body mass (g)	Source
Heterocephalus glaber	3.56	2.8	32	Hislop and Buffenstein 1994
Cryptomys hottentotus	4.51	5.4	102	Haim and Fairall 1986
Cryptomys damarensis	1.94	4.1	123	Hislop and Buffenstein 1994
Nannospalax ehrenbergi n = 60	3.3	2.1	134	Haim et al. 1983

[a]Non-shivering thermogenic capacity, expressed as the ratio between maximal oxygen consumption and minimal metabolic rate, after and prior to noradrenaline stimulation with the concomitant change in body temperature (Tb).

In summary, most subterranean rodents encounter similar environmental conditions and challenges below ground, regardless of aboveground climatic conditions. It is not surprising, therefore, that they share a common pattern of energy flux. In all cases, they are predominantly herbivorous, with similar gut morphology and highly efficient gastrointestinal function. In terms of energy output, their low basal metabolic rates and body temperatures, coupled with high rates of thermal conductance, are generalized features considered well suited to overcoming the problems encountered below ground. However, even within a single subterranean rodent family, vast differences in thermoregulatory ability are evident. Considerable work is needed in order to understand the mechanisms employed at the cellular level and the direct effects of these mechanisms on other biological systems.

Responses to Hypoxia and Hypercapnia

In addition to problems of energy flux, subterranean rodents must contend with a number of physiological challenges imposed by the atmospheric conditions found in underground burrows. Specifically, the prevalence of hypoxic and hypercapnic atmospheres (Withers 1978; Maclean 1981) may pose a suite of physiological problems for these animals that are not faced by their surface-dwelling counterparts. With the exception of species living at high altitudes, surface-dwelling rodents are not well adapted to atmospheres containing low partial pressures of oxygen and high partial pressures of carbon dioxide; under these conditions, cardiovascular function, ventilation, respiration, and acid-base balance may all be adversely affected (Boggs, Kilgore, and Birchard 1984). Subterranean rodents appear to be far more tolerant of hypoxic and hypercapnic conditions (Faleschini and Whitten 1975; Tucker et al. 1976; Arieli et al. 1984; Arieli 1990; Arieli and Nevo 1991), as might be expected of animals adapted to life in poorly ventilated tunnels. Thus subterranean species are able to survive and achieve high levels of metabolic activity (e.g., during digging: Vleck 1979; Lovegrove 1989) under atmospheric conditions in which laboratory rats cannot maintain even a "normal" BMR (Arieli, Ar, and Shkolnik 1977).

Subterranean rodents display a remarkable suite of morphological adaptations to hypoxic and hypercapnic environments. These adaptations include increased capillary densities in skeletal, cardiac, and pulmonary tissues, as well as increased fractional volumes of mitochondria in muscle tissue (Arieli and Ar 1981a; Widmer et al. 1997). In addition, elevated myoglobin concentrations have been reported for some taxa (Lechner 1976; Ar, Arieli, and Shkolnik 1977). Collectively, these features should result in reduced diffusion distances during gas exchange and enhanced oxygen transfer from blood to metabolically active tissues. Thus they may complement the physiological responses to the gaseous environment in the burrow.

Ventilatory Responses

One of the most basic adaptive responses to hypoxia is to increase breathing frequency and depth (Davies and Schadt 1989), thereby increasing the rate of gas exchange between the lungs and the environment. This response is attenuated in subterranean rodents. *T. bottae* and *N. ehrenbergi* display ventilation rates that are 20% and 30–40% lower than expected, respectively, when breathing standard laboratory air (Arieli and Ar 1979; Boggs, Kilgore, and Birchard 1984). Low ventilation rates are linked to the low BMR of chthonic species (see table 2.2) and should result in reduced levels of carbon dioxide production by these animals. Similarly, the lung morphological features of both *H. glaber* and *T. splendens* are consistent with low BMR and concomitantly low oxygen requirements. Although few studies of the lung morphology of subterranean rodents have been conducted (Stein, chap. 1, this volume), data from *H. glaber* and *T. splendens* (Maina, Maloiy, and Makanya 1992) suggest that lung development in these animals may be less complete than in surface-dwelling taxa, with conspicuous neotenic features.

If surface-dwelling mammals encounter hypercapnic conditions, they respond to the altered partial pressure of carbon dioxide by increasing both ventilation rate and the depth of inhalation, since arterial carbon dioxide concentration is the main regulator of ventilation and the entire respiratory control system (West 1995). In contrast, reduced ventilatory responses to increased concentrations of carbon dioxide have been reported in three orders of burrowing mammals as well as in some burrowing birds (Boggs, Kilgore, and Birchard 1984). Not surprisingly, the greatest reduction in sensitivity to carbon dioxide content is exhibited by strictly subterranean rodents (fig. 2.3) such as *T. bottae* (Darden 1972), *N. ehrenbergi* (Arieli and Ar 1979; Boggs, Kilgore, and Birchard 1984), and *C. hottentotus* (G. N. Bronner, W. van Aardt, and R. Buffenstein, unpublished data). Even semisubterranean mammals, such as marsupial insectivores (e.g., *Tachyglossus aculeatus*) and burrowing rodents (e.g., *Tamias striatus* and *Marmota monax*), show considerably attenuated ventilatory responses to elevated carbon dioxide concentrations when compared with laboratory rats or humans (Darden 1972; Boggs, Kilgore, and Birchard 1984). This convergent respiratory response to subterranean atmospheric conditions is considered highly adaptive in that it reduces the energetic respiratory costs associated with the deeper and more rapid breathing induced in a hypercapnic environment.

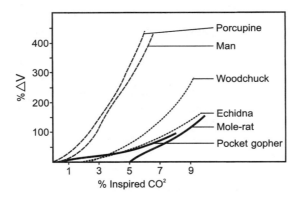

FIGURE 2.3. Ventilatory response to inspired carbon dioxide in fully subterranean (solid line), semi-subterranean (dotted line), and non-subterranean mammals (dashed line). Subterranean rodents appear relatively insensitive to changing carbon dioxide content. (From Boggs, Kilgore, and Birchard 1984)

Cardiac Responses

Associated with adaptations in ventilation rate are adaptations in both resting heart rate and heart mass. Specifically, resting heart rate may be lower and heart mass may be greater in subterranean species than in surface-dwelling taxa. Both tendencies have been reported for a number of semi-subterranean rodents such as ground squirrels and chipmunks (Burlington et al. 1970; Jones and Wang 1976). Among the taxa considered in this volume, studies of cardiac function have been conducted only for *N. ehrenbergi* (Arieli and Ar 1981b; Storier, Wollberg, and Ar 1981; Arieli et al. 1986; Edoute, Arieli, and Nevo 1988). Under "normal" burrow conditions, the heart rate of this species is approximately 72% lower than that predicted on the basis of body size (Arieli et al. 1986). This reduced rate is attributed to high innervation from the parasympathetic nervous system, which typically slows ventricular contraction, and low innervation from the sympathetic nervous system, which characteristically increases both heart rate and the force of contraction and counteracts the predominant parasympathetic control (Storier, Wollberg, and Ar 1981).

Reduced heart rates may be indicative of greater cardiac reserves, which may provide the capacity for increased cardiac output when an individual is subjected to either greater metabolic demands or decreased atmospheric oxygen availability. *N. ehrenbergi* appears to respond to hypoxic conditions by elevating cardiac output. This response is accomplished by increasing both heart rate and stroke volume per ventricular contraction, both of which result in augmented coronary flow and improved cardiac efficiency. Enhanced cardiac output, in turn, facilitates the adequate delivery of oxygen to tissues despite the reduced availability of atmospheric and pulmonary oxygen, enabling continued and unimpaired functioning of other body processes during hypoxia (Edoute, Arieli, and Nevo 1988). Cardiac function has yet to be examined in other subterranean taxa, but, given the ability of these animals to persist in hypoxic environments, similar responses to low partial pressures of oxygen seem likely.

Blood Transport

Once oxygen has entered the alveolar capillary beds, it must be transported to all portions of the body. As indicated above, this transport relies in part upon cardiac function. Oxygen transport is also influenced by the carrying capacity of the blood, as well as by processes at the cellular level that are responsible for offloading oxygen carried in the bloodstream. Thus efficient oxygen transport under hypoxic conditions requires modifications of a number of physiological traits in addition to cardiac output.

The oxygen transport properties of blood vary markedly among subterranean rodents (fig. 2.4; table 2.6). For example, *T. bottae* exhibits an increased oxygen carrying capacity that is facilitated by elevated hemoglobin concentrations, increased red blood cell counts, and increased hematocrits (Chapman and Bennett 1975; Lechner 1976; Arieli et al. 1986). Although hematocrits and hemoglobin concentrations are also high in *C. hottentotus* and in some subspecies of *N. ehrenbergi* (G. N. Bronner, W. van Aardt and R. Buffenstein, unpublished data), in *H. glaber* (Johansen et al. 1976) and other subspecies of *N. ehrenbergi* (Arieli et al. 1986), these parameters are no different from those found in surface-dwelling rodents (fig. 2.4; table 2.7). Interestingly, both *H. glaber* and the latter group of *N. ehrenbergi* subspecies live in arid habitats, where gas diffusion rates through soil may be greater, with the result that the degree of hypoxia may be lower than that encountered by subterranean species living in more mesic areas. Additional studies are needed to determine whether this apparent ecological correlation represents a general trend among subterranean rodents.

Perhaps the most common blood transport adaptation to hypoxia is the presence of hemoglobin with a high intrinsic affinity for oxygen (Boggs, Kilgore, and Birchard 1984). High oxygen affinities have been reported for *T. bottae, H. glaber, C. hottentotus,* and *N. ehrenbergi* (fig. 2.4; table 2.6) as well as for some semi-subterranean rodents such as black-tailed prairie dogs (Boggs, Kilgore, and Birchard 1984; Arieli et al. 1986).

Hemoglobin affinity for oxygen is influenced by the concentration of 2,3-diphosphoglycerate (2,3-DPG) in red blood cells. When 2,3-DPG concentra-

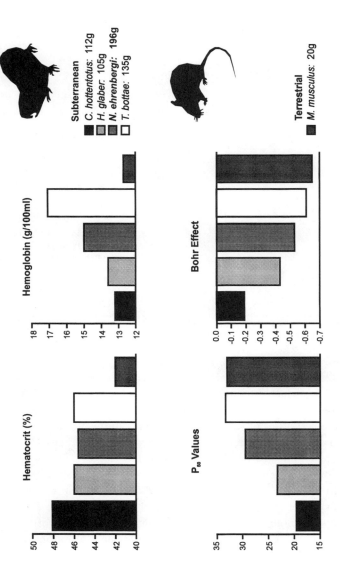

FIGURE 2.4. Oxygen transport properties of blood in four subterranean and one terrestrial rodent species. Hematocrits are higher in subterranean than in surface-dwelling rodents. Hemoglobin concentrations are also higher in subterranean rodents, but this effect is more variable. The P_{50} value (the partial pressure of oxygen at which 50% of the blood is saturated) is generally lower in subterranean rodents, indicating that their blood has a higher oxygen affinity and is able to become fully saturated at lower partial pressures of oxygen. Bohr effects are less pronounced than in most surface dwellers, implying a degree of insensitivity to changes in blood pH. (Data from G. N. Bronner, W. van Aardt, and R. Buffenstein, unpublished data; Lechner 1976; and Ar, Arieli, and Shkolnik 1977)

TABLE 2.6 BLOOD OXYGEN AFFINITY AND BUFFERING CAPACITY OF SUBTERRANEAN RODENTS

Species	Mass (g)	P_{50} (Torr)	P_{50} (predicted)	$\dfrac{\Delta \log PCO_2}{\Delta PH}$	$\dfrac{\Delta \log P_{50}}{\Delta pH}$	Source
Heterocephalus glaber	106	23.3	35–39	—	−0.43	Johansen et al. 1976
Thomomys bottae	135	33.3	39	−2.67	−0.61	Lechner 1976
Nannospalax ehrenbergi	196	29.5	37.9	−1.32	−0.53	Ar, Arieli, and Shkolnik 1977

Source: Modified from Boggs, Kilgore, and Birchard 1984.

TABLE 2.7 HEMATOCRIT AND HEMOGLOBIN CONCENTRATION OF SUBTERRANEAN RODENTS FROM DIFFERENT HABITATS

Species	Habitat	Hematocrit (%)	Hemoglobin (g/dl)	Source
Nannospalax ehrenbergi				
$n = 52$	Mesic	52	16.0	Arieli et al. 1986
$n = 54$	Mesic	51	16.3	
$n = 58$	Mesic	51	16.6	
$n = 60$	Arid	48	14.7	
Cryptomys hottentotus	Mesic	48	14	R. Buffenstein, W. van Aardt, and G. N. Bronner, unpublished data
Heterocephalus glaber	Arid	46	13.6	Johansen et al. 1976
Thomomys bottae	Arid	46	17.1	Lechner 1976
Nannospalax ehrenbergi	Arid	45.6	15.0	Ar, Arieli, and Shkolnik 1977

tions are high, the binding affinity of hemoglobin for oxygen is decreased, resulting in reduced oxygen transport but enhanced release of oxygen at the tissue level. Although surface-dwelling animals are well known to respond to hypoxic conditions by increasing the concentration of 2,3-DPG within their red blood cells, the responses of subterranean rodents under similar atmospheric conditions are more variable. In some burrowing mammals (e.g., the nine-banded armadillo, *Dasypus novemcinctus*), low 2,3-DPG levels are thought to contribute to high oxygen affinities (Dhindsa, Hoversland, and Metcalfe 1971). In contrast, *H. glaber* exhibits 2,3-DPG levels similar to those found in laboratory mice (Johansen et al. 1976), while *C. hottentotus, T. bottae,* and the European mole *(Talpa europaea)* exhibit elevated levels of this compound (W. van Aardt, G. N. Bronner, and R. Buffenstein, unpublished data; Lechner 1976; Jelkmann et al. 1981).

Although the precise relationships between hypoxia, 2,3-DPG levels, and oxygen affinity remain to be determined for subterranean rodents, it is possible that, in species with high hemoglobin affinities for oxygen, substantial increases in 2,3-DPG levels serve to bring oxygen affinities into the "normal" functional range. Thus, elevated 2,3-DPG levels may help to prevent oxygen deprivation of tissues by lowering the oxygen affinity of hemoglobin. This change in affinity, in accordance with the Bohr effect, would shift oxygen equilibrium curves to the right (fig. 2.5), thus promoting unloading of oxygen at the tissues where it is needed. Thus species-typical oxygen affinities may be modified by changing 2,3-DPG levels to allow individuals to function in environments with varying degrees of hypoxia.

Acid-Base Balance

Hypercapnic environments, commonly encountered below ground, alter the amount of carbon dioxide inhaled and thus the partial pressure of carbon dioxide in blood. Because acid-base balance is inextricably linked to the partial pressure of carbon dioxide in blood, subterranean rodents may

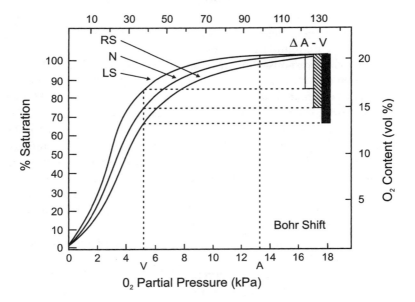

FIGURE 2.5. The Bohr effect has adaptive consequences for oxygen delivery to the tissues. Oxygen equilibrium curves for the blood of a hypothetical normal surface-dwelling animal (N) and of animals exhibiting a left-shifted (LS) and a right-shifted (RS) oxygen equilibrium curve are shown, with the normal arterial (A) and venous (V) partial pressures of oxygen indicated. Left- and right-shifted curves indicate changes in oxygen affinity, which may be induced by changes in pH or may be dependent upon the inherent blood properties of the particular species. The arterial-venous oxygen difference (ΔA-V) is greatest in the right-shifted curve, enabling greater offloading of oxygen in the tissues without substantially impairing loading. (Adapted from Withers 1992)

encounter problems in regulating blood pH. Elevated carbon dioxide concentrations in the blood may result in a concomitant increase in carbonic acid, which may, in turn, result in respiratory acidosis, which can create potentially serious problems for enzymatic activity and other pH-sensitive physiological processes (West 1995). Counteracting these negative consequences of a hypercapnic environment may require physiological adaptations enabling greater acid tolerance or enhanced buffering capacity. Regulation of blood pH is primarily accomplished by variation in respiration rate and depth and counter-ion exchange in the kidney and gastrointestinal tract.

Alterations of acid-base balance in response to hypercapnia have been detected in *C. hottentotus* (R. Buffenstein et al., unpublished data) and *N. ehrenbergi* (Ar, Arieli, and Shkolnik 1977), but not in *T. bottae* (Lechner 1976). In the former two species, unaltered buffering capacities and greater tolerance of variations in pH may enable shifts in oxygen affinity curves (see fig. 2.5), thus facilitating oxygen exchange, particularly when offloading oxygen to tissues. In contrast, *T. bottae* exhibits an elevated non-carbonic buffering capacity, and thus a greater ability to regulate pH. Lechner (1976) attributes this ability to higher concentrations of hemoglobin and organic phosphates than are found in surface-dwelling rodents.

The attenuated responses to hypercapnia and hypoxia and high tolerance of carbon dioxide observed in both *C. hottentotus* (R. Buffenstein et al., unpublished data) and *N. ehrenbergi* (Ar, Arieli, and Shkolnik 1977) differ considerably from what one would expect if surface-dwelling animals were exposed to similar gaseous conditions (Boggs, Kilgore, and Birchard 1984; West 1995). Boggs, Kilgore, and Birchard (1984) suggested that nitrogenous wastes, and in particular, potentially high ammonia concentrations within the burrow, may buffer pH and play a pivotal role in facilitating this tolerance. To date, however, few studies have reliably measured true burrow gaseous conditions, let alone when animals are residing in and active within the burrow. As such, it is not surprising that physiological responses to the gaseous conditions within burrows are poorly understood and still require considerable exploration.

In summary, considerable convergence exists among subterranean rodents in their physiological responses to and tolerance of hypoxia and hypercapnia, and many of these features are also shared with surface-dwelling animals living at high altitudes (Boggs, Kilgore, and Birchard 1984; Nevo 1979). Specialized respiratory responses are evident even in captive animals de-acclimated to burrow environments and housed under normal atmospheric conditions (Ar, Arieli, and Shkolnik 1977; Nevo 1979). This finding suggests that genetic factors determine these physiological characteristics (Ar 1986). Specialized ventilatory, blood, and tissue adaptations contribute to this adaptive response to gaseous conditions underground; these adaptations include increased ventilation, a high oxygen affinity in the blood, resistance to acidosis, and tolerance of tissues to low partial pressures of oxygen and high partial pressures of carbon dioxide.

Water Flux

In addition to the energetic and atmospheric challenges imposed by the subterranean habitat, rodents that live in underground burrows face specific problems of water balance. Subterranean rodents, like many small mammals, do not drink free water, but instead obtain moisture through the foods that they consume. As a result, these animals rely on the high water content of vegetation and on water liberated during metabolic oxidation processes to maintain an appropriate level of hydration (fig. 2.6).

Water balance is achieved primarily by adjusting water loss to match water intake. Neither pulmocutaneous evaporative water loss nor fecal water loss is typically tightly controlled in subterranean rodents. Fecal water loss is dependent primarily upon the digestibility of the diet. Because the proportion of water in fecal samples is relatively constant, fecal water loss is dependent upon the quantity of feces voided. Respiratory water loss from the lungs and nasal passages varies with the rate of gas exchange, while cutaneous water loss is primarily dependent upon the vapor pressure gradient between the animal and its environment. The high humidity within subterranean burrows, coupled with the relatively low temperature differential between the animal and the burrow environment, results in a comparatively low vapor pressure gradient and concomitantly low rates of pulmocutaneous evapora-

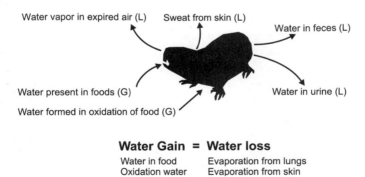

FIGURE 2.6. Water balance in a subterranean rodent.

TABLE 2.8 MAXIMUM URINE CONCENTRATION AND URINE/PLASMA RATIOS FOR RODENTS INDEPENDENT OF EXOGENOUS WATER

Species	Urine concentration (mmol/kg)	Urine/plasma ratio	Source
Notomys alexis	9,370	18	MacMillen and Lee 1967
Desmodillus auricularis	5,507	14	Buffenstein, Campbell, and Jarvis 1985
Meriones unguiculatus	3,500	10	Edwards and Peters 1988
Rattus norvegicus	3,250	8	MacMillen and Lee 1967
Acomys subspinosus	3,784	7	Buffenstein, Campbell, and Jarvis 1985
Nannospalax			
$n = 52$	1,102	—	Nevo et al. 1989
$n = 54$	1,154		
$n = 58$	1,213		
$n = 60$	1,522		
Heterocephalus glaber	1,520	5	Urison and Buffenstein 1994a

tive water loss for subterranean rodents (Gettinger 1984; Buffenstein and Yahav 1991b).

Urine production is the one avenue of water loss that can be tightly regulated. This is achieved primarily by altering the concentration of the urine voided. To date, urine concentrating ability has been investigated only for bathyergid and spalacine mole-rats. The urine concentrating abilities of both *N. ehrenbergi* and *H. glaber* are moderate. Maximal urine concentrations for both species, predicted from kidney morphology and measured during water deprivation experiments, range from 1,100 to 1,650 mmol/kg (Nevo et al. 1989; Yahav, Simson, and Nevo 1990; Urison and Buffenstein 1994a). These values are approximately five times higher than the concentration of plasma. This degree of renal efficiency is most similar to that of mesic, rather than desert-dwelling, semi-fossorial rodents (Baverstock 1976; table 2.8).

Neither *N. ehrenbergi* nor *H. glaber* can remain in water balance under even moderately stressful laboratory conditions. Both species exhibit high rates of weight loss even though the volume of urinary water loss is dramatically diminished (Yahav, Simson, and Nevo 1989, 1990; Urison and Buffenstein 1994a). Their minimum urinary water loss of 0.01 ml/g/day, although similar to that for mesic rodents (Baverstock 1976), is more than twice the amount reported for other African desert rodents (Buffenstein, Campbell, and Jarvis 1985) and contributes substantially to high rates of water turnover in these species. Not surprisingly, therefore, daily rates of water turnover in *N. ehrenbergi* are more than double those for desert rodents (Gettinger 1984; Yahav, Simson, and Nevo 1990; Degen 1997). These data suggest that although many subterranean rodents successfully inhabit arid regions, water balance in these animals is not facilitated by superior renal efficiency or other specialized mechanisms of water retention. Rather, the combination of high humidity and low radiant heat load associated with burrow microclimates may minimize cutaneous water loss so that only a moderate kidney concentrating ability is required to achieve water balance. This reliance on the burrow environment to maintain water balance may ultimately limit the distributions of subterranean species. For example, Nevo et al. (1989) have postulated that its moderate kidney concentrating ability may be the limiting factor restricting the distribution of *N. ehrenbergi* to areas where annual rainfall exceeds 100 mm.

In addition to daily patterns of water flux, seasonal shifts in water balance may occur, with lower rates of water intake during periods when the water content of food is reduced. Under these circumstances, the relative salt and protein content of the diet may be increased, which may affect water balance in subterranean rodents, none of which consume free water. Mild salt or protein loading is well tolerated by both *N. ehrenbergi* and *H. glaber*, although more severe salt loading leads to salt-induced diuresis, in which approximately 1.3 times the normal volume of urine is voided, leading to dehydration (Yahav, Simson, and Nevo 1990; Urison and Buffenstein 1994a). Although mild salt loading seems likely to occur under field conditions, pronounced salt loading of the type used in laboratory studies of water regulation seems unlikely to be a real constraint on the distributions of subterranean species. Low burrow water vapor content resulting from a combination of surface aridity and soil conditions seems more likely to limit the distributions of these taxa.

Mineral Homeostasis

Since Hess and Gutman (1922) demonstrated conclusively that rickets could be cured by exposure to ultraviolet light, it has been widely accepted that sunlight plays an integral role in the regulation of calcium stores in the body. The "vitamin D_3 endocrine system" is the key regulator of calcium concentrations. Vitamin D_3 and bone mineral metabolism are thus closely linked. However, both vitamin D_3 and calcium have many other vital functions in the body, and adequate amounts of both of these substances are considered essential for life.

Calcium is a wide-ranging physiological regulator of numerous body functions, including neural transmission and excitability, cardiovascular function, muscle contraction, the stability and permeability of cell membranes, and intracellular cell signaling, over and above its crucial role in bone structure and support. As such, it is imperative that extracellular calcium concentrations be regulated. This regulation is achieved primarily by manipulating calcium fluxes between the plasma and the gut, bone, and kidney (Mundy 1990).

The active hormone in the vitamin D_3 endocrine system, 1,25 dihydroxyvitamin D_3, plays a key role in regulating these calcium fluxes. However, this light-dependent hormone also has numerous other functions unrelated to calcium regulation. These functions include the modulation of DNA replication, cell differentiation, and cell proliferation, the regulation of the immune system, and the control of peptide hormone secretion (for review see Reichel, Koeffler, and Norman 1989; Henry and Norman 1990). This hormone is therefore considered a comprehensive somatotropic regulator and modulator.

Animals leading a strictly chthonic existence live in a perpetually dark environment, devoid of ultraviolet radiation, and are therefore incapable of using the known biochemical pathways to synthesize vitamin D_3. In the absence of sunlight, vitamin D cannot be synthesized endogenously, although it may be obtained by the consumption of fat from animals previously exposed to sunlight. Because subterranean rodents are herbivorous, they have no obvious source for this essential substance. These observations raise intriguing questions as to whether and how subterranean rodents can acquire vitamin D_3 in these circumstances, and whether they have responded to this environmental stress by modifying either their vitamin D_3 or their mineral metabolism.

Vitamin D_3

Vitamin D_3 is synthesized (from 7-dehydrocholesterol) by photolytic processes in the skin. Vitamin D_3 is considered inert, and undergoes two further hydroxylation processes to form the active hormone, 1,25 dihydroxy-vitamin D_3 [$1.25(OH)_2D_3$]. The first hydroxylation takes place in the liver, producing the principal circulating metabolite, 25 hydroxy-vitamin D_3, while the final hydroxylation, facilitated by 1-α hydroxylase located in the kidney, results in the formation of the active hormone. Other hydroxylated metabolites of vitamin D_3 are also formed in the kidney, the functions of which are mainly unknown. One metabolite, 24,25 dihydroxy-vitamin D_3, is thought to be an inert by-product of the process, and through its synthesis, rather than that of $1.25(OH)_2D_3$, the amount of active hormone available is regulated. In animals with an adequate supply of vitamin D_3, this metabolite is formed preferentially, resulting in raised levels of the 24-hydroxylase enzyme, whereas with vitamin D_3 deficiency, 24-hydroxylase activity declines, while 1-α hydroxylase activity is augmented (Henry and Norman 1990).

To date, vitamin D_3 metabolism has been studied only in bathyergid subterranean rodents. The six bathyergid species examined all exhibit low lev-

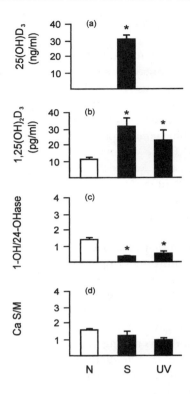

FIGURE 2.7. Changes in (a) plasma concentration of 25 hydroxy-vitamin D_3, (b) plasma concentration of 1,25 dihydroxy-vitamin D_3, (c) the ratio of 1-α hydroxylase activity to that of 24-hydroxylase activity, and (d) the mode of intestinal uptake of calcium, as shown by serosal/mucosal ratio of labeled calcium uptake (ratios of less than 2 indicate a passive mode of uptake) in Damara mole-rats subjected to three different treatments: N, freshly field-caught mole-rats receiving no vitamin D supplements and kept in the dark; S, a physiological dose of an oral vitamin D_3 supplement; UV, at least 30 minutes exposure to UV light daily for 10 days. Freshly caught animals exhibit a typical vitamin D_3-deficient profile, with undetectable levels of $25(OH)D_3$, low concentrations of $1,25(OH)_2D_3$, elevated 1-α hydroxylase activity, and passive intestinal uptake. Oral supplementation with vitamin D_3 increased concentrations of the two vitamin D_3 metabolites, decreased 1-α hydroxylase activity, and did not change the mode of intestinal uptake. Similar but less pronounced effects occurred with sunlight exposure, although $25(OH)D_3$ was still below the level of sensitivity for our assay. (From Buffenstein and Pitcher 1996)

els of vitamin D_3, as indicated by low circulating levels of its principal metabolites (25 hydroxy-vitamin D_3 at <5 ng/ml and $1.25(OH)_2D_3$ at <20 pg/ml) (fig. 2.7) and the absence of duodenal vitamin D_3-dependent calcium binding proteins (Buffenstein et al. 1994). At the same time, levels of 1-α hydroxylase are elevated (as commonly occurs with vitamin D_3 deficiency), enabling maximum utilization of any vitamin D_3 substrate and the production of as much of the active hormone as possible.

In bathyergid mole-rats, both oral vitamin D_3 supplementation and exposure to sunlight simultaneously suppress 1-α hydroxylase activity and enhance 24-hydroxylase activity. These physiological changes in response to vitamin D_3 supplementation indicate that the amount of circulating active hormone is tightly regulated (Pitcher, Sergeev, and Buffenstein 1994b). Indeed, two different biochemical pathways (protein kinase A and protein kinase C) are involved in the down-regulation of this light-dependent endocrine system (Buffenstein, Sergeev, and Pettifor 1993). As a result, even exposure to ultraviolet light leads to rapid activation of 24-hydroxylase, promoting the formation of the inert form of vitamin D_3 and the maintenance of low concentrations of both the active hormone and the principal circulating metabolites (see fig. 2.7; Pitcher, Sergeev, and Buffenstein 1994b). Despite this clear indication that their plasma concentrations of $1.25(OH)_2D_3$ are tightly regulated, bathyergids appear to be extremely tolerant of both low and pathologically high plasma concentrations of the vitamin D_3 metabolites (Skinner, Moodley, and Buffenstein 1991; Buffenstein et al. 1995).

The mechanisms of vitamin D_3 regulation in other subterranean rodents have not yet been investigated, and thus it is not yet known to what extent vitamin D_3 metabolism differs among the other taxa considered in this volume. In particular, it is not known whether regulation of vitamin D_3 metabolites is correlated with the extent to which different subterranean species are active above ground. Furthermore, the source of vitamin D_3 in subterranean rodents remains to be elucidated. Given the fact that vitamin D_3 is highly conserved and is produced by all animals studied to date (from microorganisms to humans), there is some speculation that symbiotic microorganisms in the gut may be able to make limited amounts of this compound in the absence of light. These symbionts may provide the vitamin D_3 substrate metabolized and utilized by these chthonic rodents.

In summary, despite their lack of any obvious source of vitamin D_3, subterranean rodents clearly do have limited access to this vital prohormone, and they possess the necessary enzymes to enable a functional vitamin D_3 endocrine system. Nevertheless, at least in the bathyergids, vitamin D_3 is clearly in short supply, with considerable evidence that these animals are in a perpetual state of vitamin D_3 deficiency. The effects of this deficiency on the plethora of physiological processes regulated by vitamin D_3 remain to be studied.

Mineral Homeostasis

While adequate vitamin D_3 levels are generally considered imperative for calcium and magnesium homeostasis in surface-dwelling mammals (Reichel, Koeffler, and Norman 1989), this generalization does not hold true for bathyergids. In subterranean rodents, mineral balance is achieved by vitamin D_3-independent processes, most notably unregulated passive intestinal absorption of calcium and magnesium. Intestinal absorption is nevertheless highly efficient, with apparent fractional absorption efficiencies exceeding 90%. There is no evidence that either facilitated diffusion or active transport plays an important role in the transport of calcium, for the rate of calcium absorption is directly dependent on the calcium content of the diet, and there is no evidence of saturation effects (fig. 2.8). Indeed, even when animals were provided with sufficient vitamin D_3 supplements to suppress 1-α hydroxylase activity (i.e., they had attained a replete vitamin D_3 status), calcium acquisition continued via unregulated passive absorption (Skinner, Moodley, and Buffenstein 1991) (see fig. 2.7d).

Renal reabsorption of calcium and magnesium from the glomerular filtrate is also highly efficient, with the result that very little of these minerals (<1%) is lost in the urine. Regardless of dietary intake of calcium or vitamin

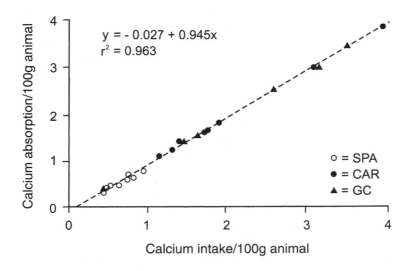

FIGURE 2.8. Daily calcium absorption is directly dependent on the dietary calcium content and concomitant daily calcium intake in Damara mole-rats fed different diets with varying dietary calcium contents. This relationship is linear with no saturation effects evident. SPA, sweet potato and apple (1.003g/kg and 0.346g/kg); CAR, carrots (2.657g/kg); GC, gemsbok cucumber (5.217g/kg). (From Pitcher et al. 1992)

D_3 supplementation, a positive mineral flux occurs, with calcium intake exceeding calcium loss (Skinner, Moodley, and Buffenstein 1991; Pitcher et al. 1992). Serum calcium concentration is nevertheless tightly regulated. This regulation is achieved primarily by the deposition of calcium in mineral reservoirs in hard tissues, namely bones and teeth (Pitcher, Sergeev, and Buffenstein 1994a). Thus calcium homeostasis in bathyergids is independent of vitamin D_3.

Mineral metabolism insensitivity to vitamin D_3 does not appear to be restricted to conditions of low vitamin concentration. When an extremely high and potentially lethal dose of a vitamin D_3 supplement was inadvertently given to *H. glaber*, they not only survived, but appeared to compensate for this overdose via a novel mechanism of calcium deposition in the skin and mammary tissue (Buffenstein et al. 1995).

Calcium requirements are assumed to be greater in chisel-tooth diggers than in scratch or head-lift diggers (Stein, chap. 1, this volume). This assumption reflects the extensive wear on teeth incurred during burrow excavation. For example, the teeth of an individual *Georychus capensis* were worn down to the gums in 3 to 5 days of extensive digging (J. U. M. Jarvis, personal communication). In natural habitats, the teeth of bathyergids, which are chisel-tooth diggers, most likely represent a sink for those elements (calcium, magnesium, and phosphorus) observed to be in a positive state of flux. This is especially true for animals consuming diets that are naturally high in calcium, such as *C. damarensis,* which consumes the calcium-rich tubers of *Acanthosicyos naudinianus* (Skinner, Moodley, and Buffenstein 1991). In short, toxic serum concentrations of calcium and other minerals are avoided by depositing these minerals in the teeth. At the same time, these minerals provide the necessary tooth strength to chew through the often hard soils in which the animals live. Studies of other subterranean taxa are needed to determine whether this mode of mineral deposition is a general characteristic of subterranean rodents or whether it occurs primarily in chisel-tooth diggers.

In summary, bathyergids appear to be extremely tolerant of life in perpetual darkness and its concomitant effect on the photoendocrine vitamin D_3 system. These animals are naturally in a vitamin D_3 impoverished state, yet they continue to regulate metabolite formation to favor the maintenance of low concentrations of the active hormone. Furthermore, despite the many indicators of vitamin D_3 deficiency, they show no indications of any of the pathologies normally associated with low vitamin D_3 concentrations.

Reproductive Physiology

Reproduction in subterranean rodents is discussed in detail by Bennett, Faulkes, and Molteno in chapter 4 of this volume, and accordingly, this topic is only briefly considered here, in relation to changes in other physiological systems that occur in response to changes in reproductive status. Reproductive females must meet the increased energetic demands of pregnancy. At the same time, changes in maternal body shape and mass during pregnancy lead to changes in heat transfer properties, and these, too, must be in balance with the physical conditions within the burrow. Very little research on the effects of pregnancy on other physiological systems has been undertaken for subterranean rodents. Those studies of reproductive physiology that have been conducted have focused largely on fluctuations in levels of reproductive hormones (Faulkes and Abbott 1993; Bennett, Faulkes, and Molteno 1996b) or on morphological, life history, and behavioral correlates of reproduction (Bennett, Faulkes, and Molteno, chap. 4, this volume).

The energetic costs of reproduction have been examined in detail for *H. glaber*, an aseasonal breeder. Body mass in this species may increase by as much as 84% during pregnancy, with a mean of 9 (and maximum of 27!) pups per litter (Jarvis 1991). During pregnancy, a female's energetic needs, as determined from changes in BMR, increase by approximately 1,300 kJ, implying that pregnant animals must double their daily food intake in order to meet the energy requirements of both themselves and their fetuses (Urison and Buffenstein 1995). A lactating female requires at least an additional 1,515 kJ per day to sustain an average litter (Urison and Buffenstein 1995).

As indicated by these perturbations in metabolic rate, pregnancy also entails changes in heat flux. While pregnant, female *H. glaber* maintain a higher body temperature than nonpregnant animals. This increase in body temperature is achieved primarily by increasing metabolic heat production. Thermal conductance is also altered as the percentage of body fat increases and insulatory properties improve (R. Woodley and R. Buffenstein, unpublished data). Behaviorally, pregnant female *H. glaber* spend considerably more time basking in warm regions of the burrow system, thereby increasing the amount of heat gained from the environment (Buffenstein et al. 1996).

Patterns of heat flux appear to change over the course of pregnancy, with female *H. glaber* becoming less sensitive to changes in ambient temperature as pregnancy progresses (fig. 2.9A). Early in pregnancy, although a female's overall rate of oxygen consumption is greater, oxygen consumption increases with increasing ambient temperature only up to 27°C. Beyond that temperature, oxygen consumption follows the typical endothermic pattern of decreasing with increasing ambient temperatures (fig. 2.9B). In late pregnancy, however, females exhibit a high degree of metabolic insensitivity to

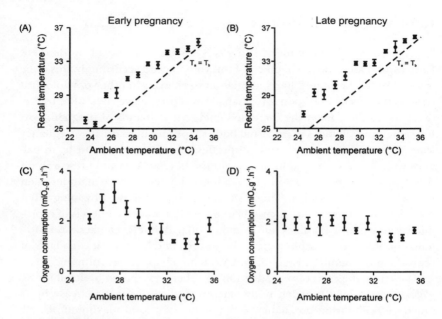

FIGURE 2.9. Effects of changes in ambient temperature on body temperature (A, B) and oxygen consumption (C, D) in early (A, C) and late (B, D) pregnancy. Regardless of stage of pregnancy or rate of oxygen consumption, body temperature varies directly with ambient temperature. Effects of ambient temperature on oxygen consumption in early pregnancy are similar to those in nonreproductive mole-rats; however, the effects of ambient temperature are attenuated in late pregnancy, such that it does not significantly influence metabolic rate. (From Urison and Buffenstein 1994b)

changing ambient temperatures (fig. 2.9C; Urison and Buffenstein 1994b), with metabolic rate remaining unchanged over a 15°C range of environmental temperatures (fig. 2.9D). This shift in thermoregulatory response may be associated with changes in rates of heat loss due to the changes in shape, size, surface area, and vascularization experienced by pregnant females. Decreased sensitivity to ambient temperatures may be advantageous in reducing thermoregulatory costs, thereby allowing females to devote more energy to the production of young. Unfortunately, these aspects of reproductive biology have not been investigated for other subterranean species, and thus the thermoregulatory consequences of pregnancy in these taxa remain unknown.

Areas for Future Research

In this review, I have attempted to draw together our limited knowledge of the physiology of subterranean rodents and the adaptive responses of these animals to the challenges imposed by underground habitats. Because ecophysiological studies of these animals have been shaped by the research interests of specific individuals and the availability of local subterranean species, we are currently faced with a mosaic of species-specific pieces of information that are difficult to assemble into a series of predictive generalizations. In some cases, our knowledge of physiological processes is based on sample sizes too small to be considered statistically representative of a single species, let alone all subterranean taxa. As a result, many of the studies currently cited as exemplars must be repeated, both to generate statistically valid sample sizes and to extend their findings to other species. On this basis alone, much remains to be done to understand the ecophysiology of subterranean rodents.

The taxonomic bias currently evident in physiological studies of these animals also needs to be rectified. While numerous researchers have investigated thermal and energy flux in bathyergids, geomyids, and spalacines, relatively little is known regarding thermoregulation and energy processing in the remaining taxa considered in this volume. Other topics have received even more restricted attention, such as vitamin D metabolism and mineral regulation, both of which have been examined for bathyergids only. Until a taxonomically more diverse data base is generated, our ability to examine convergent and divergent physiological responses to subterranean habitats remains extremely limited.

Ecophysiological studies of subterranean rodents are also limited by the quality of field data regarding the ecological and environmental conditions in which these animals live. Many published studies of the physiology of these animals are based on assumptions that have not been properly tested in the field. For example, arguments regarding the metabolic consequences of group living are based on assumptions regarding the degree of hypoxia in subterranean nest chambers, a parameter that has not been quantified under field conditions. A thorough understanding of the actual environmental conditions experienced by subterranean rodents must be attained before meaningful inferences can be drawn regarding the effects of those conditions on physiological processes.

At the same time, greater caution in interpreting the results of physiological studies conducted under laboratory conditions may be warranted. Acclimation to laboratory conditions has been reported in one study (Bennett, Taylor, and Aguilar 1993a), and, although the sample sizes were small, this finding should alert researchers to the potential for erroneous results if test animals are held in captivity for extended or unequal periods prior to data collection. This concern may be particularly relevant to studies based on repeated measures made over extended time periods. Differences in the duration and extent of acclimation may help to explain the sometimes diverse physiological responses obtained from conspecifics. Experiments conducted under simulated burrow conditions, as well as field studies, are greatly needed to determine the accuracy of many of our current perceptions of the physiology of subterranean rodents.

Finally, the range of physiological processes examined needs to be expanded to include many topics that have not yet been addressed. For example, what physiological responses are required during exercise, such as prolonged digging? How are individuals able to accommodate the tremendous energy expenditures associated with this activity? What specific features of cardiovascular, respiratory, and muscle physiology allow prolonged digging? Do these features differ from the mechanisms employed by surface-dwelling animals to accommodate energetically expensive activities? Only after we have compiled a considerably more complete picture of the abilities of subterranean taxa can we truly begin to assess the degree of physiological convergence and divergence involved in "the remarkable adaptive evolution of these subterranean mammals" (Nevo 1995).

Summary and Future Directions

The subterranean environment, although characterized by relative ecological stability, is physiologically demanding. Both its physical and its biotic features differ markedly from those encountered by surface-dwelling species, necessitating a suite of complex physiological adaptations to life underground. These adaptations include the following:

1. High energetic costs associated with burrow excavation and foraging, coupled with the relatively low energetic returns mandated by poor-quality diets, have led to a suite of morphological and physiological digestive adaptations that serve to increase the efficiency of energy extraction. These adaptations include increased hindgut length and microbial fermentation, both of which enhance the digestibility of the vegetation consumed.
2. Low metabolic rates, large thermoneutral ranges, low body temperatures, and high rates of thermal conductance reduce both the energy required to sustain life and the quantity of metabolic heat that must be dissipated to the environment. These reductions, in turn, serve to minimize rates of

gas exchange and pulmocutaneous water loss. Because relative humidity within burrows is typically high, evaporative water loss is expected to be minimal, suggesting that specialized urine concentrating mechanisms are not required, even by species that inhabit relatively arid regions.

3. Relaxed thermoregulation, reduced basal metabolic rates, and enhanced oxygen carrying capacities of blood are thought to represent responses to the hypoxic and hypercapnic conditions typical of poorly ventilated subterranean burrows.

4. Passive absorption of minerals such as calcium may facilitate homeostasis under conditions in which acquisition of vitamin D_3 is limited by the absence of light. Although naturally deficient in vitamin D_3, subterranean rodents appear to be capable of acquiring and storing sufficient quantities of calcium and other vital minerals.

With regard to future research, much remains to be done to determine the extent of physiological convergence and divergence among subterranean species. At the same time, because many of the physiological processes considered here are shared by surface-dwelling mammals, a better understanding of the physiological mechanisms employed by subterranean species should improve our understanding of similar phenomena in all mammals. Thus understanding the characteristics that allow some animals to exploit the subterranean niche has considerable relevance for understanding how all mammals meet the physiological challenges imposed by the environments in which they live.

Acknowledgments

Gary Bronner, Sally Hoskins, Eileen Lacey, and Jenny Jarvis are sincerely thanked for their critical comments on this manuscript.

Literature Cited

Altuna, C. A., L. D. Bacigalupe, and S. Corte. 1998. Food handling and feces reingestion in *Ctenomys pearsoni* (Rodentia: Ctenomyidae). Acta Theriologica 43(4):433–37.

Ar, A. 1986. Physiological adaptations to underground life in mammals. In *Comparative physiology of environmental adaptation*, edited by P. Dejours, 208–11. Basel: Karger.

Ar, A., R. Arieli, and A. Shkolnik. 1977. Blood-gas properties and function in the fossorial mole rat under normal and hypoxic hypercapnic atmosphere conditions. *Respiration Physiology* 30:201–18.

Argamaso, S. M., A. C. Froehlich, M. A. McCall, E. Nevo, I. Provencio, and R. G. Foster. 1995. Photopigments and circadian systems of vertebrates. *Biophysical Chemistry* 56:3–11.

Arieli, R. 1979. The atmospheric environment of a fossorial mole rat *(Spalax ehrenbergi):* Effect of season, soil texture, rain, temperature and activity. *Comparative Biochemistry and Physiology* 63A:569-75.

———. 1990. Adaptation of the mammalian gas transport system to subterranean life. In *Evolution of subterranean mammals at the organismal and molecular levels,* edited by E. Nevo and O. A. Reig, 251-68. Progress in Clinical and Biological Research, vol. 335. New York: Wiley-Liss.

Arieli, R., and A. Ar. 1979. Ventilation of a fossorial mammal *(Spalax ehrenbergi)* in hypoxic and hypercapnic conditions. *Journal of Applied Physiology* 47:1011-17.

———. 1981a. Blood capillary density in heart and skeletal muscles of the fossorial mole rat. *Physiological Zoology* 54:22-27.

———. 1981b. Heart rate responses of the mole-rat *(Spalax ehrenbergi)* in hypercapnic, hypoxic and cold conditions. *Physiological Zoology* 54:14-21.

Arieli, R., A. Ar, and A. Shkolnik. 1977. Metabolic responses of a fossorial rodent *(Spalax ehrenbergi)* to simulated burrow conditions. *Physiological Zoology* 50:61-75.

Arieli, R., M. Arieli, G. Heth, and E. Nevo. 1984. Adaptive respiratory variation in 4 chromosomal species of mole rats. *Experientia* 40:512-14.

Arieli, R., G. Heth, E. Nevo, Y. Zamir, and O. Neutra. 1986. Adaptive heart and breathing frequencies in 4 ecologically differentiating chromosomal species of mole-rats in Israel. *Experientia* 42:131-33.

Arieli, R., and E. Nevo. 1991. Hypoxic survival differs between two mole rat species *(Spalax ehrenbergi)* of humid and arid habitats. *Comparative Biochemistry and Physiology* 100A: 543-45.

Aschoff, J. 1981. Thermal conductance in mammals and birds: Its dependence on body size and circadian phase. *Comparative Biochemistry and Physiology* 69A:611-19.

Baverstock, P. R. 1976. Water balance and kidney function in four species of *Rattus* from ecologically diverse environments. *Australian Journal of Zoology* 24:7-17.

Bennett, N. C. 1990. Behaviour and social organization in a colony of the Damaraland mole-rat *Cryptomys damarensis. Journal of Zoology, London* 220:225-48.

Bennett, N. C., G. H. Aguilar, J. U. M. Jarvis, and C. G. Faulkes. 1994. Thermoregulation in three species of Afrotropical subterranean mole-rats (Rodentia: Bathyergidae) from Zambia and Angola and scaling within the genus *Cryptomys. Oecologia* 97:222-27.

Bennett, N. C., B. C. Clarke, and J. U. M. Jarvis. 1992. A comparison of metabolic acclimation in two species of social mole-rats (Rodentia, Bathyergidae) in southern Africa. *Journal of Arid Environments* 22:189-98.

Bennett, N. C., F. P. D. Cotterill, and A. C. Spinks. 1996. Thermoregulation in two populations of the Matabeleland mole-rat *(Cryptomys hottentotus nimrodi)* and remarks on the general thermoregulatory trends within the genus *Cryptomys* (Rodentia: Bathyergidae). *Journal of Zoology, London* 239:17-27.

Bennett, N. C., C. G. Faulkes, and A. J. Molteno. 1996. Reproductive suppression in subordinate, non-breeding female Damaraland mole-rats: Two components to a lifetime of socially-induced infertility. *Proceedings of the Royal Society of London* B 263:1599-1603.

Bennett, N. C., and J. U. M. Jarvis. 1995. Coefficients of digestibility and nutritional values of geophytes and tubers eaten by southern African mole-rats (Rodentia: Bathyergidae). *Journal of Zoology, London* 236:189-98.

Bennett, N. C., J. U. M. Jarvis, G. H. Aguilar, and E. J. McDaid. 1991. Growth and development in six species of African mole-rats (Rodentia: Bathyergidae). *Journal of Zoology, London* 225:13-26.

Bennett, N. C., J. U. M. Jarvis, and F. P. D. Cotterill. 1993. Poikilothermic traits and thermoregulation in the Afrotropical social subterranean Mashona mole-rat *(Cryptomys hottentotus darlingi)* (Rodentia: Bathyergidae). *Journal of Zoology, London* 231:179-86.

Bennett, N. C., J. U. M. Jarvis, and K. C. Davies. 1988. Daily and seasonal temperatures in the burrows of African rodent moles. *South African Journal of Zoology* 23:189–195.

Bennett, N. C., P. J. Taylor, and G. H. Aguilar. 1993. Thermoregulation and metabolic acclimation in the Natal mole-rat *(Cryptomys hottentotus natalensis)* (Rodentia: Bathyergidae). *Zeitschrift für Säugetierkunde* 58:362–67.

Ben Shlomo, R., E. Nevo, U. Ritte, S. Steinlechner, and G. Klante. 1996. 6-Sulphatoxymelatonin secretion in different locomotor activity types of the blind mole rat *Spalax ehrenbergi*. *Journal of Pineal Research* 21:243–50.

Ben Shlomo, R., U. Ritte, and E. Nevo. 1995. Activity pattern and rhythm in the subterranean mole rat superspecies *Spalax ehrenbergi*. *Behavioral Genetics* 25:239–45.

Boggs, D. F., D. L. Kilgore, and G. F. Birchard. 1984. Respiratory physiology of burrowing mammals and birds. *Comparative Biochemistry and Physiology* 77A:1–7.

Bradley, W. G., J. S. Miller, and M. K. Yousef. 1974. Thermoregulatory patterns in Pocket gophers. *Physiological Zoology* 47:172–79.

Bradley, W. G., and M. K. Yousef. 1975. Thermoregulatory responses in the Plains pocket gopher, *Geomys bursarius*. *Comparative Biochemistry and Physiology* 52A:35–38.

Brett, R. A. 1991. The ecology of naked mole-rat colonies: Burrowing, food and limiting factors. In *The biology of the naked mole-rat*, edited by P. W. Sherman, J. U. M. Jarvis, and R. D. Alexander, 137–84. Princeton, NJ: Princeton University Press.

Buffenstein, R. 1996. Ecophysiological responses to a subterranean habitat: A bathyergid perspective. *Mammalia* 60:591–605.

Buffenstein, R., W. Campbell, and J. U. M. Jarvis. 1985. Identification of crystalline allantoin in the urine of African Cricetidae (Rodentia) and its role in the water economy. *Journal of Comparative Physiology* B 155:211–18.

Buffenstein, R., J. U. M. Jarvis, L. A. Opperman, M. Cavaleros, F. P. Ross, and J. M. Pettifor. 1994. Subterranean mole-rats naturally have an impoverished calciol status, yet synthesize calciol metabolites and calbindins. *European Journal of Endocrinology* 130:402–9.

Buffenstein, R., M. T. Laundy, T. Pitcher, and J. M. Pettifor. 1995. Vitamin D_3 intoxication in naked mole-rats *(Heterocephalus glaber)* leads to hypercalcaemia and increased calcium deposition in teeth with evidence of abnormal skin calcification. *General and Comparative Endocrinology* 99:35–40.

Buffenstein, R., and T. Pitcher. 1996. Calcium homeostasis in mole-rats by manipulation of teeth and bone calcium reservoirs. In *The comparative endocrinology of calcium regulation*, edited by C. Dacke, J. Danks, I. Caple, and G. Flik, 177–82. Bristol: The Society for Endocrinology.

Buffenstein, R., I. N. Sergeev, and J. M. Pettifor. 1993. Vitamin D hydroxylases and their regulation in a naturally vitamin D deficient subterranean mammal, the naked mole-rat *(Heterocephalus glaber)*. *Journal of Endocrinology* 138:59–64.

Buffenstein, R., N. T. Urison, L. A. Van der Westhuizen, R. Woodley, and J. U. M. Jarvis. 1996. Temperature changes during pregnancy in the subterranean naked mole-rat *(Heterocephalus glaber):* The role of altered body composition and basking behaviour. *Mammalia* 60:617–26.

Buffenstein, R., and S. Yahav. 1991a. The effect of diet on microfaunal population and function in the caecum of a subterranean mole-rat, *Heterocephalus glaber*. *British Journal of Nutrition* 65:249–58.

———. 1991b. Is the naked mole-rat, *Heterocephalus glaber*, a poikilothermic or poorly thermoregulating endothermic mammal? *Journal of Thermal Biology* 16:227–32.

———. 1994. Fibre utilization by Kalahari dwelling subterranean Damara mole-rats *(Cryptomys damarensis)* when fed their natural diet of gemsbok cucumber tubers *(Acanthosicyos nuadinianus)*. *Comparative Biochemistry and Physiology* 109A:431–36.

Bunning, E. 1973. *The physiological clock*. New York: Springer.

Burda, H., V. Bruns, and A. M. Müller. 1990. Sensory adaptations in subterranean mammals. In *Evolution of subterranean mammals at the organismal and molecular levels*, edited by E. Nevo and O. A. Reig, 269–93. Progress in Clinical and Biological Research 335. New York: Wiley-Liss.

Burlington, R. F., B. K. Whitten, C. M. Sidel, M. A. Posiviata, and I. A. Salkovitz. 1970. Effect of hypoxia on glycolysis in perfused hearts from rats and ground squirrels *(Citelus lateralis)*. *Comparative Biochemistry and Physiology* 35:403–14.

Carneheim, C., J. Nedergaard, and B. Cannon. 1984. Beta-adrenergic stimulation of lipoprotein lipase in rat brown adipose tissue during acclimation to cold. *American Journal of Physiology* 246:E327–E333.

Chapman, R. C., and A. F. Bennett. 1975. Physiological correlates of burrowing in rodents. *Comparative Biochemistry and Physiology* 51A:599–603.

Comparatore, M. V., M. S. Cid, and C. Busch. 1995. Dietary preferences of two sympatric subterranean rodent populations in Argentina. *Revista Chilena de Historia Natural* 68: 197–206.

Contreras, L. C., and B. K. McNab. 1990. Thermoregulation and energetics in subterranean mammals. In *Evolution of subterranean mammals at the organismal and molecular levels*, edited by E. Nevo and O. A. Reig, 231–50. Progress in Clinical and Biological Research, vol. 335. New York: Wiley-Liss.

Cooper, H. M., M. Herbin, and E. Nevo. 1993a. Ocular regression conceals adaptive progression of the visual system in a blind subterranean mammal. *Nature* 361:156–59.

———. 1993b. The visual system of a micro-ophthalmic mammal: The blind mole-rat *Spalax ehrenbergi*. *Journal of Comparative Neurobiology* 328:313–50.

Cossins, A. R., and K. Bowler. 1987. *Temperature biology of animals*. London: Chapman and Hall.

Darden, T. R. 1972. Respiratory adaptations of a fossorial mammal, the pocket gopher, *Thomomys bottae*. *Journal of Comparative Physiology* 78:121–37.

Davies, D. G., and J. C. Schadt. 1989. Ventilatory responses of the ground squirrel, *Spermophilus tridecemlineatus*, to various levels of hypoxia. *Comparative Biochemistry and Physiology* A 92:255–57.

Degen, A. A. 1997. *Ecophysiology of small desert mammals*. Berlin: Springer.

de Villiers, M. S. 1993. A comparison of subterranean herbivory by fossorial and terrestrial mammals. *Transactions of the Royal Society of South Africa* 48:257–64.

Dhindsa, D. S., A. S. Hoversland, and J. Metcalfe. 1971. Comparative studies on the respiratory functions of mammalian blood. VII. Armadillo *(Dasypus novemcinctus)*. *Respiratory Physiology* 5:221–33.

du Toit, J. J., J. U. M. Jarvis, and G. N. Louw. 1985. Nutrition and burrowing energetics of the Cape mole-rat *Georychus capensis*. *Oecologia* 66:81–87.

Edoute, Y., R. Arieli, and E. Nevo. 1988. Evidence for improved myocardial oxygen delivery and function during hypoxia in the mole-rat. *Journal of Comparative Physiology* B 158:575–82.

Edwards, B. A., and A. Peters. 1988. Water balance during saline inhibition in the Mongolian gerbil *(Meriones unguiculatus)*. *Comparative Biochemistry and Physiology* 90A:93–98.

Faleschini, R. J., and B. K. Whitten. 1975. Comparative hypoxic tolerance in Sciuridae. *Comparative Biochemistry and Physiology* 52A:217–21.

Faulkes, C. G., and D. H. Abbott. 1993. Social control of reproduction in breeding and non-breeding naked mole-rats, *Heterocephalus glaber*. *Journal of Reproduction and Fertility* 93:427–35.

Feder, M. E., A. F. Bennett, W. W. Burggren, and R. B. Huey. 1987. *New directions in ecological physiology*. Cambridge: Cambridge University Press.

Foster, D. O. 1984. Quantitative contribution of brown adipose tissue thermogenesis to overall metabolism. *Canadian Journal of Biochemistry and Cell Biology* 62:618–22.

Frahm, H. D., G. Rehkämper, and E. Nevo. 1997. Brain structure volumes in the mole rat, *Spalax ehrenbergi* (Spalacidae, Rodentia) in comparison to the rat and subterrestrial insectivores. *Journal of Brain Research* 38:209–22.
French, A. R. 1993. Physiological ecology of the Heteromyidae: Economics of energy and water utilization. In *Biology of the Heteromyidae*, edited by H. H. Genoways and J. H. Brown, 509–38. Special publication 10. Provo, Utah: American Society of Mammalogists, Brigham Young University.
Gettinger, R. D. 1984. Energy and water metabolism of free-ranging pocket gophers, *Thomomys bottae*. *Ecology* 65:740–51.
Gilad, E., U. Shanas, J. Terkel, and N. Zisapel. 1997. Putative melatonin receptors in the blind mole rat Harderian gland. *Journal of Experimental Zoology* 277:435–41.
Haim, A., and N. Fairall. 1986. Physiological adaptations to the subterranean environment by the mole rat *Cryptomys hottentotus*. *Cimbebasia* 8A:49–53.
Haim, A., G. Heth, H. Pratt, and E. Nevo. 1983. Photoperiodic effects on thermoregulation in a "blind" subterranean mammal. *Journal of Experimental Biology* 107:59–64.
Hart, J. S. 1971. Rodents. In *Comparative physiology of thermoregulation*, edited by G. C. Whittow, 1–149. New York: Academic Press.
Hayssen, V., and R. C. Lacy. 1985. Basal metabolic rates in mammals: Taxonomic differences in the allometry of BMR and body mass. *Comparative Biochemistry and Physiology* 81:741–54.
Heffner, R. S., and H. E. Heffner. 1992. Hearing and sound localizations in blind mole-rats, *Spalax ehrenbergi*. *Hearing Research* 62:206–16.
———. 1993. Degenerate hearing and sound localisation in naked mole-rats (*Heterocephalus glaber*) with an overview of central auditory structures. *Journal of Comparative Neurobiology* 331:418–33.
Heffner, R. S., H. E. Heffner, C. Contos, and D. Kearns. 1994. Hearing in prairie dogs: Transition between surface and subterranean rodents. *Hearing Research* 73:185–89.
Heldmaier, G., S. Klaus, and H. Wiesinger. 1990. Seasonal adaptation of thermoregulatory heat production in small mammals. In *Thermoreception and temperature regulation*, edited by J. Bligh and K. Voigt, 235–43. Berlin: Springer Verlag.
Henry, H. L., and A. W. Norman. 1990. Vitamin D: Metabolism and mechanism of action. In *Primer on the metabolic bone diseases and disorders of mineral metabolism*, edited by M. J. Favus, 47–52. Kelseyville, CA: The American Society of Bone and Mineral Research.
Herbin, M., J. Reperant, and H. M. Cooper. 1994. Visual system of the fossorial mole-lemmings, *Ellobius talpinus* and *Ellobius lutescens*. *Journal of Comparative Neurobiology* 346:253–75.
Hess, A. F., and M. B. Gutman. 1922. Cure of infantile rickets by sunlight. *Journal of the American Medical Association* 78:29–30.
Heth, G., E. Frankenberg, and E. Nevo. 1986. Adaptive optimal sound for vocal communication in tunnels of a subterranean mammal, *Spalax ehrenbergi*. *Experientia* 42:1287–89.
Heth, G., E. Frankenberg, H. Pratt, and E. Nevo. 1991. Vibrational communication in the blind subterranean mole-rat: Patterns of head thumping and their detection in the *Spalax ehrenbergi* superspecies in Israel. *Journal of Zoology, London* 224:633–38.
Heth, G., E. Frankenberg, A. Raz, and E. Nevo. 1987. Vibrational communication in subterranean mole rats (*Spalax ehrenbergi*). *Behavioral Ecology and Sociobiology* 21:31–33.
Heth, G., E. M. Goldenberg, and E. Nevo. 1989. Foraging strategy in a subterranean rodent, *Spalax ehrenbergi*: A test case for optimal foraging theory. *Oecologia* 79:496–505.
Hickman, G. C. 1980. Locomotory activity of captive *Cryptomys hottentotus* (Mammalia: Bathyergidae), a fossorial rodent. *Journal of Zoology, London* 192:225–35.
Hislop, M. S., and R. Buffenstein. 1994. Noradrenaline induces non-shivering thermogenesis in both the naked mole-rat (*Heterocephalus glaber*) and the Damara mole-rat (*Crypto-*

mys damarensis) despite very different modes of thermoregulation. *Journal of Thermal Biology* 19:25–32.

Jarvis, J. U. M. 1973. Activity patterns in the mole-rats *Tachyoryctes splendens* and *Heliophobius argenteocinereus*. *Zoologica Africana* 8:101–19.

———. 1991. Reproduction of naked mole-rats. In *The biology of the naked mole-rat*, edited by P. W. Sherman, J. U. M. Jarvis, and R. D. Alexander, 384–426. Princeton, NJ: Princeton University Press.

Jarvis, J. U. M., and N. C. Bennett. 1990. The evolutionary history, population biology and social structure of African mole-rats: Family Bathyergidae. In *Evolution of subterranean mammals at the organismal and molecular levels*, edited by E. Nevo and O. A. Reig, 97–128. Progress in Clinical and Biological Research, vol. 335. New York: Wiley-Liss.

———. 1991. Ecology and behavior of the family Bathyergidae. In *The biology of the naked mole-rat*, edited by P. W. Sherman, J. U. M. Jarvis, and R. D. Alexander, 66–97. Princeton, NJ: Princeton University Press.

Jarvis, J. U. M., N. C. Bennett, and A. C. Spinks. 1998. Food availability and foraging by wild colonies of Damaraland mole-rats *(Cryptomys damarensis):* Implications for sociality. *Oecologia* 113:290–98.

Jarvis, J. U. M., and J. B. Sale. 1971. Burrowing and burrow patterns of East African mole-rats *Tachyoryctes, Heliophobius* and *Heterocephalus*. *Journal of Zoology, London* 163:451–79.

Jelkmann, W., W. Oberthur, T. Kleinschmidt, and G. Braunitzer. 1981. Adaptation of hemoglobin function to subterranean life in the mole *Talpa europaea*. *Respiratory Physiology* 46:7–16.

Johansen, K., G. Lykkeboe, R. E. Weber, and G. M. Maloiy. 1976. Blood respiratory properties in the naked mole-rat, *Heterocephalus glaber*, a mammal of low body temperature. *Respiratory Physiology* 28:303–14.

Jones, D. L., and L. C. H. Wang. 1976. Adaptive cardiovascular modifications in the western chipmunks, genus *Eutamias*. *Journal of Comparative Physiology* 112:307–315.

Kenagy, G. J., and D. F. Hoyt. 1980. Reingestion of feces in rodents and its daily rhythmicity. *Oecologia* 44:403–9.

Klauer, G., H. Burda, and E. Nevo. 1997. Adaptive differentiation of the skin of the head in a subterranean rodent, *Spalax ehrenbergi*. *Journal of Morphology* 233:53–66.

Koteja, P. 1991. On the relation between basal and field metabolic rates in birds and mammals. *Functional Ecology* 5:56–64.

Krebs, C. J. 1978. *Ecology: The experimental analysis of distribution and abundance*. 2d edition. New York: Harper and Row.

Lechner, A. J. 1976. Respiratory adaptations in burrowing pocket gophers from sea level and high altitude. *Journal of Applied Physiology* 41:168–73.

Lee, W. B., and D.C. Houston. 1993. The role of coprophagy in digestion in voles *(Microtus agrestis* and *Clethrionomys glareolus)*. *Functional Ecology* 7:427–32.

Loeb, S. C., R. G. Schwab, and M. W. Demment. 1991. Responses of pocket gophers *(Thomomys bottae)* to changes in diet quality. *Oecologia* 86:542–51.

Lovegrove, B. G. 1986a. The metabolism of social subterranean rodents: Adaptation to aridity. *Oecologia* 69:551–55.

———. 1986b. Thermoregulation of the subterranean rodent genus *Bathyergus* (Bathyergidae). *South African Journal of Science* 21:283–88.

———. 1989. The cost of burrowing of the social mole rats (Bathyergidae) *Cryptomys damarensis* and *Heterocephalus glaber:* The role of soil moisture. *Physiological Zoology* 62:449–69.

Lovegrove, B. G., G. Heldmaier, and T. Ruf. 1993. Circadian activity rhythms in colonies of 'blind' mole-rats, *Cryptomys damarensis* (Bathyergidae). *South African Journal of Science* 28:46–55.

Lovegrove, B. G., and J. U. M.. Jarvis. 1986. Coevolution between mole-rats (Bathyergidae) and a geophyte, *Micranthus* (Iridaceae). *Cimbebasia* 8A:79–95.

Lovegrove, B. G., and A. Muir. 1996. Circadian body temperature rhythms of the solitary Cape mole rat, *Georychus capensis* (Bathyergidae). *Physiology and Behavior* 60:991–98.

Lovegrove, B. G., and C. Wissel. 1988. Sociality in mole-rats: Metabolic scaling and the role of risk sensitivity. *Oecologia* 74:600–606.

MacLean, G. S. 1981. Factors influencing the composition of respiratory gases in mammal burrows. *Comparative Biochemistry and Physiology* 69A:373–83.

MacMillen, R. E., and A. K. Lee. 1967. Australian desert mice: Independence of exogenous water. *Science* 158:383–85.

Maina, J. N., G. M. O. Maloiy, and A. N. Makanya. 1992. Morphology and morphometry of the lungs of two East African mole rats, *Tachyoryctes splendens* and *Heterocephalus glaber* (Mammalia, Rodentia). *Zoomorphology* 112:167–79.

Mann, M. D., G. Rehkämper, H. Reinke, H. D. Frahm, R. Necker, and E. Nevo. 1997. Size of the somatosensory cortex and of somatosensory thalamic nuclei of the naturally blind mole rat *Spalax ehrenbergi*. *Journal für Hirnforschung* 38:47–59.

Marhold, S., and A. Nagel. 1995. The energetics of the common mole-rat *Cryptomys*, a subterranean eusocial rodent from Zambia. *Journal of Comparative Physiology* B 164: 636–45.

Mayer, W. 1955. The protective value of a burrow system to the hibernating Arctic ground squirrel *(Spermophilus undulatus)*. *Anatomical Record* 122:437–38.

McNab, B. K. 1966. The metabolism of fossorial rodents: A study of convergence. *Ecology* 47:712–33.

———. 1979. The influence of body size on the energetics and distribution of fossorial and burrowing mammals. *Ecology* 60:1010–21.

———. 1980. On estimating thermal conductance in endotherms. *Physiological Zoology* 53:145–56.

Miller, R. S. 1964. Ecology and distribution of pocket gophers (Geomyidae) in Colorado. *Ecology* 45:256–72.

Müller, M., and H. Burda. 1989. Restricted hearing range in a subterranean rodent, *Cryptomys hottentotus*. *Naturwissenschaften* 76:134–35.

Mundy, G. R. 1990. *Calcium homeostasis: Hypercalcemia and hypocalcemia*. London: Martin Dunitz.

Narins, P. M., E. R. Lewis, J. U. M. Jarvis, and J. Oriain. 1997. The use of seismic signals by fossorial southern African mammals: A neuroethological gold mine. *Brain Research Bulletin* 44:641–46.

Necker, R., G. Rehkämper, and E. Nevo. 1992. Electrophysiological mapping of body representation in the cortex of the blind mole rat. *Neurological Report* 3:505–58.

Nevo, E. 1979. Adaptive convergence and divergence of subterranean mammals. *Annual Review of Ecology and Systematics* 10:269–308.

———. 1995. Mammalian evolution underground: The ecological-genetic-phenetic interfaces. *Acta Theriologica*, Supplement 3:9–31.

Nevo, E., G. Heth, and H. Pratt. 1991. Seismic communication in a blind subterranean mammal: A major somatosensory mechanism in adaptive evolution underground. *Proceedings of the National Academy of Sciences, USA* 88:1256–60.

Nevo, E., and A. Shkolnik. 1974. Adaptive metabolic variation of chromosome forms of mole rats, *Spalax*. *Experientia* 30:724–26.

Nevo, E., S. Simson, A. Beiles, and S. Yahav. 1989. Adaptive variation in structure and function of kidneys of speciating subterranean mole rats. *Oecologia* 79:366–71.

Olszewski, J. L., and S. Skoczen. 1965. The airing of burrows of the mole, *Talpa europaea* Linnaeus 1758. *Acta Theriologica* 10:181–93.

Parra, R. 1978. Comparison of foregut and hindgut fermentation in herbivores. In *The ecology of arboreal folivores*, edited by G. G. Montgomery, 205–29. Washington, DC: Smithsonian Institution Press.

Pearson, O. P. 1959. Biology of subterranean rodents, *Ctenomys*, in Peru. *Memorias del Museo de Historia Natural "Javier Prado"* 9:1–56.

Perrin, M. R., and B. A. Curtis. 1980. Comparative morphology of the digestive system of 19 species of South African myomorph rodents in relation to diet evolution. *South African Journal of Zoology* 15:22–33.

Pevet, P., G. Heth, A. Haim, and E. Nevo. 1984. Photoperiod perception in the blind mole-rat (*Spalax ehrenbergi* Nehring): Involvement of the Harderian gland, atrophied eyes and melatonin. *Journal of Experimental Zoology* 231:41–50.

Pitcher, T., R. Buffenstein, J. D. Keegan, G. P. Moodley, and S. Yahav. 1992. The effect of dietary calcium content on calcium balance and mode of uptake in a subterranean mammal, the Damara mole-rat. *Journal of Nutrition* 122:108–14.

Pitcher, T., I. N. Sergeev, and R. Buffenstein. 1994a. The effect of dietary calcium content and vitamin D_3 supplementation on mineral homeostasis in a subterranean mole-rat, *Cryptomys damarensis*. *Bone and Mineral* 27:145–57.

———. 1994b. Vitamin D_3 metabolism in the Damara mole-rat is altered by sunlight yet mineral metabolism is unaffected. *Journal of Endocrinology* 143:367–74.

Porter, A. 1957. Entozoa and endophyta of the naked mole-rat. *Proceedings of the Zoological Society of London* 128:515–27.

Quay, W. B. 1981. Pineal atrophy and other neuroendocrine and circumventricular features of the naked mole-rat, *Heterocephalus glaber* (Ruppell), a fossorial equatorial rodent. *Journal of Neural Transmission* 52:107–15.

Rado, R., H. Gev, B. D. Goldman, and J. Terkel. 1991. Light and circadian activity in the blind mole-rat. In *Photobiology*, edited by E. Riklis, 581–89. New York: Plenum Press.

Rado, R., N. Levi, H. Hauser, J. Witcher, N. Adler, N. Intrator, Z. S. Wolberg, and J. Terkel. 1987. Seismic signalling as a means of communication in a subterranean mammal. *Animal Behaviour* 35:1249–66.

Rado, R., U. Shanas, I. Zuri, and J. Terkel. 1993. Seasonal activity in the blind mole-rat (*Spalax ehrenbergi*). *Canadian Journal of Zoology* 71:1733–37.

Rehkämper, G., R. Necker, and E. Nevo. 1994. Functional anatomy of the thalamus in the blind mole rat *Spalax ehrenbergi:* An architectonic and electrophysiologically controlled tracing study. *Journal of Comparative Neurobiology* 347:570–84.

Reichel, H., H. P. Koeffler, and A. W. Norman. 1989. The role of the vitamin D endocrine system in health and disease. *New England Journal of Medicine* 320:980–91.

Reichman, O. J., and J. U. M. Jarvis. 1989. The influence of three sympatric species of fossorial mole-rats (Bathyergidae) on vegetation. *Journal of Mammalogy* 70:763–71.

Reichman, O. J., and S. C. Smith. 1985. Impact of pocket gophers on overlying vegetation. *Journal of Mammalogy* 66:720–25.

Reiter, R. J., M. N. Reiter, A. Hattori, K. Yaga, D.C. Barlow, and L. Walden. 1994. The pineal melatonin rhythm and its regulation by light in a subterranean rodent, the valley pocket gopher *(Thomomys bottae)*. *Journal of Pineal Research* 16:145–53.

Reiter, R. J., B. A. Richardson, S. A. Mathews, S. J. Lane, and B. N. Ferguson. 1983. Rhythms in immunoreactive melatonin in the retina and Harderian glands of rats: Persistence after pinealectomy. *Life Science* 32:1229–36.

Sanyal, S., H. G. Jansen, W. G. De Grip, E. Nevo, and W. W. De Grip. 1990. The eye of the blind mole rat *Spalax ehrenbergi:* Rudiment with hidden function. *Investigative Ophthalmology and Visual Science* 31:1398–1401.

Schmidt-Nielsen, K. 1975. Desert rodents: Physiological problems of desert life. In *Rodents in desert environments*, edited by I. Prakash and P. K. Ghosh, 379–88. The Hague: Dr W. Junk.

Seagrave, R. C. 1971. Heat transfer in living systems. In *Biomedical applications of heat and mass transfer*, edited by R. C. Seagrave, 93–113. Ames: Iowa State University Press.

Skinner, D.C., G. P. Moodley, and R. Buffenstein. 1991. Is vitamin D_3 essential for mineral metabolism in the Damara mole-rat *(Cryptomys damarensis)*? *General and Comparative Endocrinology* 81:500–505.

Smolen, M. J., H. H. Genoways, and R. J. Baker. 1980. Demographic and reproductive characteristics of the yellow-cheeked pocket gopher *(Pappogeomys castanops)*. *Journal of Mammalogy* 61:224–36.

Speakman, J. R., R. M. McDevitt, and K. R. Cole. 1993. Measurement of basal metabolic rates: Don't lose sight of reality in the quest for comparability. *Physiological Zoology* 66: 1045–49.

Storier, D., Z. Wollberg, and A. Ar. 1981. Low and nonrhythmic heart rate of the mole rat *(Spalax ehrenbergi)*: Control by the autonomic nervous system. *Journal of Comparative Physiology* 142:533–38.

Stuebe, M. M., and D.C. Anderson. 1985. Nutritional ecology of a fossorial herbivore: Protein, nitrogen and energy value of winter caches made by the northern pocket gopher, *Thomomys talpoides*. *Canadian Journal of Zoology* 63:1101–5.

Towe, A. L., and M. D. Mann 1995. Habitat-related variations in brain and body size of pocket gophers. *Journal für Hirnforschung* 36:195–201.

Tucker, C. E., W. E. James, M. A. Berry, C. J. Johnstone, and R. F. Glover. 1976. Depressed myocardial function in the goat at high altitude. *Journal of Applied Physiology* 41:356–61.

Urison, N. T., and R. Buffenstein. 1994a. Kidney concentrating ability of a subterranean xeric rodent, the naked mole-rat *(Heterocephalus glaber)*. *Journal of Comparative Physiology* B 163:676–81.

———. 1994b. Shifts in thermoregulatory patterns with pregnancy in the poikilothermic mammal—the naked mole-rat *(Heterocephalus glaber)*. *Journal of Thermal Biology* 19: 365–71.

———. 1995. Metabolic and body temperature changes during pregnancy and lactation in the naked mole-rat *(Heterocephalus glaber)*. *Physiological Zoology* 68:402–20.

van Jaarsveld, A., M. C. Mhatve, and R. J. Reiter. 1989. Porphyrin levels in the Harderian glands of female and male Syrian hamsters: Early changes following gonadectomy or light deprivation and lack of a circadian rhythm. *Biomedical Research* 10:1–8.

van Soest, P. J. 1982. Gastrointestinal fermentation. In *Nutritional ecology of the ruminant*, edited by P. J. van Soest, 152–229. Corvallis, OR: O&B Books.

Vleck, D. 1979. The energy costs of burrowing in the pocket gopher *Thomomys bottae*. *Physiological Zoology* 64:871–84.

———. 1981. Burrow structure and foraging costs in the fossorial rodent *Thomomys bottae*. *Oecologia* 49:391–96.

West, J. B. 1995. *Respiratory physiology: The essentials*. 5th edition. Baltimore: Williams & Wilkins.

Widmer, H. R., H. Hoppeler, E. Nevo, C. R. Taylor, and E. R. Weibel. 1997. Working underground: Respiratory adaptations in the blind mole rat. *Proceedings of the National Academy of Sciences, USA* 94:2062–67.

Withers, P. C. 1978. Models of diffusion mediated gas exchange in animal burrows. *American Naturalist* 112:1101–12.

———. 1992. *Comparative animal physiology*. Fort Worth, TX: Saunders College Publishing.

Yahav, S., and R. Buffenstein. 1991a. The effect of temperature on caecal fermentation processes in a poikilothermic mammal, *Heterocephalus glaber*. *Journal of Thermal Biology* 16:345–49.

———. 1991b. Huddling behaviour facilitates homeothermy in *Heterocephalus glaber*. *Physiological Zoology* 64:871–84.

———. 1992. Caecal function provides energy without liberating heat in the poikilothermic mammal, *Heterocephalus glaber*. *Journal of Comparative Physiology* B 162:216–18.

Yahav, S., A. Carlston, and R. Buffenstein. 1993. Changes in food intake with ambient temperature alter hindgut fermentation in the Damara mole-rat *Cryptomys damarensis*. *Comparative Biochemistry and Physiology* 104A:357–60.

Yahav, S., and I. Choshniak. 1990. Response of digestive tract to low quality food in the fat jird *(Meriones crassus)* and in the Levant vole *(Microtus guentheri)*. *Journal of Arid Environments* 19:209–15.

Yahav, S., S. Simson, and E. Nevo. 1988. Adaptive energy metabolism in four chromosomal species of subterranean mole rats. *Oecologia* 77:533–36.

———. 1989. Total body water and adaptive water turnover rate in four chromosomal species of subterranean mole rats of the *Spalax ehrenbergi* superspecies in Israel. *Journal of Zoology, London* 218:461–469.

———. 1990. The effect of protein and salt loading on urinary concentrating ability in four chromosomal species of *Spalax ehrenbergi*. *Journal of Zoology, London* 222:341–47.

CHAPTER THREE

Sensory Capabilities and Communication in Subterranean Rodents

Gabriel Francescoli

Animal communication has long intrigued organismal biologists. Interest in this topic stems from the conspicuous nature of many animal signals as well as the realization that understanding these signals may shed light on both the social environment and the sensory capabilities of the individuals employing them. Indeed, studies of animal communication are fundamental to any research program that attempts to paint a complete picture of a species' biology.

Currently, several theoretical issues drive research on animal communication. Signal meaning is an essential component of any form of communication and is a topic that has been addressed by numerous investigators (e.g., Hauser 1996; Smith 1982). Signal referentiality is the ability of animals to use signals to inform conspecifics about abstract events, from which inferences can be drawn regarding the cognitive abilities of individuals. The intentionality of signals is also of interest due to the potential for individuals to use signals to manipulate the actions of others (reviewed by Hauser 1996). Finally, efforts to understand signal design (Endler 1993) focus on the neurobiological, physical, and ecological factors (Dusenbery 1992) influencing the nature of communication among individuals.

Clearly, these conceptual issues are interrelated, as signal structure and function may be shaped by multiple factors. For example, the structure of aggressive acoustic signals may be influenced by the context in which they occur; because fighting ability is frequently associated with large body size, aggressive vocalizations tend to be lower in frequency than acoustic signals produced in other settings, such as greeting a conspecific (Morton 1977).

At the same time, the nature of aggressive acoustic signals may be influenced by the sensory capabilities of the species and the physical conditions of the environment, as well as by possible acoustic competition with signals from other species that live in the same habitat. Thus animal signals represent compromises between multiple selective pressures generated by the varied and sometimes conflicting functions of animal communication.

Even on a more descriptive level, animal communication remains a significant research problem. Basic questions regarding animal communication that have yet to be adequately addressed for many taxa include the following: What set of signals does a species use? Which sensory modalities are favored for communication? How have specific signals been shaped by selection? What is the evolutionary origin of these signals? How has a specific meaning become associated with a given signal? Ecological and evolutionary studies of communication systems attempt to answer these questions by relating trends in signal structure and function to patterns of habitat use and social behavior. A comparative research program that makes use of available paleontological data and that is framed in terms of multiple levels of analysis (e.g., Tinbergen 1963) should prove particularly useful in addressing questions about animal communication (Hauser 1996).

Subterranean rodents may face particular communication challenges that are directly related to the habitats in which they live. Life in a dense, semisolid medium may favor certain modes of communication while limiting the use of others. At the same time, the tendency for adults to be solitary (Lacey, chap. 7, this volume) may place greater emphasis on long-distance signals than is typical of gregarious species or species in which spatial overlap between solitary adults is more readily achieved. At present, our ability to generalize about patterns of communication in subterranean rodents is limited by the absence of even descriptive studies of the signals used by most species. Nevertheless, relationships between signal structure, function, and context are beginning to emerge, and for some taxa, we are now able to formulate testable hypotheses about the ecological and evolutionary correlates of specific communication patterns. These data suggest that communication in subterranean rodents reflects a mix of convergent and divergent responses to the challenges imposed by life in underground burrows.

In this chapter, I review our current understanding of communication in subterranean rodents. Available data regarding signal characteristics, the sensory capabilities of individuals, and the behavioral and ecological contexts in which signals occur are summarized for each of the taxa considered in this volume. To place studies of communication in subterranean rodents in a more general conceptual framework, this review is organized around the disciplines of sensory physiology and sensory ecology. The various modes of communication open to subterranean rodents are considered and evaluated in terms of their potential to convey meaningful information in underground habitats. By adopting a comparative approach to signal evolution, I underscore similarities in the communication systems used by different subterranean species while illustrating the ways in which signal generation, reception, and interpretation vary among these taxa.

Communication and the Constraints of Subterranean Life

Communication assumes many forms. The ways in which conspecifics communicate (e.g., olfaction, vision, mechanoreception) are greatly influenced by the species' *Umwelt*. This concept, developed by the German "natural phylosopher" Jakob von Uexküll, refers to "the environment around the subject" and implies a relationship between the animal, its environment, and its ability to extract information from that environment (Riba 1990). Because the types of information extracted are determined by an animal's sensory capabilities, knowledge of a species' sensory physiology should yield considerable insight into the relative importance of different communication channels. Given the role of sensory physiology in shaping intraspecific communication, this review considers patterns of signal reception as well as patterns of signal generation.

All animals have evolved multiple modes of communication. Among mammals, the primary modes of communication are sight, sound, and olfaction. These forms of communication may be used singly or in conjunction with one another, although the emphasis given to each varies with the sensory capabilities and environment of the species in question (Wilson 1980; Smith 1982). Visual, auditory, and olfactory signals are used for a variety of purposes, including eliciting specific behavioral responses, eliciting contact with conspecifics, identifying individuals or species, indicating social or reproductive status, soliciting food, warning against predators, and coordinating reproductive efforts (Wilson 1980). Subterranean rodents, like other mammals, must communicate to fulfill these goals. The specialized *Umwelt* of subterranean species, however, may significantly affect the communication channels they use for these purposes.

Not surprisingly, vision is an important mode of communication among diurnal rodents, and signals employed by these animals frequently contain ritualized postures and movements. In contrast, among nocturnal rodents and rodents with reduced visual abilities, olfaction and audition play greater roles in long-distance communication (Eisenberg and Kleiman 1977). Chemical signals include urine, feces, and the products of specialized scent glands; auditory signals include both vocalizations and mechanical sounds such as hisses, tooth chatters, and the thumping of body parts against the substrate (podophony). During close-range interactions, tactile signals, including touching, pelage grasping, and allogrooming, may be important.

Life underground simultaneously constrains the use of some communication channels and allows for the elaboration of others. Vision, for example, is effectively useless in dark tunnels. Although more useful than visual signals, chemical signals are confined to the tunnels within a burrow system, and tactile signals are appropriate only in close-contact situations. In contrast, vibrational signals—in particular, seismic signals produced via podophony—may be particularly effective because they extend beyond the confines of the burrow system and can, in fact, be enhanced by the solid medium that surrounds burrows. As evident from table 3.1, however, the use of seismic signals varies among subterranean rodents; although chemical, vocal, and tactile signals are widely used by these taxa, not all genera studied emit seismic signals. The reasons for such interspecific and intergeneric differences in communication are explored in detail in the following sections of this chapter.

Studying communication in subterranean rodents is challenging due to the logistic difficulties of monitoring behaviors that take place underground. Laboratory studies of communication are useful for characterizing signal repertoires, but such studies frequently fail to provide accurate information regarding the contexts in which communication occurs. This problem may explain why data regarding modes of communication are lacking for so many subterranean taxa. This lack of information makes it difficult to generate a comprehensive picture of communication in these rodents, which in turn limits our ability to determine whether similarities in morphology (Stein, chap. 1), physiology (Buffenstein, chap. 2), or ecology (Busch et al., chap. 5; Cameron, chap. 6, this volume) have produced parallel similarities in patterns of signal generation and reception. A primary goal of this chapter is to examine convergent and divergent patterns of communication in subterranean rodents. As part of this exercise, I will identify those conceptual issues and taxa in need of further study in order to understand the evolution of communication systems in subterranean rodents.

Modes of Communication

A fundamental component of any study of animal communication is characterizing the signal modalities that are used. Like other mammals, subterranean rodents exploit four primary modes of communication, namely, olfaction, vision, touch, and audition (fig. 3.1). Each of these forms of communication is reviewed below, beginning with olfactory, or chemical, communication.

TABLE 3-1 SOME SENSORY, COMMUNICATIVE, AND SOCIAL CHARACTERISTICS OF SUBTERRANEAN RODENT TAXA

Genus	Social	Leaves burrow	Sensory abilities			Modes of communication		
			Visual	Auditory	Olfactory	Vocal	Seismic	Tactile
Ctenomys	No?[a,d,e]	Yes[f]	Good[a]	?	Yes(?)	Yes[b]	No?[c]	Yes
Spalacopus	Yes[g]	Yes[g]	Good[a]	?	?	Yes[g]	?	?
Tachyoryctes	No[i]	Yes[h]	Medium[a]	?	?	Yes[h]	?	?
Nannospalax	No[k]	No[k]	Blind[a]	Poor[j]	Yes	Yes[k]	Yes[l]	Yes
Heterocephalus	Yes[p]	No[p]	Poor[m]	Poor[m]	Yes	Yes[o]	No[o]	Yes
Cryptomys hottentotus	Yes[m]	No[m]	Reduced[m]	Poor[q]	?	Yes[m]	Yes[r]	Yes
Cryptomys damarensis	Yes[m]	No[m]	Reduced[m]	?	?	Yes[m]	Yes[r]	Yes
Cryptomys mechowi	Yes[cc]	No[aa]	Blind[bb]	?	?	Yes[z]	No[z]	Yes(?)
Cryptomys sp.	Yes[aa]	No[aa]	Blind?[dd]	Medium[dd]	?	Yes[z]	No[z]	Yes(?)
Georychus	No[m]	No[m]	Reduced[m]	?	?	Yes[r]	Yes[s]	Yes(?)
Bathyergus	No[m]	No[m]	Poor[m]	?	?	Yes[m]	Yes[m]	Yes(?)
Heliophobius	No[m]	No[t]	Poor[m]	?	?	?	No[bb]	?
Thomomys	No[v]	?	?	?	Yes(?)	Yes[u]	No[w]	?
Geomys	No[v]	No[v]	Poor[a]	Poor[x]	?	No[v]	No[v]	?
Ellobius	Yes[y]	Yes[y]	Reduced[y]	?	?	?	?	?

References: a, Burda, Bruns, and Müller 1990; b, Francescoli 1992; c, C. A. Altuna and G. Francescoli, personal observations; d, Pearson and Christie 1985; Lacey, Braude, and Wieczorek 1997; e, Reig et al. 1990; f, Altuna 1991; g, Reig 1970; h, Flynn 1990; i, Jarvis and Sale 1971; j, Heffner and Heffner 1992b; k, Heth, Frankenberg, and Nevo 1988; l, Heth et al. 1991; m, Jarvis and Bennett 1991; n, Heffner and Heffner 1993; o, Pepper et al. 1991; p, Sherman, Jarvis, and Alexander 1991; q, Müller and Burda 1989; r, Bennett and Jarvis 1988b; s, Narins et al. 1992; t, Jarvis and Bennett 1990; u, Andersen 1978; Schramm 1961; v, Hansen and Reid 1973; w, Reichman, Whitham, and Ruffines 1982; x, Heffner and Heffner 1990; y, Herbin, Reperant, and Cooper 1994; z, Credner, Burda, and Ludescher 1997; aa, H. Burda, personal communication; bb, H. Burda, personal communication; cc, Burda 1993; dd, Brückmann and Burda 1997; H. Burda, personal communication.

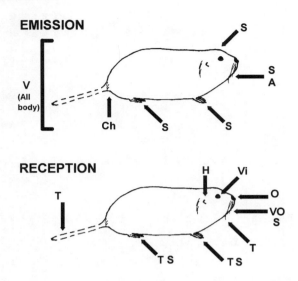

FIGURE 3.1. Schematic representation of a subterranean rodent (based on *Ctenomys*), showing the different body parts and organs involved in emission and reception of different kinds of signals. Emission: A, acoustic; Ch, chemical; S, seismic; V, visual. Reception: H, hearing; O, olfactory; S, somatosensory (seismic); T, tactile; Vi, visual; VO, vomerolfaction. The tail is represented as a dashed line because it is absent in many species of subterranean rodents.

Chemical Communication

Chemical signals are encoded as pheromones—chemicals or mixtures of chemicals that are released to the exterior by an organism and that, upon reception by a conspecific, stimulate one or more specific reactions (Shorey 1977). Pheromones are present in urine, feces, vaginal secretions, and the secretions of specialized glands, and may be deposited on the substrate or dispersed into the air. These signals can elicit either immediate, observable behavioral responses or more subtle physiological changes in the receiver. Pheromones deposited on the substrate (e.g., scent marks) can remain active over long periods of time (Eisenberg and Kleiman 1977), but pheromones expelled into the air are usually active over only relatively short time periods.

Subterranean rodents are expected to employ numerous chemical signals, both because they are mammals (Eisenberg and Kleiman 1977) and because life underground deprives them of the opportunity to use visual signals, which otherwise represent an important mode of communication among rodents. Unfortunately, little is known about chemical communication in subterranean rodents. Only scattered data are available in the literature, most of which are from studies of *Nannospalax ehrenbergi* (Heth and Todrank 1995, 1997; Heth, Nevo, and Todrank 1996). Other species for which evidence of chemical communication is available are *Heterocephalus glaber* (Faulkes and Abbott 1991, 1993) and two species of *Ctenomys: C. pearsoni* and *C. rionegrensis* (Altuna and Corte 1987). As the brevity of this list suggests, considerably more research is needed to assess the true nature and distribution of chemical communication in subterranean rodents.

CHEMICAL RECEPTION. In mammals, chemical receptors are generally present at three locations: the nasal mucosae, the tongue, and the vomeronasal organ. Of these, chemical reception by the nasal mucosae has been studied in the greatest detail. The sense of smell appears to be well developed in at least some subterranean rodents, as evidenced by the size of the morphological structures associated with olfaction. In particular, Pirlot and Nevo (1989) reported that the rhinencephalon and olfactory bulbs are well developed in *Nannospalax*, being larger than the mean sizes of these structures in surface-dwelling taxa.

In contrast, the olfactory brain centers in some subterranean taxa appear to be reduced in size, leading to speculation regarding the olfactory abilities of these animals. Animals with reduced olfactory centers include *Cryptomys* (Burda, Bruns, and Müller 1990) and *Ctenomys mendocinus* (Bee de Speroni 1995). The rhinencephalon and olfactory bulbs of the latter species have a progression index (PI) value (defined as the subject's brain weight divided by the expected brain weight of an insectivore of the same size) of about half that reported for *N. ehrenbergi*, leading to speculation that olfactory cues are less important in this species than in *N. ehrenbergi*. Because PI values may reflect structural differences that have arisen for a variety of reasons (e.g., phylogenetic history), however, extrapolating directly from PI values to sensory physiology may not provide an accurate picture of the olfactory abilities of subterranean rodents.

Even less is known about the function of the chemoreceptors on the tongue or in the vomeronasal organ. The taste capabilities of subterranean rodents are virtually unstudied, although Burda, Bruns, and Müller (1990) refer to largely unpublished data indicating that taste is highly developed in *Cryptomys hottentotus*. More controversial is the function of the vomeronasal organ (VNO). Although vomeronasal receptors are present in several mammalian groups, including rodents, their role in olfaction remains largely unknown. Data from *Mus* indicate that the epithelium of the VNO is involved in response to chemical stimuli, and studies of ungulates and carnivores suggest that this organ aids in the detection of sexual pheromones present in the urine and the vaginal secretions of females (Eisenberg and Kleiman 1972). Although some authors have argued that the functions of the VNO and olfactory mucosae do not differ, others, such as Berghard, Buck, and Liman (1996), maintain that the mammalian VNO and olfactory mucosae represent different pathways of chemical signal detection, with the VNO being specialized for the detection of pheromones (i.e., chemicals with a communicative function).

No studies of subterranean rodents have explicitly addressed the role of the VNO in chemical communication. Although accounts of anogenital contact during courtship in *Ctenomys* (Altuna, Francescoli, and Izquierdo 1988) and *H. glaber* (Lacey et al. 1991) are suggestive of pheromonal signals, it is not known whether such signals are employed, or whether the VNO is involved in detecting those signals. Similarly, pheromonal communication may occur in *N. ehrenbergi* (Heth and Todrank 1995), although the role of the VNO in these exchanges is also undocumented. Clearly, additional research is needed to determine how the VNO participates in chemical communication, as well as to identify functional differences between this structure and the olfactory mucosae. Improved understanding of the receptors involved in the detection of chemical signals should yield a better understanding of the significance of this mode of communication among subterranean taxa.

FUNCTIONAL SIGNIFICANCE OF CHEMICAL SIGNALS. Despite confusion regarding the receptor system(s) used in chemical communication, there is considerable evidence to suggest that subterranean rodents employ chemical signals. Chemical communication has been studied extensively in *N. ehrenbergi*. This species occurs in a number of different chromosomal forms in Israel and surrounding areas (Lacey, Patton, and Cameron, Introduction, this volume). Nevo, Bodmer, and Heth (1976) have demonstrated that estrous females can identify males of their own chromosomal form when provided with either soiled bedding or, in particular, male urine. These findings are corroborated by Heth et al. (1991), who demonstrated that individual *N. ehrenbergi* are capable of distinguishing between enantiomers (variations of the same compound differing only in their optical rotation) of inert substances.

Todrank and Heth (1996) have shown that *N. ehrenbergi* is capable of discriminating between individuals (of the same or different chromosomal form) based on odor cues in urine. Adult *N. ehrenbergi* do not mark their tunnels with urine, but instead use odors from urine deposited in latrine areas to deter intrusions by conspecifics (Heth and Todrank 1997). In many subterranean taxa, latrine areas (areas of heavy urine deposition) are located near nest chambers. In *N. ehrenbergi*, however, latrines are located at tunnel junctions, where they are most likely to be encountered by neighboring animals that enter the burrow system. Heth, Nevo, and Todrank

(1996) reported that the response of *N. ehrenbergi* to the urine of conspecifics varies between the breeding and nonbreeding seasons, suggesting that odor-based information may be associated with both sexual behavior and avoidance of agonistic encounters.

Work by Heth and Todrank (1995) suggests that the chemical recognition abilities of *N. ehrenbergi* may differ between contexts. When exposed to urine odors, both males and females of this species are indifferent to, or are slightly attracted to, the urine of same-sex conspecifics. If, however, the animals are first allowed to touch the urine, they avoid the odor of same-sex individuals. After contact with the urine, its odor alone is sufficient to elicit this avoidance response, although a similar avoidance is not elicited by exposure to the subject's own urine or to the urine of guinea pigs. Heth and Todrank (1995) interpreted these findings as evidence that the vomeronasal organ may play a role in interactions among conspecifics. They argued that because the VNO makes contact with molecules through openings in the palate, relatively direct exposure to pheromone-containing substances (e.g., urine) may be required for chemical signals to be detected.

More generally, Heth and Todrank (1995) proposed that *N. ehrenbergi* and other rodents use the VNO to mediate innate responses to chemical signals but use the olfactory system to mediate learned responses. This assertion seems to be contradicted, however, by the authors' work indicating that *N. ehrenbergi* avoids predator urine odors without first touching or contacting the urine. Assuming that predator avoidance is an innate response and that direct contact is required to stimulate the VNO, the ability to detect predator odors in the absence of direct contact is puzzling. Other subterranean rodents (e.g., *Thomomys talpoides:* Sullivan et al. 1990) exhibit similar abilities, raising questions regarding the chemical receptor pathways employed in predator avoidance. Comparative studies of responses in predator-naive and predator-experienced animals may help to clarify the chemosensory channels used in this context.

Recently, Shanas and Terkel (1997) have reported an alternative source of chemosensory signals. These authors have demonstrated that in *N. ehrenbergi*, Harderian glands located around the eyes produce pheromones that are deposited on the fur during grooming. Secretions from the Harderian glands of males are attractive to both sexes, although secretions from the glands of females are attractive only to males. The specific function of these secretions is unknown, but Shanas and Terkel (1997) argue that this chemosensory signal, in conjuction with the grooming behaviors that distribute it, may act as an appeasement signal during agonistic interactions between males.

In contrast to *Nannospalax,* the role of chemical communication in other subterranean taxa is virtually unknown. In some ctenomyids (e.g., *Ctenomys pearsoni* and *C. rionegrensis*), the perineal gland may play a role in scent marking (Altuna and Corte 1987). This gland, which is present in all adults of these species, consists of sebaceous (holocrine) acini connected to a keratinized tube that ends at the anal papilla. Other hystricomorph rodents with homologous structures use the secretions from this gland to scent-mark while performing a behavior termed the "perineal drag." Although this behavior has not been observed in *Ctenomys,* the presence of similar glandular structures suggests that chemical signals may be added to fecal pellets. Because *C. pearsoni* and *C. rionegrensis* pack feces into the soil used to plug burrow entrances, any odor transferred to fecal pellets could be present in burrow plugs, and could serve as a territorial warning to possible intruders (Altuna and Corte 1987). Whether similar behaviors occur in other subterranean groups remains to be determined.

In solitary species such as *N. ehrenbergi,* chemical communication is expected to occur primarily in two contexts: in long-range communication between animals in different burrow systems, and in short-range communication between animals engaged in sexual encounters. Although chemical communication may also serve these functions in social species, chemical signals appear to assume additional roles in taxa in which burrow systems are shared by multiple adults. For example, in *H. glaber,* colony odors appear to be important in maintaining colony cohesion and preventing intrusions by individuals from other colonies (O'Riain and Jarvis 1997). Colony-specific signals are apparently an amalgam of the odors of individual colony members, which are deposited in their urine; individuals may be exposed to and may acquire this collective colony scent when they visit communal latrine areas. The ability to recognize a colony-wide scent may facilitate the animals' ability to detect and repel intruders from other colonies (O'Riain and Jarvis 1997).

Chemical signals also appear to play an important role in maintaining the reproductive division of labor within colonies of social bathyergids. The role of chemicals in regulating sexual activity in naked mole-rats has been well studied (Bennett, Faulkes, and Molteno, chap. 4, this volume; Faulkes and Abbott 1991, 1993; Faulkes, Abbott, and Jarvis 1991; Faulkes et al. 1991). Although direct aggressive interactions between individuals are critical to maintaining reproductive differences among colony members, exposure to putative pheromones in the urine of the breeding female also appears to play a role in determining the reproductive status of individuals of both sexes (Bennett, Faulkes, and Molteno, chap. 4, this volume; Faulkes and Abbott 1991).

A novel form of chemical signaling that has recently been proposed for *H. glaber* is the use of scent trails to direct foraging colony members to food sources. Judd and Sherman (1996) demonstrated that colony members recruited to new food sources always choose the pathway used by the scout animal. Their experiments confirmed that recruits use no behavioral or spatial information from the scout animal, and that recruits are unable to decide which pathway to follow if the tube used by the scout is replaced by a clean tube. These results suggest that some kind of scent marking is a critical cue for recruitment. Collectively, these data from bathyergid mole-rats suggest that chemical communication may assume a number of specialized functions in social species in which individuals interact regularly with one another.

CHEMICAL COMMUNICATION IN SUBTERRANEAN HABITATS. Relative to other modes of communication, chemical signals may provide an inexpensive way to convey information to conspecifics. Pheromones are easily synthesized and may be derived from waste products that are otherwise excreted (Shorey 1977). Pheromones are also relatively long-lasting signals that may elicit extended interactions among conspecifics, as evidenced by patterns of reproductive suppression in *H. glaber* and other social bathyergids (Bennett, Faulkes, and Molteno, chap. 4, this volume). Despite these benefits, chemical signals do not appear to dominate communication among either solitary or social subterranean rodents. Instead, as outlined below, vibrational signals, including both acoustic and seismic signals, represent a sizable component of communication in these animals. Even allowing for chemical signals that have yet to be discovered, it is clear that this mode of communication is only one of several employed by subterranean species. Further research into the nature and functional significance of chemical signals in these animals should help to clarify the contexts in which such signals are preferred.

Visual Communication

A second mode of communication employed by subterranean rodents is vision. Two sources of visual signals are available to animals: bioluminescence and sunlight reflected by other objects. The former is not known to occur in subterranean rodents, and the latter seems unlikely to be of much use to animals that spend most of their lives in underground burrows. Despite the absence of light in subterranean burrows, some species show displays and/or colors that, in a surface-dwelling species, would be interpreted as visual signals. An example is the conspicuous orange coloration of rhizomyine and ctenomyid incisors (Stein, chap. 1, this volume). Among ctenomyids, this coloration is readily visible during sound emission associated with aggressive encounters (C. A. Altuna and G. Francescoli, unpublished data) and thus would likely be interpreted as a visual signal in surface-dwelling species. To date, however, no studies of the communicative function of these displays have been published.

The majority of research on vision in subterranean rodents has focused on the apparent reduction of visual capabilities in these animals (e.g., Cooper, Herbin, and Nevo 1993a,b). This emphasis is not surprising, given both the absence of light in subterranean burrows and the noticeable reduction in eye size in some species (Stein, chap. 1, this volume). However, some subterranean rodents are routinely active above ground (Busch et al., chap. 5, this volume), suggesting that reduction in visual ability is not characteristic of all taxa. Relationships between extent of surface activity, degree of visual reduction, and behavioral or morphological evidence of visual signaling provide a logical framework for exploring this form of communication in subterranean species.

One means of assessing the importance of visual communication in subterranean species is to consider the visual abilities of these animals. In general, *Spalacopus, Ctenomys, Tachyoryctes,* and *Geomys* are thought to have reasonably good visual acuity, meaning that these animals are able to distinguish different light intensities and to detect moving objects (Burda, Bruns, and Müller 1990). Perhaps not surprisingly, these animals tend to emerge from their burrows with some regularity to forage or perform other vital activities (Reig 1970; Altuna 1991; Hansen and Reid 1973; but see Heffner and Heffner 1990 on the frequency with which *Geomys* leaves its burrow). In contrast, the bathyergids and *Nannospalax* do not often leave their burrows (Heth, Frankenberg, and Nevo 1988; Jarvis and Bennett 1990, 1991; Sherman, Jarvis, and Alexander 1991), and these animals are thought to have markedly reduced visual abilities.

Detailed quantitative analyses of visual acuity have been completed for only a few species, notably *N. ehrenbergi*. In this species, which lacks external eyes (Stein, chap. 1, this volume), early development of the visual system appears normal, but eye size becomes greatly reduced, and the eye itself becomes buried beneath the skin, as ontogeny progresses (Cooper, Herbin, and Nevo 1993b). Because retinal development proceeds relatively normally, adults possess functional, albeit reduced, visual receptors (Sanyal et al. 1990). These receptors send projections to the brain, but because the thalamic visual pathway is reduced, an estimated 50% fewer neural projections reach the superior colliculus than in surface-dwelling rodents (Cooper, Herbin, and Nevo 1993a).

In contrast to this reduction in image-related projections, projections related to photoperiodic information are increased and, compared with other rodents, the associated neural pathways are hypertrophied in *Nannospalax* (Cooper, Herbin, and Nevo 1993a; Cooper et al. 1995). These structural modifications, together with the limited ecological need for visual processing, appear to have led to a reduction of the visual cortex in this species (Cooper et al. 1995; Nevo 1996). The visual cortex, in turn, appears to have subsequently been invaded by either the auditory cortex (Bronchti et al. 1989) or somatosensory cortex (Necker, Rehkämper, and Nevo 1992). Recently published data (Mann et al. 1997) indicate that the somatosensory cortex is relatively larger in *Nannospalax* than in *Rattus,* providing apparent support for Necker, Rehkämper, and Nevo's (1992) hypothesis that reduced visual ability is associated with enhanced tactile sensitivity. In summary, studies of the visual system of *Nannospalax* suggest that these animals cannot perceive images, but that they are capable of detecting light-dark cycles, which may help to entrain circadian patterns of activity and thermoregulation (Nevo 1996; Buffenstein, chap. 2, this volume).

Among subterranean rodents, *Nannospalax* is morphologically extreme in that it lacks external eyes, leading to speculation regarding the visual abilities of other subterranean species that have external eyes. The only other taxa for which detailed studies of visual perception have been conducted are *Ellobius talpinus* and *E. lutescens,* both of which have relatively reduced external eyes (Herbin, Reperant, and Cooper 1994; Stein, chap. 1, this volume). In these animals, reduced eye size is associated with a marked loss in retinal magnification and spatial resolution, although central visual projections are not substantially modified compared with surface-dwelling species. These data suggest that *Ellobius* is anatomically equipped for binocular vision (Herbin, Reperant, and Cooper 1994), but anecdotal observations of these animals suggest that they do not respond behaviorally to either stationary or moving objects, casting doubt on the importance of vision to members of this genus.

Data on the visual abilities of the remaining subterranean taxa are largely anecdotal and are often based on inferences drawn from measurements of eye size or behavioral responses to presumed visual stimuli. For example, bathyergid mole-rats generally possess minute external eyes and only poorly developed visual brain regions, findings that have been interpreted as evidence that the animals do not perceive images (Jarvis and Bennett 1991). Behaviorally, some *Cryptomys* fail to respond to changes in light intensity (Burda, Bruns, and Müller 1990). In contrast, *Tachyoryctes* and other rhizomyines, all of which have small eyes, are responsive to daylight (Flynn 1990) and can perceive objects located up to several feet away (Burda, Bruns, and Müller 1990).

Eye size is generally thought to be somewhat greater among geomyid pocket gophers (Burda, Bruns, and Müller 1990; Stein, chap. 1, this volume). In *Geomys bursarius,* the eyes are somewhat reduced in size, but have an apparently normal retinal structure that resembles the retinal structures of diurnal and crepuscular rodents (Feldman and Phillips 1984). Although not conclusive, these observations suggest that *G. bursarius* may have greater visual abilities than either *Nannospalax* or the bathyergid mole-rats. Eye sizes in *Spalacopus* and *Ctenomys* are not much reduced, and the visual acuity of these taxa is thought to be good (Burda, Bruns, and Müller 1990; Reig 1970; Pearson 1959; G. Francescoli, personal observation). The markedly dorsal location of the eyes in these genera suggests that vision is used when the animals are active at burrow entrances.

As this discussion suggests, eye size is generally thought to correlate with visual ability, despite the lack of neurological studies of vision in subterranean rodents. Eye size, in turn, is thought to be related to the proportion of time that each species spends above ground (but see Stein, chap. 1, this volume). Burda, Bruns, and Müller (1990) have argued that the retention of "normal-sized" eyes by some subterranean species must be due to selection pressures favoring visual acuity, such as selection for the ability to detect predators visually. Although this conclusion is logically appealing, additional research is needed to evaluate the forces favoring maintenance or reduction of the visual systems of subterranean species. A better understanding of the visual abilities of these animals should also allow us to explore in greater detail the relative importance of different sensory systems in species that are frequently versus only rarely active above ground.

Tactile Communication

A third mode of communication that is employed by subterranean rodents is tactile communication. Tactile signals are received when specialized mechanoreceptors are deformed during contact with an object (Dusenbery 1992). These mechanoreceptors are widely distributed over the body; certain body parts, such as the digits, the nose, and the lips, contain large numbers of these receptors and are therefore particularly sensitive to tactile stimulation. Specialized hairs such as vibrissae are also involved in tactile reception; these structures tend to be highly innervated and are capable of transmitting detailed information to the brain regarding the amplitude, direction, velocity, duration, and frequency of their deflection by an object or other stimulus (Dusenbery 1992).

Given the physical configuration of subterranean burrows, tactile signals are expected to affect subterranean rodents in numerous ways. Simply navigating through dark tunnels may involve considerable tactile input as the animals use contact with tunnel walls to guide their movements (Lacey et al. 1991). Because tactile signals require contact between individuals, some forms of tactile communication may be limited to specialized contexts, such as mating, or to social species in which animals are in frequent physical contact with burrowmates. In particular, the close contact required by tactile signals suggests that this form of communication will not be used extensively in dangerous situations such as predator detection or aggressive encounters with conspecifics; in these circumstances, other modes of communication that do not require direct physical contact should be favored.

Apparent use of tactile cues has been reported for a number of subterranean rodents. In some genera, specialized hairs on the tail appear to be used during navigation. The role of the tail as a navigational tool has been suggested for *Ctenomys* and *Geomys;* when active on the surface, members of both genera use their tails to probe for an open tunnel entrance when retreating backward into their burrows (Dubost 1968; Orofino et al. 1997; G. Francescoli, personal observation). The utility of specialized tail hairs as signal receptors is also suggested by their presence in *H. glaber,* which otherwise lacks pelage (Burda, Bruns, and Müller 1990).

Tactile communication during courtship and mating has been suggested for *C. pearsoni,* in which extensive allogrooming and other bodily contact takes place between prospective mates. During copulation, the male grasps the female with his forepaws and gently bites her on the back, probably as a form of appeasement (Altuna, Francescoli, and Izquierdo 1991). Penile contact with the female's genital tract may play an important role in inducing ovulation in this species (Altuna, Francescoli, and Izquierdo 1991; Weir 1974). Similar forms of tactile communication during mating have been postulated for *H. glaber, C. damarensis, C. hottentotus, Georychus capensis,* and *Bathyergus suillus* (Lacey et al. 1991; Burda 1989; Jarvis and Bennett 1991; Bennett, Faulkes, and Molteno, chap. 4, this volume), and the list of examples seems likely to grow as descriptions of reproductive behavior in subterranean rodents increase. To date, however, studies aimed at quantifying the importance of these forms of physical contact have not been conducted for any of these species. As a result, it is not yet possible to determine how the tactile sensitivities of subterranean rodents compare with those of surface-dwelling taxa (Eisenberg and Kleiman 1977; Burda, Bruns, and Müller 1990).

Vibrational Communication

The final mode of communication considered here is vibrational communication. The term "vibrational communication" encompasses a variety of signal types, all of which consist of waves that are transmitted through the environment. Because modes of both signal generation and signal reception can differ, vibrational communication represents a particularly challenging topic for students of animal communication. At the same time, vibrational communication includes some of the most unusual and intriguing signals used by subterranean rodents, and thus this form of communication has received particular attention from biologists interested in these animals.

Vibrational communication comprises two primary types of signals:

acoustic signals and seismic signals. Acoustic signals are those that are transmitted as sound waves and that are received and decoded using specialized sound receptors, such as ears. In contrast, seismic signals consist of substratum-borne vibrations that are received as somatosensory (tactile) signals. Thus, vibrational signals that are received via the substratum represent "true" seismic signals, but those that are received as sounds should be considered acoustic signals. This distinction is critical because certain modes of signal generation (e.g., thumping the head against the burrow ceiling) produce both seismic and acoustic signals that can function independently of each other. Differences in signal propagation are also important because they may lead to distinct receptor specializations in animals that employ vibrational communication (see Narins et al. 1992).

ACOUSTIC SIGNALS. Most studies of vibrational communication in subterranean rodents have focused on acoustic signals, which may be generated either as vocalizations or as vibrations produced by knocking or grinding together different body parts; examples of the latter include tooth chattering and tooth grinding. Acoustic signal production has been reported for almost all taxa studied, including *Tachyoryctes, Cryptomys, Georychus, Bathyergus,* and *Thomomys* (see table 3.1) (Flynn 1990; Jarvis and Bennett 1991; Bennett and Jarvis 1988a; Andersen 1978). Detailed studies of acoustic signal production have been completed for *N. ehrenbergi* (Capranica, Moffat, and Nevo 1973; Heth, Frankenberg, and Nevo 1986, 1988; Nevo et al. 1987), *H. glaber* (Pepper et al. 1991), *Cryptomys* sp. and *C. mechowi* (Credner, Burda, and Ludescher 1997), *S. cyanus* (Eisenberg 1974; S. Begall, personal communication), and *Ctenomys pearsoni* (Francescoli 1992, 1995, 1997; Francescoli and Altuna 1996).

The first study of acoustic communication in *N. ehrenbergi* (Capranica, Moffat, and Nevo 1973) identified six distinct vocalizations, each of which occurred in a different behavioral context. Of these, only the courtship call has been subject to more extensive investigation. This vocalization, which resembles a purring sound, is produced as a series of pulses (23 pulses/second) with a primary frequency of 569 Hz (Heth, Frankenberg, and Nevo 1988). This call is emitted primarily by males during courtship and copulation and appears to reduce aggressive behavior by the female (Heth, Frankenberg, and Nevo 1988). Four variants of this call have been identified, each corresponding to a different chromosomal form of *N. ehrenbergi;* work by Nevo et al. (1987) has revealed that females consistently prefer the calls of males from their own chromosomal form, providing a potential mechanism of reproductive isolation among chromosomal variants.

Another subterranean rodent whose vocal repertoire has been well studied is *H. glaber* (Pepper et al. 1991). This species employs at least seventeen different vocalizations, five of which are produced exclusively by juveniles. Eleven of these vocalizations are frequency-modulated sounds that range from 1 to 9 kHz; the remaining vocalizations are broadband "noisy" sounds ranging from 0.2 to 40 kHz. These signals are produced in a variety of behavioral contexts, including urination, digging, aggressive encounters, sexual behavior, allocoprophagy, and colony alarm (Pepper et al. 1991). Another bathyergid whose acoustic signals have been studied is an unnamed species of *Cryptomys*, which has a repertoire of thirteen vocalizations. These vocalizations, as a group, contain components ranging from 0.63 to 6.3 kHz, with the main energy of the sounds always below 10 kHz. Contact, mating, aggressive, distress, and juvenile calls were identified for these animals. Mating and aggressive calls were also identified for *C. mechowi;* frequencies for these vocalizations occur primarily in the same range as those for the unnamed *Cryptomys* mentioned above (Credner, Burda, and Ludescher 1997).

Vocalizations have also been characterized for *S. cyanus,* which emits at least nineteen different sounds, seven of which are produced by juveniles (Eisenberg 1974; S. Begall, personal communication). These vocalizations are produced by both sexes and range in frequency from approximately 0.2 to 20 kHz. Among ctenomyids, *C. pearsoni* displays a repertoire of five vocalizations, two of which are emitted by juveniles; frequencies for these sounds range from 50 to 3,300 Hz (Francescoli 1992, 1995; Francescoli and Altuna 1996; G. Francescoli, unpublished data) (fig. 3.2).

More qualitative, anecdotal accounts of acoustic signals have been published for *Thomomys, Cryptomys,* and *Georychus.* The majority of these sounds are produced during courtship. Specifically, Andersen (1978) noted that female *T. talpoides* emit soft calls while being mounted, and Schramm (1961) described a series of guttural sounds and high-pitched squeaks that are produced by females during copulation. Hickman (1982) reported that female *C. hottentotus* emit high-pitched vocalizations prior to copulation, although Burda (1989) indicated that not all individuals produce these sounds during courtship. Bennett and Jarvis (1988b) reported that both female *C. damarensis* and female *G. capensis* produce high-frequency calls or squeaks during courtship, with females of the former species also emitting guttural sounds prior to copulation. Given the apparent widespread use of acoustic signals by subterranean rodents, it seems likely that as studies of the behavior of these animals proceed, both the number of species known to use acoustic signals and the variety of contexts in which such signals are produced will increase dramatically.

ACOUSTIC RECEPTORS. Among mammals, the ear pinnae represent an essential part of the acoustic signal reception apparatus. Like their other extremities, the pinnae of subterranean rodents tend to be reduced in size relative

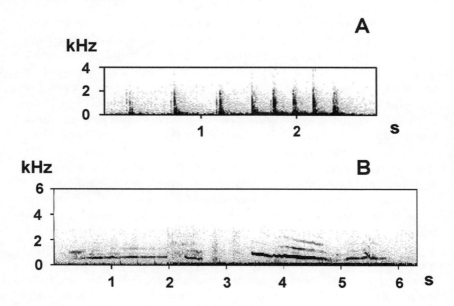

FIGURE 3.2. Sonograms of some signals emitted by *Ctenomys pearsoni*. A: S signal series, used for territorial advertisement. B: C signal notes, emitted by the female before copulation.

to those of surface-dwelling taxa (Stein, chap. 1, this volume). The effects of this structural modification on signal reception, however, are unknown. On the one hand, the absence of pinnae may make it more difficult for subterranean taxa to capture acoustic signals, as the pinnae are thought to help funnel such signals toward the auditory canal. On the other hand, given that acoustic signals must travel through the relatively confined spaces of tunnels, signal capture may not represent a substantial problem except when animals are active outside the burrow. For species that are often active at the surface, pinna size reflects a compromise between selection for reduced pinnae (e.g., to avoid injury while moving through tunnels: Stein, chap. 1, this volume) and selection for pinnae that are large enough to be functional for sound localization outside the burrow.

The structure of the inner ear varies among subterranean rodents, particularly with regard to the degree of muscle development and the structural relationship between the incus and malleus (Burda, Bruns, and Müller 1990). The auditory bullae vary in size, which may reflect interspecific differences in the nature of the acoustic signals employed; in general, bulla size is thought to be greater in species that employ low-frequency signals (Hooper 1968; Lay 1972; Webster and Webster 1975; Randall 1994). Although the converse relationship between high-frequency signals and small bullae has not been as thoroughly examined, it seems to be true for some *Ctenomys* species (Pearson and Christie 1985; Cook, Anderson, and Yates 1990; Francescoli 1992).

The cochlea of subterranean rodents tends to be tightly coiled, with a small scala tympani. The basilar membrane is long and relatively wide and maintains a nearly constant width along its length (Burda, Bruns, and Müller 1990). In contrast, other mammals tend to have basilar membranes that are narrow at the origin and wider near the apex (Burda, Bruns, and Müller 1990). The apex of the basilar membrane is the site for low-frequency signal transduction (Pickles 1988), suggesting that a uniformly wide membrane is an adaptation for low-frequency sound detection.

Among subterranean species, the organ of Corti contains more outer and fewer inner hair cells per unit area than are found in other mammals (Burda, Bruns, and Müller 1990). The "active model of cochlear mechanics" proposed by Pickles (1988) argues that this difference allows subterranean rodents to further fine-tune the sensitivity of the inner hair cells via the action of the outer hair cells (Ashmore and Russell 1983). Specifically, this model states that the inner hair cells are responsible for sound transduction, and that the outer hair cells, which contact the tectorial membrane more tightly and are receptive to nervous stimulation, act as modulators of the "stiffness" of the organ of Corti relative to a sound traveling through it (Pickles 1988). The "stiffer" the organ of Corti is, the more it will react to low-frequency sounds during transduction.

More detailed studies of ear structure and function have focused on *N. ehrenbergi*. Compared with other rodents, this species possesses an enlarged eardrum, an enlarged incus, and a structurally simple malleus that is not attached to the tympanic ring (Burda, Bruns, and Nevo 1989). The cochlea are also somewhat modified and consist of two distinct regions: a basal subsystem that is similar to the apical region of the cochlea in *Rattus*, and an apical subsystem that contains the organ of Corti and is similar to the basilar papilla of non-mammalian tetrapods (Bruns et al. 1988). Both modifications of the cochlea are thought to improve sensitivity to low-frequency sounds. Low-frequency detection is also thought to be enhanced by the relatively low density of inner hair cells on the organ of Corti and the increase in hair cell density as one moves from the base to the apex of the cochlea (Burda, Bruns, and Nevo 1989).

Audiograms of *N. ehrenbergi* based on evoked potentials or cochlear microphonics indicate that this species detects signals between 0.1 and 10 kHz, with maximum sensitivity at 0.5 and 1 kHz (Bruns et al. 1988). These values are lower than those typically reported for mammals, suggesting that the auditory system of *N. ehrenbergi* is specialized for the detection of low-frequency sounds. Under typical environmental conditions, blind mole-rats have relatively limited auditory abilities, being unable to detect sounds above 5.9 kHz (at 60 dB) or shorter than 0.5 seconds (Heffner and Heffner 1992b); this represents the poorest high-frequency sensitivity known among mammals. Experimental analyses indicate that, within a burrow, sounds produced at about 400 Hz propagate farthest. This value approximates the frequency of maximum sensitivity in this species, suggesting that its receptors are adapted to match the most effective signals produced. Middle ear structures of *N. ehrenbergi* vary across chromosomal forms; this variation has not yet been shown to correlate with intraspecific differences in the acoustic signals used by these animals, but Burda, Nevo, and Bruns (1990) assert that this correlation is likely.

Auditory ability has also been examined for *C. hottentotus*, which has an auditory range of 0.1 to 5 kHz, with maximal sensitivity between 0.5 and 1 kHz (Müller and Burda 1989). Müller and Burda (1989: 135) assert that these animals should be less sensitive to airborne sounds than other rodents, but that they should be very sensitive to "weak occasional nearby low-frequency sounds." A second bathyergid, *H. glaber*, also displays reduced hearing ability relative to other rodents, with an auditory range of 65 Hz to 12.8 kHz (at 60 dB) and a maximum sensitivity of 35 dB at 4 kHz (Heffner and Heffner 1993). Although the major auditory nuclei are present in the brainstem (Heffner and Heffner 1993), this species has poor sound localization abilities and is unable to localize sounds lasting less than 400 ms. Although the auditory abilities of the remaining bathyergids have yet to be studied, these data suggest that, as a group, these animals are not well adapted to the detection of auditory signals.

Similarly, *G. bursarius* has only limited auditory sensitivity, with a hearing range of 350 Hz to 8.7 kHz (at 60 dB) and maximal sensitivity at 2 kHz (Heffner and Heffner 1990). Its ability to localize sounds is also greatly reduced, being limited to sounds lasting more than 0.5 seconds (Heffner and Heffner 1990). The auditory nuclei in *G. bursarius* are qualitatively similar to those found in other mammals, and detailed analysis of these nuclei has revealed no morphological features that correlate with the restricted hearing abilities of this species (Heffner and Heffner 1990).

In summary, subterranean rodents are characterized by reduced auditory sensitivity compared with other mammals. More specifically, subterranean species are especially sensitive to low-frequency sounds and almost insensitive to high-frequency signals (fig. 3.3). In some species, this bias in auditory sensitivity is associated with structural changes in the auditory apparatus (Burda, Bruns, and Müller 1990), although in other species such structural changes are not apparent. The shift toward detection of low-frequency sounds is generally consistent with expectations regarding the physics of

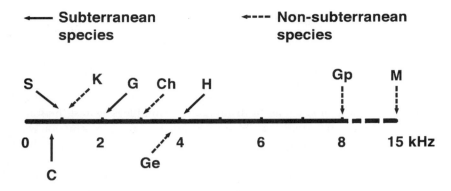

FIGURE 3.3. Frequencies for maximal hearing capabilities in different rodents. C, *Cryptomys hottentotus* (Burda 1989); Ch, chinchilla; G, *Geomys bursarius;* Ge, gerbil *(Meriones unguiculatus);* Gp, guinea pig; H, *Heterocephalus glaber* (Heffner and Heffner 1993); K, kangaroo rat; M, mouse (Ehret 1989); S, *Nannospalax ehrenbergi*. Subterranean species plus those that live in deserts (gerbil, kangaroo rat) tend to have better hearing abilities in low frequencies. The differences between subterranean and surface-dwelling species are even more obvious if we also take into account the total hearing range (not represented here). (Data from Heffner and Heffner 1990 unless another reference is specified)

acoustic signal production in subterranean habitats, but further research is needed to relate signal properties to the auditory sensitivity of the receiver.

SEISMIC SIGNALS. The production of substrate-borne seismic signals has been reported for *N. ehrenbergi* (Rado et al. 1987; Heth et al. 1987, 1991; Nevo, Heth, and Pratt 1991) and *G. capensis* (Narins et al. 1992). Based on unpublished data and personal communications with other researchers, Burda, Bruns, and Müller (1990) suggest that *Bathyergus* and *Ctenomys* also emit seismic signals. Thumping of the hind feet against the ground has been reported for *Bathyergus* spp. and *G. capensis* by Jarvis and Bennett (1991), although no published data confirm the presence of seismic signaling in *Ctenomys*.

Seismic signals may be generated in a variety of ways. In *N. ehrenbergi*, the flattened, bony, anterodorsal portion of the head is thumped against the tunnel ceiling. Thumps are typically produced in sets of four and result in a signal of less than 100 Hz (Heth et al. 1987). Animals tested together may duet, with both individuals producing thumps alternately (Rado et al. 1987). Heth et al. (1987) reported that head thumping patterns vary among the different chromosomal forms of this species, particularly with respect to the number of thumps performed in succession. Although they originally postulated that these differences represent a reproductive isolating mechanism, subsequent research (Heth et al. 1991) has revealed that variation in thumping occurs within each chromosomal form and appears to be associated with territory size and population density, suggesting that seismic signaling may function in individual spacing and recognition rather than reproductive isolation.

Narins et al. (1992) examined vibrational communication in *G. capensis*, which produces seismic signals by drumming the hind legs against the tunnel floor. These authors suggested that such signals contain information regarding individual or sexual identity, based on observations that, during the breeding season, the foot-drumming rate is sexually dimorphic and males and females alternate drumming bouts (Bennett and Jarvis 1988a; Narins et al. 1992). The propagation of artificially generated signals was tested in both soil and air. These experiments revealed (1) that acoustic signals attenuate more rapidly than seismic signals by at least an order of magnitude and (2) that vertically polarized surface seismic waves propagate with less attenuation than horizontally polarized surface waves. Narins et al. (1992) suggested that surface seismic waves are used to assess territory status and, possibly, sexual or individual identity.

SEISMIC RECEPTORS. To date, the process of detecting seismic signals has been examined only in *N. ehrenbergi*. Studies by Rado et al. (1987) provided the first demonstration that subterranean rodents can detect substrate-borne vibrations. By placing individuals in separate plastic tubes, Rado et al. discovered that the animals used head thumping to respond to each other's signals if the tubes were in physical contact; if the tubes were separated by as little as 1 mm, no mutual signaling occurred, even though the animals in different tubes could hear and smell each other. Nevo, Heth, and Pratt (1991) conducted a similar experiment in which the authors tapped on artificial tunnels; although the experimental animal responded if the tubes were in contact, tubes separated by 2 cm produced no response, even if the sound produced by tapping was audible to human observers.

N. ehrenbergi appeared to detect seismic signals by placing the cheek and lower jaw against the wall of the tube (Rado et al. 1987). This behavior suggests a bone conduction system, perhaps via full contact between the processus condylaris of the mandible and the tympanic bulla. A similar system for detecting vibration has been suggested for *Geomys* because this genus exhibits a similar pattern of mandibular-bullar mechanical contact (Rado et al. 1989; Burda, Bruns, and Müller 1990), although these animals are not known to employ seismic communication (see table 3.1). More specifically, the presence in *Nannospalax* of a ligament connecting the short process of the incus to the periotic bony lamina suggests that vibrations travel from the skull to the middle ear ossicles via this ligament. This reception mechanism contrasts with that for acoustic signals, which produce a vibration of the tympanic membrane that is transmitted to the ossicles in the middle ear and then to the cochlea. Acoustic signal reception relies upon the lever effect of the ossicles to increase vibration transmission to the cochlea. Thus signal power may be greater for acoustic signals that enter via the eardrum than for seismic signals that enter directly via the incus and bypass the amplification generated by the ossicles. Based on these findings, Rado et al. (1989) postulated that the inner ear of *N. ehrenbergi* contains two auditory systems: one for detecting airborne signals and one for detecting substrate-borne signals. While the first is used primarily for short-distance (acoustic) communication, the latter is used primarily for long-distance (seismic) communication.

A contrasting interpretation was offered by Nevo, Heth, and Pratt (1991), who argued that seismic signal reception is somatosensory, meaning that vibrations are received by mechanoreceptors not related to the ear. By using a white noise generator to mask possible acoustic cues, these authors demonstrated that *N. ehrenbergi* responds to tapping vibrations even if auditory reception is effectively precluded, suggesting that seismic signal detection does not depend on the ear.

Recently, Rado, Terkel, and Wollberg (1998) published new information on this issue. Based on experiments in which anaesthetized and deafened mole-rats were placed in plastic tubes with their lower jaws contacting the substratum and subjected to artificially produced vibrations, the researchers found that the principal component of the neural response to tapping was auditory, even if some little concomitant somatosensory response existed. The deafened animals did not show normal responses to seismic signals produced by conspecifics, even if a negligible somatosensory component was present. The authors' speculative explanation for the obvious discordance between their results and those of Nevo, Heth and Pratt (1991) lies in the possible use by the latter authors of too high a stimulation rate, which could have caused a saturation of the acoustic system and led to the rise of the somatosensory component.

No other studies supporting either position have been published to date, so this issue is far from resolved. Both the mode of analysis of seismic signals by mole rats and the mechanism of signal reception (bone conduction? general somatosensory reception?) remain unclear. Further studies of this issue, involving additional taxa, are needed to clarify the structures and processes used in the reception of seismic signals.

TAXONOMIC DISTRIBUTION AND FUNCTIONAL SIGNIFICANCE OF SEISMIC SIGNALS. What determines whether a species uses acoustic signals, seismic signals, or both? As this review indicates, some species appear to use only acoustic signals (e.g., *H. glaber:* Pepper et al. 1991). Others use a combination of short-range acoustic signals and long-range seismic signals (e.g., *N. ehrenbergi:* Heth, Frankenberg, and Nevo 1986; Heth et al. 1987). To date, however, there have been no reports of subterranean species that employ only seismic communication (see table 3.1). Seismic signals have arisen in at least two distinct lineages of subterranean rodents (Bathyergidae and Spalacinae), indicating that this mode of communication cannot be explained solely on the basis of phylogenetic history. Thus understanding the ecological and evolutionary contexts in which each signal type has been preferred remains a significant challenge to biologists studying communication in subterranean rodents.

Some authors have attempted to link different signal types to specific biological parameters. Heffner and Heffner (1992a, 1993) suggested that both auditory acuity and the ability to localize sounds are synergistically related to visual performance. Thus the use of seismic signals should be greatest in species with poor visual abilities. Alternatively, Narins et al. (1992) emphasized the importance of social behavior to the type(s) of vibrational signals used by subterranean rodents, suggesting that solitary taxa are more likely to exhibit seismic communication than social taxa. Because seismic signals propagate better than vocal signals, the former should be favored in solitary species, in which communication occurs primarily between different burrow systems (see also Burda, Bruns, and Müller 1990; Nevo, Heth, and Pratt 1991).

Given that auditory and visual acuity work together in the location and identification of external objects (e.g., rivals, predators: Heffner and Heffner 1993), it is interesting that subterranean rodents known to have poor visual acuity rarely leave their burrows and commonly utilize seismic signals (see table 3.1). The only exception to this generalization is the highly social *H. glaber,* which apparently lacks seismic communication. In contrast, subterranean species that leave their burrows with greater frequency (e.g., *S. cyanus:* Reig 1970; some *Ctenomys:* Altuna 1991; *T. splendens:* Flynn 1990; Busch et al., chap. 5, this volume) have better visual acuity and are not known to use seismic signals. Although the auditory abilities of these species have not been documented, anecdotal accounts of responses to noise suggest that at least *Spalacopus* and *Ctenomys* are quite good at localizing sounds (Pearson 1959; Reig 1970; G. Francescoli, personal observation).

Although the social systems of subterranean rodents vary, most researchers employ a dichotomous classification scheme that categorizes a species as either solitary or social (Lacey, chap. 7, this volume). According to Narins et al. (1992), social species such as *H. glaber, C. damarensis,* and *C. hottentotus* should rely primarily upon acoustic signals, or vocalizations, to communicate with conspecifics. As evident from table 3.1, seven of the eight subterranean taxa known to contain social species use vocal signals (data for *Ellobius* are lacking). In particular, *H. glaber,* the social *Cryptomys,* and *S. cyanus* have remarkably large acoustic repertoires and, apparently, do not use seismic signals. Seismic signals also appear to be lacking in *Ctenomys* and *C. mechowi.* Anecdotal reports of thumping exist for *C. hottentotus* and *C. damarensis,* although no studies of these species have demonstrated the use of seismic signals. Thus social subterranean species do appear to have extended vocal repertoires, as predicted by Narins et al. (1992). Not all solitary species, however, use seismic communication; some of these species use vocal signals that can be heard outside their tunnels (i.e., *C. pearsoni:* Francescoli 1992; G. Francescoli, unpublished data). These observations suggest that social system alone is not sufficient to explain the taxonomic distribution of seismic signals.

Undoubtedly, different sensory systems interact to determine the mix of signals used by a particular species (Jay and Sparks 1984; Knudsen and Knudsen 1985). For example, a well-developed ability to localize sounds that evolves for reasons independent of intraspecific communication (e.g., predator detection) may predispose a species to use acoustic signals. Assuming that each mode of communication entails some cost to the organism, the need to maintain acute auditory skills may constrain the elaboration of other signal modes, including seismic signaling.

When the available information is synthesized and data on social organization, visual and auditory acuity, type and size of vibrational signal repertoires, and frequency of aboveground activity are combined, the following patterns appear to emerge:

1. Social species use acoustic communication because of their reduced need for long-distance communication.
2. Solitary species that rarely, if ever, leave their burrows use acoustic signals for short-distance communication and seismic signals for long-distance communication.
3. Solitary species that are often active above ground have good visual and hearing abilities and, as a by-product of the need to maintain these sensory skills, use vocal communication.

These generalizations lead to the following predictions:

1. *H. glaber, Ctenomys sociabilis, S. cyanus,* social *Cryptomys,* and *Ellobius* use vocal communication.
2. *Nannospalax ehrenbergi* and perhaps *Tachyoryctes, Georychus, Bathyergus,* and *Heliophobius* use a mix of vocal and seismic signals.
3. The majority of *Ctenomys* and, possibly, *Thomomys* rely upon vocal communication.

Given these patterns and predictions, a primary goal of studies of interspecific differences in signal modality should be the elucidation of the selective pressures affecting communication in these animals. Only after these pressures have been enumerated can we truly begin to make sense of the functional significance and taxonomic distribution of seismic and acoustic communication.

Overview: Is a General Picture Possible?

In this review, I have attempted to sketch a general picture of communication in subterranean rodents. Clearly, much additional research is needed to characterize this aspect of the biology of subterranean species. This research should include descriptive studies of the signals employed by different taxa, as well as quantitative analyses of the sensory abilities of these animals. These data, in turn, will provide the basis for comparative studies of the ecological and evolutionary conditions that shape a species' signal repertoire.

Based on current evidence, it appears that tactile and vibrational signals are important modes of communication in subterranean rodents. Tactile and vibrational communication can be viewed as endpoints along a "distance continuum," with tactile signals used for close-range interactions and vibrational signals (acoustic or seismic) for longer-range interactions. Although less well characterized, tactile signals are likely to be more widespread, since all subterranean species engage in close physical contact during sexual encounters. In contrast, chemical signals seem to be rare, and specialized glands appear to be underused, in comparison to other mammals.

With regard to vibrational communication, it appears that seismic signals are used by only some taxa, making generalizations about this form of communication more difficult. Interestingly, in all subterranean rodents in which seismic signals are produced by thumping a body part against the tunnel walls or ceiling, the body part used is the same as that employed during digging (Nevo 1991; G. Francescoli and C. A. Altuna, unpublished data), suggesting that seismic signals may have evolved as an offshoot of digging mode. Not all species that use seismic signals dig in the same way (Stein, chap. 1, this volume), and thus this issue deserves more research.

Acoustic signals are widespread among subterranean rodents, and at least some species (e.g., *H. glaber*) possess very rich vocal repertoires. Strikingly, agonistic information (e.g., territory ownership) is conveyed via low-frequency, repetitive rhythmic signals, either calls (*C. pearsoni:* Francescoli 1992; *N. ehrenbergi:* Heth, Frankenberg, and Nevo 1988; *H. glaber:* Pepper et al. 1991) or thumps (*N. ehrenbergi:* Heth et al. 1987, 1991; *Georychus capensis:* Narins et al. 1992). The correlation between low-frequency sounds and ago-

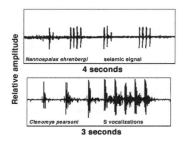

FIGURE 3.4. Comparison of oscillograms representing seismic and vocal signals. For *N. ehrenbergi*, the oscillogram shows bursts of head thumping, with every "spike" representing a single head thump (from Rado et al. 1987). For *Ctenomys*, the oscillogram shows an S signal series vocalization, in which every "spike" represents a single vocal note.

nistic messages was noted by Morton (1977), and its origin was related to the relationship between body size, resonant structure size, and sound frequency. Alternatively, the use of low-frequency rhythmic signals may be a product of the propagation conditions of the environment, which may make rhythm more reliable than frequency modulation in a spatio-aggressive scenario. The latter explanation may account for the interspecific structural similarities observed among some acoustic and seismic signals (fig. 3.4).

One topic that has not yet been addressed in this review is signal redundancy. At present, investigation of this topic is hampered by the limited number of species for which detailed data on all modes of communication are available. Nevertheless, it seems likely that multiple signal types are used simultaneously in some contexts, such as during courtship and mating. Although each signal may bear unique information that elicits a distinct response, it is also possible that some redundancy in signal content occurs, particularly for signals used in critical behavioral contexts such as mating. The reasons for this redundancy are likely to be closely linked to the functional significance of the signals in question, and thus both issues represent important topics for future research.

Summary and Future Directions

In short, almost all aspects of communication in subterranean rodents require further study. Evidence acquired to date suggests that tactile and vibrational signals are the predominant modes of communication in these animals, and that visual and chemical signals are relatively less important. Species using acoustic signals appear to outnumber those using both acoustic and seismic signals. Why seismic signals are not used more widely remains to be determined, although both social system and extent of aboveground activity have been identified as contributing to the taxonomic distribution of seismic communication.

These generalizations are based on data from only a few well-studied species, such as *N. ehrenbergi* and *H. glaber.* To assess the validity of these generalizations, a more complete picture of communication in subterranean rodents is needed. Complete signal repertoires should be compiled for all subterranean rodents, including complete listings of the visual, olfactory, tactile, and vibrational signals used by the species considered in this volume.

Once this descriptive foundation has been laid, the following more specialized questions can be addressed:

1. What are the relative selective benefits of the different modes of communication used by subterranean species? Can consistent, cross-taxon correlations be identified that shed light on the ecological and evolutionary factors favoring these different forms of communication?
2. To what extent have the physical and ecological conditions present in subterranean burrows favored one mode of communication over another? Although the role of the subterranean niche in constraining visual communication is relatively clear, the effects of subterranean burrows on other modes of communication are less evident. How has life underground influenced these other modes of communication, and has it led to signal or receptor specializations not evident in surface-dwelling taxa?
3. To what extent are different modes of communication interrelated? In other words, does specialization for one form of communication inhibit specialization for other forms, or do certain modes of communication tend to evolve in concert?
4. What ecological and other factors underlie the apparent interspecific differences in communication among subterranean rodents? Given that these animals share a number of environmental challenges associated with life underground, why do the signals used in communication differ so greatly among species? In other words, what has caused the divergence in signal repertoires evident in these animals?

Addressing these questions represents a significant challenge that will require the research efforts of numerous biologists. Given our current paucity of knowledge regarding communication in subterranean rodents, almost any of the species considered in this volume would provide an appropriate starting point. No doubt many fascinating aspects of animal communication will be revealed in our attempts to answer the broad-scale, comparative questions outlined above.

Acknowledgments

I wish to thank the editors of this volume for the invitation to contribute to it, and especially Eileen Lacey for her support, English improvement, and strong editing help. I also wish to thank Carlos Altuna for helping with literature search and for discussion of some ideas.

The discussion of the taxonomic distribution of seismic signaling is an abridged version of a manuscript in preparation by Francescoli and Altuna.

Literature Cited

Altuna, C. A. 1991. Microclima de cuevas y comportamientos de homeostasis en una población del grupo *Ctenomys pearsoni* del Uruguay. *Boletín de la Sociedad Zoológica del Uruguay* (2a. época) 6:35–46.
Altuna, C. A., and S. Corte. 1987. La glándula perineal de *Ctenomys pearsoni* y *Ctenomys rionegrensis* (Rodentia, Octodontidae) del Uruguay. *Brenesia* 28:33–39.
Altuna, C. A., G. Francescoli, and G. Izquierdo. 1988. Análisis preliminar del comportamiento copulatorio de *Ctenomys pearsoni* (Rodentia, Octodontidae) del Uruguay. Abstracts, V Reunión Iberoamericana de Conservación y Zoología de Vertebrados, 62.
———. 1991. Copulatory pattern of *Ctenomys pearsoni* (Rodentia, Octodontidae) from Balneario Solís, Uruguay. *Mammalia* 55:316–18.
Andersen, D.C. 1978. Observations on reproduction, growth, and behavior of the northern pocket gopher *(Thomomys talpoides)*. *Journal of Mammalogy* 59:418–22.
Ashmore, J. F., and I. J. Russell. 1983. The physiology of hair cells. In *Bioacoustics: A comparative approach,* edited by B. Lewis, 149–80. London: Academic Press.
Bee de Speroni, N. 1995. Encefalización y tamaño relativo de los componentes encefálicos en *Ctenomys mendocinus* Philippi 1869 (Rodentia: Ctenomyidae). *Mastozoología Neotropical* 2:31–38.
Bennett, N. C., and J. U. M. Jarvis. 1988a. The reproductive biology of the Cape mole-rat, *Georychus capensis* (Rodentia, Bathyergidae). *Journal of Zoology, London* 214:95–106.
———. 1988b. The social structure and reproductive biology of colonies of the mole-rat *Cryptomys damarensis* (Rodentia, Bathyergidae). *Journal of Mammalogy* 69:293–302.
Berghard, A., L. B. Buck, and E. R. Liman. 1996. Evidence for distinct signaling mechanisms in two mammalian olfactory sense organs. *Proceedings of the National Academy of Sciences, USA* 93:2365–69.
Bronchti, G., P. Heil, H. Scheich, and Z. Wollberg. 1989. Auditory pathway and auditory activation of primary visual targets in the blind mole rat *(Spalax ehrenbergi):* I. 2-Deoxyglucose study of subcortical centers. *Journal of Comparative Neurology* 284:253–74.
Brückmann, G., and H. Burda. 1997. Hearing in blind subterranean Zambian mole-rats (*Cryptomys* sp.): Collective behavioural audiogram in a highly social rodent. *Journal of Comparative Physiology* A 181:83–88.
Bruns, V., M. Müller, W. Hofer, G. Heth, and E. Nevo. 1988. Inner ear structure and electrophysiological audiograms of the subterranean mole rat, *Spalax ehrenbergi*. *Hearing Research* 33:1–10.
Burda, H. 1989. Reproductive biology (behavior, breeding, and postnatal development) in subterranean mole-rats, *Cryptomys hottentotus* (Bathyergidae). *Zeitschrift für Säugetierkunde* 54:360–76.
Burda, H., V. Bruns, and M. Müller. 1990. Sensory adaptations in subterranean mammals.

In *Evolution of subterranean mammals at the organismal and molecular levels,* edited by E. Nevo and O. Reig, 269–93. Progress in Clinical and Biological Research, vol. 335. New York: Wiley-Liss.

Burda, H., V. Bruns, and E. Nevo. 1989. Middle ear and cochlear receptors in the subterranean mole–rat, *Spalax ehrenbergi. Hearing Research* 39:225–30.

Burda, H., E. Nevo, and V. Bruns. 1990. Adaptive differentiation of ear structures in subterranean mole-rats of the *Spalax ehrenbergi* superspecies in Israel. *Zoologische Jahrbücher, Abteilung für Systematik, Ökologie, und Geographie der Tiere* 117:369–82.

Capranica, R. R., A. J. M. Moffat, and E. Nevo. 1973. Vocal repertoire of a subterranean rodent *(Spalax).* Abstract, Acoustical Society of America Annual Meeting, October.

Cook, J. A., S. Anderson, and T. L. Yates. 1990. Notes on Bolivian mammals. 6. The genus *Ctenomys* (Rodentia, Ctenomyidae) in the Highlands. *American Museum Novitates* 2980:1–27.

Cooper, H. M., M. Herbin, and E. Nevo. 1993a. Ocular regression conceals adaptive progression of the visual system in a blind subterranean mammal. *Nature* 361:156–59.

———. 1993b. Visual system of a naturally microphthalmic mammal: The blind mole rat, *Spalax ehrenbergi. Journal of Comparative Neurology* 328:313–50.

Cooper, H. M., M. Herbin, E. Nevo, and J. Negroni. 1995. Neuroanatomical consequences of microphthalmia in mammals. In *Les Séminaires ophtalmologiques d'IPSEN,* tome 6, *Vision et adaptation,* edited by Y. Christen, M. Doly, and M. T. Droy-Lefaix, 127–39. Paris: Elsevier.

Credner, S., H. Burda, and F. Ludescher. 1997. Acoustic communication underground: Vocalization characteristics in subterranean social mole-rats (*Cryptomys* sp., Bathyergidae). *Journal of Comparative Physiology* A 180:245–55.

Dubost, G. 1968. Les mammifères souterrains. *Revue d'écologie et de biologie du sol* 5:99–197.

Dusenbery, D. B. 1992. *Sensory ecology.* New York: W. H. Freeman.

Ehret, G. 1989. Hearing in the mouse. In *The comparative psychology of audition,* edited by R. J. Dooling and S. H. Hulse, 3–32. Hillsdale, NJ: Lawrence Erlbaum Associates.

Eisenberg, J. F. 1974. The function and motivational basis of hystricomorph vocalizations. In *The biology of hystricomorph rodents,* edited by I. W. Rowlands and B. J. Weir, 211–47. Symposia of the Zoological Society of London, 34. London: Academic Press.

Eisenberg, J. F., and D. G. Kleiman. 1972. Olfactory communication in mammals. *Annual Review of Ecology and Systematics* 3:1–32.

———. 1977. Communication in lagomorphs and rodents. In *How animals communicate,* edited by T. A. Sebeok, 634–54. Bloomington: Indiana University Press.

Endler, J. A. 1993. Some general comments on the evolution and design of animal communication systems. *Philosophical Transactions of the Royal Society of London* B 340:215–25.

Faulkes, C. G., and D. H. Abbott. 1991. Social control of reproduction in breeding and non- breeding male naked mole-rats *(Heterocephalus glaber). Journal of Reproduction and Fertility* 93:427–35.

———. 1993. Evidence that primer pheromones do not cause social suppression of reproduction in male and female naked mole-rats *(Heterocephalus glaber). Journal of Reproduction and Fertility* 99:225–30.

Faulkes, C. G., D. H. Abbott, and J. U. M. Jarvis. 1991. Social suppression of reproduction in male naked mole-rats *Heterocephalus glaber. Journal of Reproduction and Fertility* 91:593–604.

Faulkes, C. G., D. H. Abbott, C. E. Liddell, L. M. George, and J. U. M. Jarvis. 1991. Hormonal and behavioral aspects of reproductive suppression in female naked mole-rats. In *The biology of the naked mole-rat,* edited by P. W. Sherman, J. U. M. Jarvis, and R. D. Alexander, 426–45. Princeton, NJ: Princeton University Press.

Feldman, J. L., and C. J. Phillips. 1984. Comparative retinal pigment epithelium and photoreceptor ultrastructure in nocturnal and fossorial rodents: The eastern woodrat, *Neo-*

toma floridana, and the Plains pocket gopher, *Geomys bursarius. Journal of Mammalogy* 65:231–45.
Flynn, L. J. 1990. The natural history of rhizomyid rodents. In *Evolution of subterranean mammals at the organismal and molecular levels*, edited by E. Nevo and O. Reig, 155–83. Progress in Clinical and Biological Research, vol. 335. New York: Wiley-Liss.
Francescoli, G. 1992. Aportes al estudio sistematizado y analítico de la comunicación acústica en el género *Ctenomys* (Rodentia, Octodontidae). Actas de las III Jornadas de Zoología del Uruguay. *Boletín de la Sociedad Zoológica del Uruguay* (2a. época) 7:47–48.
———. 1995. Las señales acústicas en *Ctenomys* (Octodontidae) de Uruguay. *X Jornadas Argentinas de Mastozoología, Resúmenes*, 94.
———. 1997. Contact vocal signals in *Ctenomys pearsoni* pups. Unpublished abstracts, XXV International Ethological Conference (Austria), 12.
Francescoli, G., and C. A. Altuna. 1996. "Acoustic lordosis" in Uruguayan *Ctenomys* (Rodentia, Octodontidae). Abstracts, Thirty-third annual meeting of the Animal Behavior Society (USA).
Hansen, R. M., and V. H. Reid. 1973. Distribution and adaptations of pocket gophers. In *Pocket gophers and Colorado mountain rangeland*, edited by G. T. Turner, R. M. Hansen, V. H. Reid, H. P. Tietjen, and A. L. Ward, 1–19. Bulletin 5545, Colorado State University Agricultural Experiment Station, Fort Collins.
Hauser, M. D. 1996. *The evolution of communication*. Cambridge, MA: MIT Press.
Heffner, R. S., and H. E. Heffner. 1990. Vestigial hearing in a fossorial mammal, the pocket gopher *(Geomys bursarius)*. *Hearing Research* 46:239–52.
———. 1992a. Evolution of sound localization in mammals. In *The evolutionary biology of hearing*, edited by D. B. Webster, R. R. Fay, and A. N. Popper, 691–715. New York: Springer Verlag.
———. 1992b. Hearing and sound localization in blind mole rats *(Spalax ehrenbergi)*. *Hearing Research* 62:206–16.
———. 1993. Degenerate hearing and sound localization in naked mole rats *(Heterocephalus glaber)*, with an overview of central auditory structures. *Journal of Comparative Neurology* 331:418–33.
Herbin, M., J. Reperant, and H. M. Cooper. 1994. Visual system of the fossorial Mole-lemmings, *Ellobius talpinus* and *Ellobius lutescens*. *Journal of Comparative Neurology* 346: 253–75.
Heth, G., E. Frankenberg, and E. Nevo. 1986. Adaptive optimal sound for vocal communication in tunnels of a subterranean mammal *(Spalax ehrenbergi)*. *Experientia* 42: 1287–89.
———. 1988. "Courtship" call of subterranean mole rats *(Spalax ehrenbergi):* Physical analysis. *Journal of Mammalogy* 69:121–25.
Heth, G., E. Frankenberg, H. Pratt, and E. Nevo. 1991. Seismic communication in the blind subterranean mole-rat: Patterns of head thumping and their detection in the *Spalax ehrenbergi* superspecies in Israel. *Journal of Zoology, London* 224:633–38.
Heth, G., E. Frankenberg, A. Raz, and E. Nevo. 1987. Vibrational communication in subterranean mole rats *(Spalax ehrenbergi)*. *Behavioral Ecology and Sociobiology* 21:31–33.
Heth, G., E. Nevo, and J. Todrank. 1996. Seasonal changes in urinary odours and in responses to them by blind subterranean mole rats. *Physiology and Behavior* 60:963–68.
Heth, G., and J. Todrank. 1995. Assessing chemosensory perception in subterranean mole rats: Different responses to smelling versus touching odorous stimuli. *Animal Behaviour* 49:1009–15.
———. 1997. Patterns of urination of a blind subterranean rodent, *Spalax ehrenbergi*. *Ethology* 103:138–48.
Hickman, G. C. 1982. Copulation of *Cryptomys hottentotus* (Bathyergidae): A fossorial rodent. *Mammalia* 46:293–98.

Hooper, E. T. 1968. Anatomy of middle-ear walls and cavities in nine species of microtine rodents. Occasional papers, Museum of Zoology, University of Michigan, no. 657:1–28.

Jarvis, J. U. M., and N. C. Bennett. 1990. The evolutionary history, population biology and social structure of African mole-rats: Family Bathyergidae. In *Evolution of subterranean mammals at the organismal and molecular levels,* edited by E. Nevo and O. Reig, 97–128. Progress in Clinical and Biological Research, vol. 335. New York: Wiley-Liss.

———. 1991. Ecology and behavior of the family Bathyergidae. In *The biology of the naked mole-rat,* edited by P. W. Sherman, J. U. M. Jarvis, and R. D. Alexander, 66–96. Princeton, NJ: Princeton University Press.

Jarvis, J. U. M., and J. B. Sale. 1971. Burrowing and burrow patterns of East African mole-rats *Tachyoryctes, Heliophobius* and *Heterocephalus. Journal of Zoology, London* 163:451–513.

Jay, M. F., and D. L. Sparks. 1984. Auditory receptive fields in primate superior colliculus shift with changes in eye position. *Nature* 309:345–47.

Judd, T. M., and P. W. Sherman. 1996. Naked mole-rats recruit colony mates to food sources. *Animal Behaviour* 52:957–69.

Knudsen, E. I., and P. F. Knudsen. 1985. Vision guides the adjustment of auditory localization in young Barn Owls. *Science* 230:545–48.

Lacey, E. A., R. D. Alexander, S. H. Braude, P. W. Sherman, and J. U. M. Jarvis. 1991. An ethogram for the naked mole-rat: Non-vocal behaviors. In *The biology of the naked mole-rat,* edited by P. W. Sherman, J. U. M. Jarvis, and R. D. Alexander, 209–42. Princeton, NJ: Princeton University Press.

Lacey, E. A., S. H. Braude, and J. R. Wieczorek. 1997. Burrow sharing by colonial tuco-tucos *(Ctenomys sociabilis). Journal of Mammalogy* 78:556–562.

Lay, D. M. 1972. The anatomy, physiology, functional significance and evolution of specialized hearing organs of Gerbillinae rodents. *Journal of Morphology* 138:41–120.

Mann, M. D., G. Rehkämper, H. Reinke, H. D. Frahm, R. Necker, and E. Nevo. 1997. Size of somatosensory cortex and of somatosensory thalamic nuclei of the naturally blind mole rat, *Spalax ehrenbergi. Journal of Brain Research* 38:47–59.

Morton, E. S. 1977. On the occurrence and significance of motivation-structural rules in some bird and mammal sounds. *American Naturalist* 111:855–69.

Müller, M., and H. Burda. 1989. Restricted hearing range in a subterranean rodent, *Cryptomys hottentotus. Naturwissenschaften* 76:134–35.

Narins, P. M., O. J. Reichman, J. U. M. Jarvis, and E. R. Lewis. 1992. Seismic signal transmission between burrows of the Cape mole-rat, *Georychus capensis. Journal of Comparative Physiology* A 170:13–21.

Necker, R., G. Rehkämper, and E. Nevo. 1992. Electrophysiological mapping of body representation in the cortex of the blind mole rat. *Neuroreport* 3:505–8.

Nevo, E. 1991. Evolution des communications vocales et vibratoires chez les rats-taupiers *Spalax:* Structure et fonction. In *Le rongeur et l'espace,* edited by M. Le Berre and L. Le Guelte, 15–34. Paris: Chabaud.

———. 1996. Adaptive eye and brain evolution of photoperiodic perception in the blind subterranean rodent, *Spalax.* Abstract, Circadian Photoreception Conference, Lyon, France.

Nevo, E., M. Bodmer, and G. Heth. 1976. Olfactory discrimination as an isolating mechanism in speciating mole rats. *Experientia* 32:1511–12.

Nevo, E., G. Heth, A. Beiles, and E. Frankenberg. 1987. Geographic dialects in blind mole rats: Role of vocal communication in active speciation. *Proceedings of the National Academy of Sciences, USA* 84:3312–15.

Nevo, E., G. Heth, and H. Pratt. 1991. Seismic communication in a blind subterranean mammal: A major somatosensory mechanism in adaptive evolution underground. *Proceedings of the National Academy of Sciences, USA* 88:1256–60.

O'Riain, M., andJ. U. M.Jarvis. 1997. Colony member recognition and xenophobia in the naked mole-rat. *Animal Behaviour* 53:487–98.
Orofino, A. G., S. M. Giannoni, V. G. Roig, and C. B. Borghi. 1997. Uso de la cola por *Ctenomys* para obtener información espacial. *XII Jornadas Argentinas de Mastozoología (Mendoza)*, 93.
Pearson, O. P. 1959. Biology of the subterranean rodents, *Ctenomys*, in Peru. *Memorias del Museo de Historia Natural "Javier Prado"* 9:3–56.
Pearson O. P., and M. I. Christie. 1985. Los tuco-tucos (género *Ctenomys*) de los Parques Nacionales Lanin y Nahuel Huapi, Argentina. *Historia Natural (Corrientes)* 5:337–43.
Pepper, J. W., S. H. Braude, E. A. Lacey, and P. W. Sherman. 1991. Vocalizations of the naked mole-rat. In *The biology of the naked mole-rat*, edited by P. W. Sherman, J. U. M. Jarvis, and R. D. Alexander, 243–74. Princeton, NJ: Princeton University Press.
Pickles, J. O. 1988. *An introduction to the physiology of hearing*. London: Academic Press.
Pirlot, P., and E. Nevo. 1989. Brain organization and evolution in subterranean mole rats. *Zeitschrift für Zoologische Systematik und Evolutionsforschung* 27:58–64.
Rado, R., M. Himelfarb, B. Arensburg, J. Terkel, and Z. Wollberg. 1989. Are seismic communication signals transmitted by bone conduction in the blind mole rat? *Hearing Research* 41:23–30.
Rado, R., N. Levi, H. Hauser, J. Witcher, N. Adler, N. Intrator, Z. Wollberg, and J. Terkel. 1987. Seismic signalling as a means of communication in a subterranean mammal. *Animal Behaviour* 35:1249–51.
Rado, R., J. Terkel, and Z. Wollberg. 1998. Seismic communication signals in the blind mole-rat *(Spalax ehrenbergi):* Electrophysiological and behavioral evidence for their processing by the auditory system. *Journal of Comparative Physiology* A 183:503–11.
Randall, J. A. 1994. Convergences and divergences in communication and social organisation of desert rodents. *Australian Journal of Zoology* 42:405–33.
Reichman, O. J., T. G. Whitham, and G. A. Ruffnes. 1982. Adaptive geometry of burrow spacing in two pocket gopher populations. *Ecology* 63:687–95.
Reig, O. A. 1970. Ecological notes on the fossorial octodont rodent *Spalacopus cyanus* (Molina). *Journal of Mammalogy* 51:592–601.
Reig, O. A., C. Busch, M. O. Ortells, and J. R. Contreras. 1990. An overview of evolution, systematics, population biology, cytogenetics, molecular biology and speciation in *Ctenomys*. In *Evolution of subterranean mammals at the organismal and molecular levels*, edited by E. Nevo and O. Reig, 71–96. Progress in Clinical and Biological Research, vol. 335. New York: Wiley-Liss.
Riba, C. 1990. *La comunicación animal: Un enfoque zoosemiótico*. Barcelona: Anthropos.
Sanyal, S., H. G. Jansen, W. J. de Grip, E. Nevo, and W. W. de Jong. 1990. The eye of the blind mole rat, *Spalax ehrenbergi:* Rudiment with hidden function? *Investigative Ophthalmology and Visual Science* 31:1398–1404.
Schramm, P. 1961. Copulation and gestation in the pocket gopher. *Journal of Mammalogy* 42:167–70.
Shanas, U., and J. Terkel. 1997. Mole-rat Harderian gland secretions inhibit aggression. *Animal Behaviour* 54:1255–63.
Sherman, P. W., J. U. M. Jarvis, and R. D. Alexander, eds. 1991. *The biology of the naked mole-rat*. Princeton, NJ: Princeton University Press.
Shorey, H. H. 1977. Pheromones. In *How animals communicate*, edited by T. A. Sebeok, 137–63. Bloomington: Indiana University Press.
Smith, W. J. 1982. *Etología de la comunicación*. Mexico: Fondo de Cultura Universitaria.
Sullivan, T. P., D. R. Crump, H. Wizner, and E. A. Dion. 1990. Response of pocket gophers *(Thomomys talpoides)* to an operational application of synthetic semiochemicals of stoat *(Mustela erminea)*. *Journal of Chemical Ecology* 16:941–49.

Tinbergen, N. 1963. On aims and methods of ethology. *Zeitschrift für Tierpsychologie* 20: 410–33.
Todrank, J., and G. Heth. 1996. Individual odours in two chromosomal species of blind, subterranean mole rat *(Spalax ehrenbergi):* Conspecific and cross-species discrimination. *Ethology* 102:806–11.
Webster, D. B., and M. Webster. 1975. Auditory systems of Heteromyidae: Functional morphology and evolution of the middle ear. *Journal of Morphology* 146:343–76.
Weir, B. J. 1974. The tuco-tuco and plains viscacha. In *The biology of hystricomorph rodents*, edited by I. W. Rowlands and B. J. Weir, 113–30. Symposia of the Zoological Society of London, 34. London: Academic Press.
Wilson, E. O. 1980. *Sociobiología*. Barcelona: Omega.

CHAPTER FOUR

Reproduction in Subterranean Rodents

Nigel C. Bennett, Chris G. Faulkes, and Andrew J. Molteno

Reproduction is central to the biology of all organisms. Reproduction is the means by which individuals perpetuate copies of their genes, and it is through differences in reproductive success that natural selection acts to shape phenotypes. Given the fundamental role of reproduction in organismal biology, there has been surprisingly little research on the reproductive biology of subterranean rodents. Basic reproductive parameters, such as duration and frequency of pregnancy, physical dimensions of reproductive organs, and patterns of follicular development and sperm production, have been quantified for the Ctenomyidae (Reig et al. 1990; Malizia and Busch 1991), Geomyidae (Dixon 1929; Miller 1946; Gunther 1956; Smolen, Genoways, and Baker 1980; Patton and Smith 1990; Villa-Cornejo and Engeman 1994, 1995), Rhizomyinae (Jarvis 1973; Flynn 1990), Bathyergidae (Jarvis 1969; van der Horst 1972; Bennett and Jarvis 1988a), and Spalacinae (Redi et al. 1986; Simson, Lavie, and Nevo 1993). However, longitudinal studies of parameters such as courtship, copulation, gestation, parturition, and pup development are considerably fewer in number (see Bennett et al. 1991 for review).

The most thorough study of reproduction in subterranean rodents has been conducted on species from the family Bathyergidae (Jarvis 1969, 1981; Bennett 1989; Bennett and Jarvis 1988a,b; Bennett, Jarvis et al. 1994; Bennett and Aguilar 1995), a number of which display marked patterns of reproductive differentiation among members of the same social group. In this chapter, we use the bathyergids to highlight the paucity of information about reproduction in other families and subfamilies. We summarize the data available for each group, with an emphasis on identifying species for which knowledge of reproductive biology is particularly lacking.

Reproduction in the Subterranean World

Reproduction in subterranean rodents is constrained by the burrow environment. Postmortem examinations have shown that many subterranean rodents exhibit a seasonal breeding pattern (Bennett and Jarvis 1988a; Bennett et al. 1991). The fossorial nature of truly subterranean animals may preclude their use of the most common proximate cue used by seasonal breeders, namely, photoperiod (but see Buffenstein, chap. 2, this volume). Environmental cues triggering the onset of breeding could be thermal, resulting from annual changes in burrow temperature (Bennett, Jarvis, and Davies 1988), or nutritional, resulting from seasonal increases in the availability of geophytes (underground storage organs of plants). Seasonal rainfall results in softening of the soil, thereby facilitating the extension of existing burrows and enabling opposite-sex conspecifics to come together and procreate before drier soils and increased energetic costs again present physical limitations to burrowing.

Social Systems and Reproduction

Subterranean rodents are typically solitary and highly xenophobic, aggressively defending their burrow systems against conspecifics (Nevo 1979). Of the seven families and subfamilies of subterranean rodents, most contain primarily solitary species (Lacey, chap. 7, this volume). In many of these solitary species, reproduction is a brief event, during which strong barriers to aggression must be overcome. Once a male has entered a female's burrow, he must court, mate, and leave her burrow system without being injured (Bennett and Jarvis 1988a). Reproductive males in social species of subterranean rodents face no such pressures, and can mate at leisure in the confines of their own relatively secure burrow. In solitary species, courtship and copulation are usually assumed to be initiated by the male (Nevo 1979; Bennett and Jarvis 1988a), although evidence for female choice exists for pocket gophers of the genus *Thomomys* (Patton and Smith 1993). In social species of subterranean rodents, courtship is often initiated by the female (Jarvis 1991; Bennett and Jarvis 1988b).

In solitary species, plural occupancy of burrows occurs only briefly during the breeding season or when the female has young (Bennett and Jarvis 1988a; Altuna, Francescoli, and Izquierdo 1991; Lacey, chap. 7, this volume). There are reports that the octodontid *Spalacopus cyanus* is colonial, with as many as fifteen individuals in a colony (Reig 1970), although it appears to have a more primitive social organization than the colonial Bathyergidae. Further research is required to dismiss the possibility of *S. cyanus* merely being gregarious. Lacey, Braude and Wieczorek (1997) have shown in their field studies of *Ctenomys sociabilis* that burrow systems are regularly inhabited by multiple adult females and a single adult male. Preliminary studies on the ctenomyid *Ctenomys peruanus* suggest plural occupancy of burrows, but usually by harems of sexually active females in the absence of a male. The male is highly territorial and maintains a strictly solitary existence (Pearson 1959; Pearson and Christie 1985). Members of the Bathyergidae display a broad spectrum of social organization, from strictly solitary to eusocial. However, in the social bathyergid species, prospective mates must locate each other in order to found a new colony. The notable exception is the naked mole-rat, *Heterocephalus glaber*, in which a new queen is typically recruited from within the colony.

Therefore, most subterranean rodent species are confronted with the problems of locating conspecifics and communicating their intention to mate in an environment devoid of visual cues and with limited possibilities for auditory communication. How such communication is achieved has been a matter of some speculation, although few, if any, data are actually available to resolve this puzzle (Francescoli, chap. 3, this volume).

Sexual Dimorphism

Among solitary species of bathyergid mole-rats, no sexual dimorphism in body mass is apparent in *Georychus capensis*. In contrast, *Bathyergus suillus* and *B. janetta* are markedly dimorphic in body size (Taylor, Jarvis, and Crowe 1985; Jarvis and Bennett 1991). Sexual dimorphism can also be extreme in pocket gophers, although in *Thomomys* the degree of dimorphism is related to habitat quality, and thus growth differentials between the sexes (Patton and Smith 1990). Reichman, Whitham, and Ruffner (1982) suggested that in *T. bottae*, only the largest and most aggressive males mate and are polygynous, a conclusion that is not supported by the few genetic paternity data available (Patton and Feder 1981; Daly and Patton 1990). It is quite plausible that male body size is important in the solitary *Bathyergus*. Indeed, *B. suillus* males are characterized by a thick (1 cm) pad of skin on the ventral surface of the neck (Davies and Jarvis 1986), which may offer protection during fights. Badly injured *B. suillus* males have been found in the field with broken incisors and deep wounds around the head indicative of incisor punctures from a conspecific (Jarvis and Bennett 1991). Additional supportive data comes from the finding of two interlocking skulls of *B. janetta*. The incisors of one mole-rat had breached the nasal bone of the other, and evidently the two combatants had died while literally locked together (J. Harrison, personal communication).

Mate Location

The typically solitary nature of subterranean rodents means that for two animals to procreate, they must first locate each other, and the strong barrier of xenophobia and aggression must then be temporarily removed.

Seismic communication plays an important role in the subterranean environment. Seismic signals propagate at least an order of magnitude better than auditory signals through the soil medium (Narins et al. 1992; Francescoli, chap. 3, this volume). Hence solitary species should utilize seismic communication to advertise their sex, status, and intention to breed (Narins et al. 1992; Francescoli, chap. 3, this volume). Seismic communication is achieved through various means. Hind foot drumming has been reported in the Geomyidae (O. J. Reichman, personal communication) and in several bathyergids (*G. capensis, B. suillus, B. janetta, Cryptomys hottentotus,* and *C. damarensis:* Bennett and Jarvis 1988a,b; Jarvis and Bennett 1990, 1991). Incisor tapping has been reported in the rhizomyine *Tachyoryctes splendens* (Jarvis 1969), during which the upper incisors are rapped on the floor of the burrow, thus announcing the animal's presence to a potential mate. In *Nannospalax ehrenbergi,* head drumming appears to be the principal means of communication. The signaling mole-rat taps the upper part of its head against the ceiling of the tunnel (Heth et al. 1987; Rado et al. 1987). The vibrations are transmitted through the soil and sensed by the receiver, who uses the lower jaw to detect them (Rado et al. 1989). There is currently no record of seismic communication in *Ctenomys* (Altuna, Francescoli, and Izquierdo 1991).

During courtship, the initial phase of agonistic behavior may involve auditory communication and so called "long-distance advertisement" before the animals meet. In *G. capensis,* hind foot drumming in the laboratory is initiated by the male. His drumming has a high frequency, with pulses 0.035 seconds in duration, compared with the slower response of the female, with pulses of 0.05 seconds (Bennett and Jarvis 1988a). In the field, the animals' drumming can be heard above the surface nearly 10 m from the source (Bennett and Jarvis 1988a). Social bathyergid species also use tactile and olfactory communication in their mating repertoire since common occupancy of a single burrow makes the use of seismic communication unnecessary.

Courtship and Copulation

Data regarding courtship and copulation are limited to studies of captive animals, due to the inherent difficulty of observing subterranean rodents in their natural environment. In *N. ehrenbergi* (Nevo 1969), *Thomomys* (Andersen 1978; Schramm 1961), *T. splendens* (Jarvis 1969), *G. capensis* (Bennett and Jarvis 1988a), and *Ctenomys pearsoni* (Altuna, Francescoli, and Izquierdo 1991), courtship is usually initiated by the male. Similarly, in the social *C. h. hottentotus*, courtship is initiated by the male when he encounters the receptive estrous reproductive female. Courtship and mating in *C. h. hottentotus* are more elaborate than in the solitary *G. capensis* (in which the observer gets the impression that the animals mate and run). In *C. h. hottentotus*, courtship involves vocalizations from both animals. The female raises her tail, exposing the genital region, which is investigated by the male. The male then takes the female's rump into his mouth and chews gently, occasionally stroking the female's side with his head. After this elaborate courtship, the male mounts the female, restraining her by biting at the back of her neck and thrusting. After a loud squeal at climax, the pair disengage and groom their respective genitalia (Bennett 1989).

In the eusocial Damara mole-rat *(C. damarensis)*, the female typically initiates courtship. In captive studies, upon encountering the reproductive male in the burrow, the reproductive female vocalizes, briefly drums with her hind feet and then mounts the head of the male. The two animals usually enter a chamber and chase one another in a head-to-tail fashion in a tight circle. The female pauses, raises her tail, and adopts the lordosis posture. The male smells her genitalia, mounts, and then mates; there is no postmating chasing (Bennett and Jarvis 1988b; Bennett 1990a,b). This mating sequence occurs frequently for about 10 days and then ceases, indicating that multiple intromissions are necessary for conception and possibly ovulation.

In *H. glaber*, the breeding female is at the top of the dominance hierarchy. She initiates courtship by vocalizing, crouching in front of a breeding male, and adopting the lordosis posture, whereupon the male mounts her (Jarvis 1991). On occasion, mutual anogenital nuzzling and sniffing may also occur prior to or during courtship (Lacey and Sherman 1991). Mating is usually confined to a few hours and occurs 10 days postpartum (Jarvis 1991); multimale paternity of litters is known to occur (Faulkes et al. 1997).

The copulatory behaviors of a number of taxonomically unrelated solitary subterranean rodents are very similar. Many of the species studied show no lock, thrusting during intromission, multiple intromissions, and multiple ejaculations (pattern 9 from Dewsbury 1975). Copulation in *G. capensis* is brief, involving multiple intromissions, during which the female is completely docile and adopts the lordosis posture. Pelvic thrusting becomes more frequent prior to climax (Bennett and Jarvis 1988a). In both *G. capensis* and *T. splendens,* the female chases away the male after mating. In all documented cases of mating described for subterranean rodents, there is repeated mounting punctuated by short periods of autogrooming, particularly around the genitalia. Andersen (1978) suggested that the extended mating sequence generally found in *N. ehrenbergi, Thomomys talpoides,* and *T. bottae* may reflect the relative safety provided by the burrow in underground breeders. However, while the protective confines of the burrow certainly allow extended courtship and copulation, the function of a prolonged mating sequence in organisms renowned for their solitary nature remains unknown.

Paternity and Multiple Matings

In the naked mole-rat and in all species of *Cryptomys* studied to date, courtship and mating are restricted to a single female and one to three male consorts. In the genus *Cryptomys*, there appears to be no aggression among males over reproductive opportunities, although such aggression does occasionally occur in the naked mole-rat. These observations might be expected, because in naked mole-rats, new breeding individuals are recruited from within the colony, whereas Damara mole-rats are obligate outbreeders, which precludes reproductive conflict within family groups (Lacey, chap. 7, this volume).

The possibility of sperm competition following multiple matings has yet to be investigated. We do not know whether litters of species within the social genus *Cryptomys* are the offspring of the founding breeding male, or whether "sneaky" matings by other males result in mixed litters. This question is currently being investigated for both solitary and social species of bathyergids. Some degree of pair-bond formation between the reproductive animals is evident in *C. h. hottentotus,* but is more developed in *C. damarensis,* in which grooming takes place between the two (Bennett 1990a). Similarly, in the naked mole-rat, mutual anogenital sniffing and nuzzling takes place (Bennett and Jarvis 1988b; Bennett 1989; Jarvis 1991). The limited genetic data available for pocket gophers provide no evidence for multiple paternity, although the territories of individual females typically abut those of more than one male (Patton and Feder 1981). Likewise, in *Ctenomys talarum,* there is no multiple paternity of pups within litters, based on DNA fingerprinting data (Zenuto, Lacey, and Busch 1999).

Ovulation

Many solitary subterranean rodents appear to be induced ovulators. Indeed, induced ovulation appears to be the rule in solitary mammals (Zarrow and

Clark 1968). Among the Ctenomyidae, *C. talarum* is known to be an induced ovulator (Weir 1974). Altuna and Lessa (1985) studied the penile morphology of a Uruguayan species of *Ctenomys* and suggested that spines on the glans penis provide cervico-vaginal stimulation necessary for induced ovulation. Shanas et al. (1995) suggested that *N. ehrenbergi* is also an induced ovulator. These authors reported ovulatory failure after a single copulation, raising the possibility that multiple copulations are needed to induce ovulation. Gazit, Shanas, and Terkel (1996) showed that *Nannospalax* displays 7–8 days of courtship behavior prior to mounting. Copulation in this species involves multiple intromissions. After copulation, however, the male displays heightened aggression toward the female. To date, in the social bathyergids, there are no data pertaining to induced ovulation, since the nonreproductive females in colonies are socially suppressed. Faulkes and colleagues (Faulkes, Abbott, and Jarvis 1990; Faulkes et al. 1990) suggested that the naked mole-rat is a spontaneous ovulator, because if a nonbreeding female is removed from a colony and housed singly, she undergoes normal estrous cycles. In solitary species there may have been selection for induced ovulation, which would allow females to take advantage of the chance event of finding a mate during the breeding season. Further investigation is needed to determine whether other solitary subterranean rodents are induced or spontaneous ovulators and whether induced ovulation facilitates fertilization in these otherwise highly xenophobic mammals.

Gestation

Members of the Geomyidae, Spalacinae, and solitary Bathyergidae generally have shorter gestation periods than the Ctenomyidae and social Bathyergidae. Data pertaining to the estimated or actual gestation periods of solitary subterranean rodents are scarce, but they appear to be extremely variable, ranging from as few as 18 days in the Geomyidae to 120 days in the Ctenomyidae (see also Busch et al., chap. 5, this volume). Among pocket gophers, the estimated gestation period is 18–19 days for *T. talpoides* (Andersen 1978; Reid 1973) and 18 (Schramm 1961) to 30 days (Scheffer 1938; Howard and Childs 1959) for *T. bottae*. Wilks (1963) estimates 4 to 5 weeks for *Geomys bursarius*, and Ikenberry (1964) calculates a gestation length of a month for *Pappogeomys castanops*. There is limited information pertaining to gestation in the tuco-tucos of South America. Weir (1974) calculated a gestation length of 93–120 days for *C. talarum*, Pearson (1959) reported a gestation length of 4 months for *C. opimus,* and Camin (cited in Busch et al., chap. 5, this volume) determined gestation length to be 95 days in *C. mendocinus*. Among the Spalacinae, Gazit, Shanas, and Terkel (1996) have calculated the gestation period for *N. ehrenbergi* to be approximately 34 days. Within the Bathyergidae, solitary species have shorter gestations (e.g., *G. capensis:* 44 days; *B. suillus:* ca. 52 days: Bennett et al. 1991) than their smaller social counterparts (e.g., *C. darlingi* and *C. hottentotus:* 56–66 days: Bennett 1989; Bennett, Jarvis, and Cotterill 1994; *C. anselli* and *C. mechowi:* 100–111 days: Bennett and Aguilar 1995; Burda 1989; table 4.1). A long gestation is characteristic of hystricomorph rodents (Weir 1974), and its occurrence in most of the Bathyergidae provides additional support for the argument that the Bathyergidae have strong hystricomorph affinities.

Seasonality of Breeding

Seasonality of reproduction is not limited to either solitary or social species and may be variable within families (see also Busch et al., chap. 5, this volume). There are some solitary species that are highly seasonal breeders, whereas others breed throughout the year; the same holds true for the social bathyergids. Reproduction may be seasonally limited in the solitary species *T. bulbivorus* (Verts and Careraway 1991), *T. talpoides* (Hansen 1960; Andersen and MacMahon 1981), *G. bursarius* (Vaughan 1962; Desy and Druecker 1979), *N. ehrenbergi* (Rado, Wollberg, and Terkel 1992), *T. splendens* (Jarvis 1969), *B. suillus* (van der Horst 1972), *G. capensis* (Bennett and Jarvis 1988a), and *Heliophobius argenteocinereus* (Jarvis 1969). Reproduction is also seasonal in *C. h. hottentotus*. Year-round reproduction has been observed in the solitary geomyid *P. castanops* (Smolen, Genoways, and Baker 1980; Williams and Baker 1976) and several of the social bathyergids, including *C. damarensis* (Bennett and Jarvis 1988b), *C. darlingi* (Bennett, Jarvis, and Cotterill 1994), *C. mechowi* (Bennett and Aguilar 1995), and *H. glaber* (Jarvis 1991).

Interestingly, in the pocket gopher *Orthogeomys hispidus hispidus*, males were found to have the potential for year-round reproduction, whereas the breeding season of females was restricted to a particular time of the year (Villa-Cornejo and Engeman 1995). However, a histological examination of the ovaries of females showed follicular development throughout the year, indicating the potential for year-round breeding. Likewise, in Merriam's pocket gopher *(Pappogeomys merriami merriami)*, the potential for year-round reproduction was demonstrated in both sexes.

Whether species breed seasonally or year-round is in part determined by environmental factors such as rainfall and food availability (King 1927; Wood 1949). The increased friability of soils following periods of rainfall facilitates burrowing, which in turn facilitates food collection as well as access to conspecifics. A sudden flush of green vegetation or increased access

TABLE 4.1 SUMMARY OF REPRODUCTIVE DATA FOR KNOWN SPECIES IN THE FAMILY BATHYERGIDAE

Species	No. in burrow	Breeding season	Gestation (days)	No. litters annually	Litter size (range)	Birth mass (g)	Sex ratio (n)	Maximum annual recruitment per colony	Source
Bathyergus suillus	1	Jul–Oct (winter)	±52	1–2	2.4 (1–4)	34	1:1 (6)	5	Jarvis 1969; Van der Horst 1972; Bennett et al. 1991
Bathyergus janetta	1	Aug–Dec	—	1–2	3.5 (1–7)	15.4	1:3.5 (10)	7	Bennett et al. 1991
Georychus capensis	1	Aug–Dec	44–48	1–2	6 (4–10)	5–12	2.5:1 (8)	12	Bennett and Jarvis 1988a
Heliophobius argenteocinereus	1	Apr–Jun	±87	?	—(2–4)	7	—	—	Jarvis 1969
Cryptomys h. hottentotus	<14	Oct–Jan	59–66	1–2	3 (1–6)	8–9	1:1 (6)	6	Bennett 1988, 1989
Cryptomys damarensis	<41	All year	78–92	1–4	3 (1–5)	8–10	1:1 (23)	12	Bennett and Jarvis 1988b; Jarvis and Bennett 1993
Cryptomys anselli	<25	All year	100	1–3	2 (1–2)	7.8–8.1	1:1 (?)	6	Burda 1989
Cryptomys darlingi	<11	All year	56–61	1–4	1.7 (1–3)	6.9–8.2	1:1	8	Bennett, Jarvis, and Cotterill 1994
Cryptomys mechowi	<11	All year	97–111	1–3	2 (1–3)	15–21	1:1	9	Bennett and Aguilar 1995
Heterocephalus glaber	<115	All year	66–74	4–5	13 (1–27)	1.8	1:1 (>110)	48	Jarvis 1991

Source: Adapted from Bennett et al. 1991

to geophytes may trigger seasonal breeding. These observations suggest that the length of the reproductive season is dictated more by habitat quality and the length of the growing season than by any other factor. For example, in California populations of *T. bottae*, reproduction may be limited to the rainy winter months of January–March in non-irrigated fallow fields, but may extend throughout the entire year in nearby irrigated alfalfa fields (Dixon 1929; Scheffer 1938; Miller 1946; Patton and Smith 1990; Loeb 1990). Moreover, populations of the same species at low elevations will breed during the winter months, while those at high elevations will not breed until the summer, after snowmelt.

Litter Sizes

Litter sizes among subterranean rodents are generally small (Bennett et al. 1991; Malizia and Busch 1991; Busch et al., chap. 5, this volume). The seasonally breeding species typically produce one or occasionally two litters annually (Jarvis 1969; De Graaff 1981; Bennett and Jarvis 1988a; Bennett 1989; Bennett et al. 1991; Malizia and Busch 1991), although the number of litters per female may vary with the length of the season and may be highly variable even among populations of a single species. For example, in the geomyid *T. bottae*, Loeb (1990) compared reproduction in irrigated alfalfa fields and non-irrigated fallow fields near Davis, California. Females in irrigated fields produced nearly twice as many litters per year (3.6–3.9) as those in non-irrigated fields (1.7), generating 15.3–20.4 young per female per year in irrigated fields compared with 6.3–7.1 in non-irrigated ones—a significant annual difference. Litter sizes were slightly higher in irrigated (4.8, range 1–12) than in non-irrigated (4.2, range 1–8) fields, but this difference was not significant.

In geomyids, litter size ranges from 2 to 7 in *T. talpoides* (Hansen 1960; Reid 1973; Andersen and MacMahon 1981), from 1 to 13 in *T. bottae* (Miller 1946; Howard and Childs 1959; Lay 1978; Patton and Brylski 1987), and from 3 to 4 in *T. monticola* (Ingles 1952). Mean litter size is 2.5 for *G. bursarius* (Kennerly 1954), 2.3 for *G. attwateri* (Williams and Cameron 1984), 4.7 for *T. bulbivorus* (Verts and Carraway 1991), and 2 for *P. castanops* (Smolen, Genoways, and Baker 1980), *Orthogeomys cherriei* (Delgado 1992) and an unidentified species of *Pappogeomys (Cratogeomys)* (Ikenberry 1964). Among the ctenomyids, mean litter size is 1.6 for *C. opimus*, 3.5 for *C. peruanus* (Pearson 1959), and 2.9 for *C. australis* (Busch et al., chap. 5, this volume) and *C. mendocinus* (Rosi et al. 1992, 1996). *C. talarum* has been calculated to have approximately 4 young per year (Malizia and Busch 1991, 1997). Among solitary bathyergids, litter size ranges are 2–10 in *G. capensis*

(Bennett and Jarvis 1988a), 1–4 in *B. suillus*, and 1–7 in *B. janetta* (Bennett et al. 1991).

Social bathyergid mole-rat colonies contain a sole breeding female and one to three reproductive males. In the desert-dwelling species *C. h. hottentotus* and *C. damarensis*, the reproductive female may produce between 1 and 6 pups per litter (Bennett and Jarvis 1988b; Bennett 1989), with a mean of 3 pups per litter. In contrast, the tropical *C. darlingi*, *C. mechowi*, and *C. anselli* produce fewer pups, with litter sizes ranging from 1 to 3 (Bennett, Jarvis et al. 1994; Bennett and Aguilar 1995; Burda 1989) (see table 4.1). These data suggest a low annual recruitment rate, with the colony effectively acting as a single reproductive unit. Thus *C. h. hottentotus*, a seasonal breeder, produces a maximum of two litters per annum, which results in a maximum annual recruitment of 6 pups. In contrast, *C. damarensis*, although only producing a mean litter size of 3, is an aseasonal breeder with the capacity of producing four litters per annum, representing an annual recruitment of 12 pups. The tropical species of *Cryptomys*, although aseasonal breeders, show essentially the same annual recruitment rate as the common mole-rat, *C. h. hottentotus*, because of their small mean litter size.

The naked mole-rat is exceptional, producing a maximum litter size of 12, the largest recorded for any rodent (Jarvis 1991). In captivity, a female may have up to 27 pups in a litter. This potential to bear unusually large litters may enable *H. glaber* to capitalize on good rainfall or unusually large resource patches. If we consider that a colony (with an average size of forty to a hundred animals) may produce 54 pups in a season, this would be equivalent to a recruitment rate of 1 to 2.7 pups per female (assuming the sex ratio in the colony is 1:1). In *C. damarensis*, which inhabits a similar environment, an average colony of fifteen to eighteen individuals produces a maximum of 12 pups, resulting in a recruitment rate of 1 to 1.5 pups per female. In marked contrast, in the solitary genera of bathyergids, such as *Bathyergus* and *Georychus*, each adult female in the population has the capacity to breed twice a year and can produce between 5 and 12 pups. These data suggest that annual per capita recruitment is higher for solitary than for social species of bathyergids, raising intriguing questions regarding the adaptive trade-offs of group living in *Heterocephalus* and *Cryptomys* (Lacey, chap. 7, this volume).

Pup Ontogeny

Information pertaining to the postnatal development of subterranean rodents is scarce (Bennett et al. 1991). The ctenomyid *C. talarum* possesses a pelage at birth, whereas pups of the solitary bathyergids, the spalacine

N. ehrenbergi, and the geomyid *T. talpoides* are naked and altricial. The pups of *C. talarum* and *N. ehrenbergi* are generally larger (8 g) than are those of *T. talpoides* (3 g). In the latter two species, the pelage arises between 3 and 14 days of age. The eyes of all species are closed at birth, and they remain in the nest for about 9–14 days. The pups of *Nannospalax* (Gazit, Shanas, and Terkel 1996) and *Thomomys* (Andersen 1978) begin eating solid foods at about 14–17 days, an age comparable to that of the bathyergid *G. capensis*. Once their eyes have opened, at 21–28 days, sparring is observed in *Thomomys* (approximately 5 weeks), and severe fighting among pups breaks out at 8 weeks *(T. talpoides)* or 11 weeks *(T. splendens)* (table 4.2).

Like those of other subterranean rodents, the pups of bathyergids are born altricial, and their development is gradual (Bennett et al. 1991). Within the Bathyergidae, there are marked behavioral differences among the pups of the various species and subspecies. However, there is no distinct ontogenetic trend as one moves from solitary to social species (Bennett et al. 1991) (table 4.3). Interestingly, the pups of the social mole-rats are more mobile after birth. Pups of social species wander out of the nest for a period longer than 2 minutes at an earlier age (day 1 to 5) than pups of solitary species (day 5 to 9) (Bennett et al. 1991).

The pups' agility (first coordinated movement along the burrow system) develops at much the same time in all of the bathyergids. Surprisingly, the pups of *C. damarensis* begin to eat solid foods at a very early age (day 6); they are remarkably precocial when compared with the remaining bathyergids and other subterranean rodents (which begin eating solid foods at day 10–22: see tables 4.2 and 4.3). The weaning age of pups among the different bathyergids is fairly uniform, although *C. hottentotus* from Zambia are weaned much later than other species (Burda 1989).

Sparring, a behavior that is important in pup development, manifests itself early in the solitary bathyergid species *B. suillus* and *B. janetta* (day 11–16; see table 4.3), but occurs later in the social *Cryptomys* and in *Heterocephalus* (day 14–21). Interestingly, sparring occurs only after 5 weeks in the solitary *Georychus*. In all solitary genera, sparring intensifies and becomes aggressive, with wounds being inflicted if pups are unable to disperse. Hind foot drumming used in territorial advertisement is initiated by the pups of *G. capensis* at around day 50, while in the two species of *Bathyergus* it occurs at around day 80. Hind foot drumming is absent in the pups of social genera, in which the pups are recruited into the natal colony. The variation in development in the Bathyergidae cannot be related to differences in litter size, because, with the exception of *H. glaber*, litter sizes for these animals are all relatively small (see table 4.1).

The solitary *G. capensis* and *B. janetta* have similar mean growth rates of 1.23 and 1.68 g per day, respectively, for the first 70–80 days of postnatal

TABLE 4.2 MORPHOLOGICAL AND BEHAVIORAL DEVELOPMENT IN NON-BATHYERGID SUBTERRANEAN RODENTS

Species	Family	Social status	Number per burrow system	Body mass at birth (g)	Pelage present at birth	Pelage produced (d.a.b.)	Pups out of nest (d.a.b.)	Eyes open (d.a.b.)	Pups eating solids (d.a.b.)	Gestation (days)	Litter size	Source
Ctenomys talarum	Ctenomyidae	Solitary	1	8.0	Yes	1	7	2–3	?	102 (93–120)	5	Malizia and Busch 1991, 1997; Weir 1974; G. H. Aguilar, unpublished data
Nannospalax ehrenbergi	Spalacinae	Solitary	1	5.0	Hairless	7–8	12–14	Eyes closed	14–21	28–34	1–5	Nevo 1961 Shanas et al. 1995 Gazit, Shanas, and Terkel 1996
Thomomys talpoides	Geomyidae	Solitary	1	2.7–3.6	Hairless	9	9	26	17	18		Andersen 1978
Tachyoryctes splendens	Rhizomyinae	Solitary	1	11–18	Hairless	2–4	20	21–28	15–21	36–41		Jarvis 1969 Jarvis 1973

Source: Adapted from Bennett et al. 1991

TABLE 4.3 MORPHOLOGICAL AND BEHAVIORAL DEVELOPMENT IN THE FAMILY BATHYERGIDAE

Species	Pup length (cm)	Pelage present at birth	Pelage produced (d.a.b.)	Eyes open (d.a.b.)	Ear meatus at birth	Pups eating solids (d.a.b.)	Weaned (d.a.b.)	Begin to spar (d.a.b.)	Disperse (d.a.b.)	Drumming (d.a.b.)	Source
Bathyergus suillus	—	Yes	At birth	10	Closed	10	±21	12–13	60–65	>80	Jarvis 1969; Van der Horst 1972; Bennett et al. 1991
Bathyergus janetta	—	No	4	15	Closed	13	±28	11–16	60–65	>80	Bennett et al. 1991
Georychus capensis	3–4	No	7	9	Closed	17	±28	35	55–60	50	Bennett and Jarvis 1988a
Heliophobius argenteocinereus	—	—	—	—	—	—	—	—	—		Jarvis 1969
Cryptomys h. hottentotus	8–9	No	8	13	Closed	10	±28	10–14	Social		Bennett 1988, 1989
Cryptomys damarensis	8–9	No	6	18	Closed	6	±28	18–25	Social		Bennett and Jarvis 1988b; Jarvis and Bennett 1993
Cryptomys anselli	5–6	No	9	24	Closed	22	±82	?	Social		Burda 1989
Cryptomys darlingi	—	No	4	14	Closed	14	±45	36	Social		Bennett, Jarvis, and Cotterill 1994
Cryptomys mechowi	—	No	2	6	Closed	20	±42	60	Social		Bennett and Aguilar 1995
Heterocephalus glaber	—	No	No pelage	30	Closed	14	±24	21	Social		Jarvis 1991

Source: Adapted from Bennett et al. 1991

growth. In *B. suillus,* a much larger mole-rat, the mean maximum growth rate is 3.3 g per day (Bennett et al. 1991) (table 4.4). The mean rates of growth in these solitary bathyergids are comparable to the values obtained for other comparably sized solitary subterranean and fossorial rodents. *Tachyoryctes ruandae,* for example, has a mean growth rate of 1.34 g per day (Rahm 1969), *T. talpoides* 2.2 g per day (Andersen 1978), and *Tatera brantsii* 1.72 g per day (Meester and Hallett 1970). The prairie dog *Cynomys ludovicianus* has a mean growth rate of 4.0 g per day (Anthony and Foreman 1951), which is comparable to that of the similarly sized Cape dune mole-rat, *B. suillus.* The two social bathyergids *C. h. hottentotus* and *C. damarensis* have similar mean rates of growth: 0.229 and 0.233 g per day, respectively. Burda (1989), using linear regression models, calculated the rate of growth in *C. anselli* from Zambia to be 0.21 g per day in females and 0.31 g per day in males for the first 80 days after birth, which is comparable to findings for other species of *Cryptomys* (table 4.4). The growth rates of both male and female *T. bottae* in the laboratory vary over the first several months of life, depending upon the nutritional quality of the diet after weaning (Patton and Brylski 1987). These data, in turn, suggest that growth rate is labile in natural populations, a hypothesis supported by data from free-living animals (Miller 1946; Hansen and Bear 1964).

The possession of a secure nest underground may well promote the slow development of pups after birth. Altricial mammals nest exclusively below ground, in caves or in tree hollows (Case 1978). The nidicolous condition of pups in the family Bathyergidae, as well as in the other families of subterranean rodents, is believed to be related to the thermally stable and relatively secure environment of the burrow (Bennett et al. 1991). Low rates of growth in the social bathyergids may be partly due to the low metabolic rates recorded in these genera (McNab 1966; Bennett, Clarke, and Jarvis 1992;

TABLE 4.4 GROWTH IN THE FAMILY BATHYERGIDAE

Species	Asymptotic weight (A)	Inflection time (days)	Mean growth rate constant (K)	Maximum growth rate ($K.A.e^{-1}$)	Source
Cryptomys darlingi	92.6	94.1	0.008	0.272	Bennett, Jarvis, and Cotterill 1994
C. h. hottentotus	42.0	12.6	0.015	0.229	Bennett et al. 1991
C. damarensis	42.5	15.6	0.015	0.233	Bennett et al. 1991
C. mechowi	90.9	26.7	0.025	0.847	Bennett and Aguilar 1995
Georychus capensis	74.6	16.9	0.04	1.227	Bennett et al. 1991
Bathyergus janetta	90.8	14.8	0.05	1.683	Bennett et al. 1991
B. suillus	217.5	22.3	0.04	3.340	Bennett et al. 1991

Bennett, Taylor, and Aguilar 1993; Bennett, Aguilar et al. 1994). Although the trends are not clear-cut, it seems that the pups of solitary genera grow and mature more rapidly than the pups born to social genera. Interestingly, it is the pups of the solitary mole-rat species that undergo relatively rapid development and eventually disperse to defend their own burrows (Bennett and Jarvis 1988a). The pups of social species are recruited into the natal colony, and hence delay dispersal in order to increase both their own and their mother's inclusive fitness.

Sexual Maturity

Subterranean rodents generally reach sexual maturity at a much later age than their surface-dwelling counterparts (Busch et al., chap. 5, this volume). In several solitary species, females may reach sexual maturity and breed in the year in which they were born. However, males apparently do not. This delayed maturation in males is not surprising, considering that they must grow to a larger body size in order to be successful in intrasexual competition during the breeding season.

In geomyids, females may mature within the year of birth, but typically only in high-quality habitats such as agricultural fields (Smolen, Genoways, and Baker 1980; Daly and Patton 1990; Patton and Brylski 1987; Loeb 1990). Otherwise, females delay reproducing until the following season, and males apparently never breed until they are at least a year old (Patton and Smith 1990). In ctenomyids, both sexes are reported to reach sexual maturity within a year of birth (Pearson 1959; Malizia and Busch 1991, 1997). Malizia and Busch (1991) have also found that females mature before males.

In bathyergids, field data pertaining to male and female sexual development are lacking. However, laboratory-reared male *G. capensis* managed to breed only in the season subsequent to their birth (N. C. Bennett, personal observation). This observation supports the contention that in solitary bathyergids, animals must be at least a year old prior to breeding. The social species of *Cryptomys* rarely breed within a year if taken from their natal colony and paired with an animal of the opposite sex. It is probable that the extremely slow rate of growth and the extended developmental period curtails sexual activity in these hystricognaths (Bennett et al. 1991). However, captive-reared *H. glaber* have reportedly bred within a year of birth (Jarvis 1991). In nonbreeding female naked mole-rats, the ovaries show a lack of follicular development, which may be viewed as a delayed puberty. It could be argued that sexual maturity occurs only when ovarian activity recrudesces following the removal of the breeding female.

Longevity of Breeding and Lifetime Reproductive Success

Many solitary subterranean rodents have relatively short periods during which they can breed. In the Geomyidae, females tend to outlive males. Thus, whereas the average life span for males of *P. castanops* and *T. bottae* may be only 31 weeks, that for females may be up to 56 weeks in *P. castanops* and 55 weeks in *T. bottae* (Smolen, Genoways, and Baker 1980; Howard and Childs 1959). Likewise, in the Ctenomyidae, the life span of breeding *C. talarum* is less than 22 months (Busch et al. 1989; Malizia and Busch 1997). There are no longevity records for *N. ehrenbergi* (Gazit, Shanas, and Terkel 1996) or for myospalacines or rhizomyines.

A more extensive data base is available for the Bathyergidae. In social bathyergids, the residency of reproductive individuals is exceptionally long for such small animals. Records from wild naked mole-rats have shown the same male and female breeders in the colony for 10 years (S. H. Braude, personal communication), and records of captive animals indicate that reproductive tenure may exceed 15 years (J. U. M. Jarvis, personal communication). Studies on *C. damarensis* have revealed that the breeding pair may reside and breed in the natal burrow for 8 years, and that nonbreeding animals may remain there for as long as 4 years (N. C. Bennett and J. U. M. Jarvis, unpublished data). Likewise, in *C. h. hottentotus,* the reproductive pair may reside in the colony for at least 4 years (N. C. Bennett, A. C. Spinks, and C. M. Rosenthal, unpublished data). Data pertaining to solitary mole-rats are effectively absent, although a female *G. capensis* produced four litters in 4 consecutive years in captivity (N. C. Bennett, unpublished data).

There is a deficiency of data relating to lifetime reproductive success in all families of subterranean rodents, and this is an area of study that requires immediate attention. Within the social bathyergids, long-term field studies on *H. glaber* have shown that less than 0.1% of nonbreeding mole-rats eventually attain reproductive status (Jarvis et al. 1994). Field studies on *C. damarensis* (Jarvis and Bennett 1993) have shown that ecological constraints limit opportunities for dispersal, and that only approximately 10% of all nonreproductives in colonies attain reproductive status.

Dispersal

In *G. capensis,* juveniles disperse either by extending the maternal burrow system or by moving short distances above ground (Jarvis and Bennett 1991). Limited data for *B. suillus* indicate that these mole-rats disperse at a similar age and in much the same fashion as *G. capensis* (E. McDaid, unpublished data). Thus it is easy to see that local populations will exhibit low vagility and consist of closely related individuals (Busch et al., chap. 5; Lacey, chap. 7, this volume). In the solitary bathyergids, the breeding female becomes intolerant of juveniles, and it is believed that it is she who ultimately "decides" when it is time for the pups to disperse. The pups of solitary mole-rats disperse from the maternal burrow at an early age to establish their own territories.

In *Ctenomys talarum,* the sex ratio at birth is 1:1, and the sex ratio of young animals does not deviate from parity (Malizia and Busch 1991). However, in the adult population, there is a skew toward females (Malizia and Busch 1991). It is suggested that when the subadults disperse, females disperse before males. Consequently, females occupy territories close to the natal burrow system, whereas males are forced to move longer distances above ground in search of territories, and incur an increased risk of predation (Malizia and Busch 1991). A similar pattern of sex-biased dispersal distance, with females settling closer to their natal burrows, has been observed in the pocket gopher *T. bottae* (Daly and Patton 1990). Furthermore, geomyids typically exhibit skewed adult sex ratios in favor of females, although the degree of bias may be positively related to population density, as in *T. bottae* (Patton and Smith 1990).

In contrast, all of the social Bathyergidae incorporate the pups into the natal colony (Lacey, chap. 7, this volume). These "recruits" are readily accepted by the reproductive and nonreproductive colony members. The pups remain in the colony until adulthood, and during this time do not breed. These animals contribute to the direct care of pups as well as working to extend the burrow system, defend the burrow system, and engage in cooperative foraging. Social *Cryptomys* have inbreeding avoidance mechanisms (Bennett, Faulkes, and Molteno 1996; Bennett, Faulkes, and Spinks 1997), and in captivity, the death of a breeder results in the reproductive quiescence of the colony. *Cryptomys damarensis* are obligate outbreeders and, in the wild, disperse to form new colonies. In contrast, *H. glaber* continuously recruit new breeders from within the colony, although there is evidence to suggest that a "dispersing morph" may result in occasional gene flow between colonies (O'Riain, Jarvis, and Faulkes 1996; Busch et al., chap. 5; Steinberg and Patton, chap. 8, this volume).

Social Suppression of Reproduction in Bathyergid Mole-rats

The extreme range of sociality exhibited by the bathyergid mole-rats is unique among mammals (Lacey, chap. 7, this volume). Species that differ in their social organization also tend to differ with respect to reproductive opportunities. In solitary species, reproductive skew (Vehrencamp 1983) is low, and all animals have essentially an equal chance to breed. In contrast, in the cooperatively breeding social species, reproductive skew is high.

Within the social Bathyergidae, a range of mechanisms are involved in maintaining reproductive skew. In *H. glaber*, an obligate inbreeder, reproductive skew is due to socially induced infertility, whereas *C. damarensis* exhibits both dominance-related infertility and incest avoidance (Bennett, Faulkes, and Molteno 1996, Bennett, Faulkes, and Spinks 1997). Furthermore, breeding and nonbreeding individuals of the less social *C. darlingi* show no physiological differences, yet exhibit incest avoidance. The mechanistic differences underlying reproductive skew in these species probably result from a combination of extrinsic and intrinsic constraints on the opportunities to disperse and found new colonies.

It has been postulated that inhibition of the reproductive axis may be achieved through the action of primer pheromones (Vandenbergh 1988). Adult female bank voles, *Clethrionomys glareolus* (Kruczek and Marchlewska-Koj 1986), produce a urinary pheromone that delays puberty in juvenile females. In microtine rodents that live in extended family groups, suppressing pheromones released by the adults may function to reduce reproductive competition by encouraging the dispersal of offspring or preventing inbreeding (Batzli, Getz, and Hurley 1977). For *H. glaber*, however, Faulkes and Abbott (1993) used bedding transfer experiments to demonstrate that a primer pheromone alone was insufficient to effect suppression of the reproductive axis. When they removed the queen from a colony and exposed other colony members to her odor, they found that previously nonbreeding females in the colony began showing evidence of reproductive activity. Faulkes and Abbott (1993) proposed that direct contact, in the form of physical shoving by the queen, is responsible for suppression of reproduction among subordinate females (Clarke and Faulkes 1997). This hypothesis is supported by the findings of Westlin, Bennett, and Jarvis (1974), who demonstrated that within colonies of *H. glaber*, nonbreeding females appeared to be suppressed to varying degrees, and that uniformity of suppression was absent close to parturition. During this period, the queen is very gravid and is unable to shove her subordinates as frequently or as easily as when she is in a nonpregnant state.

In both captive and wild colonies of *H. glaber*, the block to reproduction in nonbreeding females results from inhibition of ovulation. Studies on the ovaries of nonbreeding females (Kayanja and Jarvis 1971; Faulkes 1990; Faulkes, Abbott, and Jarvis 1990) revealed that they resemble prepubescent ovaries, since little follicular development takes place. The lack of follicular development appears to be due to inadequate secretion of LH from the pituitary as a result of pituitary insensitivity to GnRH (Faulkes et al. 1990; Faulkes, Abbott, and Jarvis 1991). It is possible to sensitize the pituitary using exogenous GnRH, which may mirror the effect of removing a nonreproductive female from the colony. Analyses of progesterone profiles suggest that ovarian cycling can begin in as little as 7 days following removal.

The reproductive female naked mole-rat has an ovarian cycle of approximately 34 days, with a follicular phase of 6 days and a luteal phase of 28 days. This unusually long luteal phase is typical of the hystricomorph rodents. In guinea pigs *(Cavia porcellus)*, the corpus luteum of an unmated, nonpregnant female remains for 15–17 days; in the cuis *(Galea musteloides)*, 22 days; in the chinchilla *(Chinchilla laniger)*, 40 days; and in the acouchi *(Myoprocta pratti)*, 30–55 days (Weir and Rowlands 1974).

Comparison of the reproductive tracts of breeding and nonbreeding male *H. glaber* from captive colonies revealed differences in the mean testis mass relative to body mass (Faulkes et al. 1994). Numbers of spermatozoa and the proportion of motile sperm are greater in reproductive males (Faulkes et al. 1994). There are, however, differences in the degree of spermatogenesis among nonreproductive males, suggesting that reproductive suppression of males is less consistent than that of females. In addition, endocrine studies showed the nonbreeding males to have depressed titers of urinary testosterone and basal plasma LH, as well as a reduced LH response to exogenous GnRH, compared with breeders (Faulkes and Abbott 1991; Faulkes, Abbott, and Jarvis 1991). These endocrine differences can be reversed by removing a male from his natal colony. Removal results in increased testis mass and enhanced production of motile sperm. Similarly, plasma LH levels and urinary testosterone titers increase significantly after removal (Faulkes and Abbott 1991).

In *C. damarensis*, no significant differences in testosterone titer or pituitary sensitivity to GnRH stimulation are evident among males, although differences in the percentage of motile sperm were found by Maswanganye et al. (1999). Reproductive males have a greater relative testis mass than nonbreeding males, a difference that corresponds favorably with the observation that these individuals characteristically possess bulging testes that are positioned in inguinal pockets (Jarvis and Bennett 1993). However, among breeding males, larger testes are not associated with a larger number of spermatozoa (Faulkes et al. 1994).

Male bathyergids characteristically possess sperm storage sacs. In reproductively active male *H. glaber*, these structures are engorged with sperm (Faulkes et al. 1994). In nonbreeding males, they enlarge after an animal is removed from its natal colony and its reproductive system becomes activated. Similar structures are present in *C. damarensis*, yet they have not been found to contain spermatozoa. It is possible that in this species, the sacs allow for increased sperm storage by reproductive males just prior to mating with an estrous female (Faulkes et al. 1994).

H. glaber is unusual in that a socially induced physiological block to reproduction occurs in males. In contrast, *C. damarensis* conforms to the typical pattern of a socially suppressed male mammal in that there is no difference in LH or testosterone titers between breeding and nonbreeding males. Reproductive suppression apparently results from a lack of mating opportunities, since the males are closely related to females in the colony and incest is avoided (Bennett, Faulkes, and Molteno 1996).

In nonreproductive *C. damarensis* females, however, a socially induced suppression of reproductive physiology is prevalent (Bennett, Jarvis, and Cotterill 1994). As in *H. glaber,* nonreproductive females are characterized by significantly lower LH concentrations than reproductives (Bennett et al. 1993). The nonreproductives show a reduced response to exogenous GnRH compared with the reproductive female (Bennett et al. 1993). The rise in their circulating LH concentrations observed when the reproductive female is removed may promote ovulation in developing follicles that would previously have formed atretic luteinized follicles. The elevation of plasma LH in the absence of the reproductive female implicates her as a suppressive factor operating on nonbreeders (fig. 4.1). Although the response to a single exogenous GnRH challenge is higher in nonreproductives housed in the absence of the reproductive female, their response does not resemble that of a reproductive female. It is possible that the presence of a potential breeding partner provides the final cue for breeding.

As one would expect, histological examination of all females from a newly caught colony of *C. damarensis* revealed that thirteen of the fourteen females were nulliparous. Although the ovaries of the nonreproductive females showed some follicular development, they lacked mature Graafian follicles (Bennett 1988; Bennett, Jarvis, and Cotterill 1994). Instead, their ovaries contained a large number of primordial and primary follicles in which the intact zona pellucida of the ova could be seen, confirming the lack of ovulation. The periphery of the ovary of each nonreproductive female was peppered with numerous primordial follicles; this was in marked contrast to the ovaries of the reproductive female, which contained far fewer primordial follicles. The ovaries of the nonreproductive females characteristically possessed luteinized unruptured follicles (LUFs), which result from the luteinization of late secondary and early tertiary follicles. The membrana granulosa and theca interna of the LUFs were packed with ovoid luteinizing secretory cells responsible for the manufacture of progesterone.

Nonreproductive females have detectable concentrations of progesterone, which may be attributable to the presence of the LUFs. The demonstrated presence in the LUFs of the enzyme 3β-hydroxysteroid dehydroge-

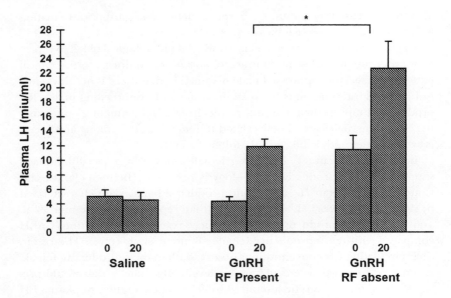

FIGURE 4.1. Concentrations of bioactive LH (mean ± SE) in intact nonreproductive female *C. damarensis* before (0) and 20 min (20) after a single injection of 200 μl of saline; a single injection of 2 μg GnRH in 200 μl of physiological saline in the presence of a reproductive female; a single injection of 2 μg GnRH in 200 μl physiological saline in the absence of a reproductive female for 3 years. (Adapted from Bennett, Faulkes, and Molteno 1996)

nase, which converts pregnenolone to progesterone via a two-step isomerase reaction, suggests that the LUFs are the structures that produce progesterone (Bennett, Jarvis et al. 1994). Bennett, Jarvis, and Wallace (1990) have shown that the number of LUFs increases with age, suggesting that LUFs accumulate over time. If the breeding female is removed from a colony, the progesterone concentrations of the nonreproductive females are significantly lowered. This drop strongly implies a reduction in the number of LUFs in the ovary. It is plausible that the progesterone produced by the LUFs induces a "pseudopregnant" condition in the nonreproductive females that restrains them from ovulating (Bennett et al. 1993b). Alternatively, it is possible that the LUFs merely reflect inadequate secretion of LH from the pituitary.

C. damarensis is an obligate outbreeder. In both the laboratory and the field, the death or experimental removal of one of the reproductive animals results in the cessation of breeding (Rickard and Bennett 1997; Jarvis and Bennett 1993). Colony formation requires an unfamiliar male or female, and new colonies are typically established by pairs or small groups of molerats that originate from different colonies (Jarvis and Bennett 1993; Jarvis et al. 1994). Behavioral studies of sibling and nonsibling pairings provide further evidence for the avoidance of breeding with familiar conspecifics. Nonsibling pairs generally engage in copulatory activity within minutes of pairing, and eventually conception takes place. In contrast, sibling pairs fail to engage in any sexual activity, and offspring are never produced (Bennett, Faulkes, and Spinks 1997). Interestingly, Burda (1995), working with the common Zambian mole-rat, *C. anselli,* found that if siblings are removed from the colony and housed separately for more than 14 days, they will mate when subsequently paired. This finding suggests that the familiarity of individuals, rather than the degree of genetic relatedness, is the important cue for these animals. Thus there appears to be a strong inhibition against breeding with familiar individuals (from the same colony), and in the field, this translates into strong incest avoidance. Studies of the suppression of reproduction in *Cryptomys* are complicated by the presence of this familiarity avoidance in addition to the physiological inhibition of fertility in the presence of the reproductive female.

In contrast to *H. glaber* and *C. damarensis,* the mesic species *C. darlingi* shows no significant difference in circulating basal or GnRH-stimulated LH levels between reproductive and nonreproductive animals of either sex (fig. 4.2). Thus neither males nor females are physiologically suppressed, and the reproductive skew appears to be maintained entirely through inhibition of reproductive behavior in these obligate outbreeders (Bennett, Faulkes, and Spinks 1997).

Cryptomys h. hottentotus, which occurs in the Cape Province, is unusual in that it is the only social bathyergid that exhibits a well-defined breeding season. The western and northern Cape is a winter rainfall region, with rain falling between May and August. At the onset of the breeding season, numerous females in a colony become perforate, despite the presence of a reproductive female (N. C. Bennett, personal observation). In contrast, males show no seasonality with respect to testicular activity, spermatogenesis, or sperm quality (motility and normal morphology). However, testis mass, testis volume, and the diameter and thickness of the seminiferous tubules are significantly smaller during the breeding season. Despite these differences in testis structure, no significant differences in sperm motility are evident between reproductive and nonreproductive males, and animals in both groups show comparable levels of sperm defects and urinary testosterone (Spinks, van der Horst, and Bennett 1997).

The maintenance of reproductive activity in males in the nonbreeding season is important, since it coincides with the period when maximal dispersal takes place. This may also explain why nonreproductive females become perforate prior to the onset of breeding; these animals may be anticipating dispersal and pair formation. The reproductive activation of dispers-

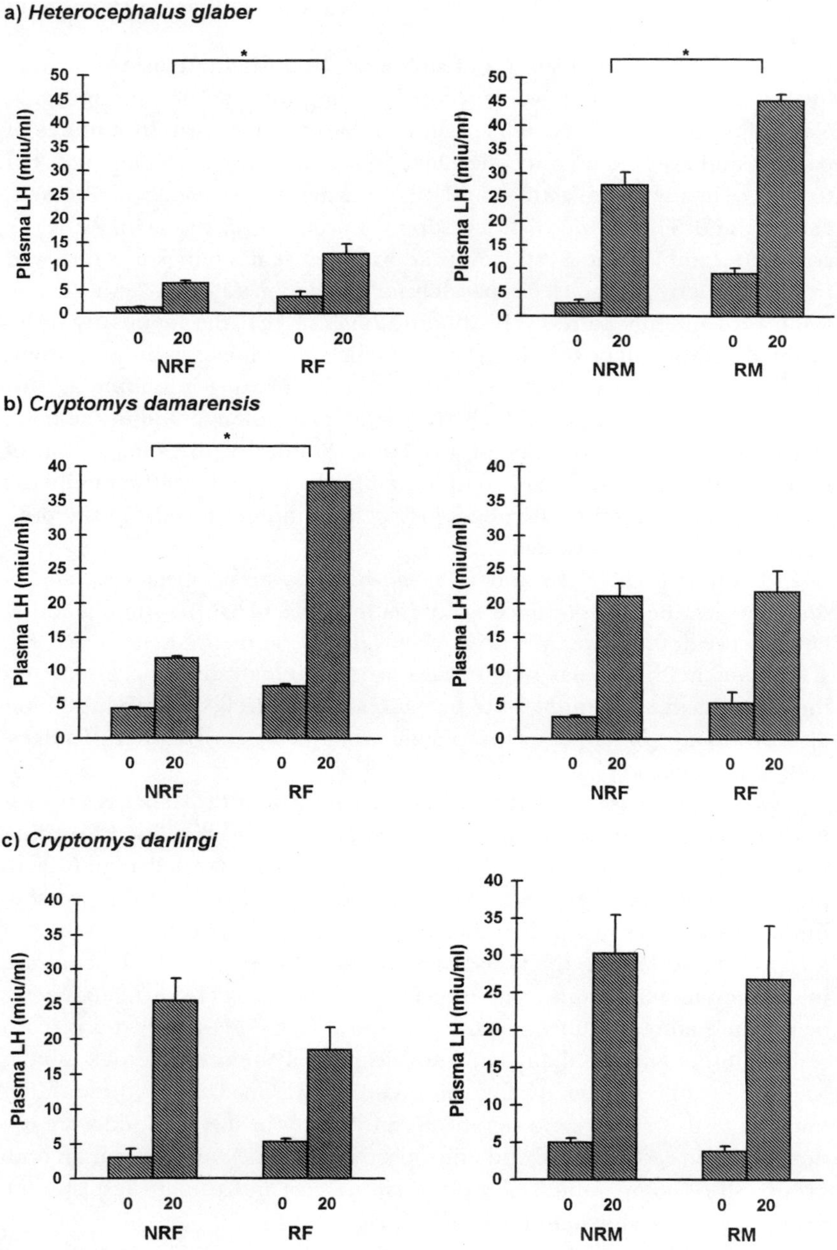

FIGURE 4.2. The socially induced infertility continuum in the mole-rats *Heterocephalus glaber*, *Cryptomys damarensis*, and *C. darlingi*. Concentrations of plasma LH (mean ± SE) in reproductive (R) and nonreproductive (NR) female and male mole-rats before (0) and 20 minutes after (20) a single subcutaneous injection of GnRH. *$P < .05$. In all cases comparisons of individual means were made using Mann-Whitney U tests.

ing males may enable intersexual recognition to occur more readily and consequently may promote pair-bond formation and subsequent colony genesis (Spinks, van der Horst, and Bennett 1997).

Thus social African mole-rats exhibit a continuum of socially induced infertility among species inhabiting regions of varying degrees of aridity (Bennett, Faulkes, and Spinks 1997) (see fig. 4.2). Sherman et al. (1995) have proposed a quantitative means of comparing the social systems of cooperatively breeding species that assesses the probability of an individual in a social group ever attaining breeding status. Differences in lifetime reproductive success are used to define the degree of reproductive division of labor. Hence, in mesic habitats, where opportunities for dispersal and reproduction are relatively great, reproductive suppression involves only an incest avoidance component. In contrast, in arid habitats, where dispersal and reproduction are impossible for most individuals, reproductive suppression includes a number of physiological components. Between these two extremes lie species in which reproductive suppression of males is behavioral, but suppression of females includes both behavioral and physiological components.

The spectrum of suppressive mechanisms evident from studies of *H. glaber*, *C. damarensis*, and *C. h. hottentotus* provides a framework for predicting the nature and extent of reproductive suppression in other social bathyergids. For example, *C. darlingi* occurs in mesic habitats, where it lives in relatively small colonies that possess only a rough social organization. Based on these attributes, we can predict that opportunities for dispersal and colony formation should be relatively common, and that more than 10% of individuals should eventually become reproductively active. Reproductive suppression within a colony should be maintained primarily via behavioral mechanisms such as familiarity and incest avoidance. Although these statements have yet to be evaluated for free-living *C. darlingi*, the comparative framework generated here provides a series of testable predictions regarding interactions between habitat, social behavior, and reproductive suppression.

Conclusions and Future Directions

Most subterranean rodents are xenophobic and highly aggressive toward conspecifics. These basic features, combined with the subterranean nature of their habitat, mean that these species do not make the easiest research subjects with which to study reproductive biology. The social bathyergids have been the focus of more detailed studies because they readily pair-bond, allowing manipulation of their reproductive state. The reproductive biology of the Ctenomyidae still requires a great deal of study. Similarly, for the Geomyidae, although there are more data pertaining to their basic reproductive biology, there are still a large number of gaps, with most species having been researched only in part. Recent studies of the Spalacinae are more complete because of success in captive breeding. However, there is still a great deal that can be undertaken. Information pertaining to reproduction in the Rhizomyinae and Myospalacinae is fragmentary, and it is for these groups in particular that further research is needed.

Although the Bathyergidae is clearly the best-studied family with regard to reproductive biology, within the group, certain species are still relatively unknown (e.g., *Heliophobius argenteocinereus*). For all bathyergids, detailed knowledge of the andrology and gynecology of these fascinating animals is absent or, at best, fragmentary. For males, quantification of sperm motility at different times of the year would enable a comparison of activity and percentage of deformity to be made. Likewise, monthly subsampling of females would allow the detection of possible times of enhanced ovarian activity. This would provide some indication of the time when reproduction occurs in these secretive species.

Future research on the reproductive biology of subterranean rodents should focus on baseline information, such as determining the length of the ovarian cycle in each species, as well as determining whether ovulation is spontaneous, induced, or both. This basic reproductive physiology would give us greater insight into the life history patterns exhibited by this fascinating array of animals. In both the solitary and social species of subterranean rodents, knowledge of the paternity of litters would also give us greater insight into the reproductive mechanisms operating in this group. Thus, in solitary animals, in which mating is generally brief, would one expect multiple paternity, or should it be singular? In the social bathyergids, multiple paternity of litters would shed light on the possible reproductive mechanisms responsible for maintaining reproductive skew in these animals. This information has important implications for answering questions on inclusive fitness benefits to colony members of varying degrees of relatedness.

However, before the above suggestions are put into practice, we should bear in mind that without the basic reproductive life histories of different species, these refinements of our knowledge of reproductive physiology would be difficult. The very nature of the subterranean niche demands that scientists researching these animals be tenacious and attentive. Indeed, if it were not for people of this makeup, we would today be all the poorer in our knowledge of the reproductive biology of these intriguing rodents.

Acknowledgments

We wish to express our deep gratitude to Eileen Lacey, Jim Patton, Guy Cameron, and Rosie Cooney for insightful comments on the manuscript. The work for this chapter was funded by grants to N. C. B. from the Universities of Pretoria and Cape Town. The Foundation for Research Development enabled much of the work on the Bathyergidae to be conducted. C. G. F. is grateful to the Wellcome Trust and the Institute of Zoology, London.

Literature Cited

Altuna, C. A., G. Francescoli, and G. Izquierdo. 1991. Copulatory pattern of *Ctenomys pearsoni* (Rodentia, Octodontidae) from Balneario Solís, Uruguay. *Mammalia* 55:440–42.

Altuna, C. A., and E. P. Lessa. 1985. Penial morphology in Uruguayan species of *Ctenomys* (Rodentia, Octodontidae). *Journal of Mammalogy* 66:483–88.

Andersen, D.C. 1978. Observations on reproduction, growth and behavior of the northern pocket gopher *(Thomomys talpoides). Journal of Mammalogy* 59:418–22.

Andersen, D.C., and J. A. MacMahon. 1981. Population dynamics and bioenergetics of a fossorial herbivore, *Thomomys talpoides* (Rodentia: Geomyidae) in a spruce fir sere. *Ecological Monographs* 51:179–202.

Anthony, A., and D. Foreman. 1951. Observations on the reproductive cycle of the black-tailed prairie dog *(Cynomys ludovicianus). Physiological Zoology* 24:242–48.

Batzli, G. O., L. L. Getz, and S. S. Hurley. 1977. Suppression of growth and reproduction of microtine rodents by social factors. *Journal of Mammalogy* 58:583–91.

Bennett, N. C. 1988. The trend towards sociality in three species of southern African mole-rats (Bathyergidae): Causes and consequences. Ph.D. thesis, University of Cape Town. 474 pp.

———. 1989. The social structure and reproductive biology of the common mole-rat, *Cryptomys h. hottentotus* and remarks on the trends in reproduction and sociality in the family Bathyergidae. *Journal of Zoology, London* 219:45–59.

———. 1990a. Behavior and social organization in a colony of the Damaraland mole-rat *Cryptomys damarensis. Journal of Zoology, London* 220:225–48.

———. 1990b. The social season is underground: The mole-rats of southern Africa. *African Wildlife* 44 (5):299–301.

Bennett, N. C., and G. H. Aguilar. 1995. The reproductive biology of the giant Zambian mole-rat, *Cryptomys mechowi* (Rodentia: Bathyergidae). *South African Journal of Zoology* 30(1) 1–4.

Bennett, N. C., G. H. Aguilar, J. U. M. Jarvis, and C. G. Faulkes. 1994. Thermoregulation in three species of Afrotropical subterranean mole-rats (Rodentia: Bathyergidae) from Zambia and Angola and scaling within the genus *Cryptomys. Oecologia* 97:222–28.

Bennett, N. C., B. C. Clarke, and J. U. M. Jarvis. 1992. A comparison of metabolic acclimation in two species of social mole-rats (Rodentia: Bathyergidae) in southern Africa. *Journal of Arid Environments* 22:189–98.

Bennett, N. C., C. G. Faulkes, and A. J. Molteno. 1996. Reproductive suppression in subordinate, non-breeding female Damaraland mole-rats: Two components to a lifetime of socially-induced infertility. *Proceedings of the Royal Society of London* B 263:1599–1603.

Bennett, N. C., C. G. Faulkes, and A. J. Spinks. 1997. LH responses to single doses of exogenous GnRH by social Mashona mole-rats: A continuum of socially-induced infertility in the family Bathyergidae. *Proceedings of the Royal Society of London* B 264:1001–6.

Bennett, N. C., and J. U. M. Jarvis. 1988a. The reproductive biology of the Cape mole-rat, *Georychus capensis* (Rodentia: Bathyergidae). *Journal of Zoology, London* 214:95–106.

———. 1988b. The social structure and reproductive biology of colonies of the mole-rat *Cryptomys damarensis* (Rodentia: Bathyergidae). *Journal of Mammalogy* 69:293–302.

Bennett, N. C., J. U. M. Jarvis, G. H. Aguilar, and E. J. McDaid. 1991. Growth rates and development in six species of African mole-rats (Rodentia: Bathyergidae) in southern Africa. *Journal of Zoology, London* B 225:13–26.

Bennett, N. C., J. U. M. Jarvis, and F. P. D. Cotterill. 1994. The colony structure and reproductive biology of the Mashona mole-rat, *Cryptomys darlingi* from Zimbabwe. *Journal of Zoology, London* 234:477–87.

Bennett, N. C., J. U. M. Jarvis, and K. C. Davies. 1988. Daily and seasonal temperatures in the burrows of African rodent moles. *South African Journal of Zoology* 23:189–95.

Bennett, N. C., J. U. M. Jarvis, C. G. Faulkes, and R. P. Millar. 1993. LH responses to single doses of exogenous GnRH by freshly captured Damaraland mole-rats, *Cryptomys damarensis*. *Journal of Reproduction and Fertility* 99:81–86.

Bennett, N. C., J. U. M. Jarvis, R. P. Millar, H. Sasano, and K. V. Ntshinga. 1994. Reproductive suppression in eusocial *Cryptomys damarensis* colonies: Socially-induced infertility in females. *Journal of Zoology, London* 234:617–30.

Bennett, N. C., J. U. M. Jarvis, and D. B. Wallace. 1990. The relative age structure and body masses of complete wild-captured colonies of two social mole-rats, the common mole-rat *Cryptomys hottentotus hottentotus* and the Damaraland mole-rat *Cryptomys damarensis*. *Journal of Zoology, London* 220:469–85.

Bennett, N. C., P. J. Taylor, and G. H. Aguilar. 1993. Thermoregulation and metabolic acclimation in the Natal mole-rat *(Cryptomys hottentotus natalensis)* (Rodentia: Bathyergidae). *Zeitschrift für Säugetierkunde* 58(6):362–67.

Burda, H. 1989. Reproductive biology (behavior, breeding and postnatal development) in subterranean mole-rats, *Cryptomys hottentotus* (Bathyergidae). *Zeitschrift für Säugetierkunde* 54:360–76.

———. 1995. Individual recognition and incest avoidance in eusocial common mole-rats rather than reproductive suppression by parents. *Experientia* 51:411–13.

Busch, C., A. I. Malizia, O. A. Scaglia, and O. A. Reig. 1989. Spatial distribution and attributes of a population of *Ctenomys talarum* (Rodentia: Octodontidae). *Journal of Mammalogy* 70:204–8.

Case, T. J. 1978. On the evolution and adaptive significance of postnatal growth rates in the terrestrial vertebrates. *Quarterly Review of Biology* 53:243–82.

Clarke, F. M., and C. G. Faulkes. 1997. Dominance and queen succession in captive colonies of the naked mole-rat, *Heterocephalus glaber*. *Proceedings of the Royal Society of London* 264:993–1000.

Daly, J. C., and J. L. Patton. 1990. Dispersal, gene flow, and allelic diversity between local populations of *Thomomys bottae* pocket gophers in the coastal ranges of California. *Evolution* 44:1283–94.

Davies, K. C., and J. U. M. Jarvis. 1986. The burrow systems and burrowing dynamics of the mole-rats *Bathyergus suillus* and *Cryptomys hottentotus* in the fynbos of the southwestern Cape, South Africa. *Journal of Zoology, London* 209:125–47.

De Graaff, G. 1981. *The rodents of Southern Africa*. Johannesburg: Butterworth.

Delgado, R. 1992. Ciclo reproductivo de la taltuza *Orthogeomys cheriei* (Rodentia: Geomyidae) en Costa Rica. *Revista de Biologia Tropical* 40:111–15.

Desy, E. A., and J. D. Druecker. 1979. The estrous cycle of the plains pocket gopher, *Geomys bursarius*, in the laboratory. *Journal of Mammalogy* 60:235–36.

Dewsbury, D. A. 1975. Diversity and adaptation in rodent copulatory behavior. *Science* 190:947–54.
Dixon, J. 1929. The breeding season of the pocket gopher in California. *Journal of Mammalogy* 10:327–28.
Faulkes, C. G. 1990. Social suppression of reproduction in the naked mole-rat, *Heterocephalus glaber.* Ph.D. thesis, University of London.
Faulkes, C. G., and D. H. Abbott. 1991. Social control of reproduction in breeding and non-breeding naked mole-rats *(Heterocephalus glaber)*. *Journal of Reproduction and Fertility* 93:427–35.
———. 1993. Evidence that primer pheromones do not cause social suppression of reproduction in male and female naked mole-rats *(Heterocephalus glaber)*. *Journal of Reproduction and Fertility* 99:225–30.
Faulkes, C. G., D. H. Abbott, and J. U. M. Jarvis. 1990. Social suppression of ovarian cyclicity in captive and wild colonies of female naked mole-rats, *Heterocephalus glaber. Journal of Reproduction and Fertility* 88:559–68.
———. 1991. Social suppression of reproduction in male naked mole-rats, *Heterocephalus glaber. Journal of Reproduction and Fertility* 91:593–604.
Faulkes, C. G., D. H. Abbott, J. U. M. Jarvis, and F. E. Sherriff. 1990. LH responses of female naked mole-rats, *Heterocephalus glaber,* to single and multiple doses of exogenous GnRH. *Journal of Reproduction and Fertility* 89:317–23.
Faulkes, C. G., D. H. Abbott, H. P. O'Brien, L. Lau, M. R. Roy, R. K. Wayne, and M. W. Bruford. 1997. Micro- and macrogeographical genetic structure of colonies of naked mole-rats *Heterocephalus glaber. Molecular Ecology* 6:615–28.
Faulkes, C. G., S. N. Trowell, J. U. M. Jarvis, and N. C. Bennett. 1994. Investigation of sperm numbers and motility in reproductively active and socially suppressed males of two eusocial African mole-rats, the naked mole-rat *(Heterocephalus glaber),* and the Damaraland mole-rat *(Cryptomys damarensis)*. *Journal of Reproduction and Fertility* 100:411–16.
Flynn, L. J. 1990. The natural history of rhizomyid rodents. In *Evolution of subterranean mammals at the organismal and molecular levels,* edited by E. Nevo and O. A. Reig, 155–83. *Progress in Clinical and Biological Research,* vol. 335. New York: Wiley-Liss.
Gazit, I., U. Shanas, and J. Terkel. 1996. First successful breeding of the blind mole-rat *(Spalax ehrenbergi)* in captivity. *Israeli Journal of Zoology* 42:3–13.
Gunther, W. C. 1956. Studies on the male reproductive system of the California pocket gopher *(Thomomys bottae navus* Merriam). *American Midland Naturalist* 55:1–40.
Hansen, R. M. 1960. Age and reproductive characteristics of mountain pocket gophers in Colorado. *Journal of Mammalogy* 41:323–35.
Hansen, R. M., and G. D. Bear. 1964. Comparison of pocket gophers from alpine, subalpine and shrub-grassland habitats. *Journal of Mammalogy* 45:636–40.
Heth, G., E. Frankenberg, A. Raz, and E. Nevo. 1987. Vibrational communication in subterranean mole-rats *(Spalax ehrenbergi)*. *Behavioral Ecology and Sociobiology* 21:31–33.
Howard, W. E., and H. E. Childs. 1959. Ecology of pocket gophers with emphasis on *Thomomys bottae mewa. Hilgardia* 29:277–358.
Ikenberry, R. D. 1964. Reproductive studies of the Mexican pocket gopher *Cratogeomys castanops perplanus.* M.Sc. thesis, Texas Technical College.
Ingles, L. G. 1952. The ecology of the mountain pocket gopher *Thomomys monticola. Ecology* 33:87–95.
Jarvis, J. U. M. 1969. The breeding season and litter size of African mole-rats. *Journal of Reproduction and Fertility* (suppl.) 6:237–48.
———. 1973. The structure of a population of mole-rats, *Tachyoryctes splendens* (Rodentia: Rhizomyidae). *Journal of Zoology, London* 171:1–14.

———. 1981. Eusociality in a mammal: Cooperative breeding in naked mole-rat colonies. *Science* 212:571–73.

———. 1991. Reproduction of naked mole-rats. In *The biology of the naked mole-rat*, edited by P. W. Sherman, J. U. M. Jarvis, and R. D. Alexander, 384–426. Princeton, NJ: Princeton University Press.

Jarvis, J. U. M., and N. C. Bennett. 1990. The evolutionary history, population biology and social structure of African mole-rats: Family Bathyergidae. In *Evolution of subterranean mammals at the organismal and molecular levels*, edited by E. Nevo and O. A. Reig, 97–128. Progress in Clinical and Biological Research, vol. 335. New York: Wiley-Liss.

———. 1991. Ecology and behavior of the Family Bathyergidae. In *The biology of the naked mole-rat*, edited by P. W. Sherman, J. U. M. Jarvis, and R. D. Alexander, 66–96. Princeton, NJ: Princeton University Press.

———. 1993. Eusociality has evolved independently in two genera of bathyergid mole-rats—but occurs in no other subterranean mammal. *Behavioral Ecology and Sociobiology* 33:353–60.

Jarvis, J. U. M., M. J. O'Riain, N. C. Bennett, and P. W. Sherman. 1994. Mammalian eusociality—a family affair. *Trends in Ecology and Evolution* 9:47–51.

Kayanja, F. I. B., and J. U. M. Jarvis. 1971. Histological observations on the ovary, oviduct and uterus of the naked mole-rat. *Zeitschrift für Säugetierkunde* 36:114–21.

Kennerly, T. E. 1954. Local differentiation in the pocket gopher *(Geomys personatus)* in southern Texas. *Texas Journal of Science* 6:297–329.

King, H. D. 1927. Seasonal variation in fertility and sex ratio of mammals with special reference to the rat. *Archiv für Entwicklungsmechanik der Organismen* 112:61–111.

Kruczek, M., and A. M. Marchlewska-Koj. 1986. Puberty delay of bank vole females in a high-density population. *Biology of Reproduction* 35:537–41.

Lacey, E. A., S. H. Braude, and J. R. Wieczorek. 1997. Burrow sharing by colonial tuco-tucos *(Ctenomys sociabilis)*. *Journal of Mammalogy* 78:556–62.

Lacey, E. A., and P. W. Sherman. 1991. Social organization of naked mole-rat colonies: Evidence for divisions of labor. In *The biology of the naked mole-rat*, edited by P. W. Sherman, J. U. M. Jarvis, and R. D. Alexander, 275–336. Princeton, NJ: Princeton University Press.

Lay, D. M. 1978. Observations on reproduction in a population of pocket gophers, *Thomomys bottae*, from Nevada. *Southwestern Naturalist* 23:375–80.

Loeb, S. C. 1990. Reproduction and population structure of pocket gophers *(Thomomys bottae)* from irrigated alfalfa fields. In *Proceedings of the 14th Vertebrate Pest Conference, University of California, Davis*, edited by L. R. Davis and R. E. Marsh, 76–81.

Malizia, A. I., and C. Busch. 1991. Reproductive parameters and growth in the fossorial rodent *Ctenomys talarum* (Rodentia: Octodontidae). *Mammalia* 55:293–305.

———. 1997. Breeding biology of the fossorial rodent *Ctenomys talarum* (Rodentia: Octodontidae). *Journal of Zoology, London* 242:463–71.

Maswanganye, K. A., N. C. Bennett, J. Brinders, and R. Cooney. 1999. Oligospermia and azoospermia in non-reproductive male Damaraland mole-rats *(Cryptomys damarensis)* (Rodentia: Bathyergidae). *Journal of Zoology, London* 248:411–18.

McNab, B. K. 1966. The metabolism of fossorial rodents: A study of convergence. *Ecology* 60:1010–21.

Meester, J., and A. F. Hallett. 1970. Notes on the postnatal development in certain southern African Muridae and Cricetidae. *Journal of Mammalogy* 51:703–11.

Miller, M. A. 1946. Reproductive rates and cycles in the pocket gopher. *Journal of Mammalogy* 27:335–58.

Narins, P. M., O. J. Reichman, J. U. M. Jarvis, and E. R. Lewis. 1992. Seismic signal transmission between burrows of the Cape mole-rat, *Georychus capensis*. *Journal of Comparative Physiology* 170:13–21.

Nevo, E. 1961. Observations on Israeli populations of the mole-rat *Spalax e. ehrenbergi* Nehring 1898. *Mammalia* 25:127–44.
———. 1969. Mole-rat *Spalax ehrenbergi:* Mating behaviour and its evolutionary significance. *Science* 163:484–86.
———. 1979. Adaptive convergence and divergence of subterranean mammals. *Annual Reviews in Ecology and Systematics* 10:269–308.
O'Riain, M. J., J. U. M. Jarvis, and C. G. Faulkes. 1996. A dispersive morph in the naked mole-rat. *Nature* 380:619–21.
Patton, J. L., and P. V. Brylski. 1987. Pocket gophers in alfalfa fields: Causes and consequences of habitat-related body size variation. *American Naturalist* 4:493–503.
Patton, J. L., and J. H. Feder. 1981. Microspatial genetic heterogeneity in pocket gophers: Non-random breeding and drift. *Evolution* 35:912–20.
Patton, J. L., and M. F. Smith. 1990. The evolutionary dynamics of the pocket gopher *Thomomys bottae*, with emphasis on California populations. *University of California Publications in Zoology* 123:1–161.
———. 1993. Molecular evidence for mating asymmetry and female choice in a pocket gopher *(Thomomys)* hybrid zone. *Molecular Ecology* 2:3–8.
Pearson, O. P. 1959. Biology of subterranean rodents, *Ctenomys*, in Peru. *Memorias del Museo de Historia Natural "Javier Prado"* 9:1–56.
Pearson, O. P., and M. I. Christie. 1985. Los tuco-tucos (genero *Ctenomys*) de los Parques Nacionales Lanin y Nahuel Huapi, Argentina. *Historia Natural* 5:337–43.
Rado, R., M. Himelfarb, B. Arensburg, J. Terkel, and Z. Wollberg. 1989. Are seismic communication signals transmitted by bone conduction in the blind mole-rat? *Hearing Research* 41:23–30.
Rado, R., N. Levi, H. Hauser, J. Witcher, N. Alder, N. Intrator, Z. Wollberg, and J. Terkel. 1987. Seismic signalling as a means of communication in a subterranean mammal. *Animal Behaviour* 35:1249–66.
Rado, R., Z. Wollberg, and J. Terkel. 1992. Dispersal of young mole-rats *(Spalax ehrenbergi)* from the natal burrow. *Journal of Mammalogy* 73:885–90.
Rahm, U. 1969. Gestation period and litter size of the mole-rat, *Tachyoryctes ruandae*. *Journal of Mammalogy* 50:383–84.
Redi, C. A., S. Garagna, G. Heth, and E. Nevo. 1986. Descriptive kinetics of spermatogenesis in four chromosomal species of the *Spalax ehrenbergi* superspecies in Israel. *Journal of Experimental Zoology* 238:81–88.
Reichman, O. J., Whitham, T. G., and Ruffner, G. A. 1982. Adaptive geometry of burrow spacing in two pocket gopher populations. *Ecology* 63:687–95.
Reid, V. 1973. Population biology of the northern pocket gopher on the Colorado mountain rangeland. In *Pocket gophers and Colorado mountain rangeland*, edited by G. T. Turner, R. M. Hansen, V. H. Reid, H. P. Tietjen, and A. L. Ward, 21–41. Bulletin 5545, Colorado State University Agricultural Experiment Station, Fort Collins.
Reig, O. A. 1970. Ecological notes on the fossorial octodontid rodent *Spalacopus cyanus* (Molina). *Journal of Mammalogy* 51:592–601.
Reig, O. A., C. Busch, M. O. Ortells, and J. R. Contreras. 1990. An overview of evolution, systematics, population biology, cytogenetics, molecular biology, and speciation in *Ctenomys*. In *Evolution of subterranean mammals at the organismal and molecular levels*, edited by E. Nevo and O. A. Reig, 71–96. Progress in Clinical and Biological Research, vol. 335. New York: Wiley-Liss.
Rickard, C. A., and N. C. Bennett. 1997. Recrudescence of sexual activity in a reproductively quiescent colony of the Damaraland mole-rat, by the introduction of a genetically unrelated male—a case of incest avoidance in "queenless" colonies. *Journal of Zoology, London* 241:185–202.
Rosi, M. I., M. I. Cona, S. Puig, F. Videla, and V. G. Roig. 1992. Estudio ecologico del

roeder subterraneo *Ctenomys mendocinus* en la precordillera de Mendoza, Argentina: Ciclo reproductivo y estructura etaria. *Revista Chilena de Historia Natural* 65:221–33.

Rosi, M. I., S. Puig, F. Videla, M. I. Cona, and V. G. Roig. 1996. Ciclo reproductivo y estructura etaria de *Ctenomys mendocinus* (Rodentia: Ctenomyidae) del Piedmonte de Mendoza, Argentina. *Ecologia Austral* 6:87–93.

Scheffer, T. H. 1938. Breeding records of Pacific Coast pocket gophers. *Journal of Mammalogy* 19:220–24.

Schramm, P. 1961. Copulation and gestation in the pocket gopher. *Journal of Mammalogy* 42:167–70.

Shanas, U., G. Heth, E. Nevo, R. Shalgi, and J. Terkel. 1995. Reproductive behaviour in the female blind mole rat *(Spalax ehrenbergi). Journal of Zoology, London* 237:195–210.

Sherman, P. W., E. A. Lacey, H. K. Reeve, and L. Keller. 1995. The eusociality continuum. *Behavioral Ecology* 6:102–8.

Simson, S., B. Lavie, and E. Nevo. 1993. Penial differentiation in speciation of subterranean mole-rats, *Spalax ehrenbergi* in Israel. *Journal of Zoology, London* 229:493–503.

Smolen, M. J., H. H. Genoways, and R. J. Baker. 1980. Demographic and reproductive parameters of the yellow-cheeked pocket gopher *(Pappogeomys castanops). Journal of Mammalogy* 61:224–36.

Spinks, A. C., G. van der Horst, and N. C. Bennett. 1997. Influence of breeding season and reproductive status on male reproductive characteristics in the common mole-rat, *Cryptomys hottentotus hottentotus. Journal of Reproduction and Fertility* 109:78–86.

Taylor, P. J., J. U. M. Jarvis, and T. M. Crowe. 1985. Age determination in the Cape mole-rat *Georychus capensis. South African Journal of Zoology* 20:261–67.

Vandenbergh, J. G. 1988. Pheromones and mammalian reproduction. In *The physiology of reproduction,* edited by E. Knobil and J. A. Neill, 1679–95. New York: Raven Press.

van der Horst, G. 1972. Seasonal effects of anatomy and histology on the reproductive tract of the male rodent mole. *Zoologica Africana* 7:491–520.

Vaughan, T. A. 1962. Reproduction in the plains pocket gopher in Colorado. *Journal of Mammalogy* 43:1–13.

Vehrencamp, S. L. 1983. Optimal degree of skew in cooperative societies. *American Zoologist* 23:327–35.

Verts, B. J., and L. N. Carraway. 1991. Summer breeding and fecundity in the camas pocket gopher, *Thomomys bulbivorus. Northwestern Naturalist* 72:61–65.

Villa-Cornejo, B., and R. M. Engeman. 1994. Reproductive characteristics of Merriam's pocket gopher *(Pappogeomys merriami merriami)* from Huitzilac, Morelos, Mexico (Rodentia: Geomyidae). *Southwestern Naturalist* 39(2):156–59.

———. 1995. Reproductive characteristics of the Hispid pocket gopher *(Orthogeomys hispidus hispidus)* in Veracruz, Mexico. *Southwestern Naturalist* 40(4):411–14.

Weir, B. J. 1974. Reproductive characteristics of hystricomorph rodents. In *The biology of hystricomorph rodents,* edited by I. W. Rowlands and B. J. Weir, 265–301. Symposia of the Zoological Society of London, 34. London: Academic Press.

Weir, B. J., and I. W. Rowlands. 1974. Functional anatomy of the hystricomorph ovary. In *The biology of hystricomorph rodents,* edited by I. W. Rowlands and B. J. Weir, 303–32. Symposia of the Zoological Society of London, 34. London: Academic Press.

Westlin, L., N. C. Bennett, and J. U. M. Jarvis. 1974. Relaxation of reproductive suppression in non-breeding naked mole-rats. In *The biology of hystricomorph rodents,* edited by I. W. Rowlands and B. J. Weir, 177–88. Symposia of the Zoological Society of London, 34. London: Academic Press.

Wilks, B. J. 1963. Some aspects of the ecology and population dynamics of the pocket gopher *(Geomys bursarius)* in southern Texas. *Texas Journal of Science* 15:241–83.

Williams, L. R., and G. N. Cameron. 1984. Demography and dispersal in Attwater's pocket gopher *(Geomys attwateri). Journal of Mammalogy* 65:67–75.

Williams, S. L., and R. J. Baker. 1976. Vagility and local movement of pocket gophers (Geomyidae: Rodentia). *American Midland Naturalist* 96:303–16.
Wood, J. E. 1949. Reproductive pattern of the pocket gopher *(Geomys breviceps brazensis)*. *Journal of Mammalogy* 30:36–44.
Zarrow, M. X., and J. H. Clark. 1968. Ovulation following vaginal stimulation in a spontaneous ovulator and its implications. *Journal of Endocrinology* 40:343–52.
Zenuto, R. R., E. A. Lacey, and C. Busch. 1999. DNA fingerprinting reveals polygyny in the subterranean rodent *Ctenomys talarum*. *Molecular Ecology* 8:1529–1532.

Part Two: Population & Community Ecology

Like other animals, subterranean rodents tend to occur in spatially distinct populations, each of which consists of numerous potentially interbreeding conspecifics. Because populations take on properties that are not evident at the level of the individual, studies of population biology are critical to our understanding of many aspects of an organism's biology. In particular, studies of population-level patterns and processes provide necessary links between individual-level traits and patterns of interspecific diversity.

As soon as multiple populations of a species arise, novel aspects of biological variation become evident. For example, the spatial distribution of conspecifics may vary among populations, as may patterns of social interactions. Morphological, physiological, and reproductive differences among populations may also become apparent, and these differences may take on new meaning when their effects on individual fitness are compared. Thus, although population-level attributes are strongly influenced by the characteristics of individuals, populations also exhibit a number of emergent properties that are not evident when considering isolated animals.

Populations of subterranean rodents, in turn, are members of larger biotic communities that include food resources and predators, as well as competitors for the subterranean niche. Members of subterranean species both influence and are influenced by other biota, and these interactions may further shape their responses to the environment. These interactions may also set the stage for long-term coevolutionary relationships such as those described in chapter 10. As a result, community-level interactions must also be considered when attempting to unravel the network of factors that affect convergence and divergence in subterranean taxa.

The chapters in part 2 of this volume consider various aspects of the population and community biology of subterranean rodents. As a result, these chapters form an integral link between the discussions of organismal biology found in the first part of the volume and the analyses of species-level diversification found in the third part. Although some aspects of population biology have long been studied by biologists working with subterranean rodents, the contents of the following chapters indicate that we still have much to learn regarding population-level patterns and processes in these taxa. In particular, the overviews provided here indicate that a number of widely held assumptions regarding the population biology of subterranean rodents may not be warranted.

Our exploration of the population biology of subterranean rodents begins with an analysis by Busch et al. (chap. 5) of the population ecology of these animals, including a discussion of the abiotic and biotic factors that determine how populations of these taxa are distributed in time and space. Busch et al. also consider demographic and life history parameters, such as natality, dispersal, and mortality, that are thought to influence population structure. In the process, Busch and colleagues reassess the validity of several long-standing generalizations regarding the foraging ecologies and life history patterns of subterranean rodents. Their conclusions indicate that, contrary to conventional wisdom, these taxa are not readily assigned to discrete ecological or life history strategies.

Cameron (chap. 6) takes these comparisons to a larger biotic scale by considering the community-level interactions between subterranean rodents and the other biotic elements of the habitats that they occupy. Subterranean rodents are known to play a vital role in many ecological communities through their effects on soil and vegetation structure. Cameron emphasizes, however, that many potential community-level interactions involving subterranean rodents have yet to be explored. Traditionally, interest in rodent-vegetation and rodent-herbivore interactions has been driven by practical concerns arising from the use of grassland habitats for agricultural purposes. As we move away from this management-oriented approach toward "pure" ecology, we are discovering that subterranean rodents inhabit complex communities that may contain numerous examples of coevolutionary interactions between these animals and other aspects of the biotic environment (see also Hafner, Demastes, and Spradling, chap. 10, this volume).

Changing focus somewhat, chapter 7 shifts from exploration of the interactions between rodent populations and their environments to consideration of the spatial and social interactions within populations of conspecifics. Although the behavior of some subterranean rodents, in particular the colonial Bathyergidae, has received considerable attention during recent years, the species studied to date represent only a few of the many spatial and social configurations possible among members of a population. Lacey considers the ecological and demographic factors thought to be associated with coloniality in these species, but also explores the importance of studies of solitary taxa to our understanding of the spatial and social biology of subterranean rodents.

As the chapters in part 2 indicate, subterranean rodents offer a wealth of opportunities for research on population- and community-level issues. Although these animals share a number of ecological and demographic characters associated with life underground, responses to these conditions vary substantially, and sometimes dramatically, among species. For each of the topics considered, data are available for only a relatively small subset of subterranean species, suggesting that additional research may reveal an even richer variety of responses to environmental conditions. Given the role of population biology as a bridge between micro- and macroevolutionary phenomena, future studies of ecology, demography, behavior, and genetic structure should serve to increase our understanding of numerous of aspects of the biology of subterranean rodents.

CHAPTER FIVE

Population Ecology of Subterranean Rodents

Cristina Busch, C. Daniel Antinuchi, J. Cristina del Valle, Marcelo J. Kittlein, Ana I. Malizia, Aldo I. Vassallo, and Roxana R. Zenuto

Interactions between animals and their environments are a fundamental aspect of the biology of any species. At the level of the individual, interactions with the environment may influence numerous traits, including aspects of morphology (chap. 1), physiology (chap. 2), and reproduction (chap. 4), as well as patterns of communication among conspecifics (chap. 3). At the level of the population, interactions with the environment determine where animals can live, as well as whether conspecifics occur alone or in groups (chap. 7). Environmental conditions may also influence population size, including patterns of natality, mortality, and dispersal. In short, ecological interactions between animals and their environment are a critical part of a species' biology.

For many mammals, including many rodents, subterranean burrows play an important role in their interactions with the environment. Burrows may be used as places of refuge and storage, as well as nest sites for young (Carter and Encarnação 1983; Ågren, Zhou, and Zhong 1989; Antinuchi and Busch 1992; Carter and Rosas 1997; Hodara, Suárez, and Kravetz 1997). Subterranean rodents often have extensive and elaborate burrows that contain multiple chambers and openings (Llanos and Crespo 1952; Yáñez and Jaksic 1978; Scheck and Fleharty 1980; Pearson 1984; Antinuchi and Busch 1992; Ojeda et al. 1996). The construction, use, and maintenance of these burrows represents a central element of the animals' lives. Given their importance to subterranean species, underground burrows are expected to substantially shape the ecology of the rodents that inhabit them.

Despite the assumption that the subterranean niche imposes similar selective pressures on all of its mammalian inhabitants, regional variation in climate, soils, and vegetation may be important in generating adaptive differences among populations and species. As a result, seemingly convergent taxa may, in fact, display different local adaptive peaks that reflect variation in local environments. Further, because adaptations to distinct environments may involve varying degrees of commitment to subterranean life (e.g., from strictly subterranean mole-rats to primarily surface-foraging tuco-tucos), differences in population ecology may reflect the combined effects of underground and surface-related selective pressures.

There is a clear need to evaluate the validity of long-accepted generalizations about the population ecology of subterranean rodents. In particular, assumptions regarding demographic parameters such as dispersal and rates of reproduction and predation need to be reconsidered in light of evidence that subterranean species are not subject to uniform selective pressures. In this chapter, we explore how differences in local environments influence the population ecology of subterranean rodents. Given the central role of the burrow system in the lives of these animals, we begin by considering how burrow structure and location are influenced by environmental conditions. We then explore the foraging ecology of subterranean rodents and consider the effects of the subterranean ecotope on the life history strategies of these animals. We conclude with a discussion of predation that underscores the prevalence of interspecific differences in response to local environments among subterranean species.

Habitat Selection by Subterranean Rodents

A critical issue that must be addressed for any species is the nature of "suitable" habitats in which the animals can live and reproduce. In other words, what determines why individuals live in some habitats but not in others? For subterranean rodents, the answer to this question appears to lie, at least in part, in the global aridization and emergence of an open-country biota at the beginning of the mid- to late Cenozoic (Nevo 1979; Cook, Lessa, and Hadly, chap. 9, this volume). These climatic changes correspond to a period of extensive diversification by subterranean taxa, and the tendency to occur in open, arid or semiarid habitats remains evident among these animals today. With the exception of *Rhizomys* and *Cannomys,* which occur in bamboo thickets or forested areas of southeastern Asia, modern subterranean rodents are found primarily in nonforest biomes such as grasslands, savannas, steppes, and deserts.

The tendency of subterranean rodents to occupy open habitats may, in part, reflect the nature of life underground. All of the species considered in this volume inhabit extensive underground burrow systems that they excavate themselves. Burrow excavation is not cheap; although the energetic costs of burrowing vary with tunnel size and shape as well as with soil characteristics such as density and cohesiveness, burrowing requires 360–3,400 times as much energy as moving the same distance across the surface (Vleck 1979; Jarvis and Bennett 1991). Thus the need to construct burrows may substantially influence the distributions of subterranean species by limiting them to habitats in which tunnel excavation is energetically feasible.

Determinants of Burrow Locations

SOIL AND VEGETATION. Most subterranean rodents tend to live in porous soils, or at least in well-drained soils of poor water-holding capacity (Contreras 1973; Brown and Hickman 1973; Vleck 1981; Reichman, Whitham, and Ruffner 1982; Cameron et al. 1988; Heth 1989; Williams and Cameron 1990; Antinuchi and Busch 1992). This tendency is thought to reflect physical and energetic limitations on digging through wet soils as well as physiological limitations imposed by the high partial pressures of CO_2 and low partial pressures of O_2 characteristic of subterranean burrows, which are exaggerated in wet soils (Buffenstein, chap. 2, this volume). A final factor thought to influence the distribution of burrow systems is soil depth: in shallow soils (e.g., soils in which underlying rock is close to the surface), the animals may not be able to avoid thermal stress by retreating to deep tunnels (Turner et al. 1977).

Where soil conditions are favorable, the absence of suitable vegetation may preclude occupation by subterranean rodents. As herbivores, these rodents feed extensively on vegetation, so appropriate plant species must be available to support the animals. In addition to providing critical food resources, vegetation may determine the distribution of subterranean rodents through its effects on patterns of ventilation. In particular, the structure and density of vegetation may influence patterns of heat flux within subterranean burrows, which may, in turn, determine whether a given habitat is suitable for underground existence. For example, Comparatore, Maceira, and Busch (1991) found that in the warm months, favorable zones for *Ctenomys talarum* were those with the greatest density and height of vegetation.

Heterogeneous distributions of appropriate soils and vegetation are thought to contribute substantially to the patchy spatial distributions characteristic of many subterranean species (Steinberg and Patton, chap. 8, this volume). For example, the local distribution of *C. talarum* is influenced by topographic, soil, and vegetation characteristics of the habitat; because areas of suitable soils and vegetation are patchily distributed, individuals also tend to be spatially clumped (Busch et al. 1989). In contrast, *C. australis,* which inhabits ecologically homogeneous sand dunes, has a relatively continuous local distribution (Zenuto and Busch 1998). These differences are significant, as the local spatial distribution of conspecifics may substantially influence patterns of dispersal, social behavior, and genetic structure.

DIGGING BEHAVIOR. The behaviors used to excavate tunnels may, in part, determine the range of soil types in which burrows can be constructed. All subterranean rodents excavate burrows by shearing soil from the wall of a tunnel, pushing the loosened soil behind them, and then moving the soil through a lateral tunnel to the surface (Vleck 1979). The specific behaviors used to accomplish these tasks, however, vary among species. Hildebrand (1985) distinguished three modes of digging: scratch digging, chisel-tooth digging, and head-lift digging. A detailed discussion of these behaviors and the associated morphological adaptations is provided in chapter 1 of this volume.

Both the mode of digging and the morphological structures used for tunnel excavation appear to be correlated with the type(s) of soil that a species inhabits. For example, Lessa and Thaeler (1989) found that *Geomys*, a scratch digger, inhabits only sandy and friable soils; in contrast, *Thomomys*, a scratch and chisel-tooth digger, occupies a much wider range of soil types, including hard soils and soils with a high clay content. This difference was attributed to the use of teeth, rather than claws, to loosen soil. Incisor enamel is the hardest material in the vertebrate body, and incisors are rooted in the skull, providing a solid base of support for the teeth during digging. In contrast, claws are formed of comparatively soft keratin and are attached to relatively flexible digits. As a result, chisel-tooth diggers are capable of handling more "difficult" soils than are scratch diggers.

Similar relationships between digging mode and habitable soil type are evident among the bathyergids and ctenomyids. Within the Bathyergidae, the scratch-digging genus *Bathyergus* inhabits soft, wind-blown sand dunes. In contrast, the chisel-tooth diggers *Heterocephalus* and *Cryptomys* inhabit compact and hard soils (Jarvis and Bennet 1991). Similarly, a recent study of two species of *Ctenomys* revealed a close association between the structures used for digging and soil hardness (Vassallo 1998). Relationships between mode of digging and soil type probably reflect the nature of the soils inhabited over evolutionary time scales. On a more immediate basis, however, the behaviors used to excavate tunnels may constrain the range of soils that a species can occupy.

One means of minimizing this limitation is to employ multiple digging modes. Some subterranean species show considerable variation in digging behavior in response to different soil types. For example, *Ctenomys pearsoni* behaves as a scratch digger when living in friable soils, but uses its incisors to dig when living in harder soils (Altuna, Izquierdo, and Tassino 1993; Giannoni, Borghi, and Roig 1996). In the plateau zokor *Myospalax baileyi*, digging speed and duration are influenced by soil hardness. With increasing soil hardness, the digging speed was decreased, but the digging duration was increased (Wang et al. 1994). As previously noted, Lessa and Thaeler (1989) reported that representatives of *Thomomys* and *Pappogeomys* loosen sandy, friable soils with their forelimbs. When confronted with more compact soils, however, these taxa make extensive use of their incisors to loosen soil. These observations indicate that, within the limits on distribution imposed by the energy costs of burrowing in particular habitats (Vleck 1981), some species may increase their range of habitable soils by adjusting their digging behavior in response to soil hardness.

OTHER DETERMINANTS OF BURROW LOCATIONS. The range of habitats occupied by subterranean rodents may also be affected by interactions between metabolic rate and body size. In general, larger-bodied species, or those with higher metabolic rates, are expected to have more stringent requirements regarding soil conditions. According to the thermal stress hypothesis, which postulates that individuals producing greater quantities of heat will settle in soils with higher heat flow (McNab 1966; MacMillen and Lee 1970), correlations between body size and soil depth and friability reflect the inverse relationship between body size and burrow temperature. Vleck (1979), however, argues that low metabolic rate and small body size are adaptations to the high cost of obtaining food by burrowing. Because this cost increases as soil friability and plant productivity decrease, smaller-bodied taxa may be able to inhabit areas that are not suitable for larger-bodied species.

As an extension of these arguments, it has been predicted that small-bodied species will inhabit deep, friable soils in regions where they have no contact with larger-bodied species, but that they will be forced into shallower, more compacted soils in areas of sympatry (McNab 1966). Recent field studies provide partial support for this hypothesis. Malizia, Vassallo, and Busch (1991) found that where distributions of *Ctenomys australis* (mean body weight = 360 g) and *C. talarum* (mean body weight = 118 g) overlap, the former inhabits areas with sparse vegetation and deep, sandy soils, but the latter inhabits areas with dense vegetation and more compact, shallow soils. Field experiments involving these species suggest that interference competition is a significant, but not the most important, factor in sustaining this pattern of habitat differentiation (Vassallo 1994). In areas of allopatry, *C. talarum* occurs in a more diverse array of habitats, ranging from deep sandy to shallow sandy-loam soils. The role of interspecific competition in shaping the biology of subterranean rodents is explored further in chapter 6 of this volume.

Numerous traits influence the locations of burrows. Among them, the most conspicuous are the physical properties of soils, the ability to dig in different soils, the metabolic costs of digging in different soils, and the ability to use different soils when faced with competition. Further studies are needed on other factors, such as climatic seasonality, that may affect burrow location and design.

The Burrow System

Extensive use of a subterranean burrow system is the trait that defines the taxa considered in this volume. Given the central role of the burrow in the ecologies of these animals, it is somewhat surprising that the architecture of burrow systems remains only poorly known. Due to the difficulty of excavating extensive tunnel systems, few studies have examined the structure of burrows in their entirety. Those studies that have characterized complete burrow systems often lack important complementary data, such as measures of soil hardness and granularity, amount of local rainfall, and plant productivity. As a result, it is difficult to make generalizations regarding interactions between environmental parameters (e.g., soil type) and burrow structure.

Despite these shortcomings, a basic blueprint for burrow system design has been identified (fig. 5.1). In general, the burrows of subterranean rodents consist of numerous shallow foraging tunnels that are connected to a single deep, central tunnel. Foraging tunnels represent as much as 80–95% of the area of the burrow system (Miller 1957; Jarvis and Bennett 1991; Antinuchi and Busch 1992; Heth 1992), with the rest consisting of one or more nest chambers, food storage chambers, and in some cases, dead-end tunnels. This general configuration has been reported for the burrow systems of bathyergids, spalacines, geomyids, and ctenomyids (Brown and Hickman 1973; Hickman 1977, 1983a,b, 1984; Vleck 1981; Andersen 1988; Antinuchi and Busch 1992; Rosi, Cona, et al. 1996), suggesting that it is a common feature of subterranean taxa.

Given the substantial energetic costs of tunnel excavation, natural selection is expected to favor burrow structures that minimize energy expenditure. Optimal burrow diameter is thought to be related to body size and soil type (Vleck 1981), with tunnel diameters rarely exceeding the minimum needed to accommodate an animal. Burrow configuration may also be optimized to minimize the costs of excavating. Studies of geomyids (Vleck 1981; Andersen 1988) and ctenomyids (Antinuchi and Busch 1992; Rosi, Cona et al. 1996) indicate that the angles at which tunnels connect, as well as the distances between tunnel junctions, are consistent with those predicted by models that minimize burrowing costs.

Burrow size and design may also be influenced by the distribution and productivity of critical plant resources (Reichman, Whitham, and Ruffner 1982). A detailed study of burrow design in *Nannospalax ehrenbergi* (Heth

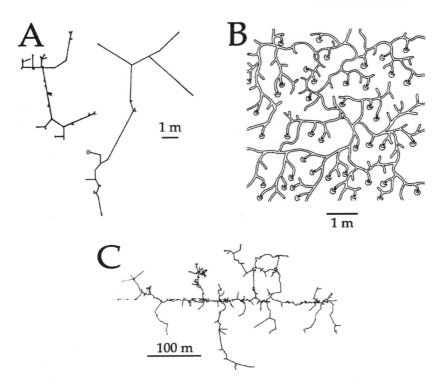

FIGURE 5.1. Burrow designs for three genera of subterranean rodents with different social systems. (A) Burrow of an adult male (right) and an adult female (left) of the solitary tuco-tuco *Ctenomys talarum*. (B) Burrow of a colony of the social *Spalacopus cyanus*. (C) Burrow of a colony of the eusocial *Heterocephalus glaber*. (A from Antinuchi and Busch 1992; B from Reig 1970; C from Brett 1991a)

1989) revealed that burrow length varied among populations as a function of both plant productivity and soil hardness. Specifically, burrows constructed in unproductive habitats tended to be longer, suggesting that they encompassed larger foraging territories. Food resource characteristics have also been implicated in the construction of the extremely large communal burrows occupied by social bathyergids. In *Heterocephalus* and *Cryptomys*, the difficulty of locating food in poor habitats (i.e., those with patchily distributed vegetation) is thought to be the primary factor favoring the evolution of sociality (Lacey, chap. 7, this volume). Extensive tunneling by colonies of bathyergid mole-rats has resulted in burrows that, although hundreds of meters long, retain a basic design and biomass ratio (number of animals per meter of burrow) similar to those of solitary species living in more productive environments (Jarvis and Bennett 1991).

An additional way in which subterranean rodents may reduce the costs of burrow excavation is by limiting digging to those portions of the year when it is least energetically expensive. Burrowing activity may vary seasonally (Andersen 1988; Cox and Hunt 1992), as has been shown for bathyergid and spalacine mole-rats; in these taxa, mound production occurs only during the rainy season (Brett 1991a; Heth 1989), when the costs of digging are presumed to be lower because the soil has been loosened by precipitation. In keeping with this hypothesis, Hickman (1983b) found that, in *Tachyoryctes,* digging through soils moistened by rain is very rapid. In contrast, Cox and Hunt (1992) found that pocket gophers tend to produce mounds at higher frequencies during early spring and late summer or autumn, and concluded that availability of forage is the most important constraint on tunneling. Further studies are needed to determine to what extent seasonality of burrow excavation is determined by precipitation as opposed to other environmental factors.

Foraging Ecology of Subterranean Rodents

Food resources have been implicated as important factors influencing both burrow location and burrow system size. The evolution of subterranean rodents has been linked to a climatic shift toward warm and xeric conditions (Nevo 1979; Cook, Lessa, and Hadly, chap. 9, this volume). These conditions are also thought to have favored the evolution of belowground storage organs in plants (Stuebe and Andersen 1985), an adaptation that may have facilitated animal exploitation of the subterranean niche by providing locally abundant food stores that could be reached via underground tunnels. As noted above, most modern subterranean animals occur in arid areas, although whether this association reflects foraging ecology or soil conditions (or both) remains to be determined.

Optimal foraging theory assumes that natural selection maximizes fitness by optimizing net energy gain per unit feeding time. As a result, at low food abundance and high foraging energy expenditures, food generalists should be favored over food specialists (Schoener 1971). Because excavating tunnels to locate food is energetically expensive (Vleck 1979, 1981; Andersen and McMahon 1981), subterranean rodents are expected to be generalists whose diets contain a large proportion of underground vegetation (Vleck 1979). Here, we examine the diet preferences and foraging strategies of subterranean rodents (table 5.1) to determine whether this general prediction of optimal foraging theory is upheld among the taxa considered in this volume.

Food Resources and Foraging Behavior

In general, geophytes and other subterranean plant structures constitute the majority of food items consumed by subterranean rodents, although the proportion of the diet consisting of aerial plant structures tends to increase as the biomass of underground vegetation in the habitat decreases. Geophytes are the main food resource for bathyergid and spalacine mole-rats that inhabit arid regions in Africa and the Middle East (Lovegrove and Jarvis 1986; Heth, Golenberg, and Nevo 1989; Jarvis and Bennett 1991; Heth 1992). The highly social *Heterocephalus glaber* consumes subterranean tubers, bulbs, and corms from a wide variety of geophytic species. Colony members cooperate to excavate tunnels leading to tubers (Jarvis 1978; Sherman, Jarvis, and Alexander 1991), and this cooperation may allow the animals to inhabit areas that are only very sparsely vegetated.

Members of the solitary bathyergid genera *Bathyergus* and *Georychus* include aerial plant parts in their diets. Foraging, however, occurs entirely below ground; the animals feed by approaching plants from underneath and then pulling vegetation down into the burrow (Davies and Jarvis 1986). In contrast, *Heliophobius*, the remaining bathyergid genus, spends limited periods of time foraging on the soil surface. Although the animals rarely leave their burrow systems entirely, individuals typically emerge to a distance of half a body length while foraging on surface-growing vegetation (Delany 1986). The foraging behavior of *N. ehrenbergi* most closely resembles that of *Bathyergus* and *Georychus* (Heth, Golenberg, and Nevo 1989), and the foraging behavior of the rhizomyine genus *Tachyoryctes* is most similar to that reported for *Heliophobius* (Delany 1986).

Foraging by geomyids and ctenomyids is more generalized in that all portions of plants may be consumed. Aerial plant parts are important diet items, although subterranean parts may constitute the majority of food consumed, particularly when vegetation is scarce (Tilman 1983; Stuebe and Andersen 1985; Cox 1989; Contreras and Gutiérrez 1991; Cox and Hunt 1992, 1994). For example, Cox (1989) determined that, on average, 21% of the diet of *Thomomys talpoides* consists of root material. For *G. bursarius*, this figure is 30% (Luce, Case, and Stubbendiek 1980), and for *G. attwateri*, subterranean plant parts constitute 60% of the diet (Williams and Cameron 1986).

Among ctenomyids, subterranean plant parts appear to represent a smaller proportion of the diet. In *C. talarum* and *C. australis*, roots never constitute more than 22% of the food ingested (Comparatore, Cid, and Busch 1995). Although ctenomyids forage within their tunnels for roots and subterranean stems, most foraging occurs above ground, as the animals venture away from their tunnels for brief periods to gather vegetation growing

TABLE 5.1 FOOD SELECTIVITY AND FEEDING STRATEGIES OF SUBTERRANEAN RODENTS

Taxon	Social status	Diet	Searching	Caches	Source
Bathyergidae					
Bathyergus	Solitary	≥60% aerial vegetation; rest roots, small geophytes	Underground; pull aerial fraction into burrow	Unknown	Brett 1991a; Delany 1986
Georychus	Solitary	<15% aerial vegetation; rest small geophytes	Underground; pull aerial fraction into burrow	Food storage for dry and hot seasons	Brett 1991a; Delany 1986
Heliophobius	Solitary	Entirely geophytes and roots	Underground	Food storage for dry and hot seasons	
Cryptomys	Colonial	Entirely geophytes and roots	Underground	Food storage for dry and hot seasons	
Heterocephalus	Colonial	Entirely geophytes and roots	Underground	Food storage for dry and hot seasons	Brett 1991a; Delany 1986
Muridae					
Rhizomyinae					
Tachyoryctes	Solitary	Wide range of roots and shoots	Underground; limited periods spent on surface	Unknown	Delany 1986
Spalacinae					
Nannospalax ehrenbergi	Solitary	61% geophytes; rest herbs and shrubs	Underground; pull aerial fraction into burrow	Food storage for gestation and lactation periods and for hot and dry season	Nevo 1991; Heth, Golenberg, and Nevo 1989

Geomyidae					
Geomys attwateri	Solitary	40% aerial vegetation, mainly perennial monocots; rest annual monocots, annual dicots, and perennial dicots	Usually underground; pull entire plants into burrow	Unknown	Williams and Cameron 1986
Thomomys talpoides	Solitary	79% aerial vegetation, mainly forbs; rest shrubs and grasses	Underground and above ground; pull entire plants into burrow	Store low-protein items in underground caches	Stuebe and Andersen 1985; Turner et al. 1977
Octodontidae					
Spalacopus cyanus	Colonial	Aerial and succulent subterranean stems and tubers of huilli (Liliceae)	Underground and above ground	Food storage	Contreras and Gutiérrez 1991
Ctenomyidae					
Ctenomys talarum	Solitary	97% aerial vegetation, mainly grasses	Above ground and underground	No food storage	Comparatore, Cid, and Busch 1995
Ctenomys australis	Solitary	97% aerial vegetation, mainly grasses	Above ground and underground	Unknown	Comparatore, Cid, and Busch 1995
Ctenomys mendocinus	Solitary	Prefer grasses to forbs, shrubs, and cacti	Above ground and underground	Food storage	Madoery 1993; Camín and Madoery 1994

on the soil surface. Although quantitative data are lacking, aerial plant parts appear to be the predominant form of vegetation consumed by *C. sociabilis* and *C. haigi* (E. Lacey, personal communication). Indeed, entrances to ctenomyid burrows are often encircled by an area of cropped vegetation that is created by the animals while foraging above ground (Reig 1970; Comparatore, Cid, and Busch 1995).

The reasons for these interspecific differences in diet have not been examined quantitatively. Williams and Cameron (1986) proposed that the proportions of subterranean plant parts consumed by geomyids reflect differences in the foraging behavior of these animals. Specifically, the proportion of aerial plant parts in the diet should be positively correlated with the amount of time that the animals spend out of their burrows. Similarly, Comparatore, Cid, and Busch (1995) suggested that the prevalence of aerial plant parts in the diet of *Ctenomys* reflects this group's tendency to forage above ground, and that interspecific differences in the diet are influenced by the nature of the available vegetation and by plant-animal interactions. Although the tendency to forage above ground provides a reasonable proximate explanation for interspecific differences in the amount of aerial vegetation consumed, this hypothesis does not explain why the tendency to forage above ground varies. The answers to this question are likely to be complex, and may reflect surface conditions (e.g., temperature, predation risk, competition) and the quality and quantity of vegetation available to the animals over evolutionary time scales.

Food Selectivity

Subterranean plant tissues may represent a more variable resource, in terms of nutritional value, than aboveground vegetation (Andersen 1987). This difference in nutritional quality may influence food selectivity in subterranean rodents. In addition to the distribution of suitable roots and tubers, the costs of foraging for these items may prevent some species from specializing on subterranean plant parts. Heth, Golenberg, and Nevo (1989) argued that subterranean herbivores cannot afford to be selective feeders because the costs of searching for preferred food items would exceed the benefits of this selectivity; as a result, subterranean rodents should utilize all foods that they encounter. In general accordance with this prediction, pocket gophers (*T. talpoides:* Stuebe and Andersen 1985; *Geomys attwateri:* Williams and Cameron 1986), tuco-tucos (*C. talarum* and *C. australis:* Comparatore, Cid, and Busch 1995; *C. mendocinus:* Madoery 1993), and bathyergid (*H. glaber:* Brett 1991a) and spalacine mole-rats (*N. ehrenbergi:* Nevo 1979) have been reported to consume a wide variety of plants.

Although most subterranean species appear to be diet generalists, this does not mean that the diets of all species are similar. Taxon-specific preferences for specific types of food have been reported for a number of species, including several species of geomyid pocket gophers (Russell and Baker 1955; Spencer et al. 1985). Several studies found high-quality perennial dicots to be the food items most frequently consumed by most geomyids, including species of *Thomomys* (Aldous 1951; Ward and Keith 1962; Miller 1964; Vaughan and Hansen 1964) and *Geomys* (Russell and Baker 1955; Schmidly 1983), as well as *Pappogeomys castanops* (Hedgal et al. 1964). Williams and Cameron (1986), however, found that *Geomys* consumes primarily lower-quality perennial monocots, which predominate in the grassland habitat they studied. Similarly, among ctenomyids, *C. talarum, C. australis,* and *C. mendocinus* select monocots over dicots (Comparatore, Cid, and Busch 1995; Madoery 1993).

There have been few cafeteria-style feeding experiments in which availabilities of potential foods were controlled and diet preferences assessed more precisely. Andersen and MacMahon (1981) tested *T. talpoides* with a large number of plant species and concluded that the animals were opportunistic, avoiding only three species and showing no clear patterns of selection among the others. Another study, however, showed that *T. monticola* was able to discriminate among the plant foods tested, harvesting roots and tops selectively (Jenkins and Bollinger 1989).

The food preferences of *C. mendocinus* (Camin and Madoery 1994) and *C. talarum* (J. C. del Valle et al., unpublished data) were also studied using cafeteria-style choice trials. Both studies showed that *Ctenomys* were capable of selective foraging in the laboratory and confirmed that these species generally prefer the aboveground parts of grasses over shrubs and roots.

In summary, although subterranean rodents are generally thought to be foraging generalists, reports of dietary preferences indicate that the animals are capable of selective foraging. Preferences for aerial versus subterranean plant parts are evident, as are preferences for particular plant species. These dietary preferences are believed to reflect gender-specific nutritional needs and reproductive status, as well as changes in the nutritional condition or palatability of plant resources (Williams and Cameron 1986; Heth, Golenberg, and Nevo 1989; Comparatore, Cid, and Busch 1995). In addition, variation in local habitats and climatic conditions may contribute to regional and seasonal dietary differences among subterranean rodents. Although the information summarized here does not address the issue of diet breadth directly, these data suggest that subterranean rodents show greater selectivity of food resources than has traditionally been attributed to them.

Foraging Strategies

CACHING FOOD. For species in which the costs of burrow excavation or the availability of food resources vary seasonally, caching food within the burrow system may provide a means of minimizing foraging during periods when this behavior is particularly expensive. Most subterranean rodents accumulate underground food stores, which they consume at times when cold or aridity precludes plant growth or tunnel excavation. Caching food in excess of that required to meet immediate nutritional needs presumably enhances fitness by eliminating the need to forage when tunnel excavation is most difficult or food resources are least abundant (Stuebe and Andersen 1985).

The food caches of *T. talpoides* are usually located in shallow, lateral chambers that are loosely sealed off from the main burrow system (Turner et al. 1977). This species also hoards vegetation in caches placed in or under the snow. Analyses of stomach contents, cache contents, and availability of plant species (Stuebe and Andersen 1985) suggest that the animals consume high-protein items as they are encountered, but cache low-protein items. The quantities of food stored are sufficient to allow individuals to endure adverse environmental conditions without entering negative energy or protein balance (Stuebe and Andersen 1985).

Caching has also been reported for *Georychus capensis* and *N. ehrenbergi* (Brett 1991a; Nevo 1991). Both species store bulbs and corms that subsequently sprout and grow, although the sprouts are not eaten. The choice of plants cached is not random. In particular, individuals appear to select large geophytes for storage (Brett 1991a). Food storage may be particularly important to female *N. ehrenbergi* during gestation and lactation, and may provide animals of both sexes with food reserves during the hot, dry summer season (Nevo 1991).

In contrast to the food caches found in the burrows of other subterranean rodents, caches in ctenomyid burrows commonly contain plant parts collected on the surface (Reig 1970). This difference in cache contents is consistent with the greater tendency of ctenomyids to forage above ground. A more conspicuous exception to the caching behavior described above occurs in *H. glaber;* rather than storing harvested food items, members of this species appear to reuse live food items by allowing tubers to regrow between successive foraging bouts at a given location (Brett 1991a).

Although food caching appears to be quite widespread among subterranean rodents, few studies have considered the energetic consequences of this behavior. If caching functions to provide individuals with food resources during periods of adverse environmental conditions, the extent to which animals cache should be positively correlated with the severity of the environmental extremes they experience. To date, however, no studies have attempted to compare caching effort by different species, and thus the ecological correlates and significance of the behavior have not been assessed. Comparative studies of cache size and energy content will yield important insights into the adaptive significance of this aspect of foraging.

LOCATING FOOD. The physical properties of soil are believed to prevent subterranean rodents from using vision or olfaction to locate food below ground (Francescoli, chap. 3, this volume). Instead, a number of authors (Williams and Cameron 1990; Brett 1991a) have suggested that subterranean rodents encounter food fortuitously while digging, rather than using sensory stimuli to determine the direction of tunnel excavation. Apparent support for this hypothesis comes from observations suggesting that *Thomomys bottae* (Williams and Cameron 1990), as well as *H. glaber* and some species of *Cryptomys* (Brett 1991a), are unable to locate edible plants that are located within a few centimeters of a tunnel.

Other studies, however, suggest that subterranean rodents possess remarkably precise abilities to locate their foraging tunnels in areas of abundant resources. The burrow systems of at least some subterranean rodents tend to be located within productive areas of the habitat, and within a burrow system, feeding tunnels are located in the most productive areas of an animal's home range. This pattern of burrow location has been reported for *Spalacopus cyanus* (Contreras and Gutiérrez 1991) and *T. bottae* (Reichman and Smith 1985). Similarly, Steuter et al. (1995) suggested that *G. bursarius* has a "fine-grained" view of its environment that allows it to locate and select compact, high-quality food items such as forb taproots.

The mechanisms by which subterranean rodents are able to place their burrows in areas of high productivity have not been determined. For species that routinely visit the surface, visual cues may play an important role in locating food resources (Andersen 1987), as has been suggested for *S. cyanus* (Contreras and Gutiérrez 1991). For species that are not active above ground, area-restricted searches for food resources may increase the probability of encountering locally abundant food items. Many edible plants have clumped distributions; once a food item has been encountered, animals that dig tunnels with numerous closely spaced turns are likely to encounter additional food resources (Huntly and Reichman 1994). This pattern of tunneling has been documented in geomyid pocket gophers (Benedix 1993; Jarvis and Sale 1971), bathyergid and spalacine mole-rats (Brett 1991a; Heth, Golenberg, and Nevo 1989), and ctenomyids (Antinuchi and Busch 1992), suggesting that subterranean herbivores search their habitat systematically as they construct foraging tunnels.

In summary, much remains to be learned about the foraging ecology of subterranean rodents. For many species, descriptive data on diet composition and foraging behavior are still needed. For well-studied species, optimality models of foraging and burrow construction must be tested via experimental manipulation of food distribution to determine how this variable affects the behavior of subterranean rodents. Cafeteria-style tests of food preferences would help to determine the nature and extent of dietary specialization. Finally, field studies of caching and foraging behavior would allow tests of optimal foraging models under the conditions faced by free-living animals.

Population Structure of Subterranean Rodents

As soon as multiple conspecifics accumulate in the same patch of habitat, patterns of population structure begin to emerge. Understanding the population structure of a species can yield important insights into ecological relationships, including ecological determinants of behavioral and genetic diversity (Lacey, chap. 7; Steinberg and Patton, chap. 8, this volume). Variables commonly used to characterize population structure include the density and spatial distribution of individuals as well as ages of individuals and the sex ratio of adults in the population.

Population Density and Spatial Distribution

Population densities for subterranean rodents tend to be lower than those for many surface-dwelling species. Although densities for the latter routinely reach 150 individuals per hectare (e.g., *Microtus:* Tamarin 1977), densities for subterranean taxa seldom exceed 80 individuals per hectare and are often considerably lower (e.g., Geomyidae: Ingles 1952; Howard and Childs 1959; Smolen, Genoways, and Baker 1980; Williams and Cameron 1984; Ctenomyidae: Pearson 1959; Gallardo and Anrique 1991; Rosi et al. 1992). Exceptions, however, do occur: more than 200 individuals per hectare have been reported for *C. talarum* (Pearson et al. 1968) and *Tachyoryctes splendens* (Jarvis 1973), suggesting that at least some subterranean species occur in very dense concentrations.

Low population densities may arise in part because of the tendency of subterranean species to be solitary, with each adult maintaining its own exclusive territory (Lacey, chap. 7, this volume). These territories are shared only by mothers and unweaned offspring and, briefly, by males and females during the breeding season. Exceptions to this pattern occur in a few species, most notably in the bathyergid genera *Heterocephalus* and *Cryptomys* (Lacey, chap. 7, this volume). The effects of sociality on population density are difficult to predict. On the one hand, burrow sharing by multiple adults may lead to higher population densities. On the other hand, although members of group-living species may be more spatially aggregated, their densities may not differ from those of solitary species if social groups are widely scattered across the habitat. To date, quantitative comparisons of population densities for solitary and social taxa have not been conducted, and thus the effects of group living on density remain to be determined.

Population densities may also be limited if environmental conditions preclude burrow construction in some portions of the habitat. Because the proportion of the habitat that can be used may vary, population density may vary both intra- and interspecifically as a function of factors such as soil heterogeneity or vegetation type. Intraspecific comparisons of population densities for several subterranean species support this hypothesis. In one study, population densities for *T. bottae* ranged from fewer than 6 individuals per hectare in desert habitats to about 60 individuals per hectare in grassland habitats (Patton 1990). In a second study, reported densities for this species varied from 5 individuals per hectare in desert scrub to 80 individuals per hectare in cultivated alfalfa fields (Patton and Brylski 1987).

This intraspecific variation in population density is not unique to pocket gophers. Nevo (1991) found that densities of *N. ehrenbergi* decreased with increasing aridity of the habitat. Similarly, among ctenomyids, densities of 15 individuals per hectare versus 55 individuals per hectare were found for populations of *C. talarum* occurring in grasslands characterized by different amounts of plant biomass (Busch et al. 1989; Malizia, Vassallo, and Busch 1991). It seems likely that additional examples of this type of variation will be reported as more studies examine multiple populations of conspecifics.

As indicated above, it is thought that populations of subterranean rodents are typically patchily distributed due to the heterogeneous distribution of suitable habitats. More uniform spatial distributions, however, were reported for *T. talpoides* with an increase in population density (Hansen and Remmenga 1961). Uniform population distributions have also been reported for *Ctenomys* species under high-density conditions and in poor habitats (Pearson et al. 1968; Gallardo and Anrique 1991; Rosi et al. 1992). A random population distribution was found for *C. australis* (Zenuto and Busch 1998) in sand dunes, which are considered an ecologically homogeneous habitat. Thus it appears that although many populations of subterranean rodents are patchily distributed, this type of spatial distribution is not obligate; changes in population density or habitat heterogeneity may lead to a more even dispersion of individuals, which may, in turn, promote changes in other behavioral or demographic parameters.

Adult Sex Ratio

The ratio of adult males to females in a population varies considerably within and among species of subterranean rodents (table 5.2). Female-biased sex ratios have been reported for a number of solitary species, although sex ratios approaching parity occur in some solitary geomyids and ctenomyids. Several of these species exhibit intraspecific variation in sex ratio, with both female-biased and unbiased sex ratios reported for different populations. To date, only a single example of a male-biased sex ratio has been reported: colonies (and therefore, presumably, populations) of *H. glaber* typically contain more males than females (Brett 1991b).

In some species, intraspecific variation in adult sex ratio appears to be density-dependent, with ratios becoming increasingly female-biased as density increases. In a high-density population of *C. talarum* studied by Busch et al. (1989) and Malizia and Busch (1991), the sex ratio at birth was even, indicating that the female-biased adult sex ratio was not due to differential conception rates or prenatal mortality. A 1:1 sex ratio was also evident among subadults in this population (Busch et al. 1989). These findings suggest that biased sex ratios arise due to differential mortality among adults.

TABLE 5.2 SEX RATIOS IN DIFFERENT POPULATIONS OF SUBTERRANEAN RODENTS

Species	Reference
Female-biased sex ratios	
Thomomys bottae	Lay 1978; Howard and Childs 1959; Patton and Feder 1981; Patton and Brylski 1987; Reichman, Whitham, and Ruffner 1982; Daly and Patton 1990
T. talpoides	Hansen 1960; Andersen 1982; Andersen and MacMahon 1981
Geomys bursarius	Wood 1949; English 1932; Vaughan 1962
Pappogeomys castanops	Ikenberry 1964; Williams and Baker 1976
Nannospalax ehrenbergi	Nevo 1982
Tachyoryctes splendens	Jarvis 1973
Orthogeomys cherriei	Delgado 1992
Ctenomys talarum	Malizia and Busch 1991
C. australis	Zenuto and Busch 1998
C. opimus	Pearson 1959
C. peruanus	Pearson 1959
Balanced sex ratios	
T. talpoides	Reid 1973
Geomys attwateri	Williams and Cameron 1990
P. castanops	Smolen, Genoways, and Baker 1980
C. talarum	Malizia and Busch 1997
C. mendocinus	Rosi, Puig et al. 1996
C. maulinus brunneus	Gallardo and Anrique 1991
Male-biased sex ratios	
Heterocephalus glaber	Brett 1991b

Specifically, as density increases, agonistic interactions among males may become more frequent, leading to greater male mortality and, as a result, an abundance of females.

Data from several species support this hypothesis. In low-density (≤ 13 individuals per ha) populations of *C. talarum,* sex ratios are close to parity, while in high-density (65 individuals per ha) populations, the sex ratio is female-biased (Busch et al. 1989; Malizia and Busch 1991, 1997). At low densities, intraspecific aggression is relatively rare, and the incidence of wounded individuals is low (Malizia and Busch 1997). Even sex ratios have also been reported for low-density populations of *C. mendocinus* (in which no wounds were found: Rosi et al. 1992; Rosi, Puig et al. 1996), and *C. peruanus* (Pearson 1959), although interpretations of these findings are more difficult because comparative data from high-density populations of conspecifics were not obtained.

Intraspecific variation in adult sex ratio also occurs in *T. bottae.* Howard and Childs (1959) reported that the ratio of males to females varied across years from 1:1 to 1:4, with female-biased sex ratios occurring when population density was high. Similarly, Patton and Brylski (1987) and Lidicker and Patton (1987) found that, at low population densities, the adult sex ratio in *T. bottae* was approximately equal, but that a significant bias toward females developed at high population densities. Although this pattern is similar to that of adult sex ratios in *C. talarum,* the reasons for female-biased ratios in high-density populations have not been explored for *T. bottae.*

In contrast to the female-biased sex ratios found in most subterranean rodents, sex ratios in *H. glaber* are male-biased (Brett 1991b; Genelly 1965). In both laboratory and free-living colonies of this species, the bias toward males may be attributable to greater female mortality, perhaps resulting from female-female competition for breeding status. As noted by Brett (1991b), the breeding tenure of males is shorter than that of females, which may mean that individual males are less motivated to compete vigorously for breeding status if a breeding animal dies or is removed. Jarvis (1991) found more males among wild-captured juveniles of *H. glaber,* although in captivity the sex ratio of pups was unbiased at birth. Jarvis (1991) reported that, in captive colonies, more females than males died before weaning, and thus female-biased pup mortality may also contribute to the biased sex ratio found in this species.

Population Age Structure

The age structure of a population describes the proportions of individuals of different ages present in the environment. Age structure is highly related to patterns of reproduction, mortality, and population growth. Among subterranean rodents, high mortality and frequent migration among subadults are believed to generate adult-biased age structures. Data from a number of subterranean species support this assertion, with the proportion of adults in a local population often approaching 70%. For example, among ctenomyids, the proportion of adults reported was 0.65 for *C. maulinus* (Gallardo and Anrique 1991), 0.55 for *C. talarum* (Busch et al. 1989), 0.75 for *C. australis* (Zenuto and Busch 1998), and 0.71 for *C. mendocinus* (Rosi et al. 1992). Data from other subterranean taxa are similar, with reported adult proportions of 0.66 for *T. splendens* (Jarvis 1973), 0.76 for *G. bursarius* (Wilks 1963), and 0.65 for *T. monticola* (Ingles 1952). Thus a predominance of adults appears to be typical of populations of subterranean rodents.

While variation in both density and sex ratio typically characterizes populations of subterranean rodents, the high proportion of adults appears to be a very stable characteristic. Except for the vulnerable period during which young animals leave the mother's home burrow and establish an independent burrow system, animals of all age groups are protected from predation by the subterranean habitat. Although adult dispersal has been observed in some species (see below), adult body size may be an impediment to most avian predators. This situation, together with a long life span, leads to the accumulation of adult individuals of different ages.

Life History Traits

Much of population ecology focuses on species-typical patterns of survival and reproduction that, collectively, are referred to as the life history of an organism. Among small mammals, there is an inverse relationship between survival and production (i.e., growth and reproduction) (French, Stoddart, and Bobek 1975). Based on this relationship, two distinct life history patterns have been recognized. The first is characterized by rapid rates of growth, maturation, and reproduction and high rates of mortality. In contrast, the second is characterized by lower rates of mortality, slow growth and maturation, and low rates of reproduction. These strategies are referred to as r-selection and K-selection, respectively (MacArthur and Wilson 1967).

Based on the assumption that subterranean habitats tend to be relatively stable and predictable, it has been suggested that the rodents that inhabit underground burrows tend to be K-selected (Nevo 1982). If this hypothesis is correct, the life histories of these animals should be characterized by small litter sizes, slow growth rates, delayed maturity and reproduction, low mortality rates, low population densities, and a relatively high proportion of breeding adults in a population. As we have seen, however, subterranean rodents display considerable variation in physiology and reproductive biology (Buffenstein, chap. 2; Bennett, Faulkes, and Molteno, chap. 4, this volume). The presence of this variation suggests that the life history strategies of these animals may not be as uniform as suspected. Here, we review the life history characteristics of subterranean rodents to determine how prevalent a *K*-selected lifestyle is among these animals.

Longevity

French, Stoddart, and Bobek (1975) concluded, based on life history data, that, according to various methods of expression of longevity or survivorship in populations of small mammals, microtines and other murids are often short-lived (e.g., life expectancy 1.8 months; survivorship 0.64/month), while the Sciuridae, Heteromyidae, Zapodidae, Insectivora, and subterranean forms are characterized by higher survivorship or longer life spans (life expectancy 7.4 to 12.5 months, survivorship 0.87–0.98/month). Cricetines and soricine shrews are intermediate (life expectancy 3.1–3.6 months, survivorship 0.64–0.73/month).

Among ctenomyids, adult *C. mendocinus* and *C. talarum* typically survive for two breeding periods, representing a maximum life span of about 2 years. More precise data for *C. talarum* indicate that longevity ranges from 20 to 22 months (Busch et al. 1989; Malizia and Busch 1997). Pocket gophers are also comparatively long-lived. Smolen, Genoways, and Baker (1980) calculated a longevity of 56 weeks for female *P. castanops,* 25% of which survived for 86 weeks after entering the trappable (sufficiently aged) population. Howard and Childs (1959) estimated a life span of 55 weeks for *T. bottae.* Hansen (1960) found that only 25% of *T. talpoides* survived for 2 years or more. Ingles (1952) calculated a longevity of 2.9 years for *T. monticola.* Puzachenko (1996) estimated a life span of 3.22 years for *Spalax microphthalmus.*

Without question, the award for longevity among subterranean rodents belongs to *H. glaber.* In captivity, individuals routinely survive for more than 15 years (Brett 1991b). Data from free-living animals suggest that life span varies with reproductive status; although reproductive females may live for close to a decade, most nonreproductive individuals disappear from their natal colony within a year or two of birth (Braude 1991). Although some of these disappearances may reflect dispersal, rather than mortality, the extremely low rates of colony formation observed suggest that most dispersal attempts end in mortality. Both the extreme life span of some individuals and the apparent dichotomy in survivorship for breeders and nonbreeders represent intriguing puzzles that, when resolved, may substantially improve our understanding of the relationships between survival and reproduction in small mammals.

Sexual Maturity and Gestation Length

All fossorial genera reach sexual maturity relatively late in life. Earlier maturation confers fitness benefits in that individuals spend less time as vulnerable juveniles, have a higher probability of surviving long enough to reach maturity, and begin reproduction sooner. In contrast, delayed maturity is associated with a longer life span and a higher initial fecundity, and is generally coupled to a longer generation time and lower survival to maturity.

Although subterranean rodents have generally been characterized as exhibiting delayed maturity, variation in this life history attribute is evident among these taxa. In solitary species with sufficiently long breeding seasons, females may reach puberty and breed during the year in which they were born, although males apparently wait until the following breeding season to begin reproduction. In other species, all animals delay reproduction until the following breeding season. For example, several authors have reported that female geomyids reach reproductive maturity during the season of their birth, but that males do not breed until they are a year old (Smolen, Genoways, and Baker 1980; Daly and Patton 1986, 1990; Patton and Brylski 1987; Nistler, Verts, and Carraway 1993). In contrast, other authors have reported that neither sex reproduces until the following breeding season, when they are about 9 months of age (Ingles 1952; Hansen 1960; Andersen 1982; Patton and Brylski 1987).

Among ctenomyids, both sexes can reach sexual maturity before they are a year old, although females typically reach puberty earlier than males (Pearson 1959; Malizia and Busch 1991, 1997; Zenuto and Busch 1998). The length of the breeding season determines whether young individuals can breed during the year of their birth. In populations of *C. talarum* from temperate regions, some females breed in the same breeding season in which they were born (Malizia and Busch 1991, 1997). In contrast, in populations of *C. mendocinus* inhabiting high Andean deserts, the breeding season is very short, and both sexes attain sexual maturity in the breeding season following their birth (Rosi, Puig et al. 1996).

Field data on age at first reproduction are not available for bathyergid mole-rats. Determining the age of sexual maturity in social bathyergids is made difficult by the reproductive division of labor evident in these taxa (Bennett, Faulkes, and Molteno, chap. 4; Lacey, chap. 7, this volume). The growth rates of captive-born individuals, however, suggest that these animals are slow to reach adult body sizes (Jarvis and Bennett 1991), implying that sexual maturity is delayed relative to surface-dwelling taxa. Both captive-born *H. glaber* and captive-born *Cryptomys* from Zambia (species identification uncertain) have bred when a year old (Burda 1989; Jarvis 1991), indi-

TABLE 5.3 GESTATION LENGTH IN SPECIES OF SUBTERRANEAN RODENTS

Taxon	Gestation length (days)	Reference
Ctenomyidae		
Ctenomys opimus	120	Pearson 1959
C. talarum	102, 95	Weir 1974; Zenuto 1999
C. mendocinus	95	S. R. Camín, personal communication
Bathyergidae		
Heterocephalus glaber	60–90	Brett, 1991b
H. glaber*	76–90	Jarvis 1991; Lacey and Sherman 1991
Cryptomys darlingi	56–61	Bennett, Jarvis, and Cotterill 1994
C. h. amatus	90–100	Burda 1989, 1990
C. mechowi	90–100	Bennett and Aguilar 1995
Geomyidae		
Georychus capensis	44	Bennett and Jarvis 1988
Thomomys talpoides	18–19	Andersen 1978; Reid 1973
T. bottae	18–30	Schramm 1961; Scheffer 1938; Howard and Childs 1959
Geomys bursarius	28–35	Scheffer 1931; Wilks 1963
Pappogeomys castanops	30	Ikenberry 1964
Spalacinae		
Nannospalax ehrenbergi	34	Gazit, Shanas, and Terkel 1996

*Captive conditions

cating that reproduction is possible at this age. In the wild, however, reproduction seems likely to be delayed for a considerably greater proportion of an individual's lifetime due to social restrictions on direct reproduction by most members of a colony.

Delayed maturity, coupled with long gestation periods, is thought to diminish the ability of subterranean rodents to undergo rapid expansions in population size. In those species for which data are available, gestation varies from 30 to 120 days (table 5.3). Among taxa, differences in gestation length may reflect phylogenetic affinities as well as ecological conditions. For example, the similarity in gestation length between bathyergids and ctenomyids may reflect the presumed close phylogenetic relationship between these taxa (Weir 1974; Honeycutt et al. 1991; Bennett et al. 1994; Bennett and Aguilar 1995; Cook, Lessa, and Hadly, chap. 9, this volume). Unlike ctenomyids, however, bathyergids produce relatively altricial pups; generally, long gestation periods are associated with the production of precocial pups, as is typical of many hystricognath rodents. Among geomyids and spalacines, gestation periods are notably shorter and more closely resemble those reported for surface-dwelling rodents.

Length of the Breeding Season

As indicated above, the length of the breeding season may substantially influence the age at which individuals begin reproducing. The length of the breeding season may also influence the numerical stability of a population. Smolen, Genoways, and Baker (1980) have suggested that a long reproductive season, coupled with a low and regular level of recruitment, gives rise to a population with a complex age structure. This complexity is thought to act as a buffer in catastrophic situations (e.g., flooding or drought). For example, *Thomomys* and *Geomys* have short annual breeding seasons during which the majority of breeders for the following year are produced (Ingles 1952; Howard and Childs 1959; Wilks 1963). In these species, an unfavorable climatic event could substantially reduce or even preclude reproductive activity during the following year. In contrast, *Pappogeomys castanops* has a longer breeding season and a more complex age structure (Smolen, Genoways, and Baker 1980), so that the loss of one year's crop of youngsters would not as dramatically influence reproduction during the following year. Also, since continuous breeding results in the lack of a discrete period of recruitment, more individuals, even young of the year, participate in breeding, thereby moderating the effects of environmental extremes.

Natality

Among subterranean rodents, the number of young per litter, or litter size, is typically low compared with that in surface-dwelling murines, microtines, and cricetines, and is comparable to that in heteromyids (French, Stoddart, and Bobek 1975). In general, subterranean rodents produce one or two litters per year, each of which contains a relatively small number of pups (table 5.4). Striking exceptions to this pattern are found in the bathyergid genera *Cryptomys* and *Heterocephalus*, members of which can bear four or five litters per year, each containing seven or more pups (Jarvis and Bennett 1991; Brett 1991b). It is not known to what extent this high reproductive output reflects the unusual social systems of these animals, which include alloparental care by nonreproductive members of the social group (Bennett, Faulkes, and Molteno, chap. 4; Lacey, chap. 7, this volume). Because the pups reared by a colony are typically the offspring of a single breeding female, however, annual recruitment at the population level may still be low relative to surface-dwelling rodents.

Although mean litter size in subterranean rodents tends to be low, variation in litter size occurs both within and among species. Given the large geographic ranges of some species, variation in litter size may reflect differences in local climate or habitat productivity. Potential support for the latter

TABLE 5.4 MEAN LITTER SIZE IN SUBTERRANEAN RODENTS

Taxon	Young/litter	Social status[a]	Reference
Ctenomys opimus	1.6	S	Pearson 1959
C. peruanus	3.5	C	Pearson 1959
C. mendocinus	2.9	S	Rosi et al. 1992; Rosi, Puig et al. 1996
C. talarum	4.09, 4.52	S	Malizia and Busch 1991, 1997
C. australis	2.9	S	Zenuto and Busch 1998
Thomomys talpoides	6.4, 4.9, 4.7, 4.4, 4.9–5.6	S	Reid 1973; Hansen 1960; Andersen and MacMahon 1981
T. bottae	4.6, 5.74, 5.7, 3.2, 5.6, 4.2	S	Howard and Childs 1959; Lay 1978; Miller 1964; Bandoli 1981
T. bulbivorus	4.68	S	Verts and Carraway 1991
T. monticola	3–4	S	Ingles, 1952
Geomys attwateri	2.3	S	Williams and Cameron 1990
G. bursarius	2.5	S	Kennerly 1954
Pappogeomys castanops	2	S	Smolen, Genoways, and Baker 1980
Pappogeomys (*Cratogeomys*)	2.01	S	Ikenberry 1964
Orthogeomys cherriei	1.9	S	Delgado 1992
Bathyergids	1–6	S,C	see review in Jarvis and Bennett 1991; Bennett et al. 1991; Bennett, Jarvis, and Cotterill 1994; Bennett and Aguilar 1995
Heterocephalus glaber*	Up to 27	C	Jarvis 1991
	Up to 12	C	Brett, 1986
Tachyoryctes splendens	1.65, 1.20	S	Jarvis 1969
Nannospalax ehrenbergi	3	S	Nevo 1961

*Captive conditions
[a]S, solitary; C, colonial

hypothesis comes from studies on *Ctenomys*. Litter sizes are larger in *C. talarum* from habitats with high productivity (Malizia and Busch 1991, 1997) than in *C. opimus* and *C. mendocinus* from low-productivity habitats (Pearson 1959; Rosi et al. 1992). Furthermore, *T. bottae* populations from alfalfa fields have larger litter sizes than populations from desert scrub. In *H. glaber,* as noted by Jarvis and Bennett (1991), litter sizes are larger in captivity, indicating that individuals have the potential to bear more offspring than they do in the wild and suggesting that resource availability or other habitat factors may limit reproduction in free-living animals. Because litter size represents a fundamental component of a species' life history strategy (Millar 1977), studies aimed at clarifying the ecological, physiological, and phylogenetic bases for variation among subterranean taxa are critical.

Mortality

Promislow and Harvey (1990) reported that, in mammals, the best predictor of life history strategy is mortality. They distinguished between two types of mortality: intrinsic and extrinsic. Intrinsic mortality is defined as the cost of reproduction, while extrinsic mortality results from factors such as predation. Species that experience high rates of extrinsic mortality may respond by increasing fecundity, while the converse is expected in species with high intrinsic mortality. In general, subterranean rodents are thought to follow the latter pattern of mortality and associated life history attributes (Nevo 1979).

Clearly, the validity of this assumption rests on our understanding of mortality patterns in subterranean taxa. Unfortunately, few quantitative data regarding mortality are available for subterranean rodents, primarily because of the difficulty of identifying instances of mortality among free-living animals. Mark-recapture studies, for example, frequently fail to distinguish mortality from dispersal. Because the subterranean niche affords considerable protection to its occupants, it has been generally assumed that extrinsic mortality among subterranean rodents is low (Nevo 1979). For most species, however, the factors causing mortality have not been fully investigated, and thus actual death rates remain unknown.

By assessing relevant ecological factors, such as time spent on the soil surface and the types of predators present in the environment, it may be possible to construct a rough, qualitative picture of relative mortality rates in different subterranean species. For example, mortality is thought to be particularly common when young animals leave the natal burrow to establish themselves elsewhere in the habitat. In keeping with this assumption, subadults in several subterranean species are characterized by high mortality rates (Howard and Childs 1959; Jarvis 1969; Smolen, Genoways, and Baker 1980), with as many as 85% of the young born each year failing to reach reproductive maturity (Patton 1990). Intersexual differences in dispersal behavior may explain the sometimes biased adult sex ratios evident in subterranean populations (see table 5.2): if mortality during dispersal is high and males tend to move farther or more often than females, increased mortality of males may lead to female-biased sex ratios among breeding adults. Thus comparisons of dispersal patterns and dispersal distances may yield some insights into general patterns of mortality in these animals.

Searching for mates may be another source of mortality if breeding individuals move about on the soil surface. Temporary contact between the burrows of neighboring males and females during the reproductive season is presented as an alternative, but little evidence of this strategy has been found in field observations. In solitary mole-rats, males and females meet by extending and joining their burrow systems or by males moving above ground in search of females (Jarvis and Bennett 1990). The latter strategy is proposed as the reason for higher mortality rates among sexually mature males versus females in *T. splendens* (Jarvis 1969).

Other sources of mortality can arise as a consequence of high densities. Under such conditions, the number and severity of aggressive interactions involving neighbors are enhanced. Mortality resulting from male-male aggression may explain deviations in adult sex ratios that favor females at high densities, as was found for *T. bottae* (Lay 1978) and *C. talarum* (Busch et al. 1989). In contrast, mortality mediated by aggression among females competing for breeding status was found for colonial naked mole-rats (Brett 1991b).

Although high subadult mortality appears to be common among subterranean rodents, patterns of mortality among adults are less clear. Among geomyids, the annual turnover of breeding adults is relatively high. In *T. bottae*, for example, 55–75% of the breeding population consists of animals born during the previous breeding season (Howard and Childs 1959; Patton 1990). Mortality is similar in *P. castanops*, in which only 20–25% of males survive to participate in two successive breeding seasons (Smolen, Genoways, and Baker 1980). Adult mortality also appears to be high in ctenomyids; populations of *C. talarum* exhibit adult turnover rates of 50–70% between breeding seasons (Busch et al. 1989; Malizia 1998), with a survival rate of only 38.5% for subadult individuals (Malizia 1998). Studies of other species, however, are needed to determine whether high adult mortality is characteristic of subterranean species.

As this discussion suggests, mortality may vary with age, as well as between the sexes. On the one hand, these differences may reflect only age- or sex-related variation in trappability that generates biased estimates of mortality. On the other hand, these differences in mortality may be real, and may reflect variation in both rates and sources of mortality. At present, studies of mortality in subterranean rodents remain largely descriptive, with virtually no information available for groups such as the myospalacines and rhizomyines. As a starting point, data on annual rates of population turnover and apparent adult mortality in these animals would be useful.

In summary, the life history strategies of subterranean rodents are not as uniform as has long been assumed. Although in most cases the reasons for inter- or intraspecific differences in traits such as longevity, rate of sexual maturation, and litter size remain unknown, the available data indicate that these parameters are variable. This variation, in turn, has significant implications for other aspects of population biology, including population stability and rates of population growth or decline. Thus, in addition to comparing life history variables for subterranean and surface-dwelling taxa, it is important that we begin considering the ecological and evolutionary explanations for life history differences among the taxa considered in this volume. By searching for correlations between ecology, phylogeny, and the life history variables discussed above, it should be possible to begin identifying potential causal relationships underlying the variations in reproduction and survival summarized here.

Dispersal

Dispersal, the permanent movement of individuals from one area to another, is a fundamental component of demography that has profound effects on the population dynamics, social behavior, and genetic structure of populations. In general, dispersal rates for subterranean rodents have been assumed to be low. This assumption, coupled with the difficulty of studying dispersal in natural populations, means that direct quantification and characterization of individual movements has been limited to only a few species (e.g., *G. attwateri:* Williams and Cameron 1984; *N. ehrenbergi:* Rado, Wollberg, and Terkel 1992; *C. talarum:* Malizia, Zenuto, and Busch 1995; *C. australis:* Zenuto and Busch 1998; *H. glaber:* O'Riain, Jarvis, and Faulkes 1996).

Who Disperses?

Among small mammals, it is generally males that disperse (Greenwood 1980). Among subterranean rodents, however, dispersal by both sexes appears to be relatively common. Inferential studies of dispersal based on population dynamics have revealed that in many subterranean species, the primary dispersers are young (juvenile and subadult) individuals of both sexes (Howard and Childs 1959; Vaughan 1962; Wilks 1963; Pearson et al. 1968; Williams and Baker 1976; Smolen, Genoways, and Baker 1980). Among geomyids, individuals may disperse either early (Vaughan 1962) or late (Howard and Childs 1959) in the breeding season, perhaps in response to factors such as population density or the probability of inbreeding. Bisexual dispersal has also been reported for *S. microphthalmus* (Wei et al. 1997).

Dispersal distances may vary from tens to hundreds of meters. Because data on individual dispersal distances are virtually nonexistent, it is not known whether animals of one sex tend to disperse farther than animals of the other sex. In general, female dispersal is thought to be mediated by resource distribution, while male dispersal is mediated by the distribution of females. Although these selective pressures are clearly linked, they may not be identical, and they lead to testable predictions regarding intersexual differences in dispersal under certain ecological conditions. Williams and Cameron (1984) initiated the first experimental study of dispersal in pocket gophers, which revealed that dispersing *G. attwateri* were young individuals of both sexes that left their natal area over the course of an extended breeding season, most probably in response to aggressive interactions with conspecifics. Although they are difficult to obtain, more data on dispersal distances would be useful in light of the consequences of dispersal for the demographic and genetic structure of a population.

Dispersal patterns may also vary with age, particularly between adult and subadult individuals. Studies of *C. talarum* and *C. australis* employing experimental removals of animals (Malizia, Zenuto, and Busch 1995; Zenuto and Busch 1998) have revealed that although the oldest individuals in a population do not disperse, the age structure of *C. talarum* dispersers is otherwise a random sample of the animals in the source population(s). The authors proposed that this pattern of dispersal, which differs from the subadult-only dispersal pattern reported for geomyids, may be related to patterns of reproductive dominance within the source population(s).

Dispersal among social bathyergids has been a topic of particular interest, as constraints on dispersal are thought to be a major factor favoring group living in these animals (Lacey, chap. 7, this volume). O'Riain, Jarvis, and Faulkes (1996) have identified what appears to be a dispersive morph within colonies of *H. glaber*. Members of this morph are morphologically, physiologically, and behaviorally distinct from other colony members: these individuals are above average in body weight, exhibit high levels of luteinizing hormone, and solicit matings only with non-colony members. The timing and mechanism of dispersal by these animals has not yet been identified, but O'Riain and colleagues argue that favorable environmental conditions promote dispersal events, which are followed by cycles of outbreeding as dispersers either invade established colonies or establish colonies of their own.

Modes of Dispersal

A second fundamental question about dispersal in subterranean rodents is the manner in which individuals move from one location to another. In short, dispersal may occur above or below ground. Dispersal is assumed to be difficult for all subterranean rodents due to the high energetic and survival costs of moving to an unfamiliar environment and the low probability of finding an appropriate site for establishing residence. Aboveground dispersal may be especially difficult for taxa such as *N. ehrenbergi*, in which limited visual acuity may place individuals at particular risk of predation. Although underground dispersal may be safer, individuals can travel only short distances before the energetic costs of digging become prohibitive. Thus either mode of dispersal is expected to present individuals with significant challenges. As a first step toward understanding how different subterranean taxa address these challenges, the method of dispersal used by each species must be determined.

In *N. ehrenbergi*, dispersal is initiated underground. Rado, Wollberg, and

Terkel (1992) reported that, during the rainy season, young mole-rats establish burrow systems adjacent to their mothers'. Being located adjacent to the maternal burrow system is thought to be advantageous in that young individuals have access to prolonged maternal care and an established food supply. Young animals found wandering above ground during the dispersal season probably have been excluded from the tunnels by their siblings; these individuals are assumed to have little chance of finding or establishing their own burrow systems. Similarly, in *G. capensis* and *B. suillus,* young animals disperse at approximately 60 days of age, either by extending the maternal burrow system underground or by moving short distances (100 m) above ground (Jarvis and Bennett 1990).

In other species, dispersing individuals reportedly travel above ground. Aboveground movement has been reported for *G. attwateri* (Williams and Cameron 1984), *C. talarum* (Malizia, Zenuto, and Busch 1995; Zenuto and Busch 1998), and *H. glaber* (Braude 1991; S. H. Braude, personal communication). It is not known whether dispersal in these species is initiated underground and then continues above ground as individuals move farther from the natal burrow. Conversely, it is not known whether belowground movements in species such as *G. attwateri* are followed by longer-distance movements on the soil surface. Dispersal is difficult to study in any small mammal, although increased use of mark-recapture and radiotelemetry techniques should improve our understanding of how dispersal occurs in subterranean species.

Dispersal Rates

Data on rates of dispersal are available for only a few species. Weekly recovery rates of marked individuals in *C. talarum* (Malizia, Zenuto, and Busch 1995), *C. australis* (Zenuto and Busch 1998), and *G. attwateri* (Williams and Cameron 1984) range from 2% to 6%, and are comparable to 4.5% for the semi-subterranean *Microtus pennsylvanicus* (Tamarin 1977), which exhibits among the lowest dispersal rates known for surface-dwelling rodents. On the one hand, these apparently low rates of dispersal may reflect the difficulties of documenting movements of individuals within and among free-living populations. On the other hand, dispersal rates in subterranean taxa may truly be low, as has been assumed based on other aspects of the biology of these animals.

Clearly, additional research is needed to quantify dispersal in subterranean rodents. At present, we do not have even basic information, such as the mode of dispersal, for many species, let alone rates of success or environmental correlates of dispersal. As molecular genetic studies of population structure increase (Steinberg and Patton, chap. 8, this volume), evidence that dispersal occurs continues to accumulate. The ecology of dispersal, however, remains virtually unknown, and thus considerable further research is needed to understand how, why, and when individuals set off in search of a new burrow system.

Predation

The final aspect of population ecology that we wish to consider here is predation. Although patterns and rates of mortality have already been discussed, predation occupies such a vital place in ecological analyses of demography and behavior that we believe it deserves particular attention. Further, detailed studies of predation have the potential to greatly revise conventional wisdom regarding the ecology of subterranean rodents.

Traditionally, it has been assumed that the subterranean ecotope is buffered against intensive predation (Nevo 1979, 1995; Reig et al. 1990). Both biotic and abiotic factors are thought to have transformed Cenozoic rodents living in exposed, open environments into inhabitants of subterranean burrows. One biotic factor that is thought to have been important in this transition is predation—in particular, predation by raptorial birds. In part, the assertion that subterranean species have been freed from predation pressure is based on the paucity of quantitative accounts of predation upon them; because predation has rarely been observed, it has been assumed to be infrequent, and subterranean rodents have been considered unlikely candidates for studies of predator-prey relationships.

As field studies of subterranean rodents have continued, however, evidence to the contrary has begun to accumulate. Rather than being freed from predation pressure, subterranean species appear to represent important diet items for a variety of aerial and, in some cases, terrestrial predators. As perceptions of predation on these animals begin to change, perceptions of the role of predation in determining population attributes are also changing. Further, as comparative data become available, researchers are now beginning to assess quantitatively long-postulated correlations between predation and other aspects of ecology, such as the amount of time spent above ground while foraging or dispersing.

Among the taxa considered in this volume, predation has been most extensively documented for geomyid pocket gophers. *Thomomys* is regularly preyed upon by several species of owls *(Bubo virginianus, Tyto alba, Asio otus, A. flammeus)* and hawks *(Buteo jamaicensis, B. regalis)* (Janes and Barss 1985; Horn and Fitch 1942; Coulombe 1971; Marti et al. 1993; Holt and Leasure 1993), and pocket gophers may constitute up to 50% of the diet of these predators. Terrestrial predators of *Thomomys* include foxes *(Urocyon cinereus)* and coyotes *(Canis latrans)*, for which gophers may constitute 10–15% of prey (Horn and Fitch 1942). Some of these predators (e.g., *T. alba*) appear to take particular size classes of gophers preferentially, while others (e.g., *B. virginianus*) appear to take prey randomly with respect to size (Janes and Barss 1985). Susceptibility to predation appears to vary among gopher species, with *T. talpoides* being particularly vulnerable, perhaps due to its relatively greater activity above ground (Proulx et al. 1995).

Predation on other geomyids has also been examined. *Geomys* species reportedly constitute 1–14% of the diet of barn owls *(T. alba)* (Otteni, Bolen, and Cottam 1972; Baker 1991). *G. attwateri* is preyed upon by coyotes (Wilks 1963), and common predators of *G. breviceps* include king snakes, great horned owls, red-tailed hawks, long-tailed weasels, and striped skunks (Sulentich, Williams, and Cameron 1991). In addition to these predators, *Pappogeomys tylorhinus* is thought to be preyed on by gopher snakes, as well as domestic dogs and cats (Cervantes et al. 1993).

Studies of ctenomyids have revealed that these animals are frequently preyed upon by an array of vertebrates. Vassallo, Kittlein, and Busch (1994; M. J. Kittlein A. I. Vassallo, and C. Busch, unpublished data) found that *C. talarum* and *C. australis* represent 16% and 2%, respectively, of the prey items of owls *(Athene cunicularia, Asio flammeus,* and *Tyto alba)*. For *C. talarum,* subadults outnumbered adults in owl pellets; only subadult *C. australis* were found in pellets. Owls took a greater proportion of male *C. talarum* than expected based on the sex ratios of local populations of tuco-tucos. This sex-biased pattern of predation was most marked during the breeding season and may reflect greater aboveground exposure of males during this portion of the reproductive cycle. In contrast, a study of predation by *Athene cunicularia* on *C. talarum* led Pearson et al. (1968) to conclude that females should be preyed upon more frequently due to their smaller body size. This apparent discrepancy suggests that predation pressures vary geographically, even among the same predator-prey species combinations.

Surveys of owl diets from several regions of Argentina suggest that predation on *Ctenomys* is more common than previously assumed. In central Argentina, *C. talarum* accounted for 18% of the prey items taken by barn owls (Massoia, Tiranti, and Torres 1987). In southern La Pampa Province, an unidentified species of *Ctenomys* represented 26.4% of the prey biomass taken by barn owls (Tiranti 1992), and in Buenos Aires Province, *C. chasiquensis* constituted 6.8% of the prey taken by short-eared owls (Massoia 1985). Farther to the west, 5.5% of the prey taken by barn owls in Nequen Province consisted of *C. haigi* (Travaini et al. 1996), while 7.6% of the diet of great horned owls from Rio Negro Province consisted of an unidentified species of *Ctenomys* (Massoia 1983). Among terrestrial carnivores, predation on *Ctenomys* has been reported for foxes *(Pseudalopex gymnocercus)*, grison

(Galictis cuja), and opossums *(Didelphis albiventris)*, as well as skunks *(Conepatus chinga)* and armadillos *(Chaetophractus vellerosus)*; the latter may dig into tunnels to capture newborn rodents. Reptilian predators include pit vipers *(Bothrops newwidii)*; adult *C. australis* have been found in the stomachs of these snakes (Contreras 1973), which apparently enter burrows to prey upon their inhabitants. Thus, overall, predation upon ctenomyids appears to be relatively common.

Little is known about predation upon other taxa of subterranean rodents. Both lower aboveground exposure and limited documentation of predators' diets suggest that they are much less frequently preyed upon by avian and mammalian predators than geomyid and ctenomyid species. As reviewed by Heth (1991), surface activity and, consequently, predation are uncommon for *Nannospalax ehrenbergi* in Israel. Low levels of predation are not due to juvenile dispersal, but rather to dispersal by subadults. Nocturnal predation by barn owls was the principal source of mortality, while predation by diurnal raptors and mammalian carnivores appeared unimportant.

Despite an extensive series of anecdotal reports describing the effects of different predators on the biology of African mole-rats and root rats (Sherman, Jarvis, and Alexander 1991), no quantitative studies of predation upon bathyergid or rhizomyine species have been conducted. In contrast to the groups of subterranean rodents discussed so far, the principal predators upon both bathyergids and rhizomyines are assumed to be snakes (Jarvis 1973; Sherman, Jarvis, and Alexander 1991). Mole snakes *(Pseudapsis cana)* and several other species of snakes (e.g., *Aspidelaps scutatus, Rhamphiophis oxyrhynchus rostratus, Mehelya capensis, Crotaphopeltis hotamboeia*) have been observed to prey upon mole-rats, often while the rodents are digging tunnels and erecting surface mounds. Only two mammalian predators are thought to be capable of catching mole-rats underground (Jarvis and Bennett 1991): the African skunk *(Ictonyx striatus)* and the striped weasel *(Poecilogale albinucha)*. In contrast, potential surface predators include birds (e.g., herons, storks, barn owls) and mammalian carnivores (e.g., caracals, jackals, mongooses), which may take mole-rats as the latter expel loose soil from their burrows. High local densities of *T. splendens* may lead to increased surface migration, augmenting exposure to aerial predation, as suggested by the large numbers of skulls of this species found in the pellets of birds (Jarvis 1973).

Based upon these data, there appears to be a marked dichotomy regarding the degree of predation on different lineages of subterranean rodents. Geomyids and ctenomyids are frequently preyed upon by a variety of birds and mammals; in contrast, bathyergids, spalacines, and rhizomyines appear to be taken much less frequently and by a more limited suite of predators. This difference may, in part, be explained by the tendency of African and

Asian species to spend much less time above ground than do their American counterparts. More direct tests, however, are needed to determine the reasons for this apparent dichotomy. In particular, comparisons of extent of surface activity and rates of predation on Old and New World species living in similar habitats are required to characterize intertaxon differences in vulnerability to predation.

The role of predation in the population ecology of subterranean rodents is a critical issue that must be addressed in future studies of these animals. Given that subterranean species commonly display low population growth rates (Nevo 1979), even modest levels of predation may have a substantial effect on the demography of these rodents. The fact that the effects of predation on population dynamics have not been evaluated highlights the traditional disregard for predation as an important aspect of the ecology of subterranean rodents. In light of growing evidence that predation is an important component of the biology of these rodents, experimental studies of predation (e.g., use of predator exclosures) are needed to determine the effects of predation on the population structures of subterranean species.

Variable Responses to the Subterranean Habitat

As this review has illustrated, the seemingly similar conditions imposed by subterranean environments are associated with ecological, demographic, and life history differences among the rodents that inhabit underground burrows. Subterranean rodents, like other mammals, display considerable phenotypic plasticity, which results in a broad repertoire of adaptive ecological responses. For example, as described above, these animals can adjust their burrowing behavior according to environmental conditions, altering not only the digging apparatus used (e.g., incisors versus claws), but also the process of burrow construction and the time and extent of burrowing activity.

Interspecific comparisons suggest that there is also variation in patterns of foraging, dispersal, reproduction, and extent of surface activity by subterranean rodents. The degree of commitment to the subterranean habitat may also influence levels of predation. These findings suggest that relationships between subterranean rodents and other members of their community are also highly diverse (Cameron, chap. 6, this volume). Thus, we conclude that, although the population biologies and ecologies of subterranean rodents are, in many ways, convergent (table 5.5), diversity is also evident within and among species that reflects environmental conditions. In other words, adaptive ecological responses to the challenges and opportunities offered by the subterranean environment may not be as constrained and unique as previously thought.

TABLE 5.5 ECOLOGICAL ATTRIBUTES FOR FOUR TAXA OF SUBTERRANEAN RODENTS

	Taxon			
Attribute	Ctenomyidae	Geomyidae	Bathyergidae	Spalacinae
Population density	Intermediate	Intermediate	Intermediate	Intermediate
Social structure	Solitary/colonial	Solitary	Solitary/colonial/social	Solitary
Age structure	Mostly adults	Mostly adults	Mostly adults	Mostly adults
Longevity	High	High	High	High
Sexual maturity	Late	Late	Late	Late
Gestation length	Long	Short	Long	Short
Reproductive rate	Low	Low	Low	Low
Diet	Mostly aerial monocots	Aerial and roots, tubers, mainly dicots	Geophytes and some aerial	Geophytes and some aerial
Digging mode	Scratch	Scratch/chisel-tooth	Scratch/chisel-tooth	Chisel-tooth/head-lift
Foraging	Above/below ground	Above/below ground	Below ground	Below ground
Dispersal	Above ground	Above ground	Below/above ground	Below ground
Predation risk	Above ground	Above ground	Below ground	Below ground

Conclusions

Obtaining ecological information on subterranean rodents is particularly challenging given that most of the animals' activities take place in underground burrows. As a result, little is known regarding the population biology of these animals, particularly when compared with many surface-dwelling rodents. The quantity of data available varies greatly among subterranean taxa. Some groups, including geomyids, some species of ctenomyids, and most bathyergids, are well represented in the literature. Considerably less is known about the remaining subterranean taxa, with an almost complete absence of data for myospalacines and rhizomyines.

Based on available data, we suggest that the population ecology of subterranean rodents can be summarized as follows:

1. The ecologies of subterranean rodents reflect numerous adaptations, of which adaptations for digging and subterranean foraging may be considered preeminent.
2. Interspecific differences in the degree of commitment to the subterranean habitat may reflect variation in surface conditions (e.g., temperature, predation risk, competition from surface-dwelling herbivores) as well as variation in the energetic gains derived from foraging underground. Prominent underground storage organs are most commonly found in plants in arid or highly seasonal environments, and thus selec-

tion for a completely subterranean life may be greatest for animals in those habitats.

3. Because subterranean rodents use resources that tend to vary both spatially and temporally, the ecologies and population biologies of these animals are also expected to vary in space and time. Thus, although subterranean rodents exhibit a general convergence in population attributes, ecological and demographic variation is evident among these animals.

Acknowledgments

We wish to dedicate this chapter to the memory of Osvaldo A. Reig. To him, and to Oliver P. Pearson, our thanks for their pioneering role in the knowledge of the fascinating underground life of tuco-tucos. Thanks are due to the editors for inviting us to contribute to this volume, and especially to Eileen Lacey for all the help.

Literature Cited

Ågren, G., Q. Zhou, and W. Zhong. 1989. Ecology and social behaviour of Mongolian gerbils, *Meriones unguiculatus,* at Xilinhot, Inner Mongolia, China. *Animal Behaviour* 37:11–27.

Aldous, C. M. 1951. The feeding habits of pocket gophers *(Thomomys talpoides)* in the high mountain ranges of Central Utah. *Journal of Mammalogy* 32:84–87.

Altuna, C., G. Izquierdo, and B. Tassino. 1993. Análisis del comportamiento de excavación en dos poblaciones del complejo *Ctenomys pearsoni* (Rodentia: Octodontidae). *Boletín de la Sociedad Zoológica del Uruguay* 8:275–82.

Andersen, D.C. 1978. Observations on reproduction, growth, and behavior of the northern pocket gopher *(Thomomys talpoides). Journal of Mammalogy* 59:418–22.

———. 1982. Observations on *Thomomys talpoides* in the region affected by the eruption of Mount St. Helens. *Journal of Mammalogy* 63(4): 652–55.

———. 1987. *Geomys bursarius* burrowing patterns: Influence of season and food patch structure. *Ecology* 68:1306–18.

———. 1988. Tunnel-construction methods and foraging path of a fossorial herbivore, *Geomys bursarius. Journal of Mammalogy* 69:565–82.

Andersen, D.C., and J. A. MacMahon. 1981. Population dynamics and bioenergetics of a fossorial herbivore, *Thomomys talpoides* (Rodentia: Geomyidae) in a spruce-fir sere. *Ecological Monographs* 51:179–202.

Antinuchi, C. D., and C. Busch. 1992. Burrow structure in the subterranean rodent *Ctenomys talarum. Zeitschrift für Säugetierkunde* 57:163–68.

Baker, R. H. 1991. Mammalian prey of the common Barn-Owl *(Tyto alba)* along the Texas coast. *Southwestern Naturalist* 36:343–47.

Bandoli, J. H. 1981. Factors influencing seasonal burrowing activity in the pocket gopher, *Thomomys bottae. Journal of Mammalogy* 62:293–303.

Benedix, J. H., Jr. 1993. Area-restricted search by plains pocket gopher *(Geomys bursarius)* in tallgrass prairie habitat. *Behavioral Ecology* 4:318–24.

Bennett, N. C., and G. H. Aguilar. 1995. The reproductive biology of the giant Zambian mole-rat, *Cryptomys mechowi* (Rodentia: Bathyergidae). *South African Journal of Zoology* 30:1–3.

Bennett, N. C., and J. U. M. Jarvis. 1988. The reproductive biology of the Cape mole-rat *(Georychus capensis)* (Rodentia: Bathyergidae). *Journal of Zoology, London* 214:95–106.

Bennett, N. C., J. U. M. Jarvis, G. H. Aguilar, and T. McDaid. 1991. Growth and development in six species of African mole-rats. *Journal of Zoology, London* 225:13–26.

Bennett, N. C., J. U. M. Jarvis, and F. P. D. Cotterill. 1994. The colony structure and reproductive biology of the Afrotropical Mashona mole-rat, *Cryptomys darlingi*. *Journal of Zoology, London* 234:477–87.

Braude, S. H. 1991. Behavior and demographics of the naked mole-rat, *Heterocephalus glaber*. Ph.D. dissertation, University of Michigan, Ann Arbor.

Brett, R. A. 1986. The ecology and behavior of the naked mole-rat *(Heterocephalus glaber ruppell)* (Rodentia: Bathyergidae). Ph.D. dissertation, University of London.

―――. 1991a. The ecology of naked mole-rat colonies: Burrowing, food, and limiting factors. In *The biology of the naked mole-rat*, edited by P. W. Sherman, J. U. M. Jarvis, and R. D. Alexander, 137–84. Princeton, NJ: Princeton University Press.

―――. 1991b. The population structure of naked mole-rat colonies. In *The biology of the naked mole-rat*, edited by P. W. Sherman, J. U. M. Jarvis, and R. D. Alexander, 97–136. Princeton, NJ: Princeton University Press.

Brown, L. N., and G. C. Hickman. 1973. Tunnel system structure of the southeastern pocket gopher. *Florida Scientist* 36:98–103.

Burda, H. 1989. Reproductive biology (behaviour, breeding, and postnatal development) in subterranean mole-rats, *Cryptomys hottentotus* (Bathyergidae). *Zeitschrift für Säugetierkunde* 54:360–76.

―――. 1990. Constraints of pregnancy and evolution of sociality in mole-rats. *Zeitschrift für zoologische Systematik und Evolutionsforschung* 28:26–39.

Busch, C., A. I. Malizia, O. A. Scaglia, and O. A. Reig. 1989. Spatial distribution and attributes of a population of *Ctenomys talarum* (Rodentia: Octodontidae). *Journal of Mammalogy* 70:204–8.

Cameron, G. N., S. R. Spencer, B. D. Eshelman, L. R. Williams, and M. J. Gregory. 1988. Activity and burrow structure of Attwater's pocket gophers *(Geomys attwateri)*, Houston. *Journal of Mammalogy* 69:667–77.

Camín, S. R., and L. A. Madoery. 1994. Feeding behavior of the tuco-tuco *(Ctenomys mendocinus)*: Its modifications according to food availability and the changes in the harvest pattern and consumption. *Revista Chilena de Historia Natural* 67:257–63.

Carter, S. K., and F. C. W. Rosas. 1997. Biology and conservation of the giant otter *Pteronura brasiliensis*. *Mammal Review* 27:1–26.

Carter, T. S., and C. D. Encarnação. 1983. Characteristics and use of burrows by four species of armadillos in Brazil. *Journal of Mammalogy* 64:103–8.

Cervantes, F. A., V. J. Sosa, J. Martínez, R. M. González, and R. C. Dowler. 1993. *Pappogeomys tylorhinus*. *Mammalian Species* 433:1–4.

Comparatore, V. M., M. S. Cid, and C. Busch. 1995. Dietary preferences of two sympatric subterranean rodent populations in Argentina. *Revista Chilena Historia Natural* 68:197–206.

Comparatore, V. M., N. Maceira, and C. Busch. 1991. Habitat relations in *Ctenomys talarum* (Caviomorpha: Octodontidae) in a natural grassland. *Zeitschrift für Säugetierkunde* 56:112–28.

Contreras, J. R. 1973. El tuco-tuco y sus relaciones con los problemas del suelo en la Argentina. *IDIA-INTA*, supl. 29:14–36.

Contreras, L. C., and J. R. Gutiérrez. 1991. Effects of the subterranean herbivorous rodent *Spalacopus cyanus* on herbaceous vegetation in arid coastal Chile. *Oecologia* 87:106–9.

Coulombe, H. N. 1971. Behavior and population ecology of the burrowing owl, *Speotyto cunicularia*, in the Imperial Valley of California. *Condor* 73:162–76.

Cox, G. W. 1989. Early summer diet and food preferences of northern pocket gophers in north central Oregon. *Northwest Science* 63:77–82.

Cox, G. W., and J. Hunt. 1992. Relation of seasonal activity patterns of valley pocket gophers to temperature, rainfall, and food availability. *Journal of Mammalogy* 73:123–34.

———. 1994. Pocket gophers, herbivory, and mortality of ocotillo on stream terrace, bajada, and hillside sites in the Colorado Desert, southern California. *The Southwestern Naturalist* 39:364–70.

Daly, J. C., and J. L. Patton. 1986. Growth, reproduction and sexual dimorphism in *Thomomys bottae* pocket gophers. *Journal of Mammalogy* 67(2):256–65.

———. 1990. Dispersal, gene flow, and allelic diversity between local populations of *Thomomys bottae* pocket gophers in the coastal range of California. *Evolution* 44:1283–94.

Davies, K. C., and J. U. M. Jarvis. 1986. The burrow systems and burrowing dynamics of the mole-rats *Bathyergus suillus* and *Cryptomys hottentotus* in the fynbos of the southwestern Cape, South Africa. *Journal of Zoology, London* 209:125–47.

Delany, M. J. 1986. Ecology of small rodents in Africa. *Mammal Review* 16:1–41.

Delgado, R. 1992. Ciclo reproductivo de la taltuza *Orthogeomys cherriei* (Rodentia: Geomyidae) en Costa Rica. *Revista de Biologia Tropical* 40:111–15.

English, P. F. 1932. Some habits of the pocket gopher, *Geomys breviceps breviceps*. *Journal of Mammalogy* 13:126–32.

French, N. R., D. M. Stoddart, and B. Bobek. 1975. Patterns of demography in small mammal populations. In *Small mammals: Their productivity and population dynamics*, edited by F. Golley, K. Petrusewicz, and L. Ryszkowski, 73–102. Cambridge: Cambridge University Press.

Gallardo, M. H., and J. A. Anrique. 1991. Populational parameters and burrow systems in *Ctenomys maulinus brunneus* (Rodentia: Ctenomyidae). *Medio Ambiente* 11:48–53.

Gazit, I., U. Shanas, and J. Terkel. 1996. First successful breeding of the blind mole rat *(Spalax ehrenbergi)* in captivity. *Israel Journal of Zoology* 42:3–13.

Genelly, R. E. 1965. Ecology of the common mole rat *Cryptomys hottentotus* in Rhodesia. *Journal of Mammalogy* 46:647–65.

Giannoni, S. M., C. E. Borghi, and V. Roig. 1996. The burrowing behavior of *Ctenomys eremophilus* (Rodentia: Ctenomyidae) in relation to substrate hardness. *Mastozoología Neotropical* 3:161–70.

Greenwood, P. J. 1980. Mating systems, philopatry and dispersal in birds and mammals. *Animal Behaviour* 28:1140–62.

Hansen, R. M. 1960. Age and reproductive characteristics of mountain pocket gophers in Colorado. *Journal of Mammalogy* 41:323–35.

Hansen, R. M., and E. E. Remmenga. 1961. Nearest neighbor concept applied to pocket gopher populations. *Ecology* 42:812–14.

Hedgal, P. L., A. L. Ward, A. M. Johnson, and H. P. Tietjen. 1964. Notes on the life history of the Mexican pocket gopher *(Cratogeomys castanops)*. *Journal of Mammalogy* 46:334–35.

Heth, G. 1989. Burrow patterns of the blind mole-rat, *Spalax ehrenbergi* in two soil types (Terra rossa and Renzina) in Mount Carmel, Israel. *Journal of Zoology, London* 217:38–56.

Heth, G. 1991. Evidence of above-ground predation and age determination of the prey in subterranean mole rats *(Spalax ehrenbergi)* in Israel. *Mammalia* 55:529–42.

———. 1992. The environmental impact of subterranean mole rats *(Spalax ehrenbergi)* and their burrows. In *The environmental impact of burrowing animals and animal burrows*, edited by P. S. Meadows and A. Meadows, 265–80. Oxford: Clarendon Press.

Heth, G., E. M. Golenberg, and E. Nevo. 1989. Foraging strategy in a subterranean rodent, *Spalax ehrenbergi:* A test case for optimal foraging theory. *Oecologia* 79:617–22.

Hickman, G. C. 1977. Burrow system structure of *Pappogeomys castanops* (Geomyidae) in Lubbock County, Texas. *American Midland Naturalist* 97:50–58.

———. 1983a. Burrow structure of the talpid mole *Parascalops breweri* from Oswego County, New York State. *Zeitschrift für Säugetierkunde* 48:265–69.

———. 1983b. Burrows, surface movement, and swimming of *Tachyoryctes splendens* (Rodentia: Rhizomyidae) during flood conditions in Kenya. *Journal of Zoology, London.* 200:71–82.

———. 1984. An excavated burrow of *Scalopus aquaticus* from Florida, with comments on Nearctic Talpid-Geomyid burrow structure. *Säugetierkundliche Mitteilungen* 31:243–49.

Hildebrand, M. 1985. Digging in quadrupeds. In *Functional vertebrate morphology*, edited by M. Hildebrand, D. Bramble, K. Liem, and D. Wake, 89–109. Cambridge, MA: Harvard University Press.

Hodara, K., O. V. Suárez, and F. O. Kravetz. 1997. Nesting and digging behavior in two rodent species *(Akodon azarae* and *Calomys laucha)* under laboratory and field conditions. *Zeitschrift für Säugetierkunde* 62:23–29.

Holt, D. W., and S. M. Leasure. 1993. Short-eared Owl. *The birds of North America: Life histories for the 21st century*, edited by A. F. Poole, P. Stettenheim, and F. B. Gill, no. 62, 1–24. Washington, DC: American Ornithologists' Union.

Honeycutt, R. L., M. W. Allard, S. V. Edwards, and D. A. Schlitter. 1991. Systematics and evolution of the family Bathyergidae. In *The biology of the naked mole-rat,* edited by P. W. Sherman, J. U. M. Jarvis, and R. D. Alexander, 46–65. Princeton, NJ: Princeton University Press.

Horn, E. E., and H. S. Fitch. 1942. The San Joaquín Experimental Range: Interrelations of rodents and other wildlife of the Range. University of California (Berkeley), Agricultural Experiment Station Bulletin 663:96–129.

Howard, W. E., and H. E. Childs, Jr. 1959. Ecology of pocket gophers with emphasis on *Thomomys bottae mewa. Hilgardia* 29:277–358.

Huntly, N., and O. J. Reichman. 1994. Effects of subterranean mammalian herbivores on vegetation. *Journal of Mammalogy.* 75:852–59.

Ikenberry, R. D. 1964. Reproductive studies of the Mexican pocket gopher, *Cratogeomys castanops perplanus*. M. S. thesis, Texas Technological College, Lubbock.

Ingles, L. G. 1952. The ecology of the mountain pocket gopher *Thomomys monticola. Ecology* 33:87–95.

Janes, S. W., and J. M. Barss. 1985. Predation by three owl species on northern pocket gophers of different body mass. *Oecologia* 67:76–81.

Jarvis, J. U. M. 1969. The breeding season and litter size of African mole-rats. *Journal of Reproduction and Fertility* (Supplement) 6:237–48.

———. 1973. The structure of a population of mole-rats, *Tachyoryctes splendens* (Rodentia: Rhizomyidae). *Journal of Zoology, London* 171:1–14.

———. 1978. Energetics of survival in *Heterocephalus glaber* (Ruppell), the naked-mole rat (Rodentia: Bathyergidae). *Bulletin of the Carnegie Museum of Natural History* 6:81–87.

———. 1991. Reproduction in naked mole-rats. In *The biology of the naked mole-rat,* edited by P. W. Sherman, J. U. M. Jarvis, and R. D. Alexander, 384–425. Princeton, NJ: Princeton University Press.

Jarvis, J. U. M., and N. C. Bennett. 1990. The evolutionary history, population biology and social structure of African mole-rats: Family Bathyergidae. In *Evolution of subterranean mammals at the organismal and molecular levels,* edited by E. Nevo and O. A. Reig, 97–128. Progress in Clinical and Biological Research, vol. 335. New York: Wiley-Liss.

———. 1991. Ecology and behavior of the family Bathyergidae. In *The biology of the naked mole-rat,* edited by P. W. Sherman, J. U. M. Jarvis, and R. D. Alexander, 66–96. Princeton, NJ: Princeton University Press.

Jarvis, J. U. M., and J. B. Sale. 1971. Burrowing and burrow patterns of East African mole-rats *Tachyoryctes, Heliophobius* and *Heterocephalus. Journal of Zoology, London* 163: 451–79.

Jenkins, S. H., and P. W. Bollinger. 1989. An experimental test of diet selection by the pocket gopher *Thomomys monticola. Journal of Mammalogy* 70:406–12.

Kennerly, T. E., Jr. 1954. Local differentiation in the pocket gopher *(Geomys personatus)* in southern Texas. Texas Journal of Science 6:297–329.

Lacey, E. A., and P. W. Sherman. 1991. Social organization of naked mole-rat colonies: Evidence for divisions of labor. In *The biology of the naked mole-rat,* edited by P. W. Sherman, J. U. M. Jarvis, and R. D. Alexander, 275–336. Princeton, NJ: Princeton University Press.

Lay, D. M. 1978. Observations on reproduction in a population of pocket gophers, *Thomomys bottae,* from Nevada. *Southwestern Naturalist* 23:375–80.

Lessa, E., and C. S. Thaeler. 1989. A reassessment of morphological specializations for digging in pocket gophers. *Journal of Mammalogy* 70:689–700.

Lidicker, W. Z., Jr., and J. L. Patton. 1987. Patterns of dispersal and genetic structure in populations of small rodents. In *Mammalian dispersal patterns,* edited by B. D. Chepko-Sade and Z. T. Halpin, 144–61. Chicago: University of Chicago Press.

Llanos, A. C., and J. A. Crespo. 1952. Ecología de la vizcacha *(Lagostomus maximus maximus Blainville)* en el nordeste de la provincia de Entre Ríos. *Revista de Investigaciones Agrícolas* 6:289–378.

Lovegrove, B. G., and J. U. M. Jarvis. 1986. Coevolution between mole rats (Bathyergidae) and a geophyte, *Micranthus* (Iridaceae). *Cimbebasia* 8A:79–85.

Luce, D. G., R. M. Case, and J. Stubbendiek. 1980. Food habits of the plains pocket gopher on western Nebraska rangeland. *Journal of Range Management* 33:129–31.

MacArthur, R. H., and E. O. Wilson. 1967. *The theory of island biogeography.* Monographs in Population Biology, 1. Princeton, NJ: Princeton University Press.

MacMillen, R. E., and A. E. Lee. 1970. Energy metabolism and pulmocutaneous water loss of Australian hopping mice. *Comparative Biochemical Physiology* 35:355–69.

Madoery, L. A. 1993. Composición botánica de la dieta del tuco-tuco *(Ctenomys mendocinus)* del piedemonte precordillerano. *Ecología Austral* 3:49–55.

Malizia, A. I. 1998. Population dynamics of the fossorial rodent *Ctenomys talarum* (Rodentia, Octodontidae). *Journal of Zoology* 244:545–51.

Malizia, A. I., and C. Busch. 1991. Reproductive parameters and growth in the fossorial rodent *Ctenomys talarum* (Rodentia: Octodontidae). *Mammalia* 55:293–305.

———. 1997. Breeding biology of the fossorial rodent *Ctenomys talarum* (Rodentia: Octodontidae). *Journal of Zoology, London* 242:463–71.

Malizia, A. I., A. I. Vassallo, and C. Busch. 1991. Population and habitat characteristics of two sympatric species of *Ctenomys* (Rodentia: Octodontidae). *Acta Theriologica* 36:87–94.

Malizia, A. I., R. R. Zenuto, and C. Busch. 1995. Demographic and reproductive attributes of dispersers in two populations of the subterranean rodent *Ctenomys talarum* (tuco-tuco). *Canadian Journal of Zoology* 73:732–38.

Marti, C. D., K. Steenhof, M. N. Kochert, and J. S. Marks. 1993. Community trophic structure: The roles of diet, body size, and activity time in vertebrate predators. *Oikos* 67:6–18.

Massoia, E. 1983. La alimentación de algunas aves del orden strigiformes en la Argentina. *El Hornero,* N° Extraordinario 125–48.

———. 1985. Análisis de regurgitados de *Asio flammeus* del arroyo Chasicó. *ACINTACNIA* 2:7–9.

Massoia, E., S. I. Tiranti, and M. P. Torres. 1987. Mamíferos pleistocenos y recientes recolectados en el arroyo Santa Catalina, Rio Cuarto, Prov. de Córdoba. *Boletín Informativo de la Asociación Paleontológica Argentina* 16:12.

McNab, B. K. 1966. The metabolism of fossorial rodents: A study of convergence. *Ecology* 47:712–33.
Millar, J. S. 1977. Adaptive features of mammalian reproduction. *Evolution* 31:370–86.
Miller, M. A. 1957. Burrows of the Sacramento Valley pocket gopher in flood-irrigated alfalfa fields. *Hilgardia* 26:431–52.
Miller, R. S. 1964. Ecology and distribution of pocket gophers (Geomyidae) in Colorado. *Ecology* 45:256–72.
Nevo, E. 1961. Observations on Israeli populations of the mole rat *Spalax e. ehrenbergi*, Nehring 1898. *Mammalia* 25:127–44.
———. 1979. Adaptive convergence and divergence of subterranean mammals. *Annual Review of Ecology and Systematics* 10:269–308.
———. 1982. Speciation in subterranean mammals. In *Mechanisms of speciation*, edited by C. Barigozzi, G. Montalenti, and M. J. D. White, 291–98. New York: Alan R. Liss.
———. 1991. Evolutionary theory and processes of active speciation and adaptive radiation in subterranean mole-rats, *Spalax ehrenbergi* superspecies, in Israel. In *Evolutionary biology*, vol. 25, edited by M. K. Hecht, B. Wallace, and R. J. MacIntyre, 1–125. New York: Plenum Publishing Corporation.
———. 1995. Mammalian evolution underground: The ecological-genetic-phenetic interfaces. *Acta Theriologica*, Supplement 3:9–31.
Nistler, D. L., B. J. Verts, and L. N. Carraway. 1993. Aging techniques, juvenile breeding, and body mass in *Thomomys bulbivorus*. *Northwestern Naturalist* 74:25–28.
O'Riain, M. J., J. U. M. Jarvis, and C. G. Faulkes. 1996. A dispersive morph in the naked mole-rat. *Nature* 380:619–21.
Ojeda, R. A., J. M. Gonnet, C. E. Borghi, S. M. Giannoni, C. M. Campos, and G. B. Díaz. 1996. Ecological observations of the red vizcacha rat, *Tympanoctomys barrerae*, in desert habitats of Argentina. *Mastozoología Neotropical* 3:183–92.
Otteni, L. C., E. G. Bolen, and C. Cottam. 1972. Predator-prey relationships and reproduction of the barn owl in southern Texas. *Wilson Bulletin* 84:434–48.
Patton, J. L. 1990. Geomyid evolution: The historical, selective, and random basis for divergence patterns within and among species. In *Evolution of subterranean mammals at the organismal and molecular levels*, edited by E. Nevo and O. A. Reig, 49–59. Progress in Clinical and Biological Research, vol. 335. New York: Wiley-Liss.
Patton, J. L., and P. V. Brylski. 1987. Pocket gophers in alfalfa fields: Causes and consequences of habitat-related body size variation. *American Naturalist* 4:493–503.
Patton, J. L., and J. H. Feder. 1981. Microspatial genetic heterogeneity in pocket gophers: Nonrandom breeding and drift. *Evolution* 35:912–20.
Pearson, O. P. 1959. Biology of the subterranean rodents, *Ctenomys* in Perú. *Memorias del Museo de Historia Natural "Javier Prado"* 9:1–56.
———. 1984. Taxonomy and natural history of some fossorial rodents of Patagonia, southern Argentina. *Journal of Zoology, London* 202:225–37.
Pearson, O. P., N. Binztein, L. Boiry, C. Busch, M. Dipace, G. Gallopin, P. Penchaszadeh, and M. Piantanida. 1968. Estructura social, distribución espacial y composición por edades de una población de tuco tucos *(Ctenomys talarum)*. *Investigaciones Zoológicas Chilenas* 13:47–80.
Promislow, D. E. L., and P. H. Harvey. 1990. Living fast and dying young: A comparative analysis of life-history variation among mammals. *Journal of Zoology, London* 220:417–37.
Proulx, G., M. J. Baldry, P. J. Cole, R. K. Drescher, A. J. Kolenosky, and I. M. Pawlina. 1995. Summer above-ground movements of northern pocket gophers, *Thomomys talpoides*, in an alfalfa field. *Canadian Field Naturalist* 109:256–58.
Puzachenko, A. Y. 1996. The demographic structure and reproduction of mole rat *Spalax microphthalmus* (Rodentia, Spalacidae). *Zoologicheskii Zhurnal* 75:271–79.

Rado, R., G. Bronchiti, Z. Wollberg, and J. Terkel. 1992. Sensitivity to light of the blind mole rat: Behavioral and neuroanatomical study. *Israel Journal of Zoology* 38:323–31.

Rado, R., Z. Wollberg, and J. Terkel. 1992. Dispersal of young mole rats *(Spalax ehrenbergi)* from the natal burrow. *Journal of Mammalogy* 73:885–90.

Reichman, O. J., and S. C. Smith. 1985. Impact of pocket gopher burrows on overlying vegetation. *Journal of Mammalogy* 66:720–25.

Reichman, O. J., T. G. Whitham, and G. A. Ruffner. 1982. Adaptive geometry and burrow spacing in two pocket gopher populations. *Ecology* 63:687–95.

Reid, V. H. 1973. Population biology of the northern pocket gopher. In *Pocket gophers and Colorado mountain rangeland*, edited by G. T. Turner, R. M. Hansen, V. H. Reid, H. P. Tietjen, and A. L. Ward, 21–41. Bulletin 5545, Colorado State University Agricultural Experiment Station, Fort Collins.

Reig, O. A. 1970. Ecological notes on the fossorial octodont rodent *Spalacopus cyanus* (Molina). *Journal of Mammalogy* 51:592–601.

Reig, O. A., C. Busch, O. Ortells, and J. L. Contreras. 1990. An overview of evolution, systematics, population biology, cytogenetics, molecular biology and speciation in *Ctenomys*. In *Evolution of subterranean mammals at the organismal and molecular levels*, edited by E. Nevo and O. A. Reig, 71–96. Progress in Clinical and Biological Research, vol. 335. New York: Wiley-Liss.

Rosi, M. I., M. I. Cona, S. Puig, F. Videla, and V. G. Roig. 1996. Size and structure of burrow systems of the fossorial rodent *Ctenomys mendocinus* in the piedmont of Mendoza province, Argentina. *Zeitschrift für Säugetierkunde* 61:352–64.

Rosi, M. I., S. Puig, F. Videla, M. I. Cona, and V. G. Roig. 1996. Ciclo reproductivo y estructura etaria de *Ctenomys mendocinus* (Rodentia: Ctenomyidae) del Piedemonte de Mendoza, Argentina. *Ecología Austral* 6:87–93.

Rosi, M. I., S. Puig, F. Videla, L. Madoery, and V. G. Roig. 1992. Estudio ecológico del roedor subterráneo *Ctenomys mendocinus* en la precordillera de Mendoza, Argentina: Ciclo reproductivo y estructura etaria. *Revista Chilena de Historia Natural* 65: 221–33.

Russell, R. J., and R. H. Baker. 1955. Geographic variation in the pocket gopher *Cratogeomys castanops* in Coahuila, Mexico. *Miscellaneous Publications, Museum of Natural History, University of Kansas* 7:591–608.

Scheck, S. H., and E. D. Fleharty. 1980. Subterranean behavior of the adult thirteen-lined ground squirrel *(Spermophilus tridecemlineatus)*. *American Midland Naturalist* 103: 191–95.

Scheffer, T. H. 1931. Habits and economic status of the pocket gophers. *U.S. Department of Agriculture Technical Bulletin* 224:1–26.

———. 1938. Breeding records of Pacific Coast pocket gophers. *Journal of Mammalogy* 19:220–24.

Schmidly, D. J. 1983. *Texas mammals east of the Balcones fault zone*. College Station: Texas A&M University Press.

Schoener, T. W. 1971. Theory of feeding strategies. *Annual Review of Ecology and Systematics* 11:369–404.

Schramm, P. 1961. Copulation and gestation in the pocket gopher. *Journal of Mammalogy* 42:167–70.

Sherman, P. W., J. U. M. Jarvis, and R. D. Alexander, eds. 1991. *The biology of the naked mole-rat*. Princeton, NJ: Princeton University Press.

Smolen, M. J., H. H. Genoways, and R. J. Baker. 1980. Demographic and reproductive characteristics of the yellow-cheeked pocket gopher *(Pappogeomys castanops)*. *Journal of Mammalogy* 61:224–36.

Spencer, S. R., G. N. Cameron, B. D. Eshelman, L. C. Cooper, and L. R. Williams. 1985. Influence of pocket gopher mounds on a Texas coastal prairie. *Oecologia* 66:111–15.

Steuter, A. A., E. M. Stenauer, G. L. Hill, P. A. Bowers, and L. L. Tieszen. 1995. Distribution and diet of bison and pocket gophers in a sandhills prairie. *Ecological Applications* 5:756–66.

Stuebe, M. M., and D.C. Andersen. 1985. Nutritional ecology of a fossorial herbivore: Protein, N_2, and energy value of winter caches made by northern pocket gopher, *Thomomys talpoides*. *Canadian Journal of Zoology* 63:1101–5.

Sulentich, J. M., L. R. Williams, and G. N. Cameron. 1991. *Geomys breviceps*. *Mammalian Species* 383:1–4.

Tamarin, R. H. 1977. Dispersal in island and mainland voles. *Ecology* 58:1044–54.

Tilman, D. 1983. Plant succession and gopher disturbance along an experimental gradient. *Oecologia* 60:285–92.

Tiranti, S. I. 1992. Barn owl prey in southern La Pampa, Argentina. *Journal of Raptor Research* 26:89–92.

Travaini, A., J. A. Donázar, O. Ceballos, A. Rodriguez, F. Hiraldo, and M. Delibes. 1996. Food habits of common barn-owl along an elevational gradient in Andean Argentine Patagonia. *Journal of Raptor Research* 31:59–64.

Turner, G. T., R. M. Hansen, U. H. Reid, H. P. Titjen, and A. L. Ward. 1977. *Pocket gophers and Colorado mountain rangeland*. Bulletin 5545, Colorado State University Agricultural Experiment Station, Fort Collins.

Vassallo, A. I. 1994. Habitat shift after experimental removal of the bigger species in sympatric *Ctenomys talarum* and *Ctenomys australis* (Rodentia: Octodontidae). *Behaviour* 127:247–63.

———. 1998. Functional morphology, comparative behaviour, and adaptation in two sympatric subterranean rodents genus *Ctenomys* (Caviomorpha: Octodontidae). *Journal of Zoology, London* 244:415–27.

Vassallo, A. I., M. J. Kittlein, and C. Busch. 1994. Owl predation on two sympatric species of tuco-tucos (Rodentia, Octodontidae). *Journal of Mammalogy* 75:725–32.

Vaughan, T. A. 1962. Reproduction in the plains pocket gopher in Colorado. *Journal of Mammalogy* 43:1–13.

Vaughan, T. A., and R. M. Hansen. 1964. Experiments on interspecific competition between two species of pocket gophers. *American Midland Naturalist* 72:444–52.

Verts, B. J., and L. N. Carraway. 1991. Summer breeding and fecundity in the camas pocket gopher *Thomomys bulbivorus*. *Northwestern Naturalist* 72:61–65.

Vleck, D. 1979. The energy cost of burrowing by the pocket gopher *Thomomys bottae*. *Physiological Zoology* 52:122–36.

———. 1981. Burrow structure and foraging cost in the fossorial rodent, *Thomomys bottae*. *Oecologia* 49:391–96.

Wang, Q., W. Zhou, Y. Zhang, and N. Fan. 1994. The observation of burrowing behavior of plateau zokor *(Myospalax baileyi)*. *Acta Theriologica Sinica* 14:203–8.

Ward, A. L., and J. O. Keith. 1962. Feeding habits of pocket gophers in mountain grasslands, Black Mesa, Colorado. *Ecology* 43:744–49.

Wei, W., Q. Wang, W. Zhou, and N. Fan. 1997. The population dynamics and dispersal of plateau zokor after removing. *Acta Theriologica Sinica* 17:53–61.

Weir, B. J. 1974. Reproductive characteristics of hystricomorph rodents. In *The biology of hystricomorph rodents*, edited by I. W. Rowland and B. J. Weir, 265–301. Symposium of the Zoological Society of London, 34. London: Academic Press.

Wilks, B. J. 1963. Some aspects of the ecology and population dynamics of the pocket gopher *(Geomys bursarius)* in southern Texas. *Texas Journal of Science* 15:241–83.

Williams, L. R., and G. N. Cameron. 1984. Demography of dispersal in Attwater's pocket gopher *(Geomys attwateri)*. *Journal of Mammalogy* 65:67–75.

———. 1986. Food habits and dietary preferences of Attwatter's pocket gopher, *Geomys attwateri*. *Journal of Mammalogy* 67:216–24.

———. 1990. Intraspecific response to variation in food resources by Attwater's pocket gopher. *Ecology* 71:797–810.

Williams, S. L., and R. J. Baker. 1976. Vagility and local movements of pocket gophers (Geomyidae: Rodentia). *American Midland Naturalist* 96:303–16.

Wood, J. E. 1949. Reproductive pattern of the pocket gopher *(Geomys breviceps brazensis)*. *Journal of Mammalogy* 30:36–44.

Yáñez, J., and F. Jaksic. 1978. Historia natural de *Octodon degus* (Molina) (Rodentia: Octodontidae). *Museo Nacional de Historia Natural Publicaciones Ocasionales* 27:3–11.

Zenuto, R. 1999. Sistema de apareamiento en *Ctenomys talarum* (Rodentia: Octodontidae). Ph.D. thesis, Universidad Nacional de Mar Del Plata, Argentina.

Zenuto, R., and C. Busch. 1998. Population biology of the subterranean rodent *Ctenomys australis* (tuco-tuco) in a coastal dunefield in Argentina. *Zeitschrift für Säugetierkunde* 60:277–85.

CHAPTER SIX

Community Ecology of Subterranean Rodents

Guy N. Cameron

Rodents that spend most of their lives underground have distinctive anatomies, morphologies, and physiologies that allow them to exploit the evolutionary challenges presented by this unique milieu. Convergence in body size and shape among subterranean rodents (Stein, chap. 1, this volume) renders subdivision of the food niche difficult for these largely herbivorous animals. This convergence represents a very different evolutionary path than that taken by organisms such as Darwin's finches, in which diversifying evolution has yielded highly specialized beak morphologies that allow each species to exploit different subsets of food resources. Hence, unlike many other rodent taxa, subterranean species often subdivide space by allopatric distributions (Steinberg and Patton, chap. 8, this volume), and, in a few cases, by vertical stratification of the habitat.

This chapter focuses on interactions between subterranean rodents and the other members of the communities in which they live. Consequently, this chapter complements the discussion of population ecology by Busch et al. in chapter 5. Subterranean rodents affect the structure of their communities in a variety of ways, including interactions with other fauna as well as direct (i.e., through feeding) and indirect (i.e., through soil movement and alteration of soil characteristics) interactions with the flora.

The concept of keystone species has become important in community ecology because of the influence of such species on a variety of other ecological processes. A related concept is that of ecosystem engineers, organisms that directly or indirectly modulate the availability of resources to other species (Lawton and Jones 1995; Jones, Lawton, and Shachak 1997). Unlike most surface-dwelling rodents, subterranean rodents are prime examples of keystone species, because they affect the diversity and productivity of the flora; they are also ecosystem engineers, because their constant modification of the environment affects resource availability to other species. This chapter explores the nature of such interactions by reviewing available data regarding the effects of subterranean rodents on soil characteristics, floral and faunal species composition, and ecosystem processes. Because they interact with, move, and alter the structure and chemical nature of soil, subterranean rodents affect a wider variety of community and ecosystem processes than surface-dwelling rodents. Consequently, we expect their effects to be more pervasive throughout the ecosystem.

Effects of Subterranean Rodents on Soil Characteristics

The effect that subterranean rodents have upon their physical surroundings provides a logical beginning for this chapter because this interaction has both direct and indirect effects on vegetation that influence not only the floral but also the faunal structures of the communities in which these rodents live. A subterranean lifestyle requires the movement of soil, and this physical modification of the environment represents one of the most important effects that subterranean rodents have on their surroundings. Although the specific effects of rodent activity may vary from one setting to another, their excavation of nest sites, feeding tunnels, latrines, and sites for caching food has numerous physical and chemical effects on soils, and as a result, on vegetation, hydrology, and nutrient cycling. Some of these effects are related directly to the movement of soil; others are a consequence of the deposition of excavated soil in new locations.

Subterranean rodents are prodigious diggers. Although all subterranean species are noted for their digging abilities (Stein, chap. 1, this volume), the amount of soil moved varies among both taxa and localities (350–430 kg per ha per year for *Heterocephalus glaber* in South Africa: Brett 1991; > 8,100 kg per ha per year for *Geomys pinetis* in north-central Florida: Kalisz and Stone 1984; 84,300–102,800 kg per ha per year for *G. attwateri* in southern Texas: Spencer et al. 1985). Most of this variation results from differences in animal density, which ranges from nine individuals per hectare in South Africa to thirty to sixty individuals per hectare in Texas (see also Busch et al., chap. 5, this volume). Excavated soil typically is deposited on the soil surface in discrete mounds, which may reach 1.5 m in diameter and which may be recognizable for weeks, months, or years. Heth (1992) summarized studies of *Nannospalax ehrenbergi* in Israel to demonstrate that mounds produced by the excavation of feeding tunnels covered up to 25% of the soil surface, and that an average of about 700 kg of soil was excavated each year in territories that averaged 70 m² in area. Grinnell (1923) estimated that pocket gophers deposited at least 8,000 tons of soil above ground in Yosemite National Park, a figure that certainly underestimated the total extent of excavations because the gophers used some soil to fill old burrows rather than depositing it on the surface.

Mima mounds (Cox and Gakahu 1985, 1986; Roig and Cox 1985; Cox and Zedler 1986; Cox and Allen 1987; Cox 1990) and similar structures (Ross, Tester, and Breckenridge 1968; Cox, Lovegrove, and Siegried 1987; Lovegrove 1991) described from populations of *Geomys* and *Thomomys* in North America, *Ctenomys* in South America, and *Cryptomys* in South Africa are perhaps the most impressive visual evidence of both the quantity of soil that subterranean rodents can move and the long-term effect of that movement in some ecosystems. These mounds, believed to have been formed by movement of soil over decades or centuries, can reach 2 m in height and more than 10 m in diameter, and can cover more than 50% of the land surface. For example, soil deposited by *Thomomys bottae* on the surface of mima mounds in southern California exceeded 8,200 kg per ha per year, with more than 20,000 kg per ha per year deposited in the subsurface (Cox 1990).

The excavation and redistribution of soil has significant effects on its physical structure (Andersen 1987; Johnson 1989) and microrelief (Mielke 1977; Inouye, Huntly, and Wasley 1997), leading to changes in its chemical composition as well as subtle changes in abiotic conditions at the soil surface. For example, in a 31-year study, Laycock and Richardson (1975) demonstrated that activity by *Thomomys talpoides* in central Utah increased soil porosity, organic matter, total nitrogen, and total phosphorus. The soil in recently excavated mounds tends to be less dense than surrounding soils and, consequently, has a higher water infiltration rate and a higher moisture-holding capacity (Ellison and Aldous 1952; Laycock and Richardson 1975; Grant, French, and Folse 1980; Reichman and Smith 1985; Andersen 1987). Nevertheless, mound soils are not always higher in moisture content (Grant, French, and Folse 1980; Stromberg and Griffin 1996), probably because of increased evaporation from the exposed soil surface in some habitats.

The nature and extent of chemical differences between mound and non-mound soils varies with soil type and mound size. Small mounds may have lower concentrations of plant nutrients because they consist of subsurface soils (Inouye et al. 1987b; Koide, Huenneck, and Mooney 1987; McDonough 1974; Spencer et al. 1985). For example, Grant and McBrayer (1981) determined that although exchangeable potassium was higher, magnesium and calcium were lower on mounds made by *G. bursarius* than on surrounding soils. In a demonstration of the effect of pocket gophers on soil nutrients, Gonzales, Saladen, and Hakonson (1995) learned that burrowing activities by *Thomomys bottae* on simulated nuclear waste landfills reduced the soil surface concentration of cesium by 43%, although this effect may have been partially the result of translocation by plants to the rhizosphere.

In contrast, where the soil is covered by a layer of nutrient-poor material, such as volcanic ash, small mounds may have a higher nutrient content than adjacent surface soils (Andersen and MacMahon 1985). Soils of mima and mima-like mounds in North America are typically richer in nutrients and may have higher organic and water contents (McGinnies 1960), higher pHs, lower carbon contents, lower bulk densities, and higher water percolation rates than surrounding soils (Ross, Tester, and Breckenridge 1968; Mielke 1977; Cox and Gakahu 1985; Cox and Zedler 1986). Mima-like mounds in South Africa have finer soils, fewer stones, a better developed A horizon, and a higher water content than intermound soils (Lovegrove 1991). These characteristics reflect the length of time over which the mounds are constructed through the repeated deposition of soil.

In addition to disturbing and redistributing soil, subterranean rodents may affect the nature of soil by feeding, caching food, and excreting wastes. Food that is gathered during the growing season and stored underground may decompose if not eaten, leaving local pockets of enriched soil. Latrines are common features of burrow systems that can also produce pockets of highly enriched soil. Zinnel (1988) found that total soil nitrogen content was greater near food caches and den sites of *Geomys bursarius,* and that these nitrogen-rich localities supported greater root biomass than other portions of the burrow system.

Future studies of relationships between subterranean rodents and the soils they inhabit must extend beyond simply documenting that these mammals redistribute and mix soils. While effects such as breaking up dense soil aggregates, increasing water infiltration, increasing aeration, and exposing bare soil to direct sunlight may all increase mineralization rates in soil (Inouye et al. 1987b), the relative importance of these factors is likely to vary among taxa, ecosystems, and seasons. For example, differences in soil temperature may be particularly important in temperate regions and during cooler parts of the growing season, while increased water infiltration or aeration may be most important in dense soils in which microbial activity is limited by lack of water or oxygen. In addition, given the variable effects on soil described above, studies comparing how different species of subterranean rodents affect the soil are needed, as are studies detailing the consequences of these effects for vegetation. For example, how is the distribution of nutrients in the soil and, concomitantly, plant growth and distribution affected by topographic changes resulting from burrowing and mound-building activity? What are the short- and long-term results of these activities? Comparative studies are needed to address the effects of soil disturbance by subterranean mammals over multiple temporal and spatial scales.

Interactions among Subterranean Herbivores

Most subterranean rodents are solitary and aggressive toward conspecifics (Lacey, chap. 7, this volume). Intraspecific aggression affects a variety of behavioral and other attributes, including the spatial distributions of individuals and their burrow systems (*Thomomys bottae:* Reichman, Whitham, and Ruffner 1982; *Geomys attwateri:* Gregory et al. 1987; *Nannospalax ehrenbergi:* Rado, Wollberg, and Terkel 1992), plural occupancy of burrows (Hansen and Miller 1959; Reichman, Whitham, and Ruffner 1982; Bandoli 1987; Lacey, chap. 7, this volume), and reproductive rates and population sizes (mole-rats in southern Africa: Bennett 1989, 1994; Mediterranean pine voles, *Microtus duodecimcostatus:* Paradis and Guedon 1993; Busch et al., chap. 5, this volume). These individual- and population-level phenomena have important implications for community- and ecosystem-level processes. For example, intraspecific aggression alters population density, which affects not only the plant biomass consumed but also patterns of energy flow and nutrient cycling. Thus, intraspecific interactions among subterranean herbivores may significantly influence community structure.

Interspecific interactions among subterranean herbivores also may substantially influence community structure. The underground habitat is a three-dimensional niche with one dimension—depth—restricted primarily to the root zone of plants (Soriguer and Amat 1980; Busch et al., chap. 5, this volume). Different species of subterranean rodents are rarely sympatric, perhaps because their morphological similarities make resource partitioning and subdivision of the subterranean niche difficult (Stein, chap. 1, this volume). Instead, species tend to have allopatric distributions, leading to the initial conclusion that interspecific competition has had a minor effect on populations of subterranean rodents (Kennerly 1959; Pearson 1984; Reig et al. 1990). Allopatric distributions occur among geomyids, between *Spalacopus* and several species of *Ctenomys*, between *Nannospalax* and *Myospalax*, and between *Rhizomys* and *Cannomys* (Pearson 1959). Bathyergids in South Africa are allopatric with one another and with *Tachyoryctes* (Ellerman 1940).

Mechanisms of Competitive Exclusion

Despite the apparent ecological equivalency of different subterranean rodent species and the general conclusions that these animals do not co-occur and thus do not partition the underground niche ecologically, a number of studies provide evidence suggesting that competitive exclusion may be an important factor regulating the distributions of subterranean rodents. Soil texture and depth are most often identified as the factors separating species (Miller 1964; Reichman and Baker 1972; Best 1973; Reichman and Smith 1990; Borghi, Giannoni, and Martinez-Rica 1994). For example, Miller (1964) reported an allopatric distribution among four species of pocket gophers in Colorado *(Geomys bursarius, Pappogeomys castanops, Thomomys bottae,* and *T. talpoides)*, which he attributed to differential preferences for soil textures. He concluded that *G. bursarius,* the species with the most specialized soil preferences, was a superior competitor that displaced the remaining species to less favorable habitats, thereby producing an allopatric distribution. Similarly, Thaeler (1968) showed that the allopatric distribution of five species of pocket gophers in northeastern California resulted from competitive exclusion. He discovered that, in sympatry, *T. bottae* was competitively superior to other species because morphological adaptations allowed it to use harder soils. Kennerly (1959) identified allopatric populations of *G. bursarius* and *G. personatus* in southern Texas that were separated by differences in soil texture. Allopatric populations of *T. bottae, P. castanops,* and *G. bursarius* in northeastern New Mexico also achieved habitat separation by adaptation to different soil types (Best 1973).

Among other subterranean lineages, sympatry has been reported for bathyergids and ctenomyids. In the former case, overlapping ranges of *Bathyergus suillus, Georychus capensis,* and *Cryptomys hottentotus* in South Africa were explained as the result of specialization for different soil microhabitats (Thaeler 1968) or interspecific differences in food resource requirements (Reichman and Jarvis 1989). In the latter case, in sympatry, *Ctenomys australis* (mean body mass 360 g) occupies areas of deep, sandy soil and sparse vegetation, whereas *C. talarum* (mean body mass 118 g) occupies more compact, shallow soils with dense vegetation (Malizia, Vassallo, and Busch 1991). Vassallo (1994) confirmed that this separation resulted from interspecific competition by removing *C. australis,* after which *C. talarum* occupied a wider range of soil types. Collectively, these data suggest that interspecific competition may be an important component of the spatial structuring of subterranean communities, and that differential use of soil types achieves the same result as differential use of different habitat types in the aboveground world.

Other mechanisms of avoiding competition include differences in life history traits and microhabitat preferences. Vaughan and Hansen (1964) concluded that differences in dispersal ability between *T. talpoides* and *T. bottae,* combined with a greater breadth of environmental tolerance by *T. talpoides,* served to separate these otherwise ecologically similar species. Niche breadth also appears to play a role in competition among these species; in contrast to Miller's (1964) finding that *T. bottae* occupied a subset of the *T. talpoides* niche, Thaeler (1968) found that *T. bottae* occupied a variety of habitats such that the niches of these two species overlapped, rather than being inclusive.

Other studies of subterranean taxa have yielded similar results. For example, Reichman and Baker (1972) found that a narrow band of sympatry (ca. 1,300 m) between *Pappogeomys castanops* and *Thomomys bottae* in the Davis Mountains of Texas was unstable because the range of *Pappogeomys* was expanding while that of *Thomomys* was contracting. These changes in range limits were attributed to adaptations by *Pappogeomys* to xeric conditions that led to its superior competitive ability. Hafner and Barkley (1984) studied a relictual population of *Zygogeomys trichopus* in Mexico and concluded that the expansion of agriculture into narrow valleys at high altitudes forced contact with *Pappogeomys gymnurus* and contracted the range of *Z. trichopus*. In this situation, they believed that the larger size and more aggressive behavior of *P. gymnurus* allowed it to be competitively superior.

Davies and Jarvis (1986) found spatial separation between the Cape dune *(Bathyergus suillus)* and common *(Cryptomys hottentotus)* mole-rats in Cape Province, South Africa, with the former favoring sand and the latter favoring consolidated soil. However, in some areas, a third species *(Georychus capensis)* overlaps vertically with the other two. *G. capensis* is excluded from highly productive foraging areas by the larger *B. suillus* and the more numerous *C. hottentotus*. *G. capensis* is able to exist in sympatry with these other species by foraging in areas of low plant biomass where the other species are not able to forage efficiently (Reichman and Jarvis 1989; Jarvis and Bennett 1990). In Natal, *G. capensis* and *C. h. natalensis* occur sympatrically because they use different microhabitats, with *G. capensis* limited by soil depth and texture (Nanni 1988).

Pearson (1984) captured the subterranean species *Ctenomys maulinus* (Ctenomyidae), *Aconaemys sagei* (Octodontidae), and *Geoxus valdivianus* (Muridae) together in a meadow in Argentina. Because both *Aconaemys* and *Ctenomys* are herbivores, Pearson speculated that these species were competitors. Pearson noted, however, that since *Aconaemys* is less adapted for subterranean life, it spends an appreciable amount of time foraging above ground, thereby minimizing direct interactions with *Ctenomys*. Pearson further noted that *Geoxus*, an insectivore, and *Chelemys macronyx*, a subterranean herbivore, coexist in forested and humid areas in southern Argentina because of dietary separation and the twofold larger body size of *Chelemys*.

A final mechanism that may contribute to competitive exclusion among subterranean rodents is aggressive behavior. Because aggressive interactions may occur between, as well as within, these species (Vaughan and Hansen 1964; Vassallo and Busch 1992), aggression may function to relegate heterospecifics to different portions of the habitat. Baker (1974) demonstrated aggressive behavior between *T. bottae* and *T. talpoides* in Colorado; *T. talpoides*, which occupies a broader niche, was more aggressive in captivity than *T. bottae*, suggesting possible competitive superiority. Similarly, Hickman (1977) documented aggressive interactions among *T. bottae, G. bursarius,* and *P. castanops*. Larger individuals were behaviorally dominant in interspecific encounters, but smaller individuals became more aggressive when cornered in blind tunnels. Among ctenomyids, the sympatric *C. australis* and *C. talarum* are aggressive toward each other in the laboratory (Vassallo and Busch 1992), with the larger-bodied *C. australis* dominating the smaller-bodied *C. talarum*, which exhibits evasive behaviors when confronted by *C. australis*.

In some habitats, competition appears to be reduced via vertical stratification of the underground niche. In southwestern and northwestern Cape Province, South Africa, *Cryptomys h. hottentotus, Bathyergus suillus,* and *B. janetta* occur micro-sympatrically. Competition among these species is minimized, however, because the animals occupy tunnels at different depths and consume different types of food. *C. h. hottentotus* have shallow tunnels and feed upon subterranean portions of plants, while *B. suillus* and *B. janetta* occupy deeper tunnels and feed upon aerial portions of plants (Davies and Jarvis 1986). Similarly, among sympatric geomyids from the New Mexico-Colorado-Oklahoma area of the western United States, *Pappogeomys castanops* uses feeding tunnels that are deeper than those of *Thomomys bottae*, a relationship that Moulton, Choate, and Bissell (1983) attribute to competitive interactions between these species. In a similar study, Hall and Villa (1949) found that burrows of *Thomomys umbrinus* in Michoacán, Mexico, were located above those of *Pappogeomys gymnurus*. In this case, an eightfold difference in body mass, resulting in behavioral domination, was believed to have contributed to the vertical separation. Finally, Hafner and Barkley (1984) concluded that *Thomomys* and *Pappogeomys* may coexist in mountain meadows in Michoacán because the former species burrows only 10 cm below the surface while the latter burrows 20–30 cm below the surface. In contrast, *Pappogeomys* and *Zygogeomys* in this same habitat are more likely to compete with one another because they burrow at the same depth.

These mechanisms of avoidance or amelioration of competition among subterranean herbivores are similar to those found in aboveground communities of rodents. The one mechanism unique to subterranean rodents is vertical separation of species, which is largely restricted in surface-dwelling rodents to tropical ecosystems with vertical layers of vegetation.

Interactions with Other Subterranean Mammals

Subterranean rodents may interact with other small mammals that inhabit underground burrows. Ecological and behavioral interactions between the largely herbivorous subterranean rodents that are the focus of this book and insectivorous subterranean mammals may, however, be relatively benign due to the marked dietary separation between these animals. For example, although *C. hottentotus* and *Amblysomus,* an insectivorous golden mole, occur in the same habitat, these animals do not appear to compete with each other (McConnell and Hickman 1985; McConnell 1986). In Argentina, the insectivores *Notiomys edwardsii* and *Geoxus valdivianus* coexist in the same burrows with the herbivores *Ctenomys haigi* and *Aconaemys fuscus,* respectively; this coexistence appears to be possible because of differences in body size and diet (Pearson 1984). Finally, Schaeffer (1945) captured moles (insectivores) and geomyid pocket gophers (herbivores) from the same runways, suggesting that these animals are also capable of coexisting.

In summary, despite the typically allopatric distributions of subterranean rodents, substantial evidence indicates that active competitive exclusion occurs among these animals and, in some cases, may be responsible for the observed spatial segregation of species. Apparent competition among subterranean herbivores raises a number of intriguing questions regarding the intensity and ecological importance of interspecific interactions. For example, how does the relative importance of specific mechanisms of competitive exclusion differ among localities and taxa? Space is an obvious resource to be partitioned, but few investigations have determined whether other resources, such as food, are also partitioned. How do interspecific interactions affect foraging behavior and nutritional ecology among sympatric subterranean rodents? How do these interactions differ from comparable patterns among surface-dwelling rodents? These and other fundamental issues regarding interspecific competition among subterranean rodents must be addressed by future studies of these animals.

Interactions with Surface Herbivores

Subterranean rodents have long been suspected of having substantial direct effects upon surface vegetation, but until recently, few studies had considered interactions between subterranean and surface-dwelling rodents. The vast majority of studies that have examined relationships between subterranean rodents and surface herbivores have considered interactions between these rodents and grazing ungulates. For example, in Colorado, *Thomomys talpoides* exhibits significant dietary overlap not only with domestic cattle, but also with native ungulates such as elk *(Cervus canadensis)*, mule deer *(Odocoileus hemionus)*, pronghorn *(Antilocapra americana)*, and bighorn sheep *(Ovis canadensis:* Paulson 1969; Ward 1973). In California, grazing by cattle significantly reduced population densities of *Thomomys bottae* in grasslands (Hobbs and Mooney 1991; Hunter 1991; Stromberg and Griffin 1996), most likely because cattle compacted the soil or otherwise reduced the suitability of available vegetation. For other geomyids, however, these interactions are positive: in Nebraska, *G. bursarius* was most active in patches of dicots, the size and placement of which were determined by the grazing of bison *(Bos bison)* (Steuter et al. 1995).

Interactions between subterranean rodents and small surface-dwelling herbivores have been less extensively documented. Data obtained to date, however, suggest that the direction of these interactions may also vary with taxa and locality. On the one hand, a positive interaction between pocket gophers and grasshoppers has been reported by Huntly and Inouye (1988), who found that the presence of *G. bursarius* significantly increased recruitment of grasshoppers by increasing soil disturbance and floral heterogeneity, which allowed grasshoppers to exploit vegetation more efficiently. On the other hand, Klaas, Danielson, and Moloney (1998) reported a negative spatial association between *G. bursarius* and *Microtus pennsylvanicus;* these authors concluded that areas with pocket gopher activity lacked sufficient vegetative cover for the vole, leading to the effective exclusion of this species. Thus interactions among different rodent taxa may be as intricate and as varied as interactions between subterranean rodents and larger herbivores.

The distinction between ecological interactions involving only subterranean species and those involving both subterranean and surface-dwelling taxa is arbitrary, and thus future studies of these interactions must consider both types of relationships to generate a complete picture of the community ecology of subterranean rodents. Only by placing such interactions in a larger, ecosystem perspective can we begin to understand phenomena such as the role of primary producers in mediating herbivore interactions or the effects of soil disturbance on the abundance of large grazing mammals. Knowledge of these interactions, particularly in grassland and xeric habitats, would enable us to assess the overall effect of plant-animal interactions on community and ecosystem functions.

Interactions with Predators

A potentially important advantage of life underground is protection against surface and aerial predators (Jarvis and Bennett 1990; Reichman and Smith 1990; Busch et al., chap. 5, this volume). Most predation by birds, mammals, and reptiles probably occurs when subterranean rodents are outside the burrow system feeding, pushing soil, or dispersing (Busch et al., chap. 5, this volume). In contrast, predation by snakes and some mammals (e.g., mustelids) probably occurs below ground. Subterranean rodents respond to such predators by walling them off or attacking them (e.g., *Heterocephalus:* Reichman and Smith 1990; Brett 1991) or by using foot-drumming and body-pumping behaviors to challenge them (e.g., *Georychus capensis, Cryptomys hottentotus:* Eloff 1951; Reichman and Smith 1990; Francescoli, chap. 3, this volume).

Much of what is known about the population-level consequences of predation on subterranean rodents is anecdotal. As Busch et al. (chap. 5, this volume) point out, little is known about the effects of predation on the demographic or life history parameters of subterranean rodents, although, in general, the rate of predation is not thought to be sufficient to regulate population size (e.g., *Thomomys:* Colorado Cooperative Pocket Gopher Project 1960; Hansen and Ward 1966). To date, quantitative studies of predation on subterranean rodents have focused primarily on the proportion of predator diets represented by different subterranean species (Busch et al., chap. 5, this volume). Considerably less work has been conducted on the effects of predation on the rodent prey.

As this discussion suggests, analyses of the population- and community-level effects of predation on subterranean rodents are lacking. While much is known about the role of predation in structuring the aboveground fauna of communities, what role do these processes play in structuring the underground fauna? Does differential predation on subterranean species affect the structure of the community? Does predation affect the diversity, evenness, or richness of the fauna? Do effects on the fauna in turn affect the floral composition of the community? Answering these questions requires that data on predation be integrated with other aspects of community ecology to gain a thorough understanding of the role of subterranean rodents in determining community structure.

Interactions with Organisms in Burrows

Entire assemblages of species may occur in the burrows of subterranean rodents. No less than twenty-two species of animals have been found in the burrows of *T. bottae, T. talpoides,* and *G. bursarius* in Colorado (Vaughan 1961); the species encountered include tiger salamanders *(Ambystoma tigrinum),* spadefoot toads (*Scaphiopus* sp.), ornate box turtles *(Terrapene ornata),* six-lined racerunners *(Cnemidophorus sexlineatus),* earless lizards *(Holbrookia maculata),* gopher snakes *(Pituophis catenifer),* prairie rattlesnakes *(Crotalus viridis*), eastern moles *(Scalopus aquaticus)*, desert cottontails *(Sylvilagus auduboni)*, ground squirrels *(Spermophilus* sp.), kangaroo rats *(Dipodomys ordii)*, deer mice *(Peromyscus maniculatus)*, meadow voles *(Microtus* sp.), and long-tailed weasels *(Mustela frenata)*. Wilks (1963) reported several amphibians *(Bufo* sp., *Scaphiopus* sp.) and reptiles *(Arizona elegans, Heterodon platyrhinos)*, but only a single mammal *(Perognathus hispidus)*, in burrows of *G. bursarius* from Texas. The silky pocket mouse *(Perognathus flavus)* and grasshopper mouse *(Onychomys leucogaster)* occupied burrow systems of *T. bottae*, but larger kangaroo rats did not because the diameter of gopher burrows was too small. A number of new species of beetles (Histeridae) have been found in burrows of *G. bursarius* in central Texas (Cartwright 1944; Ross 1944; Blume and Aga 1979) and Florida (Ross 1940). Hubbell and Goff (1940) note that nearly eighty species of insects and other arthropods have been collected in burrows of *Geomys* sp. from Florida.

Hickman (1977) noted that *T. bottae, G. bursarius,* and *P. castanops* generally ignore amphibians and reptiles in their burrows, and cited a case of a pocket gopher picking up and moving a tiger salamander that was in the way of the gopher's excavations. This reaction is in marked contrast to the responses to subterranean predators described above. While subterranean rodents may be relatively unaffected by some of the residents of their burrows, this physical relationship may be of considerable importance to the other species involved. For example, of the eighty species of insects and other arthropods reported by Hubbell and Goff (1940), fifteen were believed to be obligate inhabitants of rodent burrows. The burrows of subterranean rodents also may harbor disease vectors, such as phlebotomine sand flies, which carry the protozoan that causes leishmaniasis, or the fleas that carry bubonic plague (Chippaux and Pajot 1983; Lechleitner, Tileston, and Kartman 1962).

The phenomenon of burrow associates is largely unique to subterranean rodents, although certain other rodents that occupy nests, houses, or underground burrows also have an assortment of associated organisms. Data regarding the burrow associates of subterranean rodents are limited primarily to geomyids. At present, little information is available on burrow associates of bathyergids, ctenomyids, octodontids, or murids. Interactions between these associates and their rodent hosts are poorly understood, although it is generally assumed that the association is one-way, with the non-excavating species dependent on the environment afforded by the burrow. Given the diverse species found in the burrows of subterranean rodents, detailed investigations of their interactions with burrow owners and other burrow associates are needed. In addition, the mechanisms used by obligate forms to disperse to new burrows are unknown. Once these data are obtained, they can be combined with data on interactions of subterranean rodents with predators, surface herbivores, and subterranean herbivores to enhance our understanding of subterranean ecosystems.

Subterranean Rodents, Community Structure, and Ecosystem Perspectives

Productivity, Species Composition, and Floral Diversity

Almost all activities of subterranean rodents have the potential to alter the surrounding vegetation. Selective feeding by these animals may disproportionately remove certain plant species from the environment. Burrowing and mound building alter the nature of the soil; deposition of soil on the ground surface can both elevate the average soil height and increase microtopographic variation, changes that are likely to influence both the establishment and subsequent growth of plants (Huntly and Inouye 1988; Reichman and Smith 1990; Inouye, Huntly, and Wasley 1997). These effects may, in turn, influence the relative abundances of plant species, as well as patterns of species richness, diversity, and productivity. Given this multitude of effects, subterranean rodents may function as keystone species (Huntly and Inouye 1988; Huntly 1991; Cox, Contreras, and Milewski 1995), and they certainly qualify as ecosystem engineers (Lawton and Jones 1995; Jones, Lawton, and Shachak 1997). This dual role makes the effect of subterranean rodents on community structure and ecosystem function more pervasive than that of surface-dwelling rodents.

In many ecosystems, the indirect effects of subterranean rodents on soil conditions and plant resources probably outweigh the direct effects of their foraging. The plant species composition of a community is strongly influenced by the movement of soil and the availability of resources such as water, light, nutrients, and germination sites (Tilman 1983, 1988). Soil moisture, texture, temperature, and nutrient content all vary in association with small-scale and large-scale disturbances. In grassland ecosystems, light intensity at the soil surface, a key factor in the germination of many plant species, is almost certainly greater on disturbed soil.

Structures created by subterranean rodents may also substantially influence soils and vegetation. Differences in texture and water-holding capacity result in more vegetation and higher productivity on mima and mima-like mounds than in areas between mounds (Gakahu and Cox 1984; Cox and Zedler 1986; Lovegrove 1991). Pocket gopher mounds and mima mounds have a dramatic effect on the structure of plant communities because they change the edaphic and abiotic structure of microsites (e.g., creating deep, loamy soils and increasing soil moisture: Cox, Contreras, and Milewski

1995), thereby creating favorable germination sites for a wide variety of plant species.

Although ecosystems in which subterranean rodents do not produce mima-like mounds may lack such concentrated patches of increased productivity, soil movement and turnover may yield similar positive correlations between rodent activity and primary production. Laycock and Richardson (1975) reported lower productivity on an experimental plot from which pocket gophers had been excluded for 31 years than on a control plot where gophers were present. Reichman and Smith (1985) reported a positive relationship between the presence of mounds of *G. bursarius* and the intensity of primary productivity, although others have not detected this relationship (Bathyergidae: Reichman and Jarvis 1989; Geomyidae: Sparks and Andersen 1988). It is not known, however, whether subterranean rodents affect productivity by their activities or whether these rodents select areas of high productivity in which to locate their burrows.

On a finer scale, the nature of the vegetation may be influenced by burrow system structure. For example, the biomass of vegetation was reduced over both geomyid (Reichman and Smith 1985) and bathyergid tunnels (Reichman and Jarvis 1989) relative to the biomass in non-tunnel areas; plants with rhizomes and fibrous roots were the least affected by the presence of rodent burrows. The mechanism of reduction in plant biomass was not clearly identified, but may involve consumption or disruption of the root zone. A similar reduction in biomass, however, was not observed above the tunnels of *G. attwateri* in Texas, suggesting that plant responses to subterranean tunnels vary between communities (M. Rezsutek and G. N. Cameron, in press).

The diversity of plant species in areas that have been disturbed by geomyids is greater than that in undisturbed areas (Andersen and MacMahon 1981; Tilman 1983; Hobbs and Mooney 1985; Williams and Cameron 1986; Hobbs and Hobbs 1987; Inouye et al. 1987b; Contreras and Gutiérrez 1991; Lovegrove 1991; Gomez-Garcia, Borghi, and Giannoni 1995). Gibson (1989) reported an apparent exception to this trend in which plant species richness was lower on mounds of *G. bursarius* than in surrounding areas. Gibson attributed the lower richness on mounds to vegetative regrowth of dominant graminoids; however, because Gibson did not measure species abundances, this study cannot be compared directly with the other studies cited.

Subterranean rodents also may affect plant species composition and abundance through their differential consumption of vegetation. The dietary preferences of subterranean rodents have been determined by examining stomach or fecal contents (e.g., Bandoli 1981; Pearson 1983; Williams and Cameron 1986; Behrend and Tester 1988; Cox 1989; Comparatore, Cid, and Busch 1995), locating food caches (Galil 1967; Andersen and Mac-

Mahon 1981; Bandoli 1981), from visual evidence of consumption of plants or plant parts (i.e., roots or tubers: Brett 1991), and through feeding trials (Radwan et al. 1982; Behrend and Tester 1988; Jenkins and Bollinger 1989; Camín and Madoery 1994). In grassland ecosystems, *Geomys* and *Thomomys* eat both shoots and roots of grasses and forbs, often preferring forbs over grasses (Ward 1973; Andersen and MacMahon 1981; Bandoli 1981; Behrend and Tester 1988; Cox 1989; Hunt 1992; Comparatore, Cid, and Busch 1995; Busch et al., chap. 5, this volume). In Nebraska, activity of *G. bursarius* was greatest in forb patches, and forbs constitute a large portion of the diet (Steuter et al. 1995). *T. bottae* selected monocots over dicots (Ellison and Aldous 1952; Martinsen, Cushman, and Whitham 1990), but *G. attwateri* selected dicots, presumably because of their higher protein content (Williams and Cameron 1986).

In more arid ecosystems, *Nannospalax ehrenbergi* feeds primarily on bulbs or corms, which are harvested during the growing season and stored for consumption throughout the year (Galil 1967). In contrast, *Heterocephalus glaber* and several species of mole-voles *(Microtus gerbei, M. duodecimcostatus,* and *M. lusitanicus)* feed on tubers and rhizomes of legumes in situ, rather than caching these food items (Brett 1991; Borghi and Giannoni 1997). *Spalacopus cyanus* feeds extensively on shoots of grasses and forbs, but primarily consumes the bulbs of geophytes (Contreras and Gutiérrez 1991).

The diet preferences of subterranean rodents may affect the structure of the aboveground flora by constraining plants to certain habitats or by restricting the growth of plant species that would otherwise be more prevalent. For example, Cox (1989) suggested that the local distributions of two species of the tuberous forb *Lomatium* were limited to areas between mima mounds by foraging by *Thomomys*. Cantor and Whitham (1989) concluded that aspen trees in the Rocky Mountains may be restricted to rocky outcrops because of predation on seedlings by *Thomomys,* and *Geomys* represents a significant source of mortality for tree seedlings in old fields in Minnesota (Inouye, Allison, and Johnson 1994). Cox and Hunt (1994) found that, in western Colorado, foraging by *T. bottae* limits the abundance of ocotillo *(Fouquieria splendens)* in certain habitats.

Subterranean rodents may also affect the distributions of particular plant species through damage associated with burrowing. For example, root pruning affects plant survivorship and biomass, and consequently flower production, in *Tragopogon dubius,* a biennial plant found in habitats occupied by *G. bursarius* (Reichman and Smith 1991). *Nannospalax* in Israel damage young forests by feeding on the roots and trunks of several tree species (Lovegrove 1991). Alternatively, the actions of subterranean rodents may stimulate plant growth and reproduction, as occurs in geophytes consumed by *Spalacopus* (Contreras and Gutiérrez 1991).

Plants, in turn, may affect the distribution and abundance of subterranean rodents. For example, Buechner (1942) observed that *Geomys breviceps* increased in abundance on overgrazed lands in Texas. Early-succession forbs represent an important part of the diet of this species, and the prevalence of such forbs on overgrazed lands may have led to increased gopher densities. One conclusion that can be drawn from the diet studies mentioned above is that subterranean rodents require both monocots and dicots. To test this hypothesis, Tietjen et al. (1967) and Keith, Hansen, and Ward (1959) experimentally removed dicots from grassland habitats and discovered that densities and reproductive rates of *T. talpoides* declined. Rezsutek and Cameron (1998) obtained similar results when they removed dicots from areas inhabited by *G. attwateri*. Such studies indicate that populations of these subterranean rodents may be limited by the availability of certain classes of food. Because forbs often constitute a larger proportion of the diet during the growing season, and contain higher concentrations of nitrogen [protein] and most other nutrients than grasses, there is likely to be a nutritional basis for this reliance on particular types of food plants (Rezsutek 1997). These nutritional requirements lead to further differential selection of forage species and exacerbate patterns of species abundance and composition caused by subterranean rodents.

Certain coevolved features of community structure result from subterranean rodent-plant interactions. For example, while most geophytes are unpalatable to subterranean rodents, geophytes of the genus *Micranthus* are quite palatable to bathyergid mole-rats, suggesting that either these rodents have evolved a tolerance for *Micranthus* or this geophyte has evolved a greater palatability that allows it to be consumed by mole-rats. The latter suggestion is not as far-fetched as it may seem; the morphology of this geophyte is unusual because the number of corm segments is less than in inedible geophytes, and, during feeding, some segments become dislodged. These morphological and possible chemical adaptations ensure dispersal of the plant (Galil 1967; Lovegrove and Jarvis 1986; Borghi and Giannoni 1997).

Despite a number of studies demonstrating correlations between floral composition and the abundance of subterranean rodents, the specific interactions between these organisms remain unclear. For example, what is the effect of subterranean rodents on primary production? How do plant responses to rodent activity vary among habitats and taxa? What role do the nutritional requirements of subterranean rodents play in these interactions? The foraging behavior of subterranean rodents (e.g., Andersen 1987; Brett 1991; Benedix 1993; Busch et al., chap. 5, this volume) may be best understood when analyzed in the context of a nutritional landscape formed by the spatial distribution of food resources. Using this approach, the effects of subterranean rodents on plant quality, quantity, and distribution can be examined, as can the effects of plants on rodent presence and abundance.

Vegetational Succession

Because their activities create disturbances on the soil surface, subterranean rodents play a prominent role in secondary succession. Disturbances caused by their deposition of soil on the surface serve as colonization sites for plants (McDonough 1974; Chase, Howard, and Roseberry 1982; Peart 1989; Cox, Contreras, and Milewski 1995). Colonizing species of grassland plants (e.g., *Penstemon grandiflorus, Berteroa incana*) grew better and reproduced sooner on bare mounds created by *G. bursarius* (Reichman 1988; Davis et al. 1991a,b). However, these plant species were unable to persist as oak savanna transformed into oak woodland, a habitat lacking pocket gopher mounds (Davis et al. 1995). Mounds of soil deposited by *T. talpoides* on the volcanic ash resulting from the 1980 eruption of Mount St. Helens allowed plants buried in ash to survive and grow through the ash layer (Andersen and MacMahon 1985), thus hastening the process of succession.

The activities of subterranean rodents may also have negative consequences for succession and subsequent plant population dynamics. For example, the deposition of mound soil by *T. bottae* retards succession; germination and establishment of native perennial grasses were reduced on pocket gopher mounds in coastal California grasslands, probably because the more xeric conditions on mounds favored exotic grasses (Stromberg and Griffin 1996). At the same time, the mounds buried and killed native perennial seedlings that were not vigorous enough to grow through the mound of soil.

Martinsen, Cushman, and Whitham (1990) noted that species diversity in a shortgrass prairie was greatest for plots characterized by pocket gopher disturbances of intermediate age; the abundance of grasses decreased and the abundance of dicots increased with increasing pocket gopher disturbance. Contreras and Gutiérrez (1991) noted that *S. cyanus* had the same effect on plant species diversity in Chile, resulting in dominance by aggressive annuals in areas heavily disturbed by coruro activity.

In grasslands in east-central Minnesota, structure and succession in plant communities on nitrogen-poor soils were affected by herbivore abundance, including the abundance of *G. bursarius*. The pocket gophers affected succession by selectively eating perennials (Tilman 1983). In addition, they moved nitrogen-poor soil to the surface, thereby affecting local species composition (Inouye et al. 1987a). Succession on these nitrogen-poor soils was retarded because the gophers' consumption of legumes slowed accumulation of nitrogen in the soil (Tilman 1985; Inouye et al. 1987a,b; Ritchie and Tilman 1995). The gophers also fed on tree seedlings, thereby retarding succession and keeping the habitat dominated by herbaceous vegetation (Inouye, Allison, and Johnson 1994).

Over longer time scales, successional changes in vegetation may influence the activity of subterranean rodents. For example, the density of *G. bursarius* mounds decreased with the age of old fields in Minnesota, and was negatively correlated with vegetation biomass and nitrogen content of vegetation (Huntly and Inouye 1987). Age-related changes in vegetation along a 50-year chronosequence of old fields included increased cover of perennial and woody species, increased cover of perennial grasses, increased species richness of plants, increased belowground biomass, and increased diversity and richness of small mammals (Inouye et al. 1987b). The decrease in mound density with field age probably reflects a decrease in the amount of excavation necessary to harvest a given biomass of plants. Even though the density of gopher mounds decreased with field age, the effect of individual mounds on vegetation was greater in older fields because differences between surface and subsurface soils increased with time, and because some plant species that were common in young fields (even in the absence of gopher disturbance) were rare in older fields except where mounds had been formed (Inouye et al. 1987a).

Compared with geomyids, little is known about the roles of other subterranean rodents in ecological succession. Even for geomyids, most of what is known comes from studies of a single community at Cedar Creek, Michigan (see studies by Huntly and Inouye). Because successional patterns may vary among localities, and because the roles of subterranean rodents in these patterns may also vary, long-term studies of other taxa and other localities are needed. Once data from a variety of taxa and ecological settings have been obtained, comparative analyses of these systems can be used to identify common elements of the successional effects of subterranean rodents.

Energy Flow and Nutrient Cycling

Although both energy flow and nutrient cycling are critical ecosystem functions, little is understood about the role that subterranean rodents play in either process. Because subterranean rodents may be major herbivores in many ecosystems, these interactions are likely to be important. To date, only a single study has analyzed nitrogen intake in pocket gophers. Gettinger (1984b) found that, despite seasonal changes in the availability of nitrogen, *T. bottae* maintained a relatively constant intake of 2,400 mg N per kg per day during summer and 2,300 mg N per kg per day in winter. This constancy of nitrogen intake appeared to result from seasonal changes in preferred food resources. Gettinger did not relate these findings to the overall nitrogen budget of the ecosystem studied, however, and thus it is unclear how annual patterns of nitrogen intake by gophers reflect community-level patterns of nitrogen use.

Mielke (1977) concluded that pocket gophers must dramatically affect the biogeochemical cycles of prairies in North America because of the substantial effect of these rodents on soil mixing, sorting, and mounding. Mielke believed that these effects were particularly important in arid habitats such as prairies because pocket gophers increased water availability at the surface. Mielke further speculated that there is a mutualistic interaction between pocket gophers and bison: bison graze and trample the dense prairie vegetation, accelerating the growth of forbs, upon which gophers feed. In turn, pocket gophers work the soil, increase soil fertility, and accelerate the growth of vegetation consumed by bison.

Few studies of the energy budgets of subterranean rodents have been undertaken. Gettinger (1984a) estimated that a population containing about forty-five *T. bottae* per hectare would consume about 2,200 MJ per ha per year in a California mountain habitat. This amount exceeds estimates for nearly all other small mammal populations (Humphreys 1979). Andersen and MacMahon (1981) estimated an average annual energy consumption of 636 MJ per ha per year (range 427–1,087 MJ per ha per year) for a population of four to sixty-six *T. talpoides* per hectare (estimated to be 30% of the annual net primary productivity allocated to belowground plant parts). The density of *S. cyanus* in Chile is similar to that of the North American pocket gophers, suggesting that their effect on energy flow may be similar (Reig 1970).

Understanding the role of subterranean rodents in nutrient cycles and energy flow requires intensive study of all taxa and habitats. At present, however, few data are available, and these come only from studies of geomyids. Nothing is known about the ecological efficiency of subterranean rodents, and little is known about their consumption or assimilation efficiencies. (Information on digestive efficiency, which, in general, tends to be high, is given by Buffenstein in chapter 2 of this volume). Soil movement, consumption of vegetation, deposition of vegetation in underground caches, and creation of latrines by subterranean rodents must dramatically affect nutrient cycling; concomitantly, these activities must affect primary productivity, although little is known about either set of interactions. Improved understanding of the foraging behavior, nutritional requirements, and population dynamics of subterranean rodents, coupled with detailed analyses of nutrient cycling and energy flow, would represent a quantum leap in our understanding of the ecosystems in which these rodents occur.

Spatial and Temporal Variation in the Ecological Effects of Subterranean Rodents

The patchy distributions of individuals and populations of subterranean rodents in space and time indicate that the ecological effects of subterranean rodents may vary. Spatially patchy distributions may reflect both biotic and abiotic factors. Predominant among biotic factors is territoriality, which results in solitary burrow occupancy and, hence, spatial separation of individuals (Reig et al. 1990). At the population level, patchy distributions of preferred habitats result in clusters of individuals (Busch et al., chap. 5, this volume).

Because burrowing is energetically expensive, the burrows of many subterranean rodents are used and maintained over the lifetimes of many individuals. This behavior produces a clustered distribution of burrows, resulting in spatial variation in the effects of subterranean rodents. For example, in grassland systems, disturbances create openings in vegetation that are likely to be colonized by forbs, some of which may be preferred food plants, leading to high local densities of pocket gophers (Laycock and Richardson 1975; Hobbs and Mooney 1985; Williams et al. 1986; Inouye et al. 1987a). Likewise, where mima-like mounds are formed, there may be strong positive feedbacks that concentrate foraging in some areas. As more soil and nutrients accumulate in mounds, and as productivity in mounds increases relative to intermound areas, foraging becomes more concentrated on or adjacent to mounds, thus exacerbating patchiness.

The productivity of the habitat may also affect the patchiness of burrow distributions. In an old field system in Minnesota, the spatial distribution of gopher mounds was less patchy in younger, less productive old fields (Huntly and Inouye 1987). One explanation for this pattern is that individuals are forced to build new feeding tunnels more frequently, with the result that mounds are produced more uniformly over a wider area in younger fields. Within a local area, burrowing activity is greater in areas of higher productivity (Hobbs and Mooney 1985; Inouye et al. 1987a; Wasley 1995; Busch et al., chap. 5, this volume), perhaps reflecting the energetic cost of burrowing (Vleck 1979, 1981) and the advantage of maintaining a high rate of return for that cost.

Following a field study that demonstrated that soil disturbance by *T. bottae* affected the population dynamics of individual plant species (Hobbs and Mooney 1985), Hobbs and Hobbs (1987) used a computer simulation to model the population processes and spatial patterns generated by this disturbance. Their model revealed that soil disturbance by gophers affected both the short-term spatial patterning and the long-term species composition of plants in this annual grassland. Addition of fertilizer to this grassland increased the abundance of grasses and the deposition of mulch (Hobbs et al. 1988); the removal of grass mulch by pocket gophers allowed invasion by forbs and affected the spatial pattern of plant species.

Abiotic cues also may strongly affect the temporal and spatial distribution of burrowing activity by subterranean rodents. Temporal variation in burrowing activity commonly reflects seasonality in primary production (Bandoli 1981), but also may be tied to precipitation events within a growing season. In relatively arid areas, where soils can be very dry and hard during part of the year, most burrowing takes place during the growing season, when soils are easier to excavate. Thus, Brett (1991) found that 73% of the mounds of naked mole-rats were produced during the rainy season, and Heth (1992) reported that *Nannospalax ehrenbergi* did most of their burrowing during the rainy winter months. Even in more mesic ecosystems, burrowing by geomyids was positively correlated with precipitation events during the growing season (Miller 1948; Cox and Hunt 1992).

In contrast, other studies of geomyids have found no correlation between precipitation and the production of mounds (Miller and Bond 1960; Hickman and Brown 1973; Bandoli 1981). These differences may reflect geographic differences in soils that influence their cohesion, and thus the likelihood that burrows will collapse. For example, in very sandy and relatively unproductive areas that lack extensive root systems, shallow burrows can be susceptible to collapse when the soil is dry, and thus burrowing may be more common after precipitation. In contrast, Anderson and MacMahon (1981) reported that pocket gophers avoided burrowing in wet soil, and suggested that this reflected attempts to reduce both metabolic costs associated with wet fur and energetic costs associated with moving heavy soil.

Annual fluctuations in temperature that affect rodent activity may also affect the temporal patterning of the community-level effects of these animals. Cox and Hunt (1992) discovered that tunnel maintenance by *T. bottae* is negatively correlated with temperature and time since last rainfall. In their study population, production of surface-access tunnels peaked at the end of the growing season in late spring-early summer; in contrast, expansion of the burrow system occurred during periods of moist soil and cool temperatures in early winter. As a result, soil disturbance tended to be greatest during winter, when soil from burrowing was deposited aboveground. This seasonality in soil movement is likely to be related to the phenology of food plants: the growth of herbaceous food plants in early winter and the availability of harvestable aboveground foods in late spring-early summer.

Biologists have just begun to explore the determinants and consequences of spatial and temporal patchiness in subterranean rodent activity. As a result, considerable further research is needed to understand how this patchiness is generated and how it affects other aspects of community structure.

Phylogenetically controlled comparisons of species exhibiting different degrees of spatial or temporal clustering may prove particularly useful in addressing this topic. Such studies are critical to evaluating the role of subterranean rodents as keystone species in ecosystem processes.

Conclusions and Future Directions

One theme of this volume is the role of the subterranean niche in shaping the biology of the rodents that inhabit it. This chapter has focused on how occupants of this niche interact with one another, with the soil medium, with other subterranean taxa, and with surface flora and fauna. By analyzing these interactions, we can begin to understand how subterranean rodents influence and are influenced by the communities in which they live. Unfortunately, the data available at present are insufficient to provide a complete picture of how adaptations to life underground contribute to community-level processes. Most of the extant information comes from studies of geomyids; virtually nothing is known about the community ecologies of myospalacines, arvicolines, or rhizomyines. Clearly, long-term ecological studies of more taxa are needed to characterize the interactions between subterranean rodents and their ecosystems.

Despite the current state of our knowledge, it is possible to identify several ways in which specialization for subterranean life contributes to community dynamics. By virtue of their inclination to construct tunnels and mounds and to create food caches and latrines, subterranean rodents may have substantially greater effects on soil texture, water content, and nutrient content than their surface-dwelling counterparts. These effects on soil structure are, in turn, likely to generate significant effects on vegetation through different growing conditions in burrow and non-burrow areas. In particular, the presence of subterranean rodents may alter patterns of plant dispersion, growth, and reproduction, which may contribute to variation in the productivity and species diversity of the floral community.

There are many types of interactions between subterranean rodents, other subterranean creatures, and surface-dwelling species. Intra- and interspecific competition may occur, leading to a variety of mechanisms intended to reduce the negative consequences of this type of interaction. Among subterranean animals, such mechanisms include diet separation and vertical stratification of the habitat. The former is a widely used mechanism in aboveground communities, but the latter is used only in those few communities with sufficient vertical stratification of vegetation. Comparatively little is known about interactions or mechanisms for avoiding competition between subterranean and surface-dwelling herbivores. Such interactions may be important in structuring communities because of the effects of both above- and belowground herbivores on plants. Similarly, while lists of predators on subterranean rodents have been compiled, our understanding of how predation affects the structure of communities containing subterranean rodents is poor. Studies that view subterranean rodents as only one component of a complex network of ecological relationships are needed to characterize the role that these animals play in community structure.

Adding another dimension—subsurface depth—to community and ecosystem studies is a fascinating but challenging endeavor. As this review illustrates, life underground involves a variety of ecological interactions that, if not unique, at least differ substantially from interactions at the surface or above. In short, an exciting ecological realm is revealed by studying the fauna that dwell in subterranean burrows. Future studies must expand our knowledge of this realm to divulge the full complexity and importance of the role that these fascinating animals play in ecological communities.

Acknowledgments

I thank E. Bolen for piquing my interest in the pocket gopher-coastal prairie ecosystem, my former graduate students, L. Williams and M. Rezsutek, for explorations of the subterranean world, L. Drawe for logistic and plant taxonomy support, and D. Andersen, E. Bryant, B. Danielson, and O. J. Reichman for discussions about pocket gophers. I particularly thank the numerous colleagues who have assisted with fieldwork: K. Bruce, H. Caravello, L. Combs (Meffert), L. Cooper, M. Descalzi, B. Eshelman, M. Gregory, K. Groue, D. Hadley, L. Hugg, B. Kincaid, F. Latteo, N. Muderrisoglu, K. Myer, K. Nolte, D. Scheel, S. Schwartz, S. Sheffield, B. Sillen-Tullberg, S. Spencer, E. Vance, T. Vincent, C. Williams, and J. Williams. Welder Wildlife Foundation, the University of Houston Coastal Center, and the Theodore Roosevelt Memorial Fund provided financial assistance for our studies of pocket gophers.

Literature Cited

Andersen, D.C. 1987. *Geomys bursarius* burrowing patterns: Influence of season and food patch structure. *Ecology* 68:1306–18.

Andersen, D.C., and J. A. MacMahon. 1981. Population dynamics and bioenergetics of a fossorial herbivore, *Thomomys talpoides* (Rodentia: Geomyidae), in a spruce-fir sere. *Ecological Monographs* 51:179–202.

———. 1985. Plant succession following the Mount St. Helens volcanic eruption: Facilitation by a burrowing rodent, *Thomomys talpoides*. *American Midland Naturalist* 114:62–69.

Baker, A. E. M. 1974. Interspecific aggressive behavior of pocket gophers *Thomomys bottae* and *T. talpoides* (Geomyidae: Rodentia). *Ecology* 55:671–73.

Bandoli, J. H. 1981. Factors influencing seasonal burrowing activity in the pocket gopher, *Thomomys bottae. Journal of Mammalogy* 62:293–303.

———. 1987. Activity and plural occupancy of burrows in Botta's pocket gopher *Thomomys bottae. American Midland Naturalist* 118:10–14.

Behrend, A. F., and J. R. Tester. 1988. Feeding ecology of the plains pocket gopher in east central Minnesota. *Prairie Naturalist* 20:99–107.

Benedix, J. H. Jr. 1993. Area-restricted search by the plains pocket gopher *(Geomys bursarius)* in tallgrass prairie habitat. *Behavioral Ecology* 4:318–24.

Bennett, N. C. 1989. The social structure and reproductive biology of the common mole-rat, *Cryptomys h. hottentotus* and remarks on the trends in reproduction and sociality in the family Bathyergidae. *Journal of Zoology, London* 219:45–59.

———. 1994. Reproductive suppression in social *Cryptomys damarensis* colonies—a lifetime of socially-induced sterility in males and females (Rodentia: Bathyergidae). *Journal of Zoology, London* 234:25–39.

Best, T. L. 1973. Ecological separation of three genera of pocket gophers (Geomyidae). *Ecology* 54:1311–19.

Blume, R. R., and A. Aga. 1979. Additional records of *Aphodius* from pocket gopher burrows in Texas (Coleoptera: Scarabaeidae). *Coleopterists Bulletin* 33:131–32.

Borghi, C. E., and S. M. Giannoni. 1997. Dispersal of geophytes by mole-voles in the Spanish Pyrenees. *Journal of Mammalogy* 78:550–55.

Borghi, C. E., S. M. Giannoni, and J. P. Martinez-Rica. 1994. Habitat segregation of three sympatric fossorial rodents in the Spanish Pyrenees. *Zeitschrift für Säugetierkunde* 59:52–57.

Brett, R. 1991. The ecology of naked mole-rat colonies: Burrowing, food, and limiting factors. In *The biology of the naked mole-rat*, edited by P. W. Sherman, J. U. M. Jarvis, and R. D. Alexander, 137–84. Princeton, NJ: Princeton University Press.

Buechner, H. K. 1942. Interrelationships between the pocket gopher and land use. *Journal of Mammalogy* 23:346–48.

Camín, S. R., and L. A. Madoery. 1994. Feeding behavior of the tuco tuco *(Ctenomys mendocinus):* Its modifications according to food availability and the changes in the harvest pattern and consumption. *Revista Chilena de Historia Natural* 67:257–63.

Cantor, L. F., and T. G. Whitham. 1989. Importance of belowground herbivory: Pocket gophers may limit aspen to rock outcrop refugia. *Ecology* 70:962–70.

Cartwright, O. 1944. New *Aphodius* from Texas gopher burrows. *Entomological News* 55:129–35, 146–50.

Chase, J. D., W. E. Howard, and J. T. Roseberry. 1982. Pocket gophers (Geomyidae). In *Wild mammals of North America: Biology, management, and economics*, edited by J. A. Chapman and G. A. Feldhamer, 239–55. Baltimore: Johns Hopkins University Press.

Chippaux, J.-P., and F.-X. Pajot. 1983. Leishmaniasis in French Guiana. 4. Preliminary note on phlebotomine sand flies from mammal burrows. *Entomological Medicine and Parasitology* 21:149–54.

Colorado Cooperative Pocket Gopher Project. 1960. Pocket gophers in Colorado. Bulletin 508S, Colorado State University Agricultural Experiment Station.

Comparatore, V. M., M. S. Cid, and C. Busch. 1995. Dietary preferences of two sympatric fossorial rodent populations in Argentina. *Revista Chilena de Historia Natural* 68:197–206.

Contreras, L. C., and J. R. Gutiérrez. 1991. Effects of the fossorial herbivorous rodent *Spalacopus cyanus* on herbaceous vegetation in arid coastal Chile. *Oecologia* 87:106–9.

Cox, G. W. 1989. Early summer diet and food preferences of northern pocket gophers in north central Oregon. *Northwest Science* 63:77–82.

———. 1990. Soil mining by pocket gophers along topographic gradients in a mima moundfield. *Ecology* 71:837–43.

Cox, G. W., and D. W. Allen. 1987. Soil translocation by pocket gophers in a mima moundfield. *Oecologia* 72:207–10.

Cox, G. W., L. C. Contreras, and A. V. Milewski. 1995. Role of fossorial animals in community structure and energetics of Pacific Mediterranean ecosystems. In *Ecology and biogeography of Mediterranean ecosystems in Chile, California, and Australia*, edited by M. T. K. Arroyo, P. H. Zedler, and M. D. Fox, 383–98. New York: Springer-Verlag.

Cox, G. W., and C. G. Gakahu. 1985. Mima mound microtopography and vegetation pattern in Kenyan savannas. *Journal of Tropical Ecology* 1:23–36.

———. 1986. A latitudinal test of the fossorial rodent hypothesis of Mima mound origin. *Zeitschrift für Geomorphologie* 30:485–501.

Cox, G. W., and J. Hunt. 1992. Relation of seasonal activity patterns of valley pocket gophers to temperature, rainfall, and food availability. *Journal of Mammalogy* 73:123–34.

———. 1994. Pocket gopher herbivory and mortality of ocotillo on stream terrace, bajada, and hillside sites in the Colorado desert, southern California. *Southwestern Naturalist* 39:364–70.

Cox, G. W., B. G. Lovegrove, and W. R. Siegried. 1987. The small stone content of mimalike mounds in the South African Cape region: Implications for mound origin. *Catena* 14:165–76.

Cox, G. W., and J. B. Zedler. 1986. The influence of mima mounds on vegetation patterns in the Tijuana Estuary salt marsh, San Diego County, California. *Bulletin of the Southern California Academy of Science* 85:158–72.

Davies, K. C., and J. U. M. Jarvis. 1986. The burrow system and burrowing dynamics of the mole-rats *Bathyergus suillus* and *Cryptomys hottentotus* in the fynbos of the southwestern Cape, South Africa. *Journal of Zoology, London* 209:125–47.

Davis, M. A., B. Ritchie, N. Graf, and K. Gregg. 1995. An experimental study of the effects of shade, conspecific crowding, pocket gophers and surrounding vegetation on survivorship, growth and reproduction in *Penstemon grandiflorus*. *American Midland Naturalist* 134:237–43.

Davis, M. A., J. Villinski, K. Banks, J. Buckman-Fifield, J. Dicus, and S. Hofman. 1991a. Combining effects of fire, mound-building by pocket gophers, root loss, and plant size on growth and reproduction in *Penstemon grandiflorus*. *American Midland Naturalist* 125:150–61.

Davis, M. A., J. Villinski, S. McAndrew, H. Scholtz, and E. Young. 1991b. Survivorship of *Penstemon grandiflorus* in an oak woodland: Combined effects of fire and pocket gophers. *Oecologia* 86:113–18.

Ellerman, J. R. 1940. *The families and genera of living rodents*. Vol. 1. London: British Museum of Natural History. 689 pp.

Ellison, L., and C. M. Aldous. 1952. Influence of pocket gophers on vegetation of subalpine grassland in central Utah. *Ecology* 33:177–85.

Eloff, G. 1951. Orientation in the mole-rat *Cryptomys*. *British Journal of Psychology* 42:134–45.

Gakahu, C. G., and G. W. Cox. 1984. The occurrence and origin of Mima mound terrain in Kenya. *African Journal of Ecology* 22:31–42.

Galil, J. 1967. On the dispersal of the bulbs of *Oxalis cerua* Thunb. by mole-rats (*Spalax ehrenbergi* Nehring). *Journal of Ecology* 55:787–92.

Gettinger, R. D. 1984a. Energy and water metabolism of free-ranging pocket gophers *Thomomys bottae*. *Ecology* 65:740–51.

———. 1984b. Seasonal patterns of nitrogen utilization by pocket gophers, *Thomomys bottae*. *Comparative Biochemistry and Physiology* 78A:657–59.

Gibson, D. J. 1989. Effects of animal disturbance on tallgrass prairie vegetation. *American Midland Naturalist* 121:144–54.

Gomez-Garcia, D., C. E. Borghi, and S. M. Giannoni. 1995. Vegetation differences caused by pine vole mound building in subalpine plant communities in the Spanish Pyrenees. *Vegetatio* 117:61–67.

Gonzales, G. J., M. T. Saladen, and T. E. Hakonson. 1995. Effects of pocket gopher burrowing on cesium-133 distribution on engineered test plots. *Journal of Environmental Quality* 24:1056–62.

Grant, W. E., N. R. French, and L. J. Folse, Jr. 1980. Effects of pocket gopher mounds on plant production in shortgrass prairie ecosystems. *Southwestern Naturalist* 25:215–24.

Grant, W. E., and J. F. McBrayer. 1981. Effects of mound formation by pocket gophers *(Geomys bursarius)* on old-field ecosystems. *Pedobiologia* 22:21–28.

Gregory, M. J., G. N. Cameron, L. M. Combs, and L. W. Williams. 1987. Agonistic behavior in Attwater's pocket gopher, *Geomys attwateri*. *Southwestern Naturalist* 32:143–46.

Grinnell, J. 1923. The burrowing rodents of California as agents in soil formation. *Journal of Mammalogy* 4:137–49.

Hafner, M. S., and L. J. Barkley. 1984. Genetics and natural history of a relictual pocket gopher, *Zygogeomys*. *Journal of Mammalogy* 65:474–79.

Hall, E. R., and B. R. Villa. 1949. An annotated checklist of the mammals of Michoacán, Mexico. *Miscellaneous Publications, Museum of Natural History, University of Kansas* 1:431–72.

Hansen, R. M., and R. S. Miller. 1959. Observations on the plural occupancy of pocket gopher burrow systems. *Journal of Mammalogy* 40:577–84.

Hansen, R. M., and A. L. Ward. 1966. Some relations of pocket gophers to rangelands on Grand Mesa, Colorado. Technical Bulletin 88, Colorado State University Agricultural Experiment Station.

Heth, G. 1992. The environmental impact of subterranean mole rats *(Spalax ehrenbergi)* and their burrows. In *The environmental impact of burrowing animals and animal burrows*, edited by P. S. Meadows and A. Meadows, 265–80. Oxford: Clarendon Press.

Hickman, G. C. 1977. Geomyid interaction in burrow systems. *Texas Journal of Science* 29:235–43.

Hickman, G. C., and L. N. Brown. 1973. Pattern and rate of mound production in the southeastern pocket gopher *(Geomys pinetis)*. *Journal of Mammalogy* 54:971–75.

Hobbs, R. J., S. L. Gulmon, V. J. Hobbs, and H. A. Mooney. 1988. Effects of fertiliser addition and subsequent gopher disturbance on a serpentine annual grassland community. *Oecologia* 75:291–95.

Hobbs, R. J., and V. J. Hobbs. 1987. Gophers and grassland: A model of vegetation response to patchy soil disturbance. *Vegetatio* 69:141–46.

Hobbs, R. J., and H. A. Mooney. 1985. Community and population dynamics of serpentine grassland annuals in relation to gopher disturbance. *Oecologia* 67:342–51.

———. 1991. Effects of rainfall variability and gopher disturbance on serpentine annual grassland dynamics. *Ecology* 72:59–68.

Hubbell, T. H., and C. C. Goff. 1940. Florida pocket-gopher burrows and their arthropod inhabitants. *Proceedings of the Florida Academy of Sciences* 4:127–66.

Humphreys, W. F. 1979. Production and respiration in animal populations. *Journal of Animal Ecology* 48:427–53.

Hunt, J. 1992. Feeding ecology of valley pocket gophers *(Thomomys bottae sanctidiegi)* on a California coastal grassland. *American Midland Naturalist* 127:41–51.

Hunter, J. E. 1991. Grazing and pocket gopher abundance in a California annual grassland. *Southwestern Naturalist* 36:117–18.

Huntly, N. 1991. Herbivores and the dynamics of communities and ecosystems. *Annual Review of Ecology and Systematics* 22:477–504.

Huntly, N., and R. S. Inouye. 1987. Small mammal populations of an old-field chronosequence: Successional patterns and association with vegetation. *Journal of Mammalogy* 68:739–45.

———. 1988. Pocket gophers in ecosystems: Patterns and mechanisms. *BioScience* 38: 786–93.
Inouye, R. S., T. B. Allison, and J. C. Johnson. 1994. Old field succession on a Minnesota sand plain: Effects of deer and other factors on invasion by trees. *Bulletin of the Torrey Botanical Club* 121:266–76.
Inouye, R. S., N.J. Huntly, D. Tilman, andJ. R. Tester. 1987a. Pocket gophers *(Geomys bursarius)*, vegetation, and soil nitrogen along a successional sere in east central Minnesota. *Oecologia* 72:178–84.
Inouye, R. S., N.J. Huntly, D. Tilman, J. Tester, M. Stilwell, and K. C. Zinnell. 1987b. Oldfield succession on a Minnesota sand plain. *Ecology* 68:12–26.
Inouye, R. S., N. Huntly, and G. A. Wasley. 1997. Effects of pocket gophers *(Geomys bursarius)* on microtopographical variation. *Journal of Mammalogy* 78:1144–48.
Jarvis, J. U. M., and N. C. Bennett. 1990. The evolutionary history, population biology and social structure of African mole-rats: Family Bathyergidae. In *Evolution of subterranean mammals at the organismal and molecular levels,* edited by E. Nevo and O. A. Reig, 97–128. Progress in Clinical and Biological Research, vol. 335. New York: Wiley-Liss.
Jenkins, S. H., and P. W. Bollinger. 1989. An experimental test of diet selection by the pocket gopher *Thomomys monticola. Journal of Mammalogy* 70:406–12.
Johnson, D. L. 1989. Subsurface stone lines, stone zones, artifact-manuport layers, and biomantles produced by bioturbation via pocket gophers *(Thomomys bottae). American Antiquity* 54:370–89.
Jones, C. G., J. H. Lawton, and M. Shachak. 1997. Positive and negative effects of organisms as physical ecosystem engineers. *Ecology* 78:1946–57.
Kalisz, P. J., and E. L. Stone. 1984. Soil mixing by scarab beetles and pocket gophers in North Central Florida. *Soil Science Society of America Journal* 48:169–72.
Kennerly, T. E., Jr. 1959. Contact between the ranges of two allopatric species of pocket gophers. *Evolution* 13:247–63.
Keith, J. O., R. M. Hansen, and A. L. Ward. 1959. Effect of 2,4-D on abundance and foods of pocket gophers. *Journal of Wildlife Management* 23:137–45.
Klaas, B. A., B. J. Danielson, and K. A. Moloney. 1998. Influence of pocket gophers on meadow voles in a tallgrass prairie. *Journal of Mammalogy* 79:942–52.
Koide, R. T., L. F. Huenneck, and H. A. Mooney. 1987. Gopher mound soil reduces growth and affects ion uptake of two annual grassland species. *Oecologia* 72:284–90.
Lawton, J. H., and C. G. Jones. 1995. Organisms as ecosystem engineers. In *Linking species and ecosystems,* edited by C. G. Jones and J. H. Lawton, 141–50. New York: Chapman and Hall.
Laycock, W. A., and B. Z. Richardson. 1975. Long-term effects of pocket gopher control on vegetation and soils of a subalpine grassland. *Journal of Range Management* 28:458–62.
Lechleitner, R., J. Tileston, and L. Kartman. 1962. Die-off of a Gunnison's prairie dog colony in central Colorado. I. Ecological observations and description of the epizootic. *Zoonoses Research* 1:185–99.
Lovegrove, B. G. 1991. Mima-like mounds (heuweltjes) of South Africa: The topographic, ecological, and economic impact of burrowing animals. In *The environmental impact of burrowing animals and animal burrows,* edited by P. S. Meadows and A. Meadows, 183–98. Oxford: Clarendon Press.
Lovegrove, B. G., andJ. U. M. Jarvis. 1986. Coevolution between mole-rats (Bathyergidae) and a geophyte, *Micranthus* (Iridaceae). *Cimbebasia* 8A:79–85.
Malizia, A. I., A. I. Vassallo, and C. Busch. 1991. Population and habitat characteristics of two sympatric species of *Ctenomys* (Rodentia: Octodontidae). *Acta Theriologica* 36:87–94.
Martinsen, G. C., J. H. Cushman, and T. G. Whitham. 1990. Impact of pocket gopher disturbance on plant species diversity in a short-grass prairie community. *Oecologia* 83: 132–38.

McConnell, C. S. 1986. Coexistence of the golden mole *Amblystomus hottentotus* (Chrysochloridae) and the mole-rat *Cryptomys hottentotus* (Rodentia). Thesis, University of Natal, Pietermaritzburg, South Africa.

McConnell, S., and G. C. Hickman. 1985. Coexistence of *Amblysomus hottentotus* (Insectivora) and *Cryptomys hottentotus* (Rodentia). *South African Journal of Science* 81:699–700.

McDonough, W. T. 1974. Revegetation of gopher mounds on aspen range in Utah. *Great Basin Naturalist* 34:267–75.

McGinnies, W. J. 1960. Effect of mima-type microrelief on herbage production of five seeded grasses in western Colorado. *Journal of Range Management* 13:231–34.

Mielke, H. W. 1977. Mound building by pocket gophers (Geomyidae): Their impact on soils and vegetation in North America. *Journal of Biogeography* 4:171–80.

Miller, M. A. 1948. Seasonal trends in burrowing of pocket gophers *(Thomomys)*. *Journal of Mammalogy* 29:38–44.

Miller, M. A., and H. E. Bond. 1960. The summer burrowing activity of pocket gophers. *Journal of Mammalogy* 41:469–75.

Miller, R. S. 1964. Ecology and distribution of pocket gophers (Geomyidae) in Colorado. *Ecology* 45:256–72.

Moulton, M. P., J. R. Choate, and S. J. Bissell. 1983. Biogeographic relationships of pocket gophers in southeastern Colorado. *Southwestern Naturalist* 28:53–60.

Nanni, R. F. 1988. The interaction of mole-rats *(Georychus capensis* and *Cryptomys hottentotus)* in the Nottingham road region of Natal. M.Sc. thesis, University of Natal, Pietermaritzburg.

Paradis, E., and G. Guedon. 1993. Demography of a mediterranean microtine: The Mediterranean pine vole, *Microtus duodecimcostatus*. *Oecologia* 95:47–53.

Paulson, H. A., Jr. 1969. Forage values on a mountain grassland-aspen range in western Colorado. *Journal of Range Management* 22:102–7.

Pearson, O. P. 1959. Biology of the fossorial rodent, *Ctenomys*, in Peru. *Memorias del Museo de Historia Natural "Javier Prado"* 9:1–56.

———. 1983. Characteristics of a mammalian fauna from forests in Patagonia, southern Argentina. *Journal of Mammalogy* 64:476–92.

———. 1984. Taxonomy and natural history of some fossorial rodents of Patagonia, southern Argentina. *Journal of Zoology, London* 202:225–37.

Peart, D. R. 1989. Species interactions in a successional grassland. III. Effects of canopy gaps, gopher mounds and grazing on colonization. *Journal of Ecology* 77:267–89.

Rado, R., Z. Wollberg, and J. Terkel. 1992. Dispersal of young mole rats *(Spalax ehrenbergi)* from the natal burrow. *Journal of Mammalogy* 73:885–90.

Radwan, M. A., G. L. Crough, C. A. Harrington, and W. D. Ellis. 1982. Terpenes of ponderosa pine and feeding preferences by pocket gophers. *Journal of Chemical Ecology* 8:241–53.

Reichman, O. J. 1988. Comparison of the effects of crowding and pocket gopher disturbance on mortality, growth and seed production of *Berteroa incana*. *American Midland Naturalist* 120:58–69.

Reichman, O. J., and R. J. Baker. 1972. Distribution and movements of two species of pocket gophers (Geomyidae) in an area of sympatry in the Davis Mountains, Texas. *Journal of Mammalogy* 53:21–33.

Reichman, O. J., and J. U. M. Jarvis. 1989. The influence of three sympatric species of fossorial mole-rats (Bathyergidae) on vegetation. *Journal of Mammalogy* 70:763–71.

Reichman, O. J., and S. C. Smith. 1985. Impact of pocket gopher burrows on overlying vegetation. *Journal of Mammalogy* 66:720–25.

———. 1990. Burrows and burrowing behavior by mammals. In *Current mammalogy*, vol. 2, edited by H. Genoways, 197–244. New York: Plenum Press.

———. 1991. Responses to simulated leaf and root herbivory by a biennial, *Tragopogon dubius*. *Ecology* 72:116–24.
Reichman, O. J., T. G. Whitham, and G. A. Ruffner. 1982. Adaptive geometry of burrow spacing in two pocket gopher populations. *Ecology* 63:687–95.
Reig, O. A. 1970. Ecological notes on the fossorial octodont rodent *Spalacopus cyanus* (Molina). *Journal of Mammalogy* 51:592–601.
Reig, O. A., C. Busch, M. O. Ortells, and J. R. Contreras. 1990. An overview of evolution, systematics, population biology, cytogenetics, molecular biology and speciation in *Ctenomys*. In *Evolution of subterranean mammals at the organismal and molecular levels*, edited by E. Nevo and O. A. Reig, 71–96. Progress in Clinical and Biological Research, vol. 335. New York: Wiley-Liss.
Rezsutek, M. 1997. The relationship of dicot availability to demography in Attwater's pocket gopher. Ph.D. dissertation, University of Houston.
Rezsutek, M., and G. N. Cameron. 1998. Influence of resource removal on demography of Attwater's pocket gopher. *Journal of Mammalogy* 79:538–50.
———. In press. Vegetation edge effects and pocket gopher tunnels. *Journal of Mammalogy*.
Ritchie, M. E., and D. Tilman. 1995. Responses of legumes to herbivores and nutrients during succession on a nitrogen-poor soil. *Ecology* 76:2648–55.
Roig, V. G., and G. W. Cox. 1985. La presencia de monticulos tipo mima en la Argentina en relacion con roedores del genero *Ctenomys*. *Ecosur* 12/13:93.
Ross, B. A., J. R. Tester, and W. J. Breckenridge. 1968. Ecology of mima-type mounds in northwestern Minnesota. *Ecology* 49:172–77.
Ross, E. S. 1940. New Histeridae (Coleoptera) from the burrows of the Florida pocket gopher. *Annals of the Entomological Society of America* 33:1–9.
———. 1944. *Onthyophilus kirni* new species, and two other noteworthy Histeridae from burrows of a Texas pocket-gopher. *Entomological News* 55:113–18.
Schaeffer, T. H. 1945. Burrow associations of small mammals. *Murrelet* 26:24–26.
Soriguer, R. C., and J. A. Amat. 1980. On the structure and function of the burrows of the Mediterranean vole *(Pitymys duodecimcostantus)*. *Acta Theriologica* 25:268–70.
Sparks, D. W., and D.C. Andersen. 1988. The relationship between habitat quality and mound building by a fossorial rodent, *Geomys bursarius*. *Journal of Mammalogy* 69:583–87.
Spencer, S. R., G. N. Cameron, B. D. Eshelman, L. C. Cooper, and L. R. Williams. 1985. Influence of pocket gopher mounds on a Texas coastal prairie. *Oecologia* 66:111–15.
Steuter, A. A., E. M. Steinauer, G. L. Hill, P. A. Bowers, and L. L. Tieszen. 1995. Distribution and diet of bison and pocket gophers in a sandhills prairie. *Ecological Applications* 5:756–66.
Stromberg, M. R., and J. R. Griffin. 1996. Long-term patterns in coastal California grasslands in relation to cultivation, gophers, and grazing. *Ecological Applications* 6:1189–1211.
Thaeler, C. S. Jr. 1968. An analysis of the distribution of pocket gopher species in northeastern California (genus *Thomomys*). University of California Publications in Zoology 86:1–46.
Tietjen, H. P., C. H. Halvorson, P. L. Hegdal, and A. M. Johnson. 1967. 2,4-D herbicide, vegetation, and pocket gopher relationships Black Mesa, Colorado. *Ecology* 48:634–43.
Tilman, D. 1983. Plant succession and gopher disturbance along an experimental gradient. *Oecologia* 60:285–92.
———. 1985. The resource-ratio hypothesis of plant succession. *American Naturalist* 125:827–52.
———. 1988. *Plant strategies and the dynamics and structure of plant communities*. Monographs in Population Biology, 26. Princeton, NJ: Princeton University Press.

Vassallo, A. I. 1994. Habitat shift after experimental removal of the bigger species in sympatric *Ctenomys talarum* and *Ctenomys australis* (Rodentia: Octodontidae). *Behaviour* 127:247–63.

Vassallo, A. I., and C. Busch. 1992. Interspecific agonism between two sympatric species of *Ctenomys* (Rodentia: Octodontidae) in captivity. *Behaviour* 120:40–50.

Vaughan, T. A. 1961. Vertebrates inhabiting pocket gopher burrows in Colorado. *Journal of Mammalogy* 42:171–74.

Vaughan, T. A., and R. M. Hansen. 1964. Experiments on interspecific competition between two species of pocket gophers. *American Midland Naturalist* 72:444–52.

Vleck, D. 1979. The energy cost of burrowing by the pocket gopher *Thomomys bottae*. *Physiological Zoology* 52:122–35.

———. 1981. Burrow structure and foraging costs in the fossorial rodent, *Thomomys bottae*. *Oecologia* 49:391–96.

Ward, A. 1973. Food habits and competition. In *Pocket gophers and Colorado mountain rangeland*, edited by G. T. Turner, R. M. Hansen, V. H. Reid, H. P. Tietjen, and A. L. Ward, 41–61. Bulletin 5545, Colorado State University Agricultural Experiment Station, Fort Collins.

Wasley, G. A. 1995. The effects of productivity on mound production and burrow geometry of the plains pocket gopher *(Geomys bursarius)* in an old-field in east central Minnesota. M. S. thesis, Idaho State University, Pocatello.

Wilks, B. J. 1963. Some aspects of the ecology and population dynamics of the pocket gopher *(Geomys bursarius)* in southern Texas. *Texas Journal of Science* 15:241–83.

Williams, L. R., and G. N. Cameron. 1986. Food habits and dietary preferences of Attwater's pocket gopher, *Geomys attwateri*. *Journal of Mammalogy* 67:489–96.

Williams, L. R., G. N. Cameron, S. R. Spencer, B. D. Eshelman, and M. J. Gregory. 1986. Experimental analysis of the effects of pocket gopher mounds on Texas coastal prairie. *Journal of Mammalogy* 67:672–79.

Zinnel, K. C. 1988. Telemetry studies of the ecology of plains pocket gophers *(Geomys bursarius)* in east-central Minnesota. Ph.D. dissertation, University of Minnesota.

CHAPTER SEVEN

Spatial and Social Systems of Subterranean Rodents

Eileen A. Lacey

Studies of social behavior represent a vital part of efforts to understand the biology of subterranean rodents. Even in species in which each adult typically inhabits a different burrow system, individuals must contact one another in order to reproduce, thereby providing an opportunity for complex social interactions to evolve. Because such interactions have the potential to influence a variety of phenotypic characters, including demography (Andelman 1986; Busch et al., chap. 5, this volume), reproductive success (Clutton-Brock 1988; Bennett, Faulkes, and Molteno, chap. 4, this volume), and population genetic structure (Steinberg and Patton, chap. 8, this volume), social behavior may be a potent contributor to patterns of evolutionary change. Thus, understanding why different patterns of social behavior have arisen among subterranean taxa may yield intriguing new insights into other aspects of phenotypic and genotypic variation in these animals.

Although no single, universally recognized definition of social behavior exists, interactions among conspecifics are central to all studies of this aspect of animal behavior (Hinde 1976; Lee 1994). Several commonly addressed themes in studies of social behavior are the tendency of conspecifics to live in groups, the number and type of individuals chosen as mates, patterns of offspring care, and patterns of dispersal and mortality (Alexander 1974; Lee 1994). As this list suggests, complete characterizations of social behavior include the spatial distribution, reproductive biology, and demography of the animals under study. The resulting composite of behavioral and other attributes is frequently referred to as a species' "social system." Studies of interspecific differences in social systems and the ecological factors underlying this diversity represent a substantial proportion of modern behavioral research.

During the past two decades, interest in the social behavior of subterranean rodents has exploded, due primarily to the discovery of remarkably complex social systems among some species of bathyergid mole-rats (Jarvis 1981; Sherman, Jarvis, and Alexander 1991; Jarvis and Bennett 1993). These findings have generated considerable interest in the ecological correlates of sociality in these animals (Jarvis et al. 1994; Lacey and Sherman 1997) and have forced a reevaluation of the extent to which vertebrate and invertebrate social systems converge (Lacey and Sherman 1991; Sherman et al. 1995). The social behavior of most subterranean rodents, however, remains undescribed, making it difficult to assess the generality of ecological and evolutionary hypotheses developed for bathyergids. Comparative studies of other social subterranean taxa are beginning (e.g., Lacey, Braude and Wieczorek 1997), but much remains to be learned regarding the nature and extent of behavioral diversity in these animals. In particular, the role of the subterranean environment in shaping rodent social systems has not been documented, and thus the behavioral consequences of this shared modus vivendi are undetermined.

In this chapter, I review the social systems of subterranean rodents. I begin by describing the methods used to characterize the behavior of these animals. I then survey available data regarding spatial and social relationships among conspecifics in order to assess the nature and extent of behavioral diversity in these taxa. Specifically, I examine the prevalence and taxonomic distribution of group living, with emphasis on comparisons of group structure among social species. I then consider data from solitary taxa and explore the utility of the current dichotomous classification system (solitary versus social) as a model for behavioral diversity in subterranean taxa. Finally, I consider the role of the subterranean environment in shaping the social systems of these animals, and I identify several promising topics and taxa for future studies of the social behavior of subterranean rodents.

Logistic Challenges: Methods of Study

One reason that the social systems of most subterranean rodents remain poorly known is the logistic difficulty of studying the behavior of small, secretive animals that inhabit underground burrows. Most studies of social behavior rely upon direct, visual observation of interactions among conspecifics to characterize a species' social system. Observational studies of behavior have been conducted for a number of fossorial rodent taxa, most notably ground squirrels of the genera *Spermophilus, Cynomys,* and *Marmota* (Murie and Michener 1984; Hoogland 1995; Barash 1989). Although they inhabit underground burrow systems, ground squirrels are active above ground during much of the day, making it relatively easy to observe numerous aspects of the animals' social behavior, including alarm calling (Sherman 1977, 1985), competition for mates (Sherman 1989; Schwagmeyer 1990), and patterns of parental care (McLean 1983; Lacey 1991). In contrast, the rodent taxa considered in this volume are rarely seen above ground, and direct visual observation of their behavior is virtually impossible. As a result, patterns of social behavior must be characterized using alternative methods that do not rely on observations of behavioral interactions among free-living animals.

A variety of techniques have been employed to study spatial and social relationships among subterranean rodents. Most commonly, kill-trapping has been used to document adult spatial distributions, including those in several species of geomyids (*Thomomys bottae:* Patton and Feder 1981; Reichman, Whitham, and Ruffner 1982; *T. talpoides:* Hansen and Miller 1959; *Geomys bursarius:* Vaughan 1962) and ctenomyids (*Ctenomys talarum:* Pearson et al. 1968; *C. peruanus:* Pearson 1959; *C. sociabilis:* Pearson and Christie 1985), as well as in *Spalacopus cyanus* (Reig 1970). In contrast, live-trapping has been favored for studies of bathyergids (e.g., *Heterocephalus glaber:* Hill et al. 1957; Jarvis 1985; *Cryptomys hottentotus:* Hickman 1979; *C. damarensis:* Bennett 1990; *C. darlingi:* Bennett et al. 1994). In addition to generating field data on spatial relationships, this method provides subjects for the extensive laboratory studies of social behavior that have been conducted on these animals (e.g., Sherman, Jarvis, and Alexander 1991; Bennett, Faulkes, and Molteno, chap. 4, this volume).

Trapping and extirpation studies have proved particularly useful in identifying species in which burrow systems are shared by multiple adults (e.g., *C. damarensis:* Bennett and Jarvis 1988; *C. mechowi:* Burda and Kawalika 1993; *C. sociabilis:* Pearson and Christie 1985; *S. cyanus:* Reig 1970). Because the accuracy of this method is contingent upon capturing all animals in a burrow system or population, various criteria have been developed to determine when trapping is complete (e.g., Braude 1991; Lacey, Braude, and Wieczorek 1997, 1998). Spatial and social relationships are dynamic, however, and thus data based upon a single, short-term trapping effort (or extirpation study) cannot fully portray the complexity of a species' social system. As a result, other methods are required to obtain a detailed understanding of the social biology of these animals.

Mark-recapture studies provide a better means of tracking temporal changes in spatial and social relationships. This approach has been used to monitor spatial dynamics and demographic patterns in *H. glaber* (Braude 1991; Brett 1991a), *C. talarum* (Busch et al. 1989), and *T. bottae* (Daly and Patton 1990). For subterranean rodents, mark-recapture programs are most effective when combined with some form of remote monitoring that can be used to follow the movements of known individuals without the disruption of repeated trapping. Currently, the most common form of remote monitoring is radiotelemetry, which has been used in studies of geomyids (Bandoli 1987), bathyergids (Brett 1991a), ctenomyids (Lacey, Braude, and Wieczorek 1997), spalacines (Rado and Terkel 1989; Zuri and Terkel 1996), and myospalacines (Zhou and Dou 1990). Alternatively, radioactive wire implants have been used to monitor the movements of individual *T. bottae* (Gettinger 1984). When used in tandem, mark-recapture and remote monitoring procedures provide essential data not only on the spatial distributions of animals, but also on temporal patterns of activity and changes in spatial affinities between individuals.

Two additional methodologies that can be used to characterize the social systems of subterranean rodents are molecular genetic surveys and observations of captive animals. Molecular genetic studies provide valuable insights into patterns of parentage and kinship within social groups, as well as estimates of dispersal (measured as gene flow) among populations (Burke 1989; Hughes 1998; Parker et al. 1998; Steinberg and Patton, chap. 8, this volume). To date, however, molecular analyses aimed at uncovering patterns of social behavior have been completed for only a small number of subterranean species, notably *H. glaber* (Reeve et al. 1990; Faulkes, Abbott, and Mellor 1990; Faulkes, Abbott et al. 1997) and *T. bottae* (Patton and Feder 1981; Patton and Smith 1993).

In contrast, studies of captive animals have been used much more extensively to document aspects of social behavior that cannot be characterized under field conditions. In particular, studies of captive bathyergids have been used to quantify patterns of circadian activity (Davis-Walton and Sherman 1994), parental care (Lacey and Sherman 1991), mating behavior (Jarvis 1991b; Lacey and Sherman 1991; Hickman 1982), dominance and aggression (Jacobs et al. 1991; Reeve and Sherman 1991; Schieffelin and Sherman 1995; Gaylard, Harrison, and Bennett 1998), and behavioral specialization among individuals (Jarvis 1981; Payne 1982; Lacey and Sherman 1991; Bennett 1990, 1992; Wallace and Bennett 1998). Studies of captive animals have also been used to document mating behavior and aggression in spalacines (Shanas et al. 1995; Gazit, Shanas, and Terkel 1996; Ganem and Nevo 1996), as well as reproductive behavior in ctenomyids (R. R. Zenuto, personal communication). As techniques for housing captive animals improve (e.g., Jarvis 1991a; Gazit, Shanas, and Terkel 1996), the number of species studied in this manner should increase.

The spatial and social systems of subterranean rodents are best understood by combining several of the approaches outlined here. Long-term field studies of individually marked animals provide a critical foundation for any study of social behavior, and thus studies of free-living populations are essential. When combined with both molecular genetic analyses and observations of captive animals, such field studies should provide a detailed picture of social relationships in these taxa. Thus, despite the logistic challenges imposed by the subterranean niche, rigorous comparative studies of social behavior are possible. As the following discussion indicates, subterranean rodents exhibit considerable variation in social behavior. Understanding why this variation occurs among taxa inhabiting superficially similar environments is an important goal of efforts to understand the biology of these animals.

Spatial Relationships among Individuals

Spatial relationships among members of a population provide the foundation for social behavior. In many cases, data on the spatial distribution of individuals yield the first insight into a species' social system. For example, data indicating that multiple adults can be captured at a single burrow entrance suggest a different pattern of social behavior than do data indicating only a single adult per burrow system; such differences may provide the impetus for more detailed comparative studies of different taxa (Lacey, Braude, and Wieczorek 1998). Spatial relationships also provide insights into the relative complexity of social interactions among individuals. For example, it seems unlikely that behavioral specializations such as the size-related division of labor exhibited by nonreproductive *H. glaber* (Jarvis 1981; Payne 1982; Isil 1983; Lacey and Sherman 1991; Jarvis, O'Riain, and McDaid 1991) would have evolved unless members of this species routinely occurred in groups composed of animals from different weight classes. Thus understanding spatial relationships among individuals represents an important first step toward understanding a species' social system.

Spatial relationships among members of subterranean species are typically divided into two categories: those in which the areas used by different adults do not overlap (fig. 7.1A) and those in which the areas used by different adults overlap almost completely (fig. 7.1C). Nonoverlapping spatial distributions are thought to be common among subterranean taxa (Nevo 1979). This type of spatial system is thought to predominate among geomyids (Patton 1993), ctenomyids (Reig et al. 1990), spalacines (Savic and Nevo 1990) and myospalacines (Topachevskii 1976), as well as some genera of bathyergids (Jarvis and Bennett 1990, 1991). Although few studies have examined spatial relationships in these taxa using radiotelemetry or other forms of remote monitoring, a number of the species that have been studied in detail (e.g., *T. bottae:* Bandoli 1987; *C. haigi:* Lacey, Braude, and Wieczorek 1998; *N. ehrenbergi:* Rado and Terkel 1989) exhibit little, if any, spatial overlap among adults. These data suggest that nonoverlapping spatial distribu-

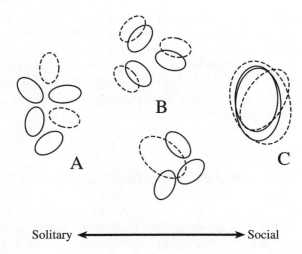

Solitary ←——————————→ Social

FIGURE 7.1. Schematic summary of spatial relationships in subterranean rodents. Within a population of subterranean rodents, adults exhibit either (A) no spatial overlap or (C) essentially complete spatial overlap associated with burrow sharing. Intermediate patterns of spatial overlap (B) have not yet been documented, although several lines of evidence suggest that, in at least some species, adults in neighboring burrow systems may interact briefly with one another at regular intervals. Spatial relationships among adults have frequently been used to infer patterns of social behavior, as indicated by the continuum of sociality depicted at the bottom of the figure.

tions are widespread among subterranean rodents, occurring in a phylogenetically diverse array of species.

In contrast, species in which the areas used by different adults overlap substantially are thought to be rare among subterranean rodents. Typically, spatial overlap occurs when adults share a burrow system, and thus burrow sharing is often taken as evidence of concordant spatial distributions. Most known examples of burrow sharing come from studies of bathyergid mole-rats, in particular *H. glaber* (Jarvis 1981; Braude 1991; Sherman, Jarvis, and Alexander 1991), *C. damarensis* (Bennett and Jarvis 1988; Jarvis and Bennett 1991), and *C. hottentotus* (Bennett 1989; Jarvis and Bennett 1991), although burrow sharing may also occur in several other species of *Cryptomys* (Jarvis and Bennett 1991; Wallace and Bennett 1998) (table 7.1). Outside of the Bathyergidae, burrow sharing has been documented in *C. sociabilis* (Pearson and Christie 1985; Lacey, Braude, and Wieczorek 1997), *S. cyanus* (Reig 1970), and *Spalax microphthalmus* (Puzachenko 1993), with anecdotal accounts of burrow sharing published for *C. peruanus* (Pearson 1959), *C. porteousi* (Contreras and Maceiras 1970), and *Prometheomys schaposchrikowi* (Grizmek 1975). Of these taxa, only *H. glaber* and *C. sociabilis* have been studied using radiotelemetry or other forms of remote sensing (Brett 1991a; Lacey,

TABLE 7.1 THE TAXONOMIC DISTRIBUTION OF BURROW SHARING IN SUBTERRANEAN RODENTS

Taxon	Social species known[a]	Philopatric sex	Group structure	References
Geomyidae				
Geomys	0			
Orthogeomys	0			
Pappogeomys	0			
Thomomys	0			
Zygogeomys	0			
Muridae: Arvicolinae				
Ellobius	0			
Prometheomys	schaposchrikowi	(M,F)	???	Grizmek 1975
Muridae: Myospalacinae				
Myospalax	0			
Muridae: Rhizomyinae				
Cannomys	0			
Rhizomys	0			
Tachyoryctes	0			
Muridae: Spalacinae				
Nannospalax	0			
Spalax	microphthalmus	M,F	Pair-based	Puzachenko 1993
Ctenomyidae				
Ctenomys	sociabilis	F	Female kin-based	Lacey, Braude, and Wieczorek 1997a
	peruanus	(F)	(Female kin-based)	Pearson 1959
	porteousi	(F)	(Female kin-based)	Contreras and Maceiras 1970
Octodontidae				
Spalacopus	cyanus	F?	Female kin-based (?)	Reig 1970 Begall, Burda, and Gallardo 1999
Bathyergidae				
Bathyergus	0			
Cryptomys	damarensis	M,F	Pair-based	Wallace and Bennett 1998[b]
	darlingi	M,F	Pair-based	Jarvis and Bennett 1991[b]
	hottentotus	M,F	Pair-based	Jarvis and Bennett 1990[b]
	mechowi	M,F	Pair-based	Burda and Kawalika 1993[b]
	sp.	M,F	Pair-based	
Georychus	0			
Heliophobius	0			
Heterocephalus	glaber	M,F	Pair-based[c]	Sherman, Jarvis, and Alexander 1991 Braude 1991

Note: Data from anecdotal reports of burrow sharing are denoted with parentheses; the spatial distribution of adults in these taxa requires quantitative documentation.
[a]Species in which burrow systems are routinely inhabited by two or more adults.
[b]Reference applies to multiple Cryptomys species.
[c]Breeding female may mate with one to three different males while in estrus.

Braude, and Wieczorek 1997). Data obtained from individuals inhabiting the same burrow system indicate that the spatial distributions of burrowmates overlap extensively, with no evidence that burrow systems are partitioned among individuals.

In addition to using the same set of tunnels, conspecifics that share burrows typically share the same nest site. Nest sharing by adults has been documented for both group-living bathyergids (Jarvis and Bennett 1991) and *C. sociabilis* (Lacey, Braude, and Wieczorek 1997). In these taxa, nest sharing occurs in burrow systems that are known to contain multiple nest sites, suggesting that overlapping adult spatial distributions reflect the formation of cohesive social groups whose members associate with one another even when spatial segregation (i.e., use of different nests) is possible. In *C. sociabilis*, multiple females with dependent young share the same nest site (Lacey, Braude, and Wieczorek 1997). Because unweaned young should be most vulnerable to infanticide or other threats from conspecifics, nest sharing by lactating females further suggests that burrowmates represent a behaviorally integrated unit whose members benefit from sharing tunnels and nest sites. Thus, in the species studied to date, burrow sharing does appear to reflect both concordant spatial distributions and the formation of cohesive social groups.

In contrast to other mammalian taxa, subterranean rodents do not appear to exhibit intermediate forms of spatial behavior in which home ranges overlap to some degree but all individuals maintain an area of exclusive use (fig. 7.1B). In particular, systems in which the home range of a male encompasses the spatially distinct home ranges of several females appear to be absent. This pattern of adult spatial distribution is relatively common among surface-dwelling and semi-subterranean rodents such as ground squirrels (Michener 1983), voles (Getz 1961; Wolff 1985), and mice (Eisenberg 1968; Ribble and Millar 1996), but has not been documented among the subterranean taxa considered in this volume. In part, this lack of "intermediate" spatial systems may reflect the paucity of data available for most species; as studies of subterranean rodents continue to generate information regarding the spatial distributions of adults, the apparent gap between overlapping and nonoverlapping spatial systems may diminish.

Alternatively, the lack of intermediate spatial systems may be a real phenomenon that reflects physical constraints on movement associated with the use of subterranean burrows. Because movement underground is typically restricted to well-defined corridors that are costly to construct (Busch et al., chap. 5, this volume), subterranean animals may be less able to travel freely through the habitat than are surface-dwelling taxa. For the ranges of individuals to overlap, animals must necessarily use the same tunnels, a requirement that may alter the costs and benefits of sharing space. For example, it may be easier for members of subterranean species to defend well-defined tunnels, thereby increasing the chances that nonresident animals will be deterred by aggressive encounters. Conversely, under circumstances in which prolonged or frequent interactions between individuals are important, the high cost of excavating tunnels may favor burrow sharing as a means of allowing contact while avoiding the costs of maintaining distinct tunnel systems.

Although subterranean rodents do not seem to exhibit intermediate patterns of spatial overlap, they display somewhat similar systems in which the burrow inhabited by a male is located adjacent to those of several females. Among surface-dwelling rodents, the size of individual home ranges is frequently sexually dimorphic: males are often active over larger areas than females, presumably because larger home ranges allow them to overlap the ranges of, or otherwise make contact with, multiple females. Sexual dimorphism in burrow system size has been reported for populations of *T. bottae* from Arizona and New Mexico; in both cases, burrow systems of males were larger than those of females (Reichman, Whitham, and Ruffner 1982; Bandoli 1987). Similarly, home range sizes for male *Ctenomys minutus* exceed those for females (Gastal 1994). Although these data do not provide direct evidence that larger burrow systems are associated with increased access to females, this dimorphism in burrow size clearly resembles the dimorphism in home range size found among surface-dwelling taxa in which males do have access to multiple females.

Home range size may not be the only mechanism by which males can achieve access to multiple females. In a number of subterranean species, home range sizes for males and females do not differ (e.g., *N. ehrenbergi*: Zuri and Terkel 1996; *C. haigi*: Lacey, Braude, and Wieczorek 1998), but burrows occupied by males appear to be located in the midst of those of several females. Maps depicting the locations of males and females are available for a number of species, including *T. bottae* (Patton and Feder 1981; Reichman, Whitham, and Ruffner 1982) and several ctenomyids (e.g., *C. talarum*: Pearson et al. 1968; *C. peruanus*: Pearson 1959; *C. haigi*: Lacey, Braude, and Wieczorek 1998). Visual inspection of these figures suggests that burrow systems of males and females are interspersed, with the result that males are likely to be located adjacent to one or more females. Because few statistical analyses of burrow locations have been conducted, it is unclear whether this pattern arises from a random distribution of males and females or from more subtle processes of burrow selection by individuals. Anecdotal observations indicating that, in *C. haigi*, some burrows are consistently inhabited by males (E. A. Lacey, unpublished data), suggest that the positioning of males and females within a population may not be random.

Reproductive interests, including access to potential mates, are not the only factors shaping the spatial distributions of adults. Other potentially important components of adult spatial patterns include kin relationships among individuals. In particular, natal philopatry (the retention of adult offspring in the natal area) may lead to the formation of clusters of related individuals. Extreme natal philopatry appears to underlie the formation of social groups in species exhibiting burrow sharing, with some individuals never leaving their natal burrow system (see below). In less extreme examples, however, individuals may remain in close proximity to their natal burrow but occupy a physically distinct tunnel system. This form of natal philopatry is common among female ground squirrels (Michener 1983), and the resulting aggregation of closely related individuals is thought to have played an important role in the evolution of ground squirrel social systems (Sherman 1977, 1981).

Few long-term mark-recapture studies have been conducted for subterranean species with nonoverlapping adult distributions, and thus the prevalence of natal philopatry and kin cluster formation in these animals is difficult to assess. Nevertheless, data from *T. bottae* in central California indicate that natal philopatry may occur. Daly and Patton (1990) reported that 63% of animals first captured as juveniles were retrapped within 40 m (roughly two home range diameters) of their natal burrow as adults. Although comparative studies of other taxa are required to assess the generality of this finding, these data suggest that some degree of natal philopatry may occur in species in which adults occupy distinct burrows. This pattern has important implications for population genetic structure (Steinberg and Patton, chap. 8, this volume) because aggregations of closely related individuals should lead to the clustering of alleles that are identical due to shared ancestry. At the same time, the formation of kin clusters may increase the probability of specific types of social interactions, such as nepotism and cooperation. Thus, understanding the spatial systems of subterranean rodents is critical to understanding other aspects of the biology of these animals.

Sociality in Subterranean Rodents

Like studies of spatial relationships, studies of social behavior have tended to divide subterranean rodents into two groups: solitary taxa, in which the spatial distributions of adults do not overlap, and social taxa, in which there is extensive spatial overlap among adults. More specifically, the term *social* has been reserved for those species that exhibit burrow sharing by adults (e.g., Jarvis and Bennett 1991; Sherman et al. 1995; Lacey and Sherman 1997). The shift from solitary to shared burrow occupancy represents a significant step in the evolution of social complexity because group living requires increased interaction among individuals, which facilitates both cooperation and competition among conspecifics. At the same time, group living sets the stage for more elaborate forms of social behavior that are not expected to occur in solitary species, including nepotism, reproductive suppression, and alloparental care. Because group living is believed to have such a fundamental effect on conspecific interactions (e.g., Alexander 1974), one of the primary goals of behavioral studies of subterranean rodents has been to understand why burrow sharing has arisen, presumably independently, in several lineages of these animals.

Within the framework of this apparent convergence, however, lies considerable diversity. Although group-living subterranean species resemble one another in that multiple adults share the same burrow system, patterns of social behavior differ markedly among these taxa. As our knowledge of the behavior of subterranean species has increased, it has become evident that we must explain not only why burrow sharing occurs, but also why it assumes so many forms. At present, attempts to compare the social systems of these animals quantitatively are hampered by a lack of data. The best-studied examples of social subterranean rodents are bathyergid mole-rats in the genera *Heterocephalus* and *Cryptomys* (Sherman et al. 1991; Jarvis et al. 1994). Considerably less is known about other social species. Nevertheless, preliminary studies of these animals suggest that their behavior differs from that of the bathyergids in several important ways, including differences in group size and structure, patterns of natal dispersal, and patterns of reproductive and other behavioral specialization within groups.

Group Size

The number of adults that share a burrow system is a fundamental property of group living that is thought to be causally linked to numerous other aspects of social behavior (Sherman et al. 1995). In particular, the degree of reproductive and behavioral specialization among individuals appears to increase with increasing group size, leading to the general perception that species with larger groups are more socially "complex." Among subterranean rodents, the number of adults that share a burrow system varies by as much as two orders of magnitude (table 7.2). The number of adults in a social group, often referred to as the "colony size," is greatest in *H. glaber*, in which as many as 250–300 adults may share a burrow system, although mean group size is less than 100 individuals (Braude 1991; Brett 1991b). In contrast, group sizes in social species of *Cryptomys* rarely exceed 40 individuals (table 7.2). The only other social species for which quantitative data are available is *C. sociabilis*, in which groups are typically composed of 6 or fewer adults (Lacey, Braude, and Wieczorek 1997).

TABLE 7.2 GROUP SIZES FOR SOCIAL SUBTERRANEAN RODENTS

Taxon	No. of adults per burrow system	References
Bathyergidae		
Heterocephalus glaber	70–80	Braude 1991
	60[a]	Brett 1991b
Cryptomys damarensis	18	Wallace and Bennett 1998
	20	Burda and Kawalika 1993
	8–25	Jarvis and Bennett 1991
Cryptomys darlingi	7	Wallace and Bennett 1998
Cryptomys hottentotus	14	Wallace and Bennett 1998
	8	Burda and Kawalika 1993
	2–14	Jarvis and Bennett 1991
Cryptomys mechowi	11	Wallace and Bennett 1998
	40	Burda and Kawalika 1993
Cryptomys sp.	20	Burda and Kawalika 1993
Ctenomyidae		
Ctenomys sociabilis	1–7	Lacey, Braude, and Wieczorek 1997
		E. A. Lacey, unpublished data
Octodontidae		
Spalacopus cyanus	15–26	Reig 1970
		Begall, Burda, and Gallardo 1999

Note: Data are from seven species for which quantitative data on group sizes are available; although burrow sharing is also thought to occur in several other species (see table 7.1), detailed studies of these taxa have not been conducted. Estimates given here are "typical" group sizes; for most species, mean colony sizes have not been calculated and colony size varies greatly within species.
[a] Mean group size for seven colonies from which all animals were captured.

Each of the species listed in table 7.2 is characterized by intraspecific variation in group size. In part, this variation may reflect the time interval since a group was founded; among bathyergids, pairs or small collections of adults are typically thought to be recently established groups that have not yet accumulated larger numbers of individuals (Braude 1991; Brett 1991b). At the same time, however, colony size in *H. glaber* fluctuates between years (Braude 1991; S. H. Braude, personal communication), suggesting that the number of adults in a group may vary in response to changes in social or environmental conditions. This variation in group size provides a potentially valuable tool for determining the ecological bases for burrow sharing by conspecifics. By comparing offspring production by groups of different sizes, the reproductive consequences of living with conspecifics can be determined (e.g., Hoogland and Sherman 1976). These findings can then be used to draw inferences regarding the reasons for burrow sharing. For example, situations in which per capita direct fitness decreases with group size suggest that groups form because opportunities for dispersal and subsequent reproduction are limited (i.e., ecological constraints arguments: Emlen 1982, 1991). Such findings may prompt efforts to identify ecological and demographic factors that constrain dispersal.

Group Structure

In addition to group size, group structure appears to vary among subterranean taxa. In all social subterranean rodents studied to date (as in many other group-living mammals), burrow sharing arises due to natal philopatry. Which sex is philopatric, however, varies among taxa. In the social bathyergids, both males and females remain in their natal burrow past the age at which reproduction is possible (Braude 1991; Brett 1991b; Jarvis and Bennett 1993; Jarvis et al. 1994). In contrast, only female *C. sociabilis* are philopatric; males disperse from their natal group between 4 and 8 months of age, with the result that social groups are composed of several related females plus an immigrant male (Lacey, Braude, and Wieczorek 1997; E. A. Lacey, unpublished data). In the social bathyergids, philopatry may continue over a significant proportion of an individual's lifetime, and reproductive animals tend to remain together for multiple years (Braude 1991). Although female *C. sociabilis* tend to live together for several years, males do not remain in the same burrow system for more than one breeding season (E. A. Lacey, unpublished data).

These differences in group structure have potentially important implications for social interactions in these species. In the social bathyergids, philopatric males and females are closely related to their burrowmates, suggesting that individuals of both sexes will receive inclusive fitness benefits from engaging in cooperative interactions with other group members. In contrast, because only females in a group of *C. sociabilis* are related, conflicts of interest between the sexes are more likely. Although male and female *C. sociabilis* share short-term interests associated with rearing young, other, more enduring forms of cooperation may be lacking. In particular, male *C. sociabilis* are not expected to engage in cooperative activities that yield primarily inclusive, rather than direct, fitness benefits. In this regard, the social system of *C. sociabilis* resembles those of ground-dwelling sciurids in that nepotism and other forms of cooperation occur primarily among members of female kin groups (Michener 1983). Thus group composition and patterns of kinship may substantially influence the nature of social interactions among conspecifics.

Reproductive Specialization

One of the most striking aspects of the social systems of group-living bathyergids is the restriction of direct reproduction to only a single female and one or a few males per social group (Lacey and Sherman 1991; Jarvis 1991b; Bennett, Faulkes, and Molteno, chap. 4, this volume). Although not uncommon among group-living mammals (Emlen 1991), this bias in direct reproduction reaches its extreme among the bathyergids—in particular, *H. glaber* (Sherman et al. 1995; Bennett, Faulkes, and Molteno, chap. 4, this volume). Despite group sizes that may exceed two hundred individuals, the vast majority of colonies captured in the field contain only a single breeding female. In a detailed study of more than thirty free-living colonies, multiple breeding females have been detected only five times (Braude 1991; S. H. Braude, personal communication). Each of these colonies was trapped over several field seasons, revealing that multiple breeding females were present in some, but not all, years. Similarly, laboratory colonies of *H. glaber* are typically characterized by only a single breeding female, although multiple breeding females are occasionally present for brief periods (Lacey and Sherman 1991; Jarvis 1991b). The number of breeding males in a colony is more difficult to determine, but behavioral observations of captive animals indicate that as many as three different males may mate with the breeding female during a single estrous period (Lacey and Sherman 1991; Jarvis 1991b).

Reproduction within groups of social *Cryptomys* appears to be similarly restricted to a single female and one or a few males (Bennett, Faulkes, and Molteno, chap. 4, this volume). In contrast, all females in a group of *C. sociabilis* appear to reproduce directly. Four lines of evidence suggest that direct reproduction is not restricted in this species. First, the number of pups reared by a group exceeds the maximum number of pups per individual, as determined from counts of placental scars (Lacey, Braude, and Wieczorek 1997). Second, all of the adult female *C. sociabilis* ($n = 10$) examined over the past 6 years have exhibited placental scars (E. A. Lacey, unpublished data), providing no evidence that some females forego direct reproduction. Third, pups reared in multi-female groups frequently vary in body size at the time of capture, with distinct weight classes evident among the young captured. Finally, all females lactate during the portion of the year when unweaned young are present in the population (Lacey, Braude, and Wieczorek 1997). These findings stand in marked contrast to morphological data from *H. glaber*, in which breeding and nonbreeding females are readily distinguished on the basis of reproductive tract condition (Kayanja and Jarvis 1971; Jarvis 1991b; Bennett, Faulkes, and Molteno, chap. 4, this volume).

The behavioral and endocrinological mechanisms underlying reproductive differences in *Heterocephalus* and *Cryptomys* have been the subject of intensive investigation and are discussed in detail by Bennett, Faulkes, and Molteno (chap. 4, this volume). Most relevant to the current discussion is the finding that these reproductive differences are reversible; under appropriate social conditions, previously nonreproductive individuals can become breeders (Lacey and Sherman 1991; Jarvis 1991b; Bennett, Faulkes, and Molteno, chap. 4, this volume). This appears to be true of both males and females in *H. glaber* and the social species of *Cryptomys* studied to date (Bennett, Faulkes, and Molteno, chap. 4, this volume). This finding has important implications for patterns of lifetime reproductive success in these species. Specifically, individuals that do not reproduce directly in their natal group may do so after dispersing and establishing themselves elsewhere in the habitat. Thus "instantaneous" estimates of reproductive differences among conspecifics may not always reflect lifetime differences in reproductive success for these animals.

Available data suggest that, in *H. glaber*, the vast majority of individuals never reproduce directly, but instead rely entirely upon indirect sources of fitness. Histograms of direct lifetime reproductive success for females in captive colonies are bimodal in distribution: a small number of females produce all the young, and a much larger number of females produce no young (Lacey and Sherman 1997). To date, efforts to quantify male reproductive success have been hampered by low levels of genetic variability in this species (Reeve et al. 1990; Faulkes, Abbott, and Mellor 1990). Nevertheless, behavioral data indicating that only one to three males per colony mate imply that even if litters are multiply sired, the percentage of males that achieve direct reproductive success is quite small. Comparable data from free-living animals are now beginning to emerge; these data suggest that less than 0.1% of individuals reproduce directly during their lifetimes (Jarvis et al. 1994), thus supporting the assertion that opportunities for direct reproduction are markedly restricted in this species.

In general, patterns of lifetime reproductive success in free-living social *Cryptomys* are assumed to be similar to those in *Heterocephalus*. In *C. damarensis*, however, colonies seem to disband more frequently than they do in *H. glaber* (Jarvis et al. 1994; see below), suggesting that individual *C. damarensis* have a greater probability of reproducing directly at some point during their lifetimes. Field studies of *C. damarensis* appear to support this hypothesis: although reproductive opportunities are constrained, an estimated 8–10% of individuals reproduce directly during their lifetimes (Jarvis and Bennett 1993). Thus the probability that an individual will breed appears to be a hundred times greater than in *H. glaber*. As a result, the number of individuals producing offspring should be larger, suggesting that histograms of lifetime reproductive success will be less markedly bimodal than they are in *Heterocephalus*. Data collected as part of ongoing field studies by J. U. M. Jarvis and colleagues will eventually provide a quantitative test of this hypothesis. As field studies of other social *Cryptomys* progress, comparative data on lifetime reproductive success in these species should also become available.

In contrast to *H. glaber* and *C. damarensis*, lifetime reproductive success in *C. sociabilis* is not expected to have a bimodal distribution. Because effectively all females that survive to adulthood produce young (Lacey, Braude, and Wieczorek 1997), individual lifetime reproductive success is expected to be relatively normal in distribution. The presence of only a single adult male per group suggests a harem-polygynous mating system in which reproductive success varies among individual males, but many males achieve at least some direct fitness (Clutton-Brock 1989; Davies 1991). Because it seems unlikely that groups of *C. sociabilis* contain a distinct subset of nonreproductive animals that achieve only indirect fitness, behaviors associated with maximizing inclusive fitness may not reach the extremes that they do in social bathyergids. Thus, patterns of lifetime reproductive success may provide important clues about the degree of alloparental care or other forms of inclusive fitness maximizing expected within a given society.

Reproductive differences among members of a social group are currently receiving considerable attention from behavioral ecologists interested in understanding why reproductive skew (defined as the difference in direct reproductive success among same-sex members of a social group) varies among social groups (e.g., Keller and Reeve 1994; Reeve, Emlen, and Keller 1998; Clutton-Brock 1998). In particular, the social dynamics leading to reproductive suppression of some group members are currently subject to debate. Among vertebrates, the degree of reproductive skew exhibited appears to reflect intragroup conflict over reproductive opportunities; to the extent that one individual can "win" such conflicts, direct reproduction will be biased in favor of that individual (Clutton-Brock 1998). Although quantitative data are limited, there does appear to be an association between group structure and degree of skew. In a review of reproductive skew in vertebrate and invertebrate societies, Reeve and Keller (1995) present data suggesting that skew is consistently greater in mother-daughter groups than it is in sibling associations. Given the variation in group structure and reproductive skew found among subterranean rodents, comparative studies of these animals may provide important insights into this and other aspects of reproductive differentiation within social groups.

Inbreeding and Breeder Replacement

The social systems of the group-living bathyergids are similar in that reproduction is typically restricted to only a single female and a few males per group. Other aspects of the reproductive structure of groups, however, differ among these species. In *H. glaber*, breeding animals are replaced by nonbreeding conspecifics from within the same social group (Braude 1991; Lacey and Sherman 1997). Following the loss of a breeding female, several nonbreeding females undergo a period of rapid weight gain, during which they may fight vigorously with one another until a single, behaviorally dominant individual has established herself as breeder (Lacey and Sherman

1991; Jarvis 1991b). In contrast, loss of a breeding male is not followed by conspicuous changes in body weight or behavior among nonbreeders (Lacey and Sherman 1991, 1997; Jarvis 1991b). Instead, which males become breeders may, in part, be determined by the breeding female: aggression between a new breeding female and her predecessor's mate(s) frequently results in the death of the male, creating an opening for a new reproductive male (Lacey and Sherman 1991, 1997).

The loss of a breeding animal has very different consequences for groups of *C. damarensis*. In this species, loss of a breeder results in the disintegration of the social group, with individuals presumably dispersing and attempting to establish new groups elsewhere in the environment (Jarvis et al. 1994). This response is expected to generate demographic patterns quite different from those found in *H. glaber*. In particular, because breeders are not replaced from within the social group, inbreeding should be reduced in *C. damarensis* relative to *H. glaber*. At the same time, because groups of the former species disband following the loss of a breeder, closely related individuals do not accumulate in the same location for extended periods of time, as they do in *H. glaber*. These demographic differences are expected to influence the genetic structure of populations (Steinberg and Patton, chap. 8, this volume) and may also affect the nature of social interactions within each species by altering the potential for long-term cooperation among closely related individuals.

Perhaps related to inbreeding, natal philopatry, and the generally insular nature of bathyergid colonies are observations suggesting that *Heterocephalus* and at least some species of group-living *Cryptomys* are xenophobic. In *H. glaber*, individuals readily distinguish unfamiliar animals from colonymates and respond aggressively to the presence of a foreign conspecific (Lacey and Sherman 1991; Sherman, Jarvis, and Braude 1992; O'Riain and Jarvis 1997). Similarly, members of *C. hottentotus* colonies respond aggressively to unfamiliar animals, although the magnitude of this response varies with sex, reproductive status, and the aridity of the environment (Spinks, O'Riain, and Polakow 1998). Gender may also be an important factor in the response of *C. damarensis* to foreign conspecifics; although the outcomes of same-sex versus opposite-sex introductions have not been compared directly, Jarvis et al. (1994) indicate that, in captivity, unfamiliar males and females show no apparent aggression when paired with a member of the opposite sex. The degree of xenophobia exhibited by social bathyergids is expected to vary as a function of the costs associated with allowing a foreign animal into the social group. Because these costs may vary with habitat, season, or colony composition, comparative studies of the type conducted by Spinks, O'Riain, and Polakow (1998) may prove useful in elucidating the relative importance of group cohesion among social bathyergids.

Alloparental Care

Individuals that do not reproduce directly frequently contribute to the care of young in other ways. Alloparental care occurs in a wide variety of mammals, including numerous rodents (Solomon and French 1997). In the social subterranean taxa considered here, alloparental care assumes a variety of forms. In *H. glaber*, nonbreeding colony members groom and transport pups, as well as feed pups both vegetation and freshly excreted feces (Lacey and Sherman 1991). Similar patterns of alloparental care have been reported for *C. damarensis* (Bennett and Jarvis 1988), and allocoprophagy between nonbreeders and pups occurs in *C. hottentotus* (Bennett 1992). The behavior of the remaining social *Cryptomys* has not been as thoroughly documented, but it seems quite likely that similar patterns of care by nonbreeding animals also occur in these species.

Even though groups of *C. sociabilis* lack a distinct subset of nonbreeding individuals, alloparental care may occur in this species. In particular, allonursing of young may take place among females in the same social group. Radiotelemetry studies of free-living animals have revealed that members of the same social group share a common nest site during the portion of the year when females are lactating (Lacey, Braude, and Wieczorek 1997), raising the possibility of communal nursing among burrowmates. At present, however, this hypothesis remains untested. Use of immunological (Glass et al. 1990, 1991) or radioactive (Hoogland, Tamarin, and Levy 1989) markers that are transferred from females to young during nursing provides the most promising means of resolving this question.

A fundamental goal of studies of alloparental care in vertebrates is to determine the reproductive consequences of such care (Emlen 1991). Among mammals, both breeders and nonreproductive alloparents may benefit from such care via a variety of proximate mechanisms (Jennions and Macdonald 1994). Typically, however, alloparental care is expected to increase breeder fitness through increased survival of offspring and to increase alloparental fitness through inclusive fitness benefits. Correlational evidence from studies of several mammalian cooperative breeders supports the assertion that alloparental care increases breeder success (e.g., *Canis mesomelas:* Moehlman 1979; *Lycaon pictus:* Malcolm and Marten 1982; *Helogale parvula:* Rood 1990). Because alloparents in most mammalian societies are closely related to the breeders that they assist, increased breeder fitness is generally interpreted as evidence of inclusive fitness benefits to alloparents.

In contrast to other mammalian taxa, the reproductive consequences of alloparental care have not been explicitly addressed for the subterranean rodents in which such care occurs. In *H. glaber* and *C. sociabilis*, alloparental care is not required, as male-female pairs can successfully rear young (Jarvis

1991b; Lacey and Sherman 1991; Lacey, Braude, and Wieczorek 1997; E. A. Lacey, unpublished data). Similarly, alloparental care does not seem to be necessary among social *Cryptomys,* as groups are thought to form from reproductive pairs (Jarvis et al. 1994). Although alloparental care may not be *required* to rear pups, it may nevertheless increase pup survival, and Sherman, Braude, and Jarvis (1999) have argued that it contributes to the extremely large litter sizes found in *H. glaber* (Bennett, Faulkes, and Molteno, chap. 4, this volume). Somewhat surprisingly, studies of captive bathyergids have not compared litter sizes or pup survival for male-female pairs with and without alloparents. Although such studies should, ideally, be conducted under field conditions, data from captive colonies may provide some insights into the reproductive consequences of alloparental care. These data are needed to bring studies of subterranean rodents to the level of research on other alloparental species of mammals.

Behavioral Specialization

One of the most distinctive features of the social system of *H. glaber* is the division of labor among nonbreeding colony members. Among nonbreeders of both sexes, behavior is correlated with body weight. Smaller nonbreeders are the primary participants in colony maintenance activities such as transporting food and nesting material and clearing the tunnels of debris (Lacey and Sherman 1991; Jarvis, O'Riain, and McDaid 1991). In contrast, larger nonbreeders are the primary participants in colony defense and, in at least some contexts, tunnel excavation (Lacey and Sherman 1991; Jarvis, O'Riain, and McDaid 1991). Factors influencing body weight and thus behavior include age, group size, and past opportunities to become a breeder. Interactions among these factors are complex, but, in the absence of changes in colony composition, individuals appear to shift from maintenance to defense activities at 30–40 months of age (Lacey and Sherman 1997).

The level of activity in a colony of *H. glaber* appears to be regulated by the breeding female (Reeve and Sherman 1991; Reeve 1992). In intact groups, the breeding female is behaviorally dominant over all other individuals, including breeding males (Rymond 1991; Reeve and Sherman 1991; Reeve 1992). The breeding female regularly patrols the burrow system and uses her muzzle to push and shove all individuals that she meets (Lacey and Sherman 1991; Reeve and Sherman 1991). This pushing and shoving is most frequently directed toward larger nonbreeders and toward individuals that are more distantly related to her (Reeve and Sherman 1991; Reeve 1992). This pattern may reflect either size-related or kinship-related conflicts of interest that arise because larger or less related individuals gain fewer benefits from engaging in risky activities such as colony defense (Lacey and Sherman 1997). If a breeding female is temporarily removed from her group, overall activity levels decline, suggesting that the female's presence—in particular, her pushing and shoving of burrowmates—plays an important role in enforcing nonbreeder behavior (Reeve 1992).

Behavioral specializations among nonbreeding *Cryptomys* are generally similar in form to those observed in *Heterocephalus*. In all species studied to date, breeding animals appear to be behaviorally dominant over other group members, although in *C. damarensis* it is the breeding male, rather than the breeding female, that is most dominant (Bennett and Jarvis 1988; Bennett 1989; Jacobs et al. 1991; Jarvis et al. 1994; Gabathuler, Bennett, and Jarvis 1996; Wallace and Bennett 1998). Among nonbreeders, behavioral differences similar to those observed in *H. glaber* have been reported for *C. damarensis* (Bennett and Jarvis 1988) and *C. mechowi* (Wallace and Bennett 1998). In *C. damarensis,* participation in colony maintenance is correlated with body weight; as in *H. glaber,* smaller nonbreeders are the primary participants in these activities (Bennett and Jarvis 1988). In contrast, both large and small nonbreeding *C. mechowi* participate extensively in colony maintenance activities (Wallace and Bennett 1998). The two other social species of *Cryptomys* that have been studied, *C. hottentotus* and *C. darlingi,* do not exhibit well-defined patterns of behavioral specialization (Rosenthal, Bennett, and Jarvis 1992; Gabathuler, Bennett, and Jarvis 1996).

Quantitative data regarding behavioral differences among group members are not yet available for *C. sociabilis.* At least two lines of evidence, however, suggest that this species is unlikely to display the type of intragroup behavioral specialization found in social bathyergids. First, all adults in a group of *C. sociabilis* appear to participate in foraging and tunnel excavation (E. A. Lacey, unpublished data). Second, because all females appear to reproduce directly, groups of *C. sociabilis* do not contain distinct subsets of reproductive and nonreproductive individuals. In the absence of marked differences in direct fitness, individuals are not expected to specialize behaviorally in ways that function primarily to increase inclusive fitness, particularly if such specializations detract from activities important to direct reproductive success. Given how little is known about this species, however, additional studies are needed to determine whether dominance hierarchies exist among females and whether other, more subtle behavioral differences occur within groups.

Trends in Sociality

As the preceding discussion has emphasized, even among group-living subterranean rodents, there is considerable variation in social behavior. Comparative studies of these animals—in particular, comparative studies of so-

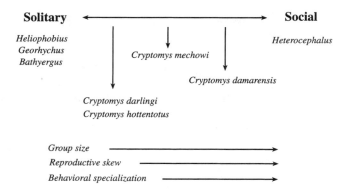

FIGURE 7.2. Schematic representation of the continuum of social systems found among bathyergid mole-rats. The relative positioning of *Cryptomys* and *Heterocephalus* is based on Wallace and Bennett (1998).

cial bathyergids—suggest that these societies comprise a spectrum of social systems that differ quantitatively with regard to a number of characteristics, including group size, degree of reproductive skew, and degree of behavioral specialization among nonbreeders (fig. 7.2). As summarized by Wallace and Bennett (1998), the group-living bathyergids can be arranged in order of increasing sociality as follows: *C. hottentotus* and *C. darlingi,* followed by *C. mechowi, C. damarensis,* and, finally, *H. glaber.* This apparent continuum of social variation closely parallels the (eu)sociality continuum proposed by Sherman et al. (1995).

As argued by Sherman et al. (1995), placing these species on a continuum of social complexity may help to identify relationships between behavioral, ecological, and demographic variables that are otherwise not readily detected. With regard to social bathyergids, apparent correlations between group size, degree of natal philopatry, and degree of reproductive skew suggest that the social systems of these animals become more complex (e.g., individuals exhibit greater reproductive and behavioral specialization) as the probability of reproducing directly declines. This realization generates a number of testable hypotheses regarding interspecific differences in demography and ecology that may influence the distribution of individual direct fitness. Efforts are already under way to examine the effects of these variables on social elaboration within group-living bathyergids (Jarvis et al. 1994; Faulkes, Bennett et al. 1997; Wallace and Bennett 1998).

In contrast to the quantitative differences among the social systems of group-living bathyergids, the social system of *C. sociabilis* may represent a qualitatively different type of society that is organized around female kin groups, rather than around a reproductive pair. Reasons for the formation of female kin-based groups, as opposed to pair-based groups, are poorly understood, and thus it is not known whether these variations in group structure represent quantitative or qualitative differences in social behavior. Both types of societies are similar in that groups of closely related individuals form due to natal philopatry. The nature and dynamics of reproductive bonds, however, may differ substantially. In societies organized around female kin groups, male tenure in a group may be brief, with only limited social bonding between reproductive partners (e.g., *Spermophilus:* Michener 1983). In contrast, pair-based societies are frequently characterized by protracted relationships between a male and female that extend beyond a single reproductive effort (e.g., *C. damarensis:* Jarvis et al. 1994). This difference has potentially important implications for other aspects of social behavior, including the relatedness of group members and the degree of reproductive skew within social groups.

In the current context, it is unclear whether *C. sociabilis* should be placed on the same continuum of social complexity used for bathyergids (see fig. 7.2), or whether a separate trajectory is needed to depict behavioral trends among species that live in female kin-based groups. Among subterranean rodents, efforts to determine whether this difference in group structure is qualitative or quantitative are hampered by the small number of social species that have been studied. To date, all examples of pair-based societies come from studies of bathyergids, with only a single example of a female-kin-based society from the Ctenomyidae. Given this taxonomic distribution, it is unclear to what extent these behavioral differences reflect phylogenetic versus current ecological conditions. Additional studies of non-bathyergids—in particular, studies of other group-living ctenomyids (see table 7.1)—are needed to assess the causal bases for the differences in group structure reported here.

ECOLOGICAL HYPOTHESES. To date, most efforts to explain sociality among subterranean rodents have focused on ecological and demographic factors. Because social bathyergids exhibit alloparental care, these explanations have frequently followed the conceptual framework applied to other cooperatively breeding vertebrates. This framework views cooperative breeding as the product of two distinct but interdependent processes: (1) the formation of social groups and (2) the elaboration of alloparental care (Emlen 1991; Koenig et al. 1992; Lacey and Sherman 1997). More specifically, ecological and demographic conditions are thought to constrain dispersal options, leading to natal philopatry and the formation of groups of closely related individuals. Within social groups, inclusive fitness and other benefits may favor alloparental behavior by group members.

Among the Bathyergidae, arid habitats and widely scattered food resources are thought to favor group living (Jarvis 1978; Jarvis et al. 1994; Faulkes, Bennett et al. 1997). Specifically, where food resources are locally abundant but patchily distributed, animals must regularly excavate additional tunnels in order to reach new feeding areas. In arid regions, opportunities for burrow expansion are limited because the energetic costs of excavation are tolerable only immediately following infrequent and unpredictable periods of rainfall. The "aridity-food distribution hypothesis" (Jarvis et al. 1994; Lacey and Sherman 1997) argues that individuals are better able to take advantage of these limited opportunities for burrow expansion if they live in groups. In short, lone individuals may not be able to dig fast enough and far enough to reach new food resources while conditions for excavation are appropriate. By living in groups, however, bathyergid mole-rats are able to complete extensive excavations very quickly, thus allowing the animals to locate new food resources before soil conditions again render the costs of tunneling prohibitive.

Within the Bathyergidae, both qualitative examinations of the geographic distributions of social species (Jarvis et al. 1994) and quantitative analyses employing independent contrasts (Faulkes, Bennett et al. 1997) have been used to test this hypothesis. The latter study, which combined a molecular genetic phylogeny of the Bathyergidae with behavioral and ecological data from five social and four solitary species, indicates that both food distribution (geophyte density) and rainfall (mean number of months with less than 25 mm of rain; unpredictability of rainfall) are significantly associated with group size. This analysis appears to resolve questions arising from qualitative analyses indicating that some social taxa (e.g., *C. mechowi*: Wallace and Bennett 1998) occur in mesic areas that, superficially, do not appear to be subject to the constraints proposed by the aridity-food distribution hypothesis. Even in apparently mesic habitats, the temporal distribution of rainfall may be such that burrowing opportunities are limited (Faulkes, Bennett et al. 1997) and thus group living may be favored. In other words, detailed quantitative studies may be needed to assess ecological patterns that are not apparent from more general, qualitative analyses. Thus, current data suggest that the spatial distribution of food resources and the temporal distribution of rainfall are associated with group living in bathyergid rodents.

On a larger scale, however, the aridity-food distribution hypothesis has not been tested for any other examples of group-living subterranean rodents. As a result, it is not yet known whether group living among non-bathyergids occurs due to similar ecological conditions, or whether group living in ctenomyids and octodontids reflects other factors. The predictions generated by the aridity-food distribution hypothesis are strongest when the ecologies of group-living species are compared with those of closely related solitary taxa (e.g., intrafamilial comparisons by Faulkes, Bennett et al. 1997). Consequently, appropriate data are needed not just from group-living ctenomyids and octodontids, but also from closely related solitary species. Comparative studies of social *(C. sociabilis)* and solitary *(C. haigi)* ctenomyids from northern Argentine Patagonia are currently under way (Lacey, Braude, and Wieczorek 1998); these analyses should yield important insights into the generality of the aridity-food distribution hypothesis as an explanation for group living in subterranean rodents.

Compared with the reasons for group living, less attention has been paid to the factors favoring alloparental care in social bathyergids (see above). In general, nonbreeding individuals are thought to gain primarily from inclusive fitness benefits associated with caring for the offspring of breeders. Because mole-rat colonies are typically composed of closely related individuals, nonbreeders should receive inclusive fitness benefits if they assist the reproductive efforts of breeders. Alloparents in other species of cooperatively breeding vertebrates have been found to receive a variety of benefits from caring for young (Emlen 1991; Jennions and Macdonald 1994). Many of these potential benefits, however, are linked to future direct reproduction by alloparents, which may render these hypotheses irrelevant for *H. glaber* and other species in which most individuals probably never reproduce directly. Although available evidence strongly suggests that inclusive fitness is the primary benefit obtained from alloparental care (Lacey and Sherman 1997), no quantitative estimates have been generated for the fitness gains to nonbreeders resulting from this care.

OTHER EXPLANATORY FACTORS. Other factors thought to contribute to sociality among bathyergid mole-rats include life history characteristics such as body size, metabolic rate, length of gestation, and litter size (Burda 1990; Lovegrove 1991; but see Sherman, Braude, and Jarvis 1999 on litter size versus sociality). Several of these traits—in particular, some reproductive variables—are generally characteristic of hystricognath rodents (Weir 1974; Burda 1989), and thus their role in the evolution of group living in bathyergids is difficult to assess. Each of these parameters may facilitate group living by altering cost-benefit relationships in favor of sociality, and, indeed, each may contribute to the relationship between burrowing costs and group living outlined above (Lacey and Sherman 1997). It seems unlikely, however, that these characteristics alone can explain group living, both because other hystricognaths are solitary and because extreme life history traits (i.e., those requiring group living) should not evolve prior to the onset of natal philopatry and group formation. In short, although life history traits may facilitate certain forms of social behavior, extrinsic factors such as resource distribution are critical to the evolution and maintenance of group living.

Predation has also been proposed as a potentially important factor in the occurrence of group living in subterranean taxa (Alexander, Noonan, and Crespi 1991). Few studies, however, have examined the effects of predation on group living in these animals. In part, this may be because predation is often difficult to quantify, making it difficult to generate comparative data on predation risk for solitary versus social species (but see Busch et al., chap. 5, this volume). Predation may affect social behavior in multiple ways. For example, increased danger of predation during dispersal may facilitate philopatry by favoring individuals that remain in the comparative safety of the natal burrow. Alternatively, group defense may be an effective means of deterring certain types of predators. With regard to bathyergids, Jarvis et al. (1994) indicate that predation may be an important factor favoring group living, but in the absence of data on rates of predation, they note that no apparent differences exist between the types of predators known to take solitary and social bathyergids (see also Busch et al., chap. 5, this volume). Although this observation implies that predation pressure is similar for both types of societies, additional studies are needed to verify this assumption. Further, apparent interspecific variation in type and rate of predation (Busch et al., chap. 5, this volume) suggests that even if predation is relatively unimportant among bathyergids, it may play a greater role in shaping the behavior of other subterranean rodents.

In contrast to hypotheses based on current ecological conditions, relatively little attention has been paid to historical hypotheses that address possible phylogenetic influences on social behavior. In particular, because most studies of group living in subterranean taxa have focused on a single monophyletic group—the Bathyergidae—the role of history in shaping the social systems of these animals has been difficult to assess. Within the Bathyergidae, *Heterocephalus* and *Cryptomys* are quite divergent (Allard and Honeycutt 1992; Faulkes, Abbott et al. 1997; Faulkes, Bennett et al. 1997); this relationship has generally been interpreted as evidence that group living has evolved twice in this family (e.g., Allard and Honeycutt 1992; Faulkes, Bennett et al. 1997). Recently, however, Burda (1997, 1998, 1999) has proposed that, given the prevalence of sociality in hystricognath taxa, group living may be ancestral in the Bathyergidae. Burda (1999) argues that, rather than asking why sociality has arisen twice in this group, comparative analyses should ask why sociality has been lost in the solitary genera *Bathyergus, Heliophobius,* and *Georychus*. At present, definitive tests of these alternatives are confounded by the lack of a clear sister group to the Bathyergidae; because placement of the bathyergids within the Hystricognathi remains problematic (Honeycutt et al. 1991), behavioral comparisons of *Heterocephalus* (the most ancestral of the bathyergids: Allard and Honeycutt 1992; Faulkes, Bennett et al. 1997) and closely related non-bathyergids have not been conducted.

Although historical hypotheses regarding the evolution of sociality deserve further consideration (Lessa, chap. 11, this volume), studies aimed at this level of analysis (Sherman 1988) should not deter complementary efforts to determine the ecological factors currently associated with group living. Regardless of whether group living has been evolutionarily gained or lost, the behavioral differences evident among bathyergids require explanation, and it seems likely that ecological conditions have played a substantial role in generating this diversity. In this context, phylogenetic data may be used for identifying and exploring behavioral convergence and divergence among subterranean rodents. As additional information becomes available regarding the behavior and ecology of the Bathyergidae and other subterranean lineages, we should begin to develop a clearer picture of how interactions between current and past pressures have shaped the social systems of these animals.

The Role of the Subterranean Niche

Given the conceptual framework for this volume, an obvious question to ask is how specialization for life underground has influenced the social behavior of subterranean rodents. The preceding discussion has already identified several ways in which the subterranean habitat may influence sociality. The high cost of tunnel excavation has been discussed with regard to group living in subterranean taxa. Other factors that may contribute to the formation of groups include the often patchy distribution of suitable habitat and the presumably high costs of aboveground dispersal (Busch et al., chap. 5, this volume), both of which may facilitate natal philopatry. These conditions are thought to apply to all subterranean rodents, however, and thus additional factors must be operating on some species to generate differences in social behavior.

Speculation as to the effects of the subterranean niche varies. On the one hand, burrow systems are well-defined, valuable resources that may be relatively easy to defend, which may favor the establishment of individual territories (Nevo 1979). On the other hand, because burrow systems are comparatively safe settings that can be expanded to meet the needs of multiple animals, subterranean rodents may be predisposed to sociality as a means of avoiding predation or other threats encountered outside of the burrow (Andersson 1984; Alexander, Noonan, and Crespi 1991); as long as food resources are adequate to support the needs of numerous individuals, young may remain in their natal burrow system rather than risk dispersal to a new location.

These hypotheses are not mutually exclusive. Instead, burrow systems that are worth defending under some conditions may, in other circumstances, be shared with conspecifics. Resolving this issue requires that critical factors such as food resources and costs of dispersal be compared quantitatively for closely related species that exhibit different patterns of spatial and social behavior. For this reason, behavioral and ecological studies of solitary taxa are critical, as are studies of species that exhibit intraspecific variation in these attributes. Comparisons based on this type of naturally occurring variation are an essential first step toward identifying aspects of the subterranean niche that are associated with variation in spatial and social relationships.

How Social Are Solitary Species?

The majority of subterranean rodents are assumed to be solitary (Nevo 1979). As with "group-living," however, the designation "solitary" may mask a wealth of variation in social behavior. Although it may seem contradictory to consider the social systems of solitary taxa, being solitary does not mean that a species is asocial. Even in societies in which adults typically live alone, individuals must come into contact to reproduce, and interactions occurring in this and related contexts may assume a variety of forms. In general, however, solitary subterranean rodents have received considerably less attention from behavioral biologists than their social counterparts, and thus, in many ways, the behavior of the former remains the greater mystery.

Some solitary species of subterranean rodents are well studied, leaving little doubt that each adult inhabits its own burrow system (e.g., *T. bottae:* Reichman, Whitham, and Ruffner 1982; Bandoli 1987; Patton 1993; *N. ehrenbergi:* Nevo 1961; Nevo et al. 1992; Zuri and Terkel 1996). Nevertheless, anecdotal accounts of plural burrow occupancy are available for some of these species—in particular, geomyid pocket gophers (e.g., *Thomomys:* Miller 1946; Hansen and Miller 1959; Howard and Childs 1959; Bandoli 1987; *Geomys:* Vaughan 1962; *Pappogeomys:* Russell 1954). The majority of these reports come from trapping studies in which multiple adults were captured in a single burrow system. In a number of cases, these accounts appear to reflect the point in the reproductive cycle at which trapping was conducted: reports of males and females captured together are more common during the breeding season and are typically attributed to sexual interactions among adults (Hansen and Miller 1959; Vaughan 1962).

Alternatively, plural burrow occupancy may reflect the methods used to monitor the distribution of individuals. Following extirpation, an individual's burrow system may be quickly reoccupied, sometimes within a matter of hours (*T. bottae:* Reichman, Whitham, and Ruffner 1982; *N. ehrenbergi:* Zuri and Terkel 1996). Thus studies reporting multiple burrow occupancy based on trapping programs conducted over several days may, in fact, be detecting the movement of neighboring animals into an empty burrow system rather than actual burrow sharing. This finding is itself intriguing, as it suggests that individuals are aware of the presence of neighboring animals and respond rapidly to the disappearance of conspecifics from nearby burrow systems (see also Reichman, Whitham, and Ruffner 1982).

Additional evidence that members of solitary species may have regular contact with one another is provided by analyses of burrow structure. Reichman, Whitham, and Ruffner (1982) reported the presence of subterranean nest chambers connecting the burrows of adjacent male and female *T. bottae*, providing a physical means by which interactions could occur. Using radiotelemetry, Bandoli (1987) found that interactions between neighboring adult *T. bottae* were brief, lasting only 10–20 minutes. These observations suggest that numerous short contacts between individuals may be missed by studies that rely on trapping or other "static" forms of data collection. Even studies employing radiotelemetry may miss such interactions unless individuals are monitored almost continuously. Clearly, detailed studies of the movements of individuals in solitary species are needed to document the nature and extent of social contact in these taxa.

Finally, support for the hypothesis that individuals monitor the presence and activity of neighboring animals comes from studies of acoustic communication in subterranean rodents. Vocal and seismic signals are used by solitary species (Francescoli, chap. 3, this volume), apparently to communicate between burrow systems. Few studies of the functional significance of these signals have been conducted, but anecdotal observations suggest that these signals may allow individuals to advertise their presence to conspecifics. For example, Narins et al. (1992) found that seismic signals produced by the solitary bathyergid *G. capensis* are sexually dimorphic and that males and females tend to alternate foot drumming bouts, leading the authors to propose that these signals contribute to gender or individual recognition. Among Patagonian tuco-tucos *(C. haigi)*, acoustic signals that are audible above ground are produced only by males, and the onset of calling by one male appears to trigger calling by other males resident in nearby burrow systems (E. A. Lacey, personal observation). Thus, in addition to direct physical contact, members of solitary species may interact indirectly via acoustic or seismic signals.

In summary, it appears that members of solitary species may interact with one another more than is typically acknowledged. Although each adult inhabits its own burrow system, individuals may monitor the presence and activity of neighboring animals through a combination of direct and indirect methods. Thus, even in solitary species, there may be well-developed patterns of social relationships among animals that have important implications for individual behavior and reproductive success. Because the behavioral complexity of solitary taxa has typically been underappreciated, few studies have explored such interactions in detail. In particular, focal animal studies (Altmann 1974) of marked individuals are needed to characterize patterns and rates of social interactions in these species.

Is There a Solitary-Social Dichotomy?

Studies of the behavioral ecology of subterranean rodents have tended to focus on taxa that exhibit extremes of spatial distribution and social behavior. At one end of this spectrum are well-studied solitary species such as *T. bottae*, in which adults clearly live alone, even if they interact regularly with conspecifics. At the other extreme are well-studied group-living species such as *H. glaber*, in which numerous conspecifics share a burrow system and interact cooperatively with one another in a variety of contexts. What lies in between these extremes, however, remains largely unknown. Do solitary and group life represent discrete behavioral categories, or are these instead endpoints along a continuum of social systems?

Most efforts to categorize social behavior recognize a series of intermediates between strictly solitary and strictly group-living taxa (see fig. 7.2; Michener 1983; Lidicker and Patton 1987; Lee 1994). Although these classification schemes use a variety of axes to characterize behavior, common themes include the degree of spatial overlap among individuals and the temporal persistence of social groupings. The apparent absence of intermediate spatial systems among subterranean rodents has been discussed above. In light of evidence suggesting routine, brief contacts among members of solitary species, however, more detailed studies of adult spatial distributions seem warranted. Such brief contacts may provide the intermediate spatial overlap (see fig. 7.1) otherwise thought to be lacking in subterranean taxa, suggesting that the distinctions between solitary (no spatial overlap) and social (complete spatial overlap) species may not be as absolute as has typically been assumed.

Although intermediate spatial and social systems are easy to propose, at present their existence among subterranean rodents remains uncertain. Clearly, additional studies of species exhibiting a variety of spatial and social patterns are needed to resolve this issue. In particular, studies of intraspecific variation in spatial and social relationships may prove useful in determining whether individuals shift from solitary to group life based on ecological or other environmental conditions. In the meantime, studies of subterranean rodents will benefit by considering interspecific trends in behavior, rather than by pigeonholing species as solitary or social and then restricting comparisons to other spatially and behaviorally similar taxa.

Mating Systems of Subterranean Rodents

As this chapter suggests, most studies of the behavioral biology of subterranean rodents have focused either directly or indirectly on the topic of group living and its consequences for other aspects of social behavior. There are, however, other important aspects of social behavior that are not typically considered as part of efforts to understand distinctions between solitary and group life. Perhaps foremost among these are the mating systems of subterranean rodents. Studies of animal mating systems constitute a vast proportion of behavioral research (Davies 1991), yet questions regarding patterns of reproductive competition, mate acquisition, and mating and reproductive success have only rarely been asked with regard to the taxa considered in this volume.

The mating systems of subterranean rodents have traditionally been categorized as "polygynous." The few quantitative studies that have been conducted indicate that males in some species do sire litters belonging to several different females (*T. bottae:* Patton and Feder 1981; *C. talarum:* Zenuto, Lacey, and Busch 1999). These data conform to a general pattern of polygyny, but fail to provide more detailed data regarding patterns of reproductive competition and mate choice. For example, do individual males have exclusive access to multiple females, or do females "choose" a single mate from among a group of several potential reproductive partners? In populations in which several males are located in close proximity to a female, are there differences in male phenotype (e.g., age, body weight) that can be used reliably to predict which male will sire the female's litter?

The questions to be addressed are even more numerous if we consider recent advances in our understanding of mammalian mating systems. As studies of reproductive behavior have increased, it has become apparent that traditional categories of mating systems are inadequate for characterizing the variation evident among free-living animals. In particular, a growing body of literature indicates that females often mate with several different males during a single period of sexual receptivity, suggesting that sperm competition is common among mammals (Ginsberg and Huck 1989; Gomendio and Roldan 1993; Birkhead and Parker 1997). As a result, data on mating success may not provide accurate indicators of paternity (see also Bennett, Faulkes, and Molteno, chap. 4, this volume). Because the outcome of sperm competition may be influenced by relatively subtle factors, such as male mating order or the interval between copulations, this phenomenon adds an additional level of complexity to patterns of male reproductive competition.

As already indicated, studies of subterranean rodents have barely scratched the surface of traditional approaches to mating system variation, let alone explored the intricacies of sperm competition and determinants of male fertilization success. As with other aspects of their behavior, our ability to study these taxa is limited by the nature of the subterranean habitat. Laboratory observations of reproductive behavior provide a partial solution to this problem. Studies of mating behavior in captive animals have been conducted for *N. ehrenbergi* (Gazit, Shanas, and Terkel 1996), *C. talarum* (R. R. Zenuto, personal communication), *S. cyanus* (Begall, Burda, and Gallardo 1999), and several species of social bathyergids (Bennett, Faulkes, and Molteno, chap. 4, this volume). Although the resulting data provide important insights into patterns of male-female interactions and copulatory behavior, it is unclear to what extent behaviors observed in these artificial environments represent naturally occurring patterns of reproductive competition and mate selection.

An alternative approach makes use of molecular genetic markers to assess patterns of paternity and reproductive success, even in the absence of behavioral data on mating behavior (Hughes 1998). For example, molecular markers have been used to quantify paternity in several species of ground squirrels that mate underground (Sherman 1989; Lacey, Wieczorek, and Tucker 1997) and to determine parentage in group-living species in which multiple adults rear young in a single communal den (e.g., dwarf mongooses, *Helogale parvula:* Keane et al. 1994). Although these types of analyses appear to hold considerable promise for behavioral studies of subterranean rodents, to date they have been applied to only three species. In *H. glaber,* generally low levels of genetic variability have effectively precluded efforts to determine which of a female's mates sire her young (Reeve et al. 1990; Faulkes, Abbott, and Mellor 1990). Paternity analyses conducted for *T. bottae* (Patton and Feder 1981) and *C. talarum* (R. R. Zenuto, E. A. Lacey, and C. Busch 1999) have been used to confirm generally polygynous patterns of male reproductive success (see above), although no instances of multiple paternity within litters have been detected in these species (see also Bennett, Faulkes, and Molteno, chap. 4, this volume).

As this discussion suggests, much additional work is needed to raise our knowledge of the mating systems of subterranean rodents to the level of understanding currently held for other taxa of small mammals. This objective will be best achieved using a combination of field studies and molecular genetic analyses of parentage and reproductive success. For example, field studies of basic attributes, such as the distributions and movements of males and females during the mating period, will provide essential information regarding the number of males competing for access to a female, as well as the frequency and duration of interactions between potential mates. When combined with genetic estimates of paternity and male reproductive success, such data should allow us to address many of the questions that currently drive mating systems research. Given the taxonomic diversity and the wealth of convergent and divergent behaviors found among subterranean rodents, detailed investigations of the mating systems of these animals should prove extremely informative.

Recurrent Themes: Directions for Future Research

Several themes have emerged from this discussion of the spatial and social systems of subterranean rodents. Perhaps the most evident of these is the lack of information available for most species. Although studies of the social behavior of subterranean rodents have, in some ways, dominated research on these animals for the past two decades, the number of taxa studied remains quite small. Indeed, the spatial and social behaviors of the vast majority of subterranean taxa remain virtually unknown. Thus one aim of future research on these topics is to provide detailed information regarding a wider range of subterranean species.

To provide a conceptual framework for future studies, it is perhaps most productive to view the spatial and social systems of different species as part of a continuum of societies. Rather than emphasizing the distinction between solitary and group-living taxa, future studies will benefit by considering the full spectrum of behavioral variation evident in subterranean rodents. Comparisons of social taxa are critical to identifying common ecological or demographic factors associated with convergent examples of group living in subterranean rodents. At the same time, however, comparisons of social and solitary taxa are needed to identify the factors underlying behavioral divergence in these animals. In short, only by working to build a single, comprehensive picture of social behavior will we truly be able to understand behavioral diversity in subterranean rodents. Studies of social species have proven extremely effective in drawing attention to subterranean rodents, but these animals represent only a small portion of the total behavioral diversity that may exist in these animals.

The theme of hidden diversity also permeates analyses of the mating systems of subterranean rodents. Here, the new frontiers are technical as well as conceptual. Applying current theoretical constructs to subterranean rodents makes it clear that much remains to be learned regarding the reproductive behavior of almost every species considered in this volume. Again, recognition that mating systems may form a continuum of variants adds impetus to efforts to study all taxa, rather than just a few select, group-living species. At the same time, studies of subterranean taxa need to partake of the technological advances that have so dramatically altered mating systems research on other organisms. There is considerable evidence to suggest that the rodents considered in this volume are just as behaviorally diverse and fascinating as other groups of small mammals—the challenge is to explore and interpret this diversity.

Acknowledgments

I would like to thank Guy Cameron and Jim Patton for inviting me to participate in this endeavor. Paul Sherman has been a major influence on my thinking about social behavior since my earliest days studying naked mole-rats in his laboratory; his efforts to improve the current manuscript are greatly appreciated. The studies of *C. sociabilis* described here could not have been conducted without the infinite patience and lightning-fast reflexes of John Wieczorek.

Literature Cited

Alexander, R. D. 1974. The evolution of social behavior. *Annual Review of Ecology and Systematics* 5:325–83.
Alexander, R. D., K. M. Noonan, and B. J. Crespi. 1991. The evolution of eusociality. In *The biology of the naked mole-rat*, edited by P. W. Sherman, J. U. M. Jarvis, and R. D. Alexander, 3–44. Princeton, NJ: Princeton University Press.
Allard, M. W., and R. L. Honeycutt. 1992. Nucleotide sequence variation in the mitochondrial 12S rRNA gene and the phylogeny of African mole-rats, Rodentia: Bathyergidae. *Molecular Biology and Evolution* 9:27–40.
Altmann, J. 1974. Observational study of behaviour: Sampling methods. *Behaviour* 49:227–67.
Andelman, S. J. 1986. Ecological and social determinants of cercopithecine mating patterns. In *Ecological Aspects of Social Evolution*, edited by D. I. Rubenstein and R. W. Wrangham, 201–16. Princeton, NJ: Princeton University Press.
Andersson, M. 1984. The evolution of eusociality. *Annual Review of Ecology and Systematics* 15:165–89.
Bandoli, J. H. 1987. Activity and plural occupancy of burrows in Botta's pocket gopher *Thomomys bottae*. *American Midland Naturalist* 118:10–14.
Barash, D. P. 1989. *Marmots: Social behavior and ecology*. Stanford, CA: Stanford University Press.
Begall, S., H. Burda, and M. H. Gallardo. 1999. Reproduction, postnatal development, and growth of social coruros, *Spalacopus cyanus* (Rodentia: Octodontidae), from Chile. *Journal of Mammalogy* 80:210–17.
Bennett, N. C. 1989. The social structure and reproductive biology of the common mole-rat, *Cryptomys h. hottentotus* and remarks on the trends in reproduction and sociality in the family Bathyergidae. *Journal of Zoology, London* 219:45–59.
———. 1990. Behaviour and social organization in a colony of the Damaraland mole-rat *Cryptomys damarensis*. *Journal of Zoology, London* 220:225–48.
———. 1992. Aspects of the social behaviour in a captive colony of the common mole-rat *Cryptomys hottentotus* from South Africa. *Zeitschrift für Säugetierkunde* 57:294–309.
Bennett, N. C., and J. U. M. Jarvis. 1988. The social structure and reproductive biology

of colonies of the mole-rat, *Cryptomys damarensis* (Rodentia, Bathyergidae). *Journal of Mammalogy* 69:293–302.

Bennett, N. C., J. U. M. Jarvis, and F. P. D. Cotterill. 1994. The colony structure and reproductive biology of the Afrotropical Mashona mole-rat, *Cryptomys darlingi. Journal of Zoology, London* 234:477–87.

Birkhead, T. R., and G. A. Parker. 1997. Sperm competition and mating systems. In *Behavioural Ecology: An Evolutionary Approach*, 4th ed., edited by J. R. Krebs and N. B. Davies, 121–53. Oxford: Blackwell Scientific Publications.

Braude, S. H. 1991. Behavior and demographics of the naked mole-rat, *Heterocephalus glaber.* Ph.D. dissertation, University of Michigan, Ann Arbor.

Brett, R. A. 1991a. The ecology of naked mole-rat colonies: Burrowing, food, and limiting factors. In *The biology of the naked mole-rat*, edited by P. W. Sherman, J. U. M. Jarvis, and R. D. Alexander, 137–84. Princeton, NJ: Princeton University Press.

———. 1991b. The population structure of naked mole-rat colonies. In *The biology of the naked mole-rat*, edited by P. W. Sherman, J. U. M. Jarvis, and R. D. Alexander, 97–136. Princeton, NJ: Princeton University Press.

Burda, H. 1989. Relationships among rodent taxa, as indicated by reproductive biology. *Zeitschrift für Zoologische Systematik und Evolutionsforschung* 27:49–57.

———. 1990. Constraints of pregnancy and evolution of sociality in mole-rats. *Zeitschrift für Zoologische Systematik und Evolutionsforschung* 28:26–39.

———. 1997. Evolution of life history strategies in African mole-rats (Bathyergidae). Abstracts, Seventh International Theriological Congress, Acapulco, Mexico.

———. 1998. Evolution of social behaviour in bathyergids and spalacids followed different paths: A reply of H. Burda to reply of G. Ganem. *Behavioral Ecology and Sociobiology* 42:369–70.

———. 1999. Syndrome of eusociality in African subterranean mole-rats (Bathyergidae, Rodentia), its diagnosis and aetiology. In *Evolutionary theory and processes: Modern perspectives, Papers in honour of Eviatar Nevo*, edited by S. P. Wasser, 385–418. Netherlands: Kluwer Academic Publishers.

Burda, H., and M. Kawalika. 1993. Evolution of eusociality in the Bathyergidae: The case of the giant mole-rat *(Cryptomys mechowi). Naturwissenschaften* 80:235–37.

Burke, T. 1989. DNA fingerprinting and other methods for the study of mating success. *Trends in Ecology and Evolution* 4:139–44.

Busch, C., A. I. Malizia, O. A. Scaglia, and O. A. Reig. 1989. Spatial distribution and attributes of a population of *Ctenomys talarum* (Rodentia: Octodontidae). *Journal of Mammalogy* 70:204–8.

Clutton-Brock, T. H., ed. 1988. *Reproductive success: Studies of individual variation in contrasting breeding systems.* Chicago: University of Chicago Press.

———. 1989. Mammalian mating systems. *Proceedings of the Royal Society of London* B 236:339–72.

———. 1998. Reproductive skew, concessions and limited control. *Trends in Ecology and Evolution* 13:288–92.

Contreras, J. R., and A. J. Maceiras. 1970. Relaciones entre el tuco-tuco y los procesos del suelo en la region semiarida del sudoeste bonaerense. *Agro* 12:3–17.

Daly, J. C., and J. L. Patton. 1990. Dispersal, gene flow, and allelic diversity between local populations of *Thomomys bottae* pocket gophers in the coastal ranges of California. *Evolution* 44:1283–94.

Davies, N. B. 1991. Mating systems. In *Behavioural ecology: An evolutionary approach*, 3d ed., edited by J. R. Krebs and N. B. Davies, 263–99. Oxford: Blackwell Scientific Publications.

Davis-Walton, J., and P. W. Sherman. 1994. Sleep arrhythmia in the eusocial naked mole-rat. *Naturwissenschaften* 81(6):272–275.

Eisenberg, J. R. 1968. Behavior patterns. In Biology of *Peromyscus* (Rodentia), edited by J. A. King, 451–95. Special publication no. 2. American Society of Mammalogists.
Emlen, S. T. 1982. The evolution of helping. I: An ecological constraints model. *American Naturalist* 119:29–39.
———. 1991. Evolution of cooperative breeding in birds and mammals. In *Behavioural ecology: An evolutionary approach,* 3d ed., edited by J. R. Krebs and N. B. Davies, 301–37. Oxford: Blackwell Scientific Publications.
Faulkes, C. G., D. H. Abbott, and A. L. Mellor. 1990. Investigation of genetic diversity in wild colonies of naked mole-rats *(Heterocephalus glaber)* by DNA fingerprinting. *Journal of Zoology, London* 221:87–97.
Faulkes, C. G., D. H. Abbott, H. P. O'Brien, L. Lau, M. R. Roy, R. K. Wayne, and M. W. Bruford. 1997. Micro- and macrogeographical genetic structure of colonies of naked mole-rats *Heterocephalus glaber. Molecular Ecology* 6:615–28.
Faulkes, C. G., N. C. Bennett, M. W. Bruford, H. P. O'Brien, G. H. Aguilar, and J. U. M. Jarvis. 1997. Ecological constraints drive social evolution in the African mole-rats. *Proceedings of the Royal Society of London* B 254:1619–27.
Gabathuler, U., N. C. Bennett, and J. U. M. Jarvis. 1996. The social structure and dominance hierarchy of the Mashona mole-rat, *Cryptomys darlingi* (Rodentia, Bathyergidae) from Zimbabwe. *Journal of Zoology, London* 240:221–31.
Ganem, G., and E. Nevo. 1996. Ecophysiological constraints associated with aggression, and evolution toward pacifism in *Spalax ehrenbergi. Behavioral Ecology and Sociobiology* 38:245–52.
Gastal, M. L. D. A. 1994. Tunnel systems and the home range of *Ctenomys minutus* Nehring, 1887 (Rodentia, Caviomorpha, Ctenomyidae). *Iheringia Serie Zoologia* 77:35–44.
Gaylard, A., Y. Harrison, and N. C. Bennett. 1998. Temporal changes in the social structure of a captive colony of the Damaraland mole-rat, *Cryptomys damarensis:* The relationship of sex and age to dominance and burrow-maintenance activity. *Journal of Zoology, London* 244:313–21.
Gazit, I., U. Shanas, and J. Terkel. 1996. First successful breeding of the blind mole-rat *(Spalax ehrenbergi)* in captivity. *Israel Journal of Zoology* 42:3–13.
Gettinger, R. D. 1984. A field study of activity patterns in *Thomomys bottae. Journal of Mammalogy* 65:76–84.
Getz, L. L. 1961. Home ranges, territoriality, and movement of the meadow vole. *Journal of Mammalogy* 42:24–36.
Ginsberg, J. R., and U. W. Huck. 1989. Sperm competition in mammals. *Trends in Ecology and Evolution* 4:74–79.
Glass, G. E., J. E. Childs, J. W. LeDuc, S. D. Cassard, and A. D. Donnenberg. 1990. Determining matrilines by antibody response to exotic antigens. *Journal of Mammalogy* 71:129–38.
Glass, G. E., G. W. Korch, J. E. Gomez, and J. E. Childs. 1991. Using exotic antigens to measure reproduction and dispersal in *Peromyscus leucopus. Canadian Journal of Zoology* 69:528–30.
Gomendio, M., and E. R. S. Roldan. 1993. Mechanisms of sperm competition: Linking physiology and behavioural ecology. *Trends in Ecology and Evolution* 8:95–104.
Grizmek, B., ed. 1975. *Grizmek's animal life encyclopedia.* Vols. 10–13, *Mammals,* I–IV. New York: Van Nostrand Publishers.
Hansen, R. M., and R. S. Miller. 1959. Observations on the plural occupancy of pocket gopher burrow systems. *Journal of Mammalogy* 40:577–84.
Hickman, G. C. 1979. Burrow system structure of the bathyergid *Cryptomys hottentotus* in Natal, South Africa. *Zeitschrift für Säugetierkunde* 44:153–62.
———. 1982. Copulation of *Cryptomys hottentotus* (Bathyergidae), a fossorial rodent. *Mammalia* 46:293–98.

Hill, W. C. O., A. Porter, R. T. Bloom, J. Seago, and M. D. Southwick. 1957. Field and laboratory studies on the naked mole-rat, *Heterocephalus glaber*. *Proceedings of the Zoological Society, London* 128:455–513.

Hinde, R. A. 1976. Interactions, relationships and social structure. *Man* 11:1–17.

Honeycutt, R. L., M. W. Allard, S. V. Edwards, and D. A. Schlitter. 1991. Systematics and evolution of the family Bathyergidae. In *The biology of the naked mole-rat*, edited by P. W. Sherman, J. U. M. Jarvis, and R. D. Alexander, 45–65. Princeton, NJ: Princeton University Press.

Hoogland, J. L. 1995. *The black-tailed prairie dog: Social life of a burrowing mammal*. Chicago: University of Chicago Press.

Hoogland, J. L., and P. W. Sherman. 1976. Advantages and disadvantages of bank swallow (*Riparia riparia*) coloniality. *Ecological Monographs* 46:33–58.

Hoogland, J. L., R. H. Tamarin, and C. K. Levy. 1989. Communal nursing in prairie dogs. *Behavioral Ecology and Sociobiology* 24:91–95.

Howard, W. E., and H. E. Childs, Jr. 1959. Ecology of pocket gophers with emphasis on *Thomomys bottae mewa*. *Hilgardia* 29:277–358.

Hughes, C. 1998. Integrating molecular techniques with field methods in studies of social behavior: A revolution results. *Ecology* 79:383–99.

Isil, S. 1983. A study of social behavior in laboratory colonies of the naked mole-rat (*Heterocephalus glaber* Rüppell; Rodentia, Bathyergidae). M. S. thesis, University of Michigan, Ann Arbor.

Jacobs, D. S., N. C. Bennett, J. U. M. Jarvis, and T. M. Crowe. 1991. The colony structure and dominance hierarchy of the Damaraland mole-rat, *Cryptomys damarensis* (Rodentia: Bathyergidae), from Namibia. *Journal of Zoology, London* 224:553–76.

Jarvis, J. U. M. 1978. Energetics of survival in *Heterocephalus glaber* (Rüppell), the naked mole-rat. *Bulletin of the Carnegie Museum of Natural History* 6:81–87.

———. 1981. Eusociality in a mammal: Cooperative breeding in naked mole-rat colonies. *Science* 212:571–73.

———. 1985. Ecological studies on *Heterocephalus glaber*, the naked mole-rat, in Kenya. *National Geographic Society Research Reports* 20:429–37.

———. 1991a. Methods for capturing, transporting, and maintaining naked mole-rats in captivity. In *The biology of the naked mole-rat*, edited by P. W. Sherman, J. U. M. Jarvis, and R. D. Alexander, 467–83. Princeton, NJ: Princeton University Press.

———. 1991b. Reproduction of naked mole-rats. In *The biology of the naked mole-rat*, edited by P. W. Sherman, J. U. M. Jarvis, and R. D. Alexander, 384–425. Princeton, NJ: Princeton University Press.

Jarvis, J. U. M., and N. C. Bennett. 1990. The evolutionary history, population biology, and social structure of African mole-rats: Family Bathyergidae. In *Evolution of subterranean mammals at the organismal and molecular levels*, edited by E. Nevo and O. A. Reig, 97–128. Progress in Clinical and Biological Research, vol. 335. New York: Wiley-Liss.

———. 1991. Ecology and behavior of the family Bathyergidae. In *The biology of the naked mole-rat*, edited by P. W. Sherman, J. U. M. Jarvis, and R. D. Alexander, 66–96. Princeton, NJ: Princeton University Press.

———. 1993. Eusociality has evolved independently in two genera of bathyergid mole-rats—but occurs in no other subterranean mammal. *Behavioral Ecology and Sociobiology* 33:253–60.

Jarvis, J. U. M., M. J. O'Riain, N. C. Bennett, and P. W. Sherman. 1994. Mammalian eusociality: A family affair. *Trends in Ecology and Evolution* 9:47–51.

Jarvis, J. U. M., M. J. O'Riain, and E. McDaid. 1991. Growth and factors affecting body size in naked mole-rats. In *The biology of the naked mole-rat*, edited by P. W. Sherman, J. U. M. Jarvis, and R. D. Alexander, 358–83. Princeton, NJ: Princeton University Press.

Jennions, M. D., and D. W. Macdonald. 1994. Cooperative breeding in mammals. *Trends in Ecology and Evolution* 9:89–93.
Kayanja, F. I. B., and J. U. M. Jarvis. 1971. Histological observations on the ovary, oviduct and uterus of the naked mole-rat. *Zeitschrift für Säugetierkunde* 36:114–21.
Keane, B., P. M. Waser, S. R. Creel, N. M. Creel, L. F. Elliott, and D. J. Minchella. 1994. Subordinate reproduction in dwarf mongooses. *Animal Behaviour* 47:65–75.
Keller, L., and H. K. Reeve. 1994. Partitioning of reproduction in animal societies. *Trends in Ecology and Evolution* 9:98–103.
Koenig, W. D., F. A. Pitelka, W. J. Carmen, R. L. Mumme, and M. T. Stanback. 1992. The evolution of delayed dispersal in cooperative breeders. *Quarterly Review of Biology* 67: 111–50.
Lacey, E. A. 1991. Male dispersal and reproductive strategies in arctic ground squirrels. Ph.D. dissertation, University of Michigan, Ann Arbor.
Lacey, E. A., and P. W. Sherman. 1991. Social organization of naked mole-rat colonies: Evidence for divisions of labor. In *The biology of the naked mole-rat,* edited by P. W. Sherman, J. U. M. Jarvis and R. D. Alexander, 275–336. Princeton, NJ: Princeton University Press.
———. 1997. Cooperative breeding in naked mole-rats: Implications for vertebrate and invertebrate sociality. In *Cooperative breeding in mammals,* edited by N. G. Solomon and J. A. French, 267–301. Cambridge: Cambridge University Press.
Lacey, E. A., S. H. Braude, and J. R. Wieczorek. 1997. Burrow sharing by colonial tuco-tucos *(Ctenomys sociabilis). Journal of Mammalogy* 78:556–62.
———. 1998. Solitary burrow use by adult Patagonian tuco-tucos *(Ctenomys haigi). Journal of Mammalogy* 79:986–91.
Lacey, E. A., J. R. Wieczorek, and P. K. Tucker. 1997. Male mating behaviour and patterns of sperm precedence in arctic ground squirrels. *Animal Behaviour* 53:767–79.
Lee, P. C. 1994. Social structure and evolution. In *Behaviour and Evolution,* edited by P. J. B. Slater and T. R. Halliday, 266–303. Cambridge: Cambridge University Press.
Lidicker, W. Z., and J. L. Patton. 1987. Patterns of dispersal and genetic structure in populations of small rodents. In *Mammalian dispersal patterns,* edited by B. D. Chepko-Sade and Z. T. Halpin, 144–61. Chicago: University of Chicago Press.
Lovegrove, B. G. 1991. The evolution of eusociality in mole-rats (Bathyergidae): A question of risks, numbers, and costs. *Behavioral Ecology and Sociobiology* 28:37–45.
McLean, I. G. 1983. Paternal behaviour and killing of young in arctic ground squirrels. *Animal Behaviour* 31:32–44.
Malcolm, J. R., and K. Marten. 1982. Natural selection and the communal rearing of pups in African wild dogs *(Lycaon pictus). Behavioral Ecology and Sociobiology* 10:1–13.
Michener, G. R. 1983. Kin identification, matriarchies, and the evolution of sociality in ground-dwelling sciurids. In *Advances in the Study of Mammalian Behavior,* edited by J. F. Eisenberg and D. G. Kleiman, 528–72. Special publication no. 7. American Society of Mammalogists.
Miller, M. A. 1946. Reproductive rates and cycles in the pocket gopher. *Journal of Mammalogy* 27:335–58.
Moehlman, P. D. 1979. Jackal helpers and pup survival. *Nature* 277:382–83.
Murie, J. O., and G. R. Michener, eds. 1984. *The biology of ground-dwelling squirrels.* Lincoln: University of Nebraska Press.
Narins, P. M., O. J. Reichman, J. U. M. Jarvis, and E. R. Lewis. 1992. Seismic signal transmission between burrows of the Cape mole-rat, *Georychus capensis. Journal of Comparative Physiology* A 170:13–21.
Nevo, E. 1961. Observations on Israeli populations of the mole rat *Spalax e. ehrenbergi* Nehring 1898. *Mammalia* 25:127–44.

———. 1979. Adaptive convergence and divergence of subterranean mammals. *Annual Review of Ecology and Systematics* 10:269–308.
Nevo, E., S. Simson, G. Heth, and A. Beiles. 1992. Adaptive pacifistic behaviour in subterranean mole-rats in the Sahara Desert, contrasting to and originating from polymorphic aggression in Israeli species. *Behaviour* 123:70–76.
O'Riain, M. J., and J. U. M. Jarvis. 1997. Colony member recognition and xenophobia in the naked mole-rat. *Animal Behaviour* 53:487–98.
Parker, P. G., A. A. Snow, M. D. Schug, G. C. Booton, and P. A. Fuerst. 1998. What molecules can tell us about populations: Choosing and using a molecular marker. *Ecology* 79:361–82.
Patton, J. L. 1993. Hybridization and hybrid zones in pocket gophers (Rodentia, Geomyidae). In *Hybrid zones and the evolutionary process*, edited by R. G. Harrison, 290–308. Oxford: Oxford University Press.
Patton, J. L., and J. H. Feder. 1981. Microspatial genetic heterogeneity in pocket gophers: Non-random breeding and drift. *Evolution* 35:912–20.
Patton, J. L., and M. F. Smith. 1993. Molecular evidence for mating asymmetry and female choice in a pocket gopher *(Thomomys)* hybrid zone. *Molecular Ecology* 2:3–8.
Payne, S. F. 1982. Social organization of the naked mole-rat *(Heterocephalus glaber)*: Cooperation in colony labor and reproduction. M. S. thesis, University of California, Santa Cruz.
Pearson, O. P. 1959. Biology of the subterranean rodents, *Ctenomys*, in Peru. *Memorias del Museo de Historia Natural "Javier Prado"* 9:1–56.
Pearson, O. P., N. Binsztein, L. Boiry, C. Busch, M. Di Pace, G. Gallopin, P. Penchaszadeh, and M. Piantanida. 1968. Estructura social, distribucion espacial y composicion por edades de una poblacion de tuco-tucos *(Ctenomys talarum)*. *Investigaciones Zoologicas Chilenas* 13:47–80.
Pearson, O. P and M. I. Christie. 1985. Los tuco-tucos (genero *Ctenomys*) de los Parques Nacionales Lanin y Nahuel Huapi, Argentina. *Historia Natural* 5:337–44.
Puzachenko, A. Y. 1993. Social organization in the mole rat population, *Spalax microphthalmus* (Rodentia, Spalacidae). *Zoologicheskii Zhurnal* 72:123–31.
Rado, R., and J. Terkel. 1989. A radio-tracking system for subterranean rodents. *Journal of Wildlife Management* 53:946–49.
Reeve, H. K. 1992. Queen activation of lazy workers in colonies of the eusocial naked mole-rat. *Nature* 358:147–49.
Reeve, H. K., S. T. Emlen, and L. Keller. 1998. Reproductive sharing in animal societies: Reproductive incentives or incomplete control by dominant breeders? *Behavioral Ecology* 9:267–78.
Reeve, H. K., and L. Keller. 1995. Partitioning of reproduction in mother-daughter versus sibling associations: A test of optimal skew theory. *American Naturalist* 145:119–32.
Reeve, H. K., and P. W. Sherman. 1991. Intracolonial aggression and nepotism by the breeding female naked mole-rat. In *The biology of the naked mole-rat*, edited by P. W. Sherman, J. U. M. Jarvis, and R. D. Alexander, 337–57. Princeton, NJ: Princeton University Press.
Reeve, H. K., D. F. Westneat, W. A. Noon, P. W. Sherman, and C. F. Aquadro. 1990. DNA "fingerprinting" reveals high levels of inbreeding in colonies of the eusocial naked mole-rat. *Proceedings of the National Academy of Sciences, USA* 87:2496–2500.
Reichman, O. J., T. G. Whitham, and G. A. Ruffner. 1982. Adaptive geometry of burrow spacing in two pocket gopher populations. *Ecology* 63:687–95.
Reig, O. A. 1970. Ecological notes on the fossorial octodont rodent *Spalacopus cyanus* (Molina). *Journal of Mammalogy* 51:592–601.
Reig, O. A., C. Busch, M. O. Ortells, and J. R. Contreras. 1990. An overview of evolution, systematics, population biology, cytogenetics, molecular biology, and speciation in *Cten-*

omys. In *Evolution of subterranean mammals at the organismal and molecular levels,* edited by E. Nevo and O. A. Reig, 71–96. Progress in Clinical and Biological Research, vol. 335. New York: Wiley-Liss.

Ribble, D. O., and J. S. Millar. 1996. The mating system of northern populations of *Peromyscus maniculatus* as revealed by radiotelemetry and DNA fingerprinting. *Ecoscience* 3: 423–28.

Rood, J. P. 1990. Group size, survival, reproduction, and routes to breeding in dwarf mongooses. *Animal Behaviour* 39:566–72.

Rosenthal, C. M., N. C. Bennett, and J. U. M. Jarvis. 1992. The changes in the dominance hierarchy over time of a complete field-captured colony of *Cryptomys hottentotus hottentotus. Journal of Zoology, London* 228:205–25.

Russell, R. J. 1954. A multiple catch of *Cratogeomys. Journal of Mammalogy* 35:121–22.

Rymond, M. A. 1991. Aggression and dominance in the naked mole-rat *(Heterocephalus glaber).* M. S. thesis, University of Michigan, Ann Arbor.

Savic, I. R., and E. Nevo. 1990. The Spalacidae: Evolutionary history, speciation, and population biology. In *Evolution of Subterranean Mammals at the Organismal and Molecular Levels,* edited by E. Nevo and O. A. Reig, 129–54. Progress in Clinical and Biological Research, vol. 335. New York: Wiley-Liss.

Schieffelin, J. S., and P. W. Sherman. 1995. Tugging contests reveal feeding hierarchies in naked mole-rat colonies. *Animal Behaviour* 49:537–41.

Schwagmeyer, P. L. 1990. Ground squirrel reproductive behavior and mating competition: A comparative perspective. In *Contemporary issues in comparative psychology,* edited by D. A. Dewsbury, 175–96. Sunderland, MA: Sinauer Associates.

Shanas, U., G. Heth, E. Nevo, R. Shalgi, and J. Terkel. 1995. Reproductive behaviour in the female blind mole-rat *(Spalax ehrenbergi). Journal of Zoology, London* 237:195–210.

Sherman, P. W. 1977. Nepotism and the evolution of alarm calls. *Science* 197:1246–53.

———. 1981. Kinship, demography, and Belding's ground squirrel nepotism. *Behavioral Ecology and Sociobiology* 8:251–59.

———. 1985. Alarm calls of Belding's ground squirrels to aerial predators: Nepotism or self-preservation? *Behavioral Ecology and Sociobiology* 17:313–23.

———. 1988. The levels of analysis. *Animal Behaviour* 36:616–18.

———. 1989. Mate guarding as paternity insurance in Idaho ground squirrels. *Nature* 338:418–20.

Sherman, P. W., S. Braude, and J. U. M. Jarvis. 1999. Litter sizes and mammary numbers of naked mole-rats: Breaking the one-half rule. *Journal of Mammalogy* 80:720–33.

Sherman, P. W., J. U. M. Jarvis, and R. D. Alexander, eds. 1991. *The biology of the naked mole-rat.* Princeton, NJ: Princeton University Press.

Sherman, P. W., J. U. M. Jarvis, and S. H. Braude. 1992. Naked mole-rats. *Scientific American* 267:72–78.

Sherman, P. W., E. A. Lacey, H. K. Reeve, and L. Keller. 1995. The eusociality continuum. *Behavioral Ecology* 6:102–8.

Solomon, N. G., and J. A. French, eds. 1997. *Cooperative breeding in mammals.* Cambridge: Cambridge University Press.

Spinks, A. C., M. J. O'Riain, and D. A. Polakow. 1998. Intercolonial encounters and xenophobia in the common mole-rat, *Cryptomys hottentotus* (Bathyergidae): The effects of aridity, sex, and reproductive status. *Behavioral Ecology* 9:354–59.

Topachevskii, V. A. 1976. *Fauna of the USSR.* Vol. 3. *Mammals.* No. 3. *Mole-rats, Spalacidae.* Washington, DC: Smithsonian Institution.

Vaughan, T. A. 1962. Reproduction in the plains pocket gopher in Colorado. *Journal of Mammalogy* 43:1–13.

Wallace, E. D., and N. C. Bennett. 1998. The colony structure and social organization of the giant Zambian mole-rat, *Cryptomys mechowi. Journal of Zoology, London* 244:51–61.

Weir, B. J. 1974. Reproductive characteristics of hystricomorph rodents. In *The biology of hystricomorph rodents,* edited by I. W. Rowlands and B. J. Weir, 265–301. Symposia of the Zoological Society of London, 34. London: Academic Press.

Wolff, J. O. 1985. Behavior. In *Biology of New World* Microtus, edited by R. H. Tamarin, 340–72. Special publication no. 8. American Society of Mammalogists.

Zenuto, R. R., E. A. Lacey, and C. Busch. 1999. DNA fingerprinting reveals polygyny in the subterranean rodent *Ctenomys talarum. Molecular Ecology* 8:1529–32.

Zhou, W., and F. Dou. 1990. Studies on activity and home range of plateau zokor. *Acta Theriologica Sinica* 10:31–39.

Zuri, I., and J. Terkel. 1996. Locomotor patterns, territory, and tunnel utilization in the mole-rat *Spalax ehrenbergi. Journal of Zoology, London* 240:123–40.

Part Three: Evolutionary Biology

Two features unify subterranean rodents: their limited mobility due to morphological modifications for digging, and a patchy distribution of local populations dictated by the soils they inhabit. These features have been identified by many workers as fundamental to evolutionary diversification in these animals. Consequently, subterranean rodents have commonly been used to exemplify the geographic conditions that generate evolutionary novelties; they have been identified as prime examples of the distribution of character variation among as opposed to within populations; and they have been held up as special examples of rapid rates of evolution, notably of speciation.

In this third part of our volume on subterranean rodents, we focus on the evolutionary themes of genetic structure, phylogenesis, coevolution, and adaptation. The four chapters contained in this part build upon those in part 2, most specifically chapter 5 on population ecology and chapter 7 on spatial and social structure.

In chapter 8, Steinberg and Patton summarize the available data on the population genetic structure of subterranean rodents and address the question of whether these animals are, in fact, as highly structured geographically as is usually presumed. These authors explore the degree to which variation is transferred from that maintained within local populations to that which culminates in differences between populations and, ultimately, species, of subterranean rodents. They then proceed to demonstrate why species boundaries become difficult to ascertain by standard criteria, and why hypotheses of population, and thus ultimately species, genealogies generated by individual genes might not reflect the true phylogeny of the taxa being examined. Finally, this chapter acknowledges the reality of hybridization between differentiated geographic units, and argues that hybridization is not simply an epiphenomenon, but is an evolutionary force of significance to the biology of many subterranean rodents.

In chapter 9, Cook, Lessa, and Hadly build upon the themes identified by Steinberg and Patton and explore a variety of important macroevolutionary issues. First, these authors review the fossil record for subterranean rodents, emphasizing the pattern of phyletic diversity at the generic and species levels. They use this record to explore the question of temporal concordance, perhaps due to global phenomena, of peaks and valleys in diversity. They then move on to a discussion of phylogenesis, and address the central question of whether or not subterranean rodents have speciated at a faster rate than their surface-dwelling relatives. The presumption of high speciation rates has been either implicit or explicit in the minds of many students of subterranean rodents, if only because of the propensity of genetic diversity to be apportioned among, as opposed to within, local populations (but see chapter 8). This presumption is tested here for the first time.

Chapter 10 examines the potential for coevolution, including cospeciation, between subterranean rodents and other organisms. Hafner, Demastes, and Spradling review the rich data on subterranean rodents and their external and internal parasites, and use their own seminal work on pocket gophers and chewing lice to document the cospeciation of hosts and their parasites. They also show that such parasites, by themselves, can offer important insights into the evolutionary and even ecological dynamics of subterranean rodents. They then argue convincingly that this is an extremely rich area for future research, one that has hardly been tapped, much less exploited. Indeed, they end by identifying more than two hundred separate research programs with subterranean rodents wherein a coevolutionary approach will yield novel insights.

The final chapter serves as both a summary of the contents of the volume and a reflection upon the current status of research on subterranean rodents. Lessa explores the utility of adaptationist versus historical approaches to the study of these animals, and proposes a unified conceptual framework that incorporates both aspects of evolutionary change. He then reviews several evolutionary issues that have traditionally served as foci for research on subterranean taxa, and considers the progress made to date in resolving the associated controversies. In briefly recapping the highlights of each part of this volume, Lessa outlines a series of goals for future research on subterranean rodents. As this and the other chapters in part 3 emphasize, the opportunities for evolutionary research on subterranean rodents are immense. Topics deserving further study range from questions regarding species-specific adaptations to the subterranean niche to questions regarding the evolutionary diversification of major lineages of these animals.

CHAPTER EIGHT

Genetic Structure and the Geography of Speciation in Subterranean Rodents: Opportunities and Constraints for Evolutionary Diversification

Eleanor K. Steinberg and James L. Patton

Studies of subterranean rodents commonly emphasize two interrelated features that underlie the evolutionary diversification of these lineages. The first is the propensity of these species to be distributed into many, relatively small populations rather than occurring as large, panmictic, and continuously distributed ones. This propensity leads to strong geographic structuring, so that a substantial portion of character variation (be it morphological, chromosomal, or molecular) is distributed among populations rather than being polymorphic within them. A few key aspects of the basic population biology of subterranean rodents underscore the causal linkage between these two features; minimally, these include the combination of exclusive territories at the individual or colony level (and therefore relatively low population sizes) and limited dispersal coupled with naturally patchy distributions due to the availability of suitable soils (see Busch et al., chap. 5; Lacey, chap. 7, this volume). Moreover, both features impose constraints on the means and direction of evolutionary divergence, as well as presenting opportunities for such divergence, at the species level and beyond. These features have also resulted in challenges to those of us who study subterranean rodents, as we attempt to decipher the record they have left behind while searching to understand the processes underlying their diversification.

Our intent in this chapter is to summarize what is known about local population genetic structure in subterranean rodents and to discuss how this structure influences the patterns of geographic differentiation that characterize these animals. While genetic structure, local and geographic, is important unto itself, an understanding of this structure also provides the context both for recognizing and for predicting broader patterns of lineage diversification and cospeciation, themes developed in chapters 9 and 10 of this volume. Consequently, beyond our descriptive summary of the extent to which genetic variation is distributed within and among local populations of subterranean rodents, we will emphasize three factors that we believe affect evolutionary opportunity in these animals. These factors are independent of the specific taxonomic lineage, degree of sociality, or continental placement, and are probably more characteristic of subterranean organisms than of their surface-dwelling sister groups. First, and perhaps most important, is the geographic setting under which new species are generated. Second, the coupling of that geographic setting with population demography places some limitations on the types of mechanisms by which reproductive isolation can be achieved. Finally, the retained capacity for hybridization offers enhancements for continued divergence in many of these groups, rather than retardation due to gene pool fusion. While we believe that each of these factors has been among the most important elements underlying the evolutionary history of subterranean rodents, they have also left a legacy of confusion in our reading of that history. With evolutionary opportunities have come constraints that compromise simple hierarchical branching sequences of relationships, and thus our ability to cleanly bound species of these organisms in nature.

Studies of Population Genetic Structure in Subterranean Rodents

The genetic structure of a population, or a series of populations, results from the interplay of multiple demographic attributes, all of which act on the pool of available genetic variation. Key aspects of demography are the longevity of animals of both sexes, the breeding sex ratio, population turnover rates, the dispersion patterns of individuals in local populations, and dispersal distances for young animals as well as established adults (see summaries in Busch et al., chap. 5; Lacey, chap. 7, this volume). From a genetic standpoint, we should be interested in parameters such as the degree to which mating within local populations is random, the variance in reproductive success of each sex, and the genetic effectiveness of short- and long-distance dispersal (i.e., gene flow). Genetic markers provide the means to measure patterns of the dispersion of genetic traits in space and time. In combination with demographic attributes, these markers give us insights into the processes that both maintain variation within local populations and distribute variation among them.

Sewall Wright developed the use of hierarchical F statistics to quantify patterns of genetic structure within and among populations. His most commonly employed statistic is the fixation index, F_{ST}, which describes how much genetic variation is partitioned at the "subpopulation" level relative to all subpopulations combined into a single "total" population (Wright 1943, 1951). For a given locus, populations with identical allele frequencies have an F_{ST} of 0; those for which alternative alleles are fixed exhibit an F_{ST} of 1. An important predictive relationship that stems from Wright's work and that is now applied extensively to molecular data, appearing in virtually every population genetics textbook (e.g., Hartl and Clark 1997) is:

$$F_{ST} = 1/(4N_e m + 1)$$

where N_e is the effective population size and m is the migration rate. These two parameters represent the two forces determining how genetic variation will be distributed among populations at equilibrium. When N_e is small, chance events control patterns of neutral genetic variation, resulting in population differentiation due to genetic drift causing random fixation of different alleles in different populations. This differentiation can be counteracted by genetic exchange among populations, which occurs when individuals migrate.

Most studies of the genetic structure of subterranean (and other) rodent populations have employed protein electrophoresis (allozymes); few have taken advantage of the diversity of the powerful DNA-based marker systems now available. For example, Faulkes et al. (1997) used multilocus DNA fingerprints to examine genetic diversity within and between colonies of *Heterocephalus glaber*. Similarly, both Faulkes et al. (1997) and Patton and Smith (1993) employed maternally inherited mitochondrial DNA to determine the extent of female bias in mating within local populations of *Heterocephalus glaber* and within species of *Thomomys*, respectively. Importantly, these studies considered data from nuclear (multilocus DNA fingerprints or allozymes, respectively) and organelle (mitochondrial) genes simultaneously, a combination that is especially informative because of differences in patterns of inheritance and recombination. Microsatellite markers have been developed for *Thomomys* (Steinberg 1999) and *Ctenomys* (Lacey et al. 1999), but no detailed studies using these markers have been published as yet. Paternally inherited genes have also yet to be exploited, despite their potential in combination with maternal mitochondrial genes for comparing intersexual differences in behavior, including dispersal and mate choice.

To date, most genetic studies of subterranean rodents have focused on calculating levels of genetic variation or on examining phylogeographic patterns. Consequently, commonly reported statistics include calculations of average individual heterozygosity, number of alleles per locus, and percentage of loci that are polymorphic (data for subterranean and surface-dwelling rodents are summarized by Nevo 1988 and Nevo, Filippucci, and Beiles 1990). Additionally, many studies include one or more measures of genetic distance among populations. While these variables are important components of population genetics, estimates of heterozygosity are not by themselves measures of population structure. Similarly, genetic distances give only a global description of interpopulation differences; they do not reveal the processes responsible for those differences. Below, we describe some of the ways that population structure can be, and has been, measured, and we direct interested readers to good texts in population genetics (e.g., Hartl and Clark 1997) and general reviews of the methods used to link demography and population genetic structure (e.g., Rockwell and Barrowclough 1987).

There have been a relatively large number of molecular genetic papers devoted to genetic differentiation in subterranean rodents (table 8.1). Most of these have used allozymes as the marker system of choice, and nearly all have been directed toward one of two research areas. Basic systematic questions, including measures of the taxonomic uniqueness of populations, studies of hybrid zone dynamics, and hierarchical analyses of species relationships, have been a major emphasis of these studies. We describe and discuss the importance of some of these in greater detail below. The studies listed in table 8.1 are meant to be representative, not exhaustive. A second area of much effort has been the measurement of within-population variation (e.g., allelic heterozygosity), which has formed part of the larger debate about selective versus stochastic influences on patterns and levels of genetic variability. We make no attempt to review this aspect of genetic diversity, but refer interested readers to the general reviews of Nevo (1990), Nevo, Filippucci, and Beiles (1994), Nevo et al. (1997), Patton (1990), and Lessa (chap. 11, this volume) for contrasting opinions.

Given the number of papers dealing with genetic population differentiation in subterranean rodents and the general interest in understanding the patterns observed, it is surprising that so few studies have included detailed examinations of local patterns of population genetic structure, particularly analyses that combine genetic and demographic perspectives. Even the rich data base on genetic and other characteristics of Middle Eastern *Nannospalax* (see reviews by Nevo, Filippucci, and Beiles 1994; Nevo et al. 1997) has largely neglected aspects of the genetic demography of local populations. The colonial bathyergid *Heterocephalus glaber* has received some recent attention regarding the genetic structure of local populations (Honeycutt et al. 1991; Faulkes et al. 1997). However, only for pocket gophers has there

TABLE 8.1 STUDIES OF THE GENETIC STRUCTURE OF SUBTERRANEAN RODENTS

Taxon	Demography[a]	Genetic marker	Spatial scale	References
Ctenomyidae				
Ctenomys maulinus	Yes	Allozymes	Local/regional	Gallardo, Köhler, and Araneda 1995
Ctenomys rionegrensis	Yes	Allozymes	Local/regional	D'Elía, Lessa, and Cook 1998
Geomyidae				
Geomys bursarius	No	Allozymes	Regional	Penney and Zimmerman 1976
	No	Allozymes	Local	Zimmerman and Gayden 1981
	No	Allozymes, nuclear and mtDNA	Local	Baker et al. 1989
Thomomys bottae	No	Allozymes	Regional	Patton and Yang 1977
	Yes	Allozymes	Local	Patton and Feder 1981
	Yes	Allozymes	Local	Daly and Patton 1990
	No	Allozymes	Regional	Nadler et al. 1990[b]
Thomomys bulbivorus	No	Allozymes	Regional	Carraway and Kennedy 1993
Thomomys talpoides	No	Allozymes	Regional	Nevo et al. 1974
Thomomys townsendii	No	Allozymes	Local	Patton et al. 1984
	No	mtDNA	Local	Patton and Smith 1993, 1994
Thomomys umbrinus	No	Allozymes	Regional	Patton and Feder 1978
	No	Allozymes	Regional	Hafner et al. 1987
Spalacinae				
Nannospalax ehrenbergi	Yes	Allozymes	Regional	Nevo, Heth, and Beiles 1982
	No	Allozymes	Regional	Nevo, Filippucci, and Beiles 1994
	No	mtDNA	Regional	Nevo et al. 1993
Bathyergidae				
Cryptomys, Bathyergus, Georychus	No	Allozymes	Regional	Nevo et al. 1987
Heterocephalus glaber	No	Allozymes, mtDNA	Local & regional	Honeycutt et al. 1991
	No	Multilocus fingerprinting, mtDNA	Local & regional	Faulkes et al. 1997

Note: This table does not include those papers that examine only genetic distance for taxonomic or systematic purposes, for which the literature is large.
[a] Density, dispersal distances, mating pattern, or other populational attributes.
[b] Inferences on gopher genetic structure based on the genetic structure of their external parasites (see also chap. 10, this volume).

been coordinated genetic and demographic study of the same populations. We use these data to highlight three important aspects of population genetic structure: the applicability of isolation by distance models, the importance of extinction and recolonization dynamics, and the drift versus selection paradigms. Each of these topics is central to an eventual understanding of the influence of genetic structure on population divergence and, ultimately, species formation, as we demonstrate in subsequent sections of this chapter.

Pocket Gophers and Isolation by Distance Models

The spatial configuration of populations influences the process of genetic differentiation such that adjacent populations will exchange migrants (and thus genes) more often than more distant ones. A formal recognition of this expectation was Sewall Wright's (1943) "isolation by distance" model of genetic differentiation. According to this model, populations should become increasingly differentiated at neutral loci with increasing geographic distance (e.g., Mayamura 1971; Rohlf and Schnell 1971; Nagylaki 1976; Slatkin 1993). This prediction has been confirmed by many empirical studies (i.e., Blouin et al. 1995; Doadrio, Perdices, and Machordom 1996; Green et al. 1996; Patton, da Silva, and Malcolm 1996).

Slatkin (1993) has developed methods to test for isolation by distance in populations that do not necessarily adhere to equilibrium conditions, a basic assumption that underlies much of population genetic theory. He measured gene flow ($\hat{M}$) between all population pairs, determined from estimates of the mean coalescence derived from pairwise population F_{ST} values, and then regresses this measure against map distances between populations. Under equilibrium conditions, a significant negative slope of $\hat{M}$ versus map distance indicates isolation by distance. Slatkin examined two models that represent alternative scenarios for how a species might expand its range. In the first, an ancestral population is hypothesized to give rise rapidly to an array of subpopulations, which thereafter exchange migrants at a fairly high rate. The second hypothesizes a population that gradually (in a stepwise fashion) comes to occupy progressively more area. Using these underlying population models as foundations for generating genetic structure, Slatkin simulated genetic exchange between populations over time, using the coalescent approach described by Hudson (1990), and showed that given these scenarios, it is possible to detect influences of isolation by distance.

To test whether geographic patterns of genetic structure found in nature could be interpreted with his model, Slatkin (1993) used a large allozyme data set for the pocket gopher *Thomomys bottae* (from Patton and Smith 1990). His choice of this species took advantage of its known restricted dispersal between readily definable populations (Busch et al., chap. 5; Lacey, chap. 7, this volume). Analyses of the allozyme data clearly demonstrated how Slatkin's isolation by distance approach might be used to elucidate possible historical scenarios for the radiation of populations. In this example, one cluster of populations of *T. bottae* was likely to have only recently colonized its extant range, while another cluster exhibited evidence of historically limited gene flow with recurrent mutation, and thus a strong isolation by distance pattern.

Pocket Gophers and Extinction-Recolonization Dynamics

The dynamics of metapopulations—arrays of interacting populations occurring in both spatially and temporally discontinuous patches—has been a focus of theoretical and empirical ecological research for many years (e.g., Huffaker 1958; Levins 1970; S. Harrison 1991; Hanski 1996, 1998; Hanski and Gilpin 1996; McCullough 1996). From this body of work, it is clear that both spatial and temporal aspects of population structure can significantly influence population dynamics. Theoretical analyses have shown that patterns of local extinction and recolonization in subdivided populations can also have important genetic consequences for population structure (Wright 1940; Slatkin 1977; Maruyama and Kimura 1980; Wade and McCauley 1988; Whitlock and McCauley 1990; McCauley 1991). Wright (1940) suggested that genetic differentiation of subdivided populations would be enhanced by localized genetic drift. Slatkin (1977, 1985) built on some of Wright's ideas, pointing out that extinction and recolonization processes can cause subdivided populations to become either differentiated or homogenized, depending on the details of colonization and the particular population model hypothesized. Subsequently, Whitlock (1992) modeled temporal variation in extinction and migration rates, as well as variation in population size in subdivided populations, and found that fluctuations in these demographic parameters generally cause higher rates of genetic differentiation than would be expected if populations are assumed to be stable. Finally, McCauley (1991) argued that our understanding of the genetic consequences of extinction and recolonization would ultimately require a melding of simple theoretical models with the complex behavior of actual natural systems. In particular, he suggested that experimental studies of population founding events would be illuminating.

The discontinuous spatial structure and low vagility of most subterranean rodent species make them good systems for experimental studies of extinction and recolonization dynamics. The individual territories of most subterranean rodents can be readily identified by mounds of excavated earth on the surface of the ground, which makes it possible to trap, mark, and enumerate entire populations. Patton and Feder (1981) took advantage of these features in a study exploring the influence of extinction and subsequent recolonization on temporal patterns of genetic variation in *T. bottae*. In their study, the location of every resident animal was mapped, and then all individuals were removed from three adjacent fields in each of two successive years. Apportionment of allozyme variation for all adults and offspring (i.e., embryos of pregnant females) sampled was then quantified. According to data averaged over eleven loci, the three populations became less differentiated after the experimental extinctions, thus documenting a clear homogenizing influence of migration. However, when the first- and second-generation post-extinction genetic structures of the populations were compared, a large increase in differentiation among populations (based on F_{ST} values) was evident. While recolonizers appeared to represent a random sample of the genetic variability present in the source populations, interpopulation variation was quickly generated, presumably due to biases in reproductive success and low numbers of founding individuals. Thus, despite significant local population differentiation due to genetic drift, local breeding patterns and migration and extinction dynamics can maintain high levels of heterozygosity in subdivided populations of this subterranean rodent.

Pocket Gophers, Drift, and Selection

Genetic drift and natural selection are the two primary evolutionary forces thought to influence patterns of diversification, and arguments regarding the relative importance of each date back to the debates of Wright and Fisher in the early 1930s (see Provine 1971). Wright (1978) believed that random drift of genes in subdivided populations played a critical role in evolution by creating opportunities for the formation of coadapted gene complexes. Fisher thought that evolutionary diversification could be explained by selection alone (Fisher 1958; Provine 1971). Over half a century later, proponents of these differing views continue to engage in discussions about the importance of these two evolutionary forces.

The relative importance of drift and selection as agents in the genetic structuring of subterranean rodent populations provides an example of this debate (Nevo, Filippucci, and Beiles 1990; Nevo et al. 1997; Patton 1990; Lessa, chap. 11, this volume). The population parameters of most subterranean rodents (including limited dispersal, low densities, biased sex ratios, nonrandom mating, and patchy distribution of local populations) predict that genetic drift will play a significant role in determining the genetic structure of local populations. At the same time, there can be little question that both stabilizing and diversifying selection have been important contributors to evolutionary diversification among these animals. One need look no further than the exceptional degree to which pelage color matches soil color in nearly all taxa to find evidence that selection has acted on this character (Ingles 1950; Patton and Smith 1990).

Analyses of hybrid zones between species of *Geomys* and *Thomomys* (reviewed by Patton 1993) also provide excellent examples of the interaction between drift and selection in the determination of population genetic structure. Hybrid zones of pocket gophers are narrow and often show evidence of selection against karyotypic intermediates (Patton 1973; Baker et al. 1989). Moreover, phenotypic intermediates are typically limited to zones of habitat overlap (Patton 1973; Heaney and Timm 1985). Both observations strongly suggest that selection has played a central role in mediating the genetic interactions between neighboring populations. However, the same geographic narrowness limits the number of interacting individuals each generation, and thus the potential for gametic "leakage" into either parental population is determined by random spatial assortment as much as by the strength of selection differentials. Indeed, despite demonstrably strong selection against hybrid individuals, genic introgression does extend into the parental populations. Moreover, models of zone dynamics based on cline theory (Endler 1977; Barton and Gale 1993) cannot reject simple neutral diffusion as an explanation for the observed geographic distribution of alleles (e.g., Hafner et al. 1983). Consequently, horizontal gene transfer due to hybridization is a likely event in the history of many subterranean taxa (Patton and Smith 1994; see below). Similarly, the possibility of selective "sweeps" of genetic traits must be recognized and evaluated in both phylogenetic studies and those that apply population genetic models based on effective neutrality (Ruedi, Smith, and Patton 1997).

The Geographic Apportionment of Genetic Variation

The underlying presumption of many studies examining the spatial array of genetic variation in subterranean rodents is that such variation, however it might be measured, is largely partitioned among, rather than contained within, populations. In this section, we address the adequacy of this presumption. Numerous studies have examined the geographic structure of subterranean rodent populations based on a variety of genetic markers (see table 8.1). As described above, protein electrophoretic, or allozyme, variants have been used most extensively, although other molecular systems are now increasingly employed (e.g., *Ctenomys:* Sage et al. 1986; Cook and Yates 1994; bathyergids: Honeycutt et al. 1987, 1991; Nevo et al. 1987; Filippucci et al. 1994; *Nannospalax:* a large number of papers by Nevo and his co-workers [recent summaries in Nevo et al. 1993, 1995]; geomyid pocket gophers: a large number of papers summarized by Patton 1990; Smith 1998). As noted above, the vast majority of these studies have focused on measures of individual variability (heterozygosity) and population and species genetic distance.

Surprisingly, little opportunity has been taken to examine the degree to which the variation measured is hierarchically structured within and among populations of subterranean rodents. Extensive data are available on Wright's fixation index (F_{ST}) for allozyme loci in pocket gophers at both the local and regional levels (reviewed in Lidicker and Patton 1987; Patton 1990; Patton and Smith 1990). Fewer data are available for other subterranean taxa, however, despite the number of published molecular genetic studies.

For the pocket gopher *Thomomys*, up to 7% of allozymic variation is apportioned among local populations found within a 1 km radius ($F_{ST} = 0.07$: Patton and Feder 1981), and regional apportionment may reach as high as 70–80% ($F_{ST} = 0.70$–0.80: Hafner et al. 1987; Patton 1990). According to the comparative criteria developed by Sewall Wright (1978), local populations of pocket gophers are thus moderate in their degree of genetic subdivision, while species exhibit extreme structure across their ranges. These high levels of geographic partitioning are even more impressive when viewed in the context of the levels of individual heterozygosity maintained within local populations, which can be as high as 24% (Patton and Smith 1990). Both hierarchical F_{ST} and spatial autocorrelation analyses indicate that genetic differentiation in *Thomomys* increases precipitously to a geographic distance of about 200 km, but beyond this threshold further differentiation is unrelated to distance per se (Patton and Smith 1990). For non-geomyids, an average F_{ST} of 0.38 was obtained for allozyme loci in Chilean *Ctenomys maulinus* populations sampled less than 150 km distant from one another (Gallardo and Köhler 1992). An average of 42% of the variation in twenty-five polymorphic allozyme loci was distributed among Israeli populations of the *Nannospalax ehrenbergi* superspecies (Nevo, Filippucci, and Beiles 1994). Finally, Faulkes et al. (1997) recently showed that 68% of the variance in mitochondrial haplotpyes was apportioned among geographic populations of the naked mole-rat, *Heterocephalus glaber*, with nearly 32% among colonies within single localities.

A high degree of geographic differentiation in subterranean rodents is equally evident in genetic distance summaries derived from allozyme comparisons. While population samples of a "typical" surface-dwelling mammal species might differ by no more than an average Rogers' distance of 0.05 (Avise 1976; Selander and Johnson 1973), many subterranean taxa exhibit considerably greater genetic distances among populations. Comparisons of these data for different species, and between subterranean and surface-dwelling mammals, are complicated by lack of control for phylogenetic effects and the inclusion of different sets of enzyme and other protein systems in the separate studies, which in themselves vary in their individual rates of change. Nevertheless, substantial differences between subterranean and surface-dwelling rodents in the degree to which populations are genetically differentiated are usually apparent. For example, the average Rogers' genetic distance between 141 sampled populations of *Thomomys bottae* is 0.261

(Patton and Smith 1990). This level is four to nine times higher than that for pocket mice of the genus *Chaetodipus* (0.070: Patton, Sherwood, and Yang 1981) or kangaroo rats of the genus *Dipodomys* (0.03: Johnson and Selander 1971), the closest surface-dwelling relatives of pocket gophers.

In some cases, measures of the differentiation among populations are complicated by the inclusion of semi-reproductively isolated chromosomal races in the analyses, and thus some of the apportionment is probably due to factors other than population structure itself. Nevertheless, the data available are consistent with the premise that a substantial proportion of the total pool of variation in subterranean rodents is typically found among, as opposed to contained within, local populations. While subterranean rodents thus clearly exhibit strong geographic structure, an unanswered question is whether they are any more structured across the landscape than similar-sized aboveground species. Data are available to address this question, but they have not been compiled in any useful way to date.

The Geography of Species Formation

With the exception of models of speciation based on the mechanics of population genetics (e.g., Templeton 1980a, 1981), the field of speciation research has been dominated by a singular emphasis on the geographic context in which divergence takes place (Mayr 1963). How important, or even likely, is divergence in sympatry (e.g., Bush 1975, 1994)? What role has parapatric differentiation (e.g., Endler 1977) played? For subterranean rodents, the geographic context of species formation seems relatively trivial. Given the empirical observation that, with very minor exceptions, different species of subterranean rodents are unable to occupy the same space, sympatric speciation, via resource specialization or any other mechanism (Bush 1975, 1994), is extremely unlikely for this group of organisms. A similar conclusion can be drawn nearly as easily for parapatric divergence. Narrow contact zones characterize many subterranean rodents (reviewed in Barton and Hewitt 1985; Patton 1993), and their association with habitat ecotones potentially provides the steep selection gradient required by the divergence-with-gene-flow model (Rice and Hostert 1993). However, these zones are clearly secondary in origin (see discussion in Hafner et al. 1983; Patton 1993), not primary, which is another requirement of the parapatric model. Parapatric divergence has, as an immediate outcome, the expectation of a sister-group relation between the derivative species, and thus the model is easily falsified by phylogenetic analyses (Patton and Smith 1992). These types of analyses do not support parapatric divergence in subterranean rodents.

If both sympatric and parapatric models are excluded from our consideration, then subterranean rodents must have undergone one of the several types of allopatric divergence (*sensu* Mayr 1963, 1982), such as classic "dumb-bell" vicariance or peripheral isolation (Mayr 1954). This conclusion seems hardly surprising, at least to us. As conventional as it might be, however, allopatric speciation does have nontrivial effects on the opportunities for evolutionary divergence. These effects include, but are not limited to, (1) the degree to which genetic change might accompany speciation, (2) the actual mechanisms of reproductive isolation, (3) the degree of retention of hybridization ability, (4) the likelihood of selection for reinforcement, (5) the pattern of phyletic relationships among daughter populations and species, (6) the rate at which new species are formed, and (7) the overall diversity of species within lineages through time. Each of these issues is central to any understanding of evolutionary diversification at the species level, and we will address most in the sections that follow. The interrelated issues of speciation rate and overall species diversity, often assumed to be especially high in subterranean rodents, are addressed by Cook, Lessa, and Hadly in chapter 9 of this volume. However, as important as these issues are unto themselves, they still beg a more central question, which is how to define and recognize species of subterranean rodents in the first place.

Genetic Divergence and Mechanisms of Speciation

Species formation in subterranean rodents can apparently happen very quickly, as evidenced by numerous cases wherein reproductively isolated taxa share close genetic identity. For example, in both the *Nannospalax ehrenbergi* superspecies in Israel (reviewed in Nevo 1989) and the *Thomomys talpoides* complex in North America (Nevo et al. 1974), Nei genetic distances average less than 0.04. This value is less than the divergence measured among local interbreeding populations of another pocket gopher, *Thomomys bottae* (Patton and Feder 1981; Patton and Smith 1990). What, then, are the mechanisms by which rapid intrinsic barriers to gene exchange are generated? And, equally importantly, are these mechanisms related to those population parameters that are responsible for the apportionment of the available genetic variation in the first place?

The combination of small and isolated populations is often argued to be ideal for speciation by genetic transilience: a rapid shift in a previously stable genetic system through founder events (Templeton 1980b; see also Carson and Templeton 1984; Carson 1987). While this pattern of geographic population structure does characterize subterranean rodents, the conditions for the kind of "genetic revolution" envisioned by Mayr (1954) may be met only rarely in these organisms. As noted above, the population characteristics of "metapopulation" model   "boundary" model

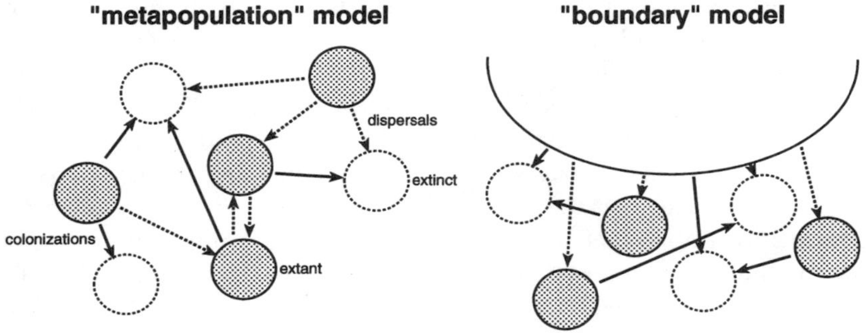

FIGURE 8.1. Two models of geographic population structure that probably characterize most subterranean rodents: (left) the "metapopulation" model and (right) the "boundary" population model. Extinct demes or vacant habitats are indicated by dashed circles; extant demes are shown by stippled circles. The solid arrows represent colonizations, and the dashed arrows indicate dispersal events (see Wade 1982).

subterranean rodents can either enhance or retard differentiation at the genic level during temporal episodes of population splitting. For simplicity, let us consider two different population models that are realistic for subterranean rodents: "metapopulations" and "boundary populations" (fig. 8.1). For species exhibiting the latter pattern, genetic revolutions in peripheral isolates are more likely, since the expected large pool of variation present in the larger source population is available for distribution, in different combinations, to the derivative isolates. However, derivative populations can exhibit decreased as well as increased differentiation when already subdivided populations are subject to random extinctions and recolonizations, as in a metapopulation model (McCauley 1991). Moreover, an array of metapopulations subject to extinction/recolonization cycles will store less genetic diversity than will a boundary population (Gilpin 1991). Consequently, the degree to which the genetic architecture of a species is open to transilient processes depends ultimately on the type of general population structure present, which in turn influences both the total pool of variation available and the pattern of apportionment of that variation prior to splitting events. If a considerable amount of variation is contained within populations, then fragmentation can increase divergence substantially through simple founder effects and drift (Wade and McCauley 1988), and the type of genetic revolution envisioned by Mayr is possible. If, on the other hand, much of the available genetic diversity is already distributed among populations, then subsequent fragmentation events will have very little or no effect on changes in among-population divergence.

Nevo (1989), following decades of detailed work on *Nannospalax* in Israel and adjacent countries, has shown that small, isolated populations are rarely different in any fundamental way from their presumptive geographic parents. His data are extensive, both in the number of population "replicates" and in the number of independent genetic systems, including chromosomal and genic, examined. Similarly, Patton and Smith (1990) have shown that isolated populations of *Thomomys bottae*, while exhibiting the loss of genic variability expected because of their size and isolation, are also more homogeneous than non-isolated populations with regard to the partitioning of the variation that does exist. In neither of these cases has range fragmentation or peripheral isolation following range expansion routinely increased genetic divergence per se. The empirical data thus suggest that the population structure typical of subterranean rodents may not enhance speciation through the types of genetic revolutions envisioned by Mayr, a conclusion that may seem counterintuitive. Rather, speciation results either from the "capture" of a post-mating reproductive barrier or, if there are significant shifts in local environmental conditions coincident with isolation, through changes in the matrix of selection differentials. The latter mechanism would provide for the "shifting balance" opportunities envisioned by Wright (summarized in Wright 1980; but see Coyne, Barton, and Turelli 1997 for a critique).

One of the most dramatic aspects of most subterranean rodent taxa, and one that characterizes all cases of "sibling" species with high degrees of genetic identity of which we are aware, is extensive chromosomal reorganization that presents a partial sterility barrier (Patton 1973; Nevo 1989; Reig 1989). Without attempting to quantify this statement, we suggest that subterranean rodents are far more chromosomally diverse than are other groups of rodents. And most investigators agree that the pattern of geographic structuring typical for these animals is ideal for the rapid fixation of strongly underdominant rearrangements, which can serve as primarily post-mating isolating barriers. For example, Nevo (1989: 221) states, in relation to the chromosomal species of *Nannospalax*, that "small isolates are indeed ... ideal sites to pass quickly through a condition of heterozygosity with lowered fitness." This is true whether chromosomal rearrangements are considered only negatively heterotic, requiring the combination of small population size and drift for fixation (Lande 1979; Walsh 1982), or whether drift is combined with positive selection, meiotic drive, or another mechanism that facilitates fixation (Hedrick 1981; Chesser and Baker 1986). While not a universal mechanism of speciation in subterranean rodents, chromosome races are a very common phenomenon, and their formation is probably facilitated by the shared demographic and population features of these animals.

Reproductive Isolation: To Reinforce or Not?

A second dramatic aspect shared by many subterranean rodent taxa is that, despite the development of severe post-mating barriers, hybridization between differentiated taxa is commonplace. Indeed, certain subterranean taxa have provided some of our best insights into the dynamics of hybrid zones, either in mammals specifically or among animals more generally (see Harrison 1993). When divergent populations produce hybrids with lowered fitness, then natural selection should favor an increase in assortative mating and thus progress toward full reproductive isolation. This simple prediction forms the basis of the reinforcement hypothesis, first elaborated by Alfred Russel Wallace (1889) over a century ago. However, the likelihood of reinforcement has been, and continues to be, a vigorously debated topic (see recent reviews by Howard 1993; Butlin 1995).

The reinforcement hypothesis, along with the question of whether subterranean rodents are good models for testing it, has been addressed in two groups, *Nannospalax* and *Thomomys*, but with opposite conclusions. Regarding *Nannospalax*, Savic and Nevo (1990) state that "the nature and extent of hybridization in the *S. ehrenbergi* complex reinforce the hypothesis that the chromosome forms are indeed young, closely related, sibling species *at early stages* of evolutionary divergence [emphasis ours]." The presumption that we are looking at the early stages of differentiation in these organisms assumes that refinement of reproductive isolation will occur with time, and that hybridization will eventually be eliminated through the process of reinforcement. Circumstantial evidence of two types supports this assertion for *Nannospalax*. First, the width of hybrid zones is negatively correlated with the degree of reproductive isolation, which in turn is positively correlated with the time since divergence. That is, more recently evolved taxa are less isolated and exhibit broader hybrid zones, with a greater range of hybrids produced, than do more distantly related ones (Nevo and Bar-El 1976; Nevo 1989). Second, the degree of assortative mating increases with the age of the taxa hybridizing, suggesting the progressive development of pre-mating ethological isolation (Beiles, Heth, and Nevo 1984).

In pocket gophers, however, Patton (1993) has argued that reinforcement is unlikely to occur, regardless of the degree of fitness deficit suffered by hybrid individuals. In the case of *T. bottae* and *T. umbrinus*, male hybrids are sterile and females exhibit an approximately 50% reduction in fertility, yet hybridization has probably persisted for centuries, and the hybrid zone itself has remained extremely stable over at least 30 years (Patton 1973; M. Ruedi, M. F. Smith, and J. L. Patton, unpublished data). Moreover, because pocket gophers have only a short annual breeding season (reviewed by

Busch et al., chap. 5, this volume), the exclusive territories of individuals limit the extent of panmixia even within a local population. Thus, the total number of potentially interbreeding individuals within hybrid zones must be quite limited. This combination of features suggests that selection for reproductive character displacement is unlikely ever to occur. An individual simply has too few potential mates, and can choose only among those with nearby territories, even if all are of the "wrong" species.

Whether reinforcement occurs or not, however, is of less importance than is the apparent length of time required for its evolution. Again, Nevo (1989 and elsewhere) stressed the gradual progression required in *Nannospalax*, as none of the chromosomal species are completely isolated reproductively, yet their divergences extend over several hundreds of thousands of years. The retention of hybridization ability, therefore, is long-standing, which increases its potential importance as a means of subsequent evolution in the interacting taxa.

Reticulation: Evolutionary Opportunism

Widely appreciated in plant genetic systems for decades, the importance of hybridization as a mechanism of evolution has been only recently recognized in animals (Harrison 1993). Its importance in subterranean rodents may be especially great. We would argue that hybridizing gene pools are unlikely to merge, even with extended time and even when hybrids exhibit total fertility. Hence, hybridization per se is unlikely to retard diversification in these animals. The narrowness of most hybrid zones, the typically fragmented nature of populations within and beyond those zones, and the low densities of the interacting taxa probably preclude genetic fusion, unless the introgressing gene is under very strong directional selection (see Ruedi, Smith, and Patton 1997). However, hybridization can be extremely important as an evolutionary force in other ways, even when hybrids exhibit reduced fertility. It can serve to introduce genes from one taxon into the other by introgression, and it can be a means to generate novel alleles, by intragenic recombination or other mechanisms (Golding and Strobeck 1983). Some introgression of alleles from one parent population into the other characterizes all subterranean rodent hybrid zones with which we are familiar, despite evidence for lowered fitness of hybrid individuals. In several particularly well studied cases in pocket gophers, gene exchange may be decidedly asymmetrical (*Thomomys:* Patton 1993; Patton and Smith 1993; *Geomys:* Baker et al. 1989; Bradley, Davis, and Baker 1991; Chesser et al. 1996). Nevertheless, evidence for introgression can be found well beyond the limits of the hybrid zone (Patton and Smith 1994; Ruedi, Smith, and Patton 1997).

Despite the lowered fitness of hybrids, the length of time over which hybridization potential persists offers sufficient opportunity for the effects of even limited introgression to be extensive over the species' ranges. It also means that the position of hybrid zones may well have changed over time, leaving a record that needs to be understood. Thus, as important as hybridization might be as a means for inputting genetic variability into subterranean rodent populations and thus influencing their subsequent evolutionary trajectories, these reticulation events can compromise our ability to use phylogenetic analyses to understand the pattern of sequential diversification.

Gene Trees and Species Histories

Two aspects of the biology of subterranean rodents impinge on our ability to hypothesize accurately the branching sequence of these species. The first is lineage sorting of ancestral polymorphisms; the second is the problem of horizontal gene transfer through reticulation events either prior to or subsequent to passage of the species boundary. An accurately resolved gene tree, of course, may not be congruent with that of the species under consideration because of lineage sorting of ancestral polymorphisms (Pamilo and Nei 1988). While mtDNA provides a rich source of data for estimating relationships among recently evolved populations and species because of its rapid rate of evolution, it provides only a single gene tree because of its clonal inheritance, and thus is especially prone to problems of lineage sorting. Nonetheless, Moore (1995) has argued that if the time between successive bifurcation events is sufficiently long so that coalescence of alleles, or haplotypes, occurs, then an mtDNA gene tree will be an accurate reflection of the species tree. This is because the probability of coalescence increases as the effective population size decreases, which for mtDNA is one-fourth that for a nuclear autosomal gene. One might expect, therefore, that descendant populations, or species, of subterranean rodents will pass through the sequence of polyphyletic to paraphyletic to eventual reciprocal monophyletic conditions (Neigel and Avise 1986) reasonably quickly.

Subterranean rodents, however, are likely to experience short internodal intervals between successive splitting events because of their propensity for population disjunction, so that coalescence will not necessarily occur between these bifurcations. Moreover, even if internodal intervals are sufficiently long and descendant populations have all reached reciprocal monophyly, the phyletic relationships depicted in the gene tree may still not be those of the species branching pattern. This situation is illustrated in figure 8.2. Here, a polymorphism ancestral to all three species was transmitted through the ancestral stem of species Sp1 and Sp2 and persists in those two

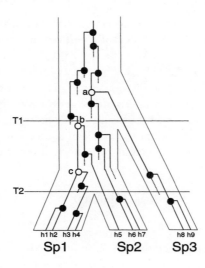

FIGURE 8.2. One possible gene tree-species tree relationship (see Moore 1995). The outer framework represents the species tree. T1 and T2 are times at which species splitting events occur. The single-trace branches within the species represent branches of the gene tree. Open circles denote ancestral haplotypes that are shared by derivative species; a polymorphism ancestral to all three species was transmitted through the ancestral stem of species Sp1 and Sp2 and persists in both. h1, h2, ... are individual haplotypes (= alleles). (Adapted from Moore 1995: figure 1A)

species (as haplotypes 1–4 in Sp1 and haplotype 5 in Sp2). If haplotypes 6 and 7 go extinct in Sp2 with time, then the gene tree will accurately reflect the organismal tree. However, if the reverse happens and haplotype 5 goes extinct in Sp2, it will become reciprocally monophyletic for the haplotype 6–7 lineage. However, Sp2 will, from that point on, display a gene tree relationship with Sp3, yet be the sister species of Sp1. In other words, the gene tree will become "fixed" and will never conform to the tree of the organisms that contain the gene.

A perfect example of this type of situation is found in pocket gophers of the *Thomomys bottae* complex in southeastern New Mexico, where two distinct races that differ in their morphology, karyotypes, and nuclear genes meet and hybridize (Patton et al. 1979). Allozyme nuclear genes give a picture of geographic relationships concordant with morphology (and thus the traditional subspecies boundaries) and chromosomes, but one quite different from that based on mitochondrial DNA (fig. 8.3). Both trees are a true reflection of the history of the divergence of these two taxa, with the mtDNA tree influenced most by a more rapid coalescent "fixation" of haplotypes (Ruedi, Smith, and Patton 1997). This study also illustrates how careful phylogenetic analyses can differentiate between selective "sweeps" of mtDNA

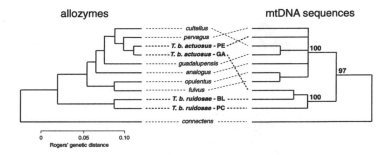

FIGURE 8.3. Geographic and phyletic relationships among populations of the pocket gopher *Thomomys bottae* in southeastern New Mexico. The cladogram on the left illustrates relationships based on nuclear markers (allozymes); that on the right is based on mitochondrial DNA. (For further details, see Ruedi, Smith, and Patton 1997.)

and historical range shifts or differential lineage sorting as mechanisms underlying discordance between mitochondrial and nuclear gene trees.

The second way in which gene trees and species trees can be confounded is through horizontal gene transfer. For example, Patton and Smith (1994) showed how hybridization between the pocket gophers *Thomomys bottae* and *T. townsendii* in the western Great Basin of the United States has affected the genetic matrix of local geographic populations, thereby complicating our understanding of hierarchical relationships among them. This horizontal gene transfer has involved mtDNA markers, but not those of the nuclear genome, as might be expected by their different times to coalescence (Moore 1995) and rates of introgression (Barton and Hewitt 1989).

Figure 8.4 illustrates the two most likely geographic scenarios for the formation of new subterranean rodent species: vicariant allopatry and peripheral isolation, due to either range expansion or contraction. Because, as we argue, subterranean species often exist naturally as a set of subdivided populations, congruence of gene and organismal trees is possible only if a splitting event is exactly coincidental with any prior phylogeographic structure. Given the probable complex combination of hierarchical and reticulate relationships, such congruence seems quite unlikely even following vicariant processes, and will always be lacking if speciation is peripatric. It should not be surprising, therefore, that paraphyletic, and even polyphyletic, relationships are commonly encountered in phylogenetic assessments of a wide range of subterranean rodents based on either nuclear or mitochondrial markers, or both. We have not tried to document this phenomenon exhaustively, but provide a range of examples in table 8.2. Since a gene tree may or may not represent hierarchical organismal relationships, patterns of divergence must be assessed from a diverse set of data and, importantly, with

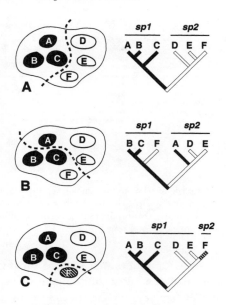

FIGURE 8.4. Relationship between phylogeographic structure and derivative species formed by vicariant allopatric (A and B) and peripatric geographic (C) processes. Species formation is concordant with prior phylogeographic structure in A, but not in either B or C (see R. G. Harrison 1991).

the recognition that clearly delineated hierarchical relationships may not even be a null expectation of organismal history. These problems, in turn, highlight the difficulty in defining species of subterranean rodents, regardless of the concept one wishes to use (see Patton and Smith 1994).

The Nature of Species Boundaries in Subterranean Rodents

Any discussion of speciation must necessarily involve the more fundamental question of just what a species is, a controversial issue with widely differing conceptual, philosophical, and practical implications (De Queiroz 1992; Cracraft 1989). This issue is especially complicated in subterranean rodents, as aspects of their biology and distribution complicate the application of any of the variety of species concepts in current vogue (reviewed by Baum and Shaw 1995). The "fixation" of gene trees that do not reflect organismal splitting events, the retention of hybridization ability between evolutionary independent lineages, and the almost universal allopatric nature of diagnosable entities means that subterranean rodent taxa, even if clearly bounded by

TABLE 8.2 EXAMPLES OF PARAPHYLETIC OR POLYPHYLETIC RELATIONSHIPS AMONG GEOGRAPHICALLY DIAGNOSABLE POPULATIONS OR SPECIES OF SUBTERRANEAN RODENTS, ESTIMATED BY EITHER NUCLEAR OR MITOCHONDRIAL GENETIC MARKERS

Taxon	Marker system	References
Geomyidae		
Thomomys bottae ruidosae and *bottae actuosus*	Allozymes, mtDNA	Ruedi, Smith, and Patton 1997
Thomomys bottae and *T. townsendii*	Allozymes, mtDNA	Patton and Smith 1989, 1994; Smith 1998
Geomys bursarius	Allozymes	Penney and Zimmerman 1976
Ctenomyidae		
Ctenomys mendocinus	Allozymes	Sage et al. 1986
Ctenomys maulinus	Allozymes	Gallardo and Köhler 1992
Ctenomys boliviensis	Allozymes	Cook and Yates 1994
Bathyergidae		
Cryptomys hottentotus	mtDNA	Honeycutt et al. 1987
Cryptomys hottentotus	Allozymes	Nevo et al. 1987; Filippucci et al. 1994
Spalacinae		
Nannospalax ehrenbergi superspecies	mtDNA	Nevo et al. 1993
Nannospalax leucodon superspecies	Allozymes	Nevo et al. 1995

one criterion, are unlikely to be similarly bounded by others. This problem is the direct and expected consequence of the general matrix of structured populations that characterizes subterranean rodents (see discussion in Patton and Smith 1994). There is, therefore, no "answer" to the problem of species identification in these organisms. The protracted debate between proponents of the "biological species concept" and the "phylogenetic species concept," for example, has contributed little of substance to our understanding of species boundaries in subterranean rodents, since both concepts, in their strictest senses, can be applied only with difficulty in these organisms (Patton and Smith 1994). Moreover, as emphasized by Avise and Wollenberg (1997), historical descent and reproductive continuity are related aspects of phylogeny that together illuminate biotic discontinuity.

Researchers of subterranean rodents have traditionally defined species in nature by a set of understandable operational criteria (for example, Patton and Smith 1989). But such operational criteria should be imbedded within a conceptual framework, and the one that we prefer has been elegantly expressed recently by Michael Ghiselin (1997: 99), who considers species to be populations "within which there is, but between which there is not, sufficient cohesive capacity to preclude indefinite divergence." Given this view, there are likely to be many more species in each of the extant subterranean rodent lineages (Cook, Lessa, and Hadly, chap. 9, this volume) than current taxonomy suggests. Certainly, as we develop and apply a consistent understanding of species boundaries in subterranean rodents, our ability to address critical macroevolutionary questions will improve.

The Future

All researchers have their biases, as each of us views the world through our own experiences. Ours come from a combined 45 years of working on, and thinking about, pocket gophers. As a result, we are prone to think of all subterranean rodents as if they were geomyids. Clearly they are not. Consequently, if the views we express above serve any purpose at all, it will be to catalyze deepening investigations into other, less well known groups, if only to show that our ideas on evolutionary constraints and opportunities are wrong! We need to know whether subterranean rodents are generally as subdivided as the limited current data suggest, which type of population structure is apparent, and the consequences of that structure with regard to the potential and process of evolutionary diversification.

Most studies of subterranean rodents employing genetic tools have focused on the systematic topics of species status and phyletic relationships, rather than on population-level patterns and processes. Until recently, these types of studies were largely or exclusively limited to the North American pocket gophers (and even then, primarily to the genus *Thomomys*) and Middle Eastern *Nannospalax*, with little attention paid to South American ctenomyids or African bathyergids. As important as these systematic topics are, we need to extend beyond the simple questions of genetic distance and phyletic relationship. We need to think about local populations as well as hierarchical relationships across geography.

We envision several areas of novel and exciting research on the evolutionary genetics of subterranean rodents, areas that promise to add greatly to our understanding of these animals as well as to contribute substantially to the empirical and theoretical literature in evolutionary biology. The first of these obviously includes continued empirical studies of all subterranean rodents within local populations and across geography, as we noted above. Intelligent sampling of natural populations and imaginative use of different marker systems (both molecular and biotic) can provide important information about the local and global genetic structuring of subterranean rodents, such as indirect estimates of dispersal distances. A recently published study that combined the use of nuclear genetic markers and chewing lice in a pocket gopher hybrid zone in New Mexico is an excellent example of such work (Hafner et al. 1998).

Subterranean rodents are also excellent model systems for investigating certain issues that are important in the evolutionary and conservation biology of natural populations of any organism, such as the effects of local extinction and recolonization on genetic structure. Consequently, a second set of studies should employ experimental manipulations of natural populations to address these general questions. Certainly, the ability to enumerate entire local populations, readily accomplished in many subterranean rodents, provides opportunities not present in most studies of surface-dwelling mammals.

A third area of research would involve the use of individual-based models (IBMs), such as that developed by Steinberg and Jordan (1997) in their study of the genetics of "virtual" pocket gopher populations. This type of approach, based on empirically determined population parameters, offers a powerful means to understand the interplay between demography, including mating system, and geographic genetic structure. For example, IBMs can be used effectively to develop expectations for empirical studies of extinction-recolonization dynamics by determining differences between complete and partial extirpations, followed by recolonization by different numbers of individuals, of different degrees of relatedness, and so forth. IBMs can also be used to assist in the optimization of experimental design by determining whether one should sample more populations, more individuals, more loci, or more variable loci to maximize the power of statistical tests. Finally, IBMs based on empirical data can be used in concert with other analytical approaches to explore the applicability of theory to natural populations.

The last research agenda we mention, though it is certainly not the only remaining one, would utilize the large series of specimens of subterranean rodents available in museum collections. For example, DNA amplified from museum specimens, including fossils, permits one to explore temporal as well as spatial shifts in patterns of genetic variation, thereby adding a direct measure of history to population differentiation (e.g., Villablanca 1994). This ability to "look back in time" thus opens a number of new doors for evolutionary investigations, such as the opportunity to evaluate the effect of gene flow on patterns of phenotypic variation. In one such study, Hadly et al. (1998) used mtDNA sequences extracted and amplified from fossil teeth of *Thomomys talpoides* to examine the interplay between climatic change and shifts in mean body size in a local population over the past 5,000 years. A critical issue in this study was whether the morphological variation found was intrinsic to the population or resulted from migration from elsewhere.

Hadly et al. (1998) examined the genetic characteristics of the temporal series in which phenotypic changes had been documented as well as those of nearby populations. They thus were able to show that the morphological shifts took place within a temporally stable local population, and were not influenced by gene flow from afar. The large collections of subterranean rodents available in museums, collections made because of earlier interests in subspeciation (e.g., Grinnell 1927; Patton and Smith 1990), offer similar opportunities to examine simultaneously both spatial and temporal patterns of variation and differentiation. For example, the pocket gopher collection housed at the University of California at Berkeley's Museum of Vertebrate Zoology includes over 18,000 individuals, representing hundreds of population samples of at least 25 individuals each. Toward this potential, Steinberg (1999) has provided useful descriptions of methods of amplifying microsatellite loci from both field and museum specimens, developed for *Thomomys mazama*.

The future holds exciting promise for studies on subterranean rodents. Our individual imaginations, and our willingness to get into the field and the laboratory, and sit before the computer, present the only limitations.

Acknowledgments

This chapter was delivered as two separate presentations at the symposium on subterranean rodents held at the Seventh International Theriological Congress. We thank Guy Cameron and Eileen Lacey for the opportunity to participate, and especially Eileen for seeing this project to fruition. J. L. P. has benefited greatly over the years from a large number of individuals who directed his thoughts concerning subterranean and other rodents, most notably Oliver Pearson, Eibi Nevo, David Wake, Peg Smith, Mark and John Hafner, Don Straney, Joanne Daly, Elizabeth Hadly, Manuel Ruedi, Ellie Steinberg, and Eileen Lacey. His research on pocket gophers has been generously supported by grants from the National Science Foundation. E. K. S.'s interest in learning all that can be known about subterranean and other rodents speaks to the immense value of inspirational teaching and mentorship, most notably that of Jim Patton, Bill Lidicker, David Ribble, and Francis Villablanca. Her research has benefited from collaborations with Jim Patton, Peg Smith, and Chris Jordan and has been supported by funding from The Nature Conservancy, the Northwestern Society for Natural History, and the University of Washington.

Literature Cited

Avise, J. C. 1976. Genetic differentiation during speciation. In *Molecular evolution*, edited by F. J. Ayala, 106–22. Sunderland, MA: Sinauer Associates.

Avise, J. C., and K. Wollenberg. 1997. Phylogenetics and the origin of species. *Proceedings of the National Academy of Sciences, USA* 94:7748–55.

Baker, R. J., S. K. Davis, R. D. Bradley, M. J. Hamilton, and R. A. Van Den Bussche. 1989. Ribosomal-DNA, mitochondrial DNA, chromosomal, and allozymic studies on a contact zone in the pocket gopher, *Geomys*. *Evolution* 43:63–75.

Barton, N. H., and K. S. Gale. 1993. Genetic analysis of hybrid zones. In *Hybrid zones and the evolutionary process*, edited by R. G. Harrison, 13–45. New York: Oxford University Press.

Barton, N. H., and G. M. Hewitt. 1985. Analysis of hybrid zones. *Annual Review of Ecology and Systematics* 16:113–48.

———. 1989. Adaptation, speciation, and hybrid zones. *Nature* 341:497–503.

Baum, D. A., and K. L. Shaw. 1995. Genealogical perspectives on the species problem. In *Experimental and molecular approaches to plant biosystematics*, edited by P. C. Hoch and A. G. Stephenson, 289–303. Monographs in Systematic Botany from the Missouri Botanical Garden, vol. 53. St. Louis: Missouri Botanical Garden.

Beiles, A., G. Heth, and E. Nevo. 1984. Origin and evolution of assortative mating in actively speciating mole rats. *Theoretical Population Biology* 26:265–70.

Blouin, M. S., C. A Yowell, C. H. Courtney, and J. B. Dame. 1995. Host movement and the genetic structure of populations of parasitic nematodes. *Genetics* 141:1007–14.

Bradley, R. D., S. K. Davis, and R. J. Baker. 1991. Genetic control of pre-mating isolating behavior: Kaneshiro's hypothesis and asymmetrical sexual selection in pocket gophers. *Heredity* 82:192–96.

Bush, G. L. 1975. Modes of animal speciation. *Annual Review of Ecology and Systematics* 6:339–64.

———. 1994. Sympatric speciation in animals: New wine in old bottles. *Trends in Ecology and Evolution* 9:285–88.

Butlin, R. K. Reinforcement: An idea evolving. *Trends in Ecology and Evolution* 10:432–34.

Carraway, L. N., and P. K. Kennedy. 1993. Genetic variation in *Thomomys bulbivorus*, an endemic to the Willamette Valley, Oregon. *Journal of Mammalogy* 74: 952–62.

Carson, H. L. 1987. The genetic system, the deme, and the origin of species. *Annual Review of Genetics* 21:405–23.

Carson, H. L., and A. R. Templeton. 1984. Genetic revolutions in relation to speciation: The founding of new populations. *Annual Review of Ecology and Systematics* 15:97–131.

Chesser, R. K., and R. J. Baker. 1986. On factors affecting the fixation of chromosomal rearrangements and neutral genes: Computer simulations. *Evolution* 40:625–32.

Chesser, R. K., R. D. Bradley, R. A. Van Den Bussche, M. J. Hamilton, and R. J. Baker. 1996. Maintenance of a narrow hybrid zone in *Geomys*: Results from contiguous clustering analysis. In *Contributions in mammalogy: A memorial volume honoring Dr. J. Knox Jones, Jr,* edited by H. H. Genoways and R. J. Baker, 35–45. Lubbock: Museum of Texas Tech University. i + 315 pp.

Cook, J. A., and T. L. Yates. 1994. Systematic relationships of the Bolivian tuco-tucos, genus *Ctenomys* (Rodentia: Ctenomyidae). *Journal of Mammalogy* 75:583–99.

Coyne, J. A., N. H. Barton, and M. Turelli. 1997. Perspective: A critique of Sewall Wright's shifting balance theory of evolution. *Evolution* 51:643–71.

Cracraft, J. 1989. Speciation and its ontology: The empirical consequences of alternative species concepts for understanding patterns and processes of differentiation. In *Speciation and its consequences,* edited by D. Otte and J. A. Endler, 28–59. Sunderland, MA: Sinauer Associates.

Daly, J. C., and J. L. Patton. 1990. Dispersal, gene flow, and allelic diversity between local populations of *Thomomys bottae* pocket gophers in the coastal ranges of California. *Evolution* 44:1283–94.

D'Elía, G., E. P. Lessa, and J. A. Cook. 1998. Geographic structure, gene flow, and maintenance of melanism in *Ctenomys rionegrensis* (Rodentia: Octodontidae). *Zeitschrift für Säugetierkunde* 63:285–96.

De Queiroz, K. 1992. Phylogenetic definitions and taxonomic philosophy. *Biology and Philosophy* 7:295–313.

Doadrio, I., A. Perdices, and A. Machordom. 1996. Allozymic variation of the endangered killifish *Aphanius iberus* and its application to conservation. *Environmental Biology of Fishes* 45: 259–71.

Endler, J. A. 1977. *Geographic variation, speciation, and clines.* Princeton, NJ: Princeton University Press.

Faulkes, C. G., D. H. Abbott, H. P. O'Brien, L. Lau, M. R. Roy, R. K. Wayne, and M. W. Brubord. 1997. Micro- and macrogeographical genetic structure of colonies of naked mole-rats *Heterocephalus glaber. Molecular Ecology* 6:615–28.

Filippucci, M. G., H. Burda, E. Nevo, and J. Kocka. 1994. Allozyme divergence and systematics of common mole-rats (*Cryptomys*, Bathyergidae, Rodentia) from Zambia. *Zeitschrift für Säugetierkunde* 59:42–51.

Fisher, R. A. 1958. *The Genetical Theory of Natural Selection.* 2d edition. New York: Dover Publications.

Gallardo, M. H., and N. Köhler. 1992. Genetic divergence in *Ctenomys* (Rodentia, Ctenomyidae) from the Andes of Chile. *Journal of Mammalogy* 73:99–105.

Gallardo, M. H., N. Köhler, and C. Araneda. 1995. Bottleneck effects in local populations of fossorial *Ctenomys* (Rodentia, Ctenomyidae) affected by vulcanism. *Heredity* 74: 638–46.

Ghiselin, M. 1997. *Metaphysics and the origin of species.* Albany: State University of New York Press. 377 pp.

Gilpin, M. 1991. The genetic effective size of a metapopulation. *Biological Journal of the Linnean Society* 42:165–75.

Golding, G. B., and C. Strobeck. 1983. Increased number of alleles found in hybrid populations due to intragenic recombination. *Evolution* 37:17–29.

Green, D. M., T. F. Sharbel, J. Kearsley, and H. Kaiser. 1996. Postglacial range fluctuation, genetic subdivision and speciation in the western North American spotted frog complex, *Rana pretiosa. Evolution* 50:374–90.

Grinnell, J. 1927. Geography and evolution in the pocket gophers of California. *Smithsonian Institution Annual Report* 2894:335–43.

Hadly, E. A., M. H. Kohn, J. A. Leonard, and R. K. Wayne. 1998. A genetic record of population isolation in pocket gophers during Holocene climatic change. *Proceedings of the National Academy of Sciences, USA* 95:6893–96.

Hafner, J. C., D. J. Hafner, J. L. Patton, and M. F. Smith. 1983. Contact zones and the genetics of differentiation in the pocket gopher *Thomomys bottae* (Rodentia: Geomyidae). *Systematic Zoology* 32:1–20.

Hafner, M. S., J. W. Demastes, D. J. Hafner, T. A. Spradling, P. D. Sudman, and S. A. Nadler. 1998. Age and movement of a hybrid zone: Implications for dispersal distance in pocket gophers and their chewing lice. *Evolution* 52:278–82.

Hafner, M. S., J. C. Hafner, J. L. Patton, and M. F. Smith. 1987. Macrogeographic patterns of genetic differentiation in the pocket gopher *Thomomys umbrinus. Systematic Zoology* 36:18–34.

Hanski, I. 1996. Metapopulation ecology. In *Population dynamics in ecological space and time,* edited by O. E. Rhodes, Jr., R. K. Chesser, and M. H. Smith, 13–43. Chicago: University of Chicago Press.

———. 1998. Metapopulation dynamics. *Nature* 395:41–49.
Hanski, I., and M. Gilpin, eds. 1996. *Metapopulation biology: Ecology, genetics and evolution.* London: Academic Press.
Harrison, R. G. 1991. Molecular changes at speciation. *Annual Review of Ecology and Systematics* 22:281–308.
———. 1993. Hybrids and hybrid zones: Historical perspective. In *Hybrid zones and the evolutionary process,* edited by R. G. Harrison, 3–12. New York: Oxford University Press.
Harrison, S. 1991. Local extinction in a metapopulation context: An empirical evaluation. *Biological Journal of the Linnean Society* 42:73–88.
Hartl, D. L., and A. G. Clark. 1997. *Principles of population genetics.* 3d edition. Sunderland, MA: Sinauer Associates.
Heaney, L. R., and R. M. Timm. 1985. Morphology, genetics, and ecology of pocket gophers (genus *Geomys*) in a narrow hybrid zone. *Biological Journal of the Linnean Society* 23:301–17.
Hedrick, P. W. 1981. The establishment of chromosomal variants. *Evolution* 35:322–32.
Honeycutt, R. L., S. V. Edwards, K. Nelson, and E. Nevo. 1987. Mitochondrial DNA variation and the phylogeny of African mole rats (Rodentia: Bathyergidae). *Systematic Zoology* 36:280–92.
Honeycutt, R. L., K. Nelson, D. A. Schlitter, and P. W. Sherman. 1991. Genetic variation within and among populations of the naked mole-rat: Evidence from nuclear and mitochondrial genomes. In *The biology of the naked mole-rat,* edited by P. W. Sherman, J. U. M. Jarvis, and R. D. Alexander, 195–208. Princeton, NJ: Princeton University Press.
Howard, D. J. 1993. Reinforcement: Origin, dynamics, and fate of an evolutionary hypothesis. In *Hybrid zones and the evolutionary process,* edited by R. G. Harrison, 46–69. New York: Oxford University Press.
Hudson, R. R. 1990. Gene genealogies and the coalescent process. In *Oxford surveys in evolutionary biology,* vol. 7, edited by D. J. Futuyma and J. Antonovics, 1–44. Oxford: Oxford University Press.
Huffaker, C. B. 1958. Experimental studies on predation: Dispersion factors and predator prey oscillations. *Hilgardia* 27:343–83.
Ingles, L. G. 1950. Pigmental variations in populations of pocket gophers. *Evolution* 4:353–57.
Johnson, W. E., and R. K. Selander. 1971. Protein variation and systematics in kangaroo rats (genus *Dipodomys*). *Systematic Zoology* 20:377–405.
Lacey, E. A., J. E. Maldonado, J. P. Clabaugh, and M. D. Matocq. 1999. Interspecific variation in microsatellites isolated from tuco-tucos (Rodentia: Ctenomyidae). *Molecular Ecology* 8:1754–56.
Lande, R. 1979. Effective deme sizes during long-term evolution estimated from rates of chromosomal rearrangement. *Evolution* 33:234–51.
Levins, R. 1970. Extinction. In *Some mathematical questions in biology: Second Symposium on Mathematical Biology,* edited by M. Gerstenhaber, 75–107. Providence, RI: American Mathematical Society.
Lidicker, W. Z., Jr., and J. L. Patton. 1987. Patterns of dispersal and genetic structure in populations of small rodents. In *Mammalian dispersal patterns,* edited by B. D. Chepko-Sade and Z. T. Halpin, 144–61. Chicago: University of Chicago Press.
Mayr, E. 1954. Change of genetic environment and evolution. In *Evolution as a process,* edited by J. Huxley, A. C. Hardy, and E. B. Ford, 157–80. London: Allen & Unwin.
———. 1963. *Animal species and evolution.* Cambridge, MA: Harvard University Press.
———. 1982. Processes of animal speciation. In *Mechanisms of speciation,* edited by D. Barigozzi, 1–19. New York: Alan R. Liss.
Maruyama, T., and M. Kimura. 1980. Genetic variability and effective population size when local extinction and recolonization of subpopulations are frequent. *Genetics* 77:6710–14.

Mayamura, T. 1971. Analysis of population structure. II. Two-dimensional stepping stone models of finite length and other geographically structured populations. *Annals of Human Genetics* 35:179–96.

McCauley, D. E. 1991. Genetic consequences of local population extinction and recolonization. *Trends in Ecology and Evolution* 6:5–8.

McCullough, D. R. 1996. *Metapopulations and wildlife conservation.* Washington, DC: Island Press.

Moore, W. S. 1995. Inferring phylogenies from mtDNA variation: Mitochondrial-gene trees versus nuclear-gene trees. *Evolution* 49:718–26.

Nadler, S. A., M. S. Hafner, J. C. Hafner, and D. J. Hafner. 1990. Genetic differentiation among chewing louse populations (Mallophaga: Trichodectidae) in a pocket gopher contact zone (Rodentia: Geomyidae). *Evolution* 44:942–51.

Nagylaki, T. 1976. The decay of genetic variability in geographically structured populations. II. *Theoretical Population Biology* 10:70–82.

Neigel, J. E., and J. C. Avise. 1986. Phylogenetic relationships of mitochondrial DNA under various demographic models of speciation. In *Evolutionary processes and theory*, edited by E. Nevo and S. Karlin, 515–34. New York: Academic Press.

Nevo, E. 1988. Genetic diversity in nature: Patterns and theory. *Evolutionary Biology* 23: 217–47.

———. 1989. Modes of speciation: The nature and role of peripheral isolates in the origin of species. In *Genetics, speciation, and the founder principle*, edited by L. V. Giddings, K. Y. Kaneshiro, and W. W. Anderson, 205–36. New York: Oxford University Press.

———. 1990. Molecular evolutionary genetics of isozymes: Patterns, theory and application. Sixth International Congress on Isozymes, Toyama, Japan, 701–42.

Nevo, E., and H. Bar-El. 1976. Hybridization and speciation in fossorial mole rats. *Evolution* 30:831–40.

Nevo, E., R. Ben-Shlomo, A. Beiles, J. U. M. Jarvis, and G. C. Hickman. 1987. Allozyme differentiation and systematics of the endemic subterranean mole rats of South Africa. *Biochemical Systematics and Ecology* 15:489–502.

Nevo, E., M. G. Filippucci, and A. Beiles. 1990. Genetic diversity and its ecological correlates in nature: Comparison between subterranean, fossorial and aboveground small mammals. In *Evolution of subterranean mammals at the organismal and molecular levels*, edited by E. Nevo and O. A. Reig, 347–66. Progress in Clinical and Biological Research, vol. 335. New York: Wiley-Liss.

———. 1994. Genetic polymorphisms in subterranean mammals (*Spalax ehrenbergi* superspecies) in the Near East revisited: Patterns and theory. *Heredity* 72:465–87.

Nevo, E., M. G. Filippucci, C. Redi, S. Simson, G. Heth, and A. Beiles. 1995. Karyotype and genetic evolution in speciation of subterranean mole rats of the genus *Spalax* in Turkey. *Biological Journal of the Linnean Society* 54:203–29.

Nevo, E., G. Heth, and A. Beiles. 1982. Population structure and evolution in subterranean mole rats. *Evolution* 36:1283–89.

Nevo, E., R. L. Honeycutt, H. Yonekawa, K. Nelson, and N. Hanzawa. 1993. Mitochondrial DNA polymorphisms in subterranean mole-rats of the *Spalax ehrenbergi* superspecies in Israel, and its peripheral isolates. *Molecular Biology and Evolution* 10:590–604.

Nevo, E., Y. J. Kim, C. R. Shaw, and C. S. Thaeler, Jr. 1974. Genetic variation, selection and speciation in *Thomomys talpoides* pocket gophers. *Evolution* 28:1–23.

Nevo, E., V. Kirzhner, A. Beiles, and A. Korol. 1997. Selection versus random drift: Long-term polymorphism persistence in small populations (evidence and modeling). *Philosophical Transactions of the Royal Society of London* B 352:381–89.

Pamilo, P., and M. Nei. 1988. Relationships between gene trees and species trees. *Molecular Biology and Evolution* 5:568–83.

Patton, J. L. 1973. An analysis of natural hybridization between the pocket gophers, *Thomomys bottae* and *Thomomys umbrinus*, in Arizona. *Journal of Mammalogy* 54:561–84.

———. 1990. Geomyid evolution: The historical, selective, and random basis for divergence patterns within and among species. In *Evolution of subterranean mammals at the organismal and molecular levels*, edited by E. Nevo and O. A. Reig, 49–69. Progress in Clinical and Biological Research, vol. 335. New York: Wiley-Liss.

———. 1993. Hybridization and hybrid zones in pocket gophers (Rodentia, Geomyidae). In *Hybrid zones and the evolutionary process*, edited by R. G. Harrison, 290–308. New York: Oxford University Press.

Patton, J. L., and J. H. Feder. 1978. Genetic divergence between populations of the pocket gopher, *Thomomys umbrinus* (Richardson). *Zeitschrift für Säugetierkunde* 43:12–30.

Patton, J. L., and J. H. Feder. 1981. Microspatial genetic heterogeneity in pocket gophers: Non-random breeding and drift. *Evolution* 43:12–30.

Patton, J. L., J. C. Hafner, M. S. Hafner, and M. F. Smith. 1979. Hybrid zones in *Thomomys bottae* pocket gophers: Genetic, phenetic, and ecologic concordance patterns. *Evolution* 33: 860–76.

Patton, J. L., S. W. Sherwood, and S. Y. Yang. 1981. Biochemical systematics of chaetodipine pocket mice. *Journal of Mammalogy* 62:477–92.

Patton, J. L., M. N. F. da Silva, and J. R. Malcolm. 1996. Hierarchical genetic structure and gene flow in three sympatric species of Amazonian rodents. *Molecular Ecology* 5: 229–38.

Patton, J. L., and M. F. Smith. 1989. Population structure and the genetic and morphologic divergence among pocket gopher species (genus *Thomomys*). In *Speciation and its consequences*, edited by D. Otte and J. A. Endler, 284–304. Sunderland, MA: Sinauer Associates.

———. 1990. The evolutionary dynamics of the pocket gopher *Thomomys bottae*, with emphasis on California populations. *University of California Publications in Zoology* 123: 1–161.

———. 1992. MtDNA phylogeny of Andean mice: A test of diversification across ecological gradients. *Evolution* 46:174–83.

———. 1993. Molecular evidence for mating asymmetry and female choice in a pocket gopher *(Thomomys)* hybrid zone. *Molecular Ecology* 2:3–8.

———. 1994. Paraphyly, polyphyly, and the nature of species boundaries in pocket gophers (genus *Thomomys*). *Systematic Biology* 43:11–26.

Patton, J. L., M. F. Smith, R. D. Price, and R. A. Hellenthal. 1984. Genetics of hybridization between the pocket gophers *Thomomys bottae* and *Thomomys townsendii* in northeastern California. *Great Basin Naturalist* 44:431–40.

Patton, J. L., and S. Y. Yang. 1977. Genetic variation in *Thomomys bottae* pocket gophers: Macrogeographic patterns. *Evolution* 31:697–720.

Penney, D. F., and E. G. Zimmerman. 1976. Genic divergence and local population differentiation by random drift in the pocket gopher genus *Geomys*. *Evolution* 30:473–83.

Provine, W. B. 1971. *The origins of theoretical population genetics*. Chicago: University of Chicago Press.

Reig, O. A. 1989. Karyotypic repatterning as one triggering factor in cases of explosive speciation. In *Evolutionary biology of transient unstable populations*, edited by A. Fontdevila, 246–89. Berlin: Springer-Verlag.

Rice, W. R., and E. E. Hostert. 1993. Laboratory experiments on speciation: What have we learned in 40 years? *Evolution* 47:1637–53.

Rockwell, R. F., and G. F. Barrowclough. 1987. Gene flow and the genetic structure of populations. In *Avian genetics*, edited by F. Cooke and P. A. Buckley, 223–55. New York: Academic Press.

Rohlf, F. J., and G. D. Schnell. 1971. An investigation of the isolation-by-distance model. *American Naturalist* 105:295–324.

Ruedi, M., M. F. Smith, and J. L. Patton. 1997. Phylogenetic evidence of mitochondrial DNA introgression among pocket gophers in New Mexico (family Geomyidae). *Molecular Ecology* 6:453–62.

Sage, R. S., J. R. Contreras, V. G. Roig, and J. L. Patton. 1986. Genetic variation in the South American burrowing rodents of the genus *Ctenomys* (Rodentia, Ctenomyidae). *Zeitschrift für Säugetierkunde* 51:158–72.

Savic, I., and E. Nevo. 1990. The Spalacidae: Evolutionary history, speciation, and population biology. In *Evolution of subterranean mammals at the organismal and molecular levels*, edited by E. Nevo and O. A. Reig, 129–43. Progress in Clinical and Biological Research, vol. 335. New York: Wiley-Liss.

Selander, R. K., and W. E. Johnson. 1973. Genetic variation among vertebrate species. *Annual Review of Ecology and Systematics* 4:75–91.

Slatkin, M. 1977. Gene flow and genetic drift in a species subject to frequent local extinctions. *Theoretical Population Biology* 12:253–62.

———. 1985. Gene flow in natural populations. *Annual Review of Ecology and Systematics* 16:393–430.

———. 1993. Isolation by distance in equilibrium and non-equilibrium populations. *Evolution* 47:264–79.

Smith, M. F. 1998. Phylogenetic relationships and geographic structure in pocket gophers in the genus *Thomomys*. *Molecular Phylogenetics and Evolution* 9:1–14.

Steinberg, E. K. 1999. Characterization of polymorphic microsatellites from current and historic populations of North American pocket gophers (genus *Thomomys*). *Molecular Ecology* 8(6):1075–76.

Steinberg, E. K. and C. E. Jordan. 1997. Using molecular genetics to learn about the ecology of threatened species: The allure and illusion of measuring genetic structure in natural populations. In *Conservation for the coming decade*, edited by P. F. Feidler and P. Kareiva, 438–60. New York: Chapman and Hall.

Templeton, A. R. 1980a. Modes of speciation and inferences based on genetic distances. *Evolution* 34:719–29.

———. 1980b. The theory of speciation via the founder principle. *Genetics* 94:1011–38.

———. 1981. Mechanisms of speciation—a population genetic approach. *Annual Review of Ecology and Systematics* 12:23–48.

Villablanca, F. X. 1994. Spatial and temporal aspects of populations revealed by mitochondrial DNA. In *Ancient DNA*, edited by B. Herrmann and S. Hummel, 31–58. New York: Springer-Verlag.

Wade, M. J. 1982. Group selection: Migration and the differentiation of small populations. *Evolution* 36:949–61.

Wade, M. J., and D. E. McCauley. 1980. Group selection: The phenotypic and genotypic differentiation of small populations. *Evolution* 34:799–812.

Wallace, A. R. 1889. *Darwinism: An exposition of the theory of natural selection with some of its applications*. London: Macmillan.

Walsh, J. B. 1982. Rate of accumulation of reproductive isolation by chromosomal rearrangements. *American Naturalist* 120:510–32.

Whitlock, M. C. 1992. Temporal fluctuations in demographic parameters and the genetic variance among populations. *Evolution* 46:608–15.

Whitlock, M. C. and D. E. McCauley. 1990. Some population genetic consequences of colony formation and extinction: Genetic correlations within founding groups. *Evolution* 44:1717–24.

Wright, S. 1940. Breeding structure of populations in relation to speciation. *American Naturalist* 74:232–48.

———. 1943. Isolation by distance. *Genetics* 28:114–38.
———. 1951. The genetic structure of populations. *Annals of Eugenics* 15:323–54.
———. 1978. *Evolution and the genetics of populations.* Vol. 4. Chicago: University of Chicago Press.
———. 1980. Genic and organismic selection. *Evolution* 34:825–43.
Zimmerman, E. G., and N. A. Gayden. 1981. Analysis of genic heterogeneity among local populations of the pocket gopher, *Geomys bursarius.* In *Mammalian population genetics,* edited by M. H. Smith and J. Joule, 272–87. Athens: University of Georgia Press.

CHAPTER NINE

Paleontology, Phylogenetic Patterns, and Macroevolutionary Processes in Subterranean Rodents

Joseph A. Cook, Enrique P. Lessa, and Elizabeth A. Hadly

Discussions of speciation processes have been frequently illustrated by the independent radiations of several subterranean rodents (e.g., Bush 1975; Givnish 1997). White's (1978: 202) classic work on chromosomal change and speciation highlighted the cases of *Nannospalax, Thomomys,* and *Ctenomys.* He concluded that species of *Ctenomys* were "undoubtedly the result of fairly recent explosive speciation." Similarly, Nevo (1979: 270) asserted that, among subterranean species, rodent "speciation underground was prolific." Reig (1989) explored processes that might contribute to high diversification rates in mammals and, again, focused on subterranean rodents. If, indeed, rapid diversification is a common theme across subterranean rodent taxa, then we need to determine whether common processes are driving speciation. Are these purported high rates of diversification due directly to the evolution of a set of innovative adaptations (e.g., morphological, physiological) that allowed the invasion of a new niche, or are they rather due to the unusual population structure (e.g., presumed low migration: Busch et al., chap. 5; Steinberg and Patton, chap. 8, this volume) associated with a life underground? Various research initiatives have focused on these issues, but few have used either the fossil record or an explicit phylogenetic framework to test them.

Key innovations allow the invasion of new habitats. These novelties are thought to spur adaptive radiations (Mayr 1963) by leading to major evolutionary shifts. Nevo (1979) noted that the "evolutionary convergence and worldwide recurrent radiations of unrelated mammals into the fossorial ecological zone seem to be related causally to the evolution of open country biota in the Cenozoic." That statement, however, does not directly tie the key innovations related to fossoriality to increased speciation. Adaptive shifts should be specified a priori when examining a purported key innovation across a number of independently derived taxa (Mitter, Farrell, and Wiegmann 1988). Further, the key innovation concept requires that arguments for a relationship between a new feature and ensuing diversification go beyond correlation to actually demonstrate evidence for causation.

One way to test the key innovation hypothesis would be to ask whether suspected adaptive shifts are repeatedly associated with accelerated diversification across independent groups (e.g., Sanderson and Donoghue 1994; but see Barrett and Graham 1997; Givnish 1997). The evolution of fossoriality in several rodent clades provides an excellent opportunity to test whether the subterranean lifestyle led repeatedly to greater diversification (i.e., greater species or generic richness). Previous assertions of increased diversity in subterranean taxa were based primarily on comparisons of standing taxonomic diversity, with all of its inherent problems (e.g., artifacts of lumpers and splitters). Indeed, the idea of prolific diversification is not without its critics. Simpson (1961: 173) mused that "*Thomomys umbrinus* has 213 currently recognized and named 'subspecies' in southwestern United States and northern Mexico, and those that enjoy that game may well go on until every little colony of those gophers sports its own Linnaean name." Subsequent reviews based on multiple kinds of data reduced the number of California subspecies of this complex from 66 to 15 (Patton and Smith 1990).

A well-supported phylogeny and a clear fossil record of species-level evolution would provide the information necessary to identify changes in diversification rate and to document when and where speciation and extinction events occurred. Unfortunately, both of these pieces of information are incomplete for our subterranean clades. In this chapter, we begin to explore issues related to common macroevolutionary patterns across these subterranean clades. Traditionally the fossil record has provided insight into these patterns; however, molecular phylogenies are increasingly being used to investigate macroevolution (Purvis 1996; Mooers and Heard 1997; Givnish and Sytsma 1997). We provide a comparative overview of taxonomic diversity, phylogenies, and the fossil record for these clades. We explore and compare patterns of origination, duration, and generic diversity in these widely dispersed, independent clades. We then examine the evidence for increased diversification rates in these taxa.

Statistical tests have been used to examine the relationship between purported key innovations and increased diversification rates. These tests have been applied to examine the evolution of sexual selection in passerine birds

(Barraclough, Harvey, and Nee 1995), phytophagy in insects (Mitter, Farrell, and Wiegmann 1988), and floral nectar spurs in flowering plants (Hodges and Arnold 1995). The tests vary in the kinds of information they require. Some need only a phylogeny, while other, more powerful tests require an estimation of the timing of diversification. The tests also differ in their ability to discriminate between alternative models (statistical power). We outline a few of the primary macroevolutionary tests, focus on those clades for which sufficient information is available, and then compare the molecular approach to patterns evident in the fossil record. Finally, we note a few of the exciting comparative studies that would benefit from the availability of a solid fossil record and phylogeny for the subterranean clades.

Overview of Extant Taxonomic Diversity and Phylogenies

Complete phylogenetic treatments are not available for any of the eight subterranean clades of rodents, although some have been studied in more detail than others. Here, we briefly outline the major attempts to reconstruct evolutionary history in these clades. Given our interest in including an estimate of at least relative evolutionary rates, we pay particular attention to phylogenies based on molecular data. Lists of all genera and species and complete synonymies are available in Wilson and Reeder 1993. We note, however, that alpha taxonomy is still not settled for several of these groups (e.g., Spalacinae: Savic and Nevo 1990; Ctenomyinae: Anderson 1997). In this vein, the current application of species concepts and definitions of species boundaries (Steinberg and Patton, chap. 8, this volume) to some of these groups is critical to estimates of species richness.

Geomyidae

Systematic treatments of the five genera *(Geomys, Orthogeomys, Pappogeomys, Thomomys, Zygogeomys)* and thirty-five species of geomyids (Patton 1993) have focused primarily on subsets of genera and species. The most extensive studies based on DNA sequences are from the mitochondrial cytochrome oxidase I (Hafner et al. 1994) and cytochrome *b* (Smith 1998) genes. Phylogenies were constructed based on morphology (Russell 1968; Akersten 1973), protein electrophoresis (e.g., Honeycutt and Williams 1982; Hafner 1982, 1991; Patton and Smith 1990), chromosomes (e.g., Thaeler 1980), and DNA (e.g., Hafner et al. 1994; Smith 1998).

Arvicolinae

Two subterranean genera *(Ellobius, Prometheomys)* with six species are known from this subfamily (Musser and Carleton 1993), but there is little evidence that these two genera are sister taxa (but see Repenning, Fejfar, and Heinrich 1990). Only the three species of *Ellobius* were included in a preliminary phylogenetic analysis (based on DNA sequence data from the *SRY* locus); however, their relationships were not resolved (Just et al. 1995). Other arvicoline species are known to be semi-subterranean (e.g., *Hyperacrius*).

Myospalacinae

There is one genus *(Myospalax)* and seven species of zokors (Musser and Carleton 1993). Lawrence's (1991) monograph on the subfamily is the most complete study to date. She constructed a cladogram for fossil and extant species of *Myospalax* based on morphological characters. None of the species has been included in a molecular analysis.

Rhizomyinae

There are three genera *(Cannomys, Rhizomys, Tachyoryctes)* and fifteen species of root rats. A phylogenetic hypothesis based on skeletal and dental characters for both extant and extinct taxa (ten genera and twenty-seven species) was proposed by Flynn (1990).

Spalacinae

The subfamily Spalacinae includes two genera *(Nannospalax, Spalax)* and approximately eight species (Musser and Carleton 1993); however, Savic and Nevo (1990) suggested that there may be more than thirty species, based on karyotypes. Evaluations of the evolutionary relationships of subsets of these taxa were based on morphology (Ognev 1963; Corbet 1978; Carleton and Musser 1984; de Bruijn 1984) and were reviewed by Savic and Nevo (1990).

Bathyergidae

The family Bathyergidae includes five genera *(Bathyergus, Cryptomys, Georychus, Heliophobius, Heterocephalus)* and about twelve species (Woods 1993). Jarvis and Bennett (1990) and Honeycutt et al. (1991) provided overviews of the systematics of the Bathyergidae, although a complete phylogeny for the family has not been attempted. Several investigations focusing on intergeneric and intrageneric relationships were based on morphology (Wood 1985), protein electrophoresis (Nevo et al. 1987; Filippucci et al. 1992), mitochondrial restriction fragment analysis (Honeycutt et al. 1987), and 12S rDNA (Honeycutt et al. 1991; Nedbal, Allard, and Honeycutt 1994) and cytochrome *b* (Faulkes et al. 1997) sequence data.

Ctenomyidae

With about fifty-six species in a single genus *Ctenomys* (Reig et al. 1990), the family Ctenomyidae constitutes the most speciose subterranean group of rodents. Early attempts to reconstruct their evolutionary relationships were hindered by limited sampling of species. Phylogenies based on protein electrophoresis (Roig and Reig 1969; Sage et al. 1986; Cook and Yates 1994) and chromosomes (e.g., Ortells 1995) are difficult to compare because they included relatively little overlap in the species examined. Phylogenies based on the mitochondrial cytochrome *b* gene (Lessa and Cook 1998; Cook and Lessa 1998; D'Elía, Lessa, and Cook 1999) now include twenty-three species.

Octodontidae

The family Octodontidae includes six genera *(Aconaemys, Octodon, Octodontomys, Octomys, Spalacopus, Tympanoctomys)* and nine species (Woods 1993). Of these, only *Spalacopus cyanus* is subterranean. Existing phylogenetic hypotheses are based on morphology of the glans penis (Spotorno 1979; Contreras et al. 1993), protein electrophoresis (Woods 1982), chromosomes (Contreras et al. 1990; Gallardo 1992), and mitochondrial DNA sequences (Lessa and Cook 1998; Cook and Lessa 1998).

Macroevolutionary Patterns Inferred from the Fossil Record

Complete fossil histories are lacking for all groups of subterranean rodents. We collected fossil data from the primary literature and from summaries wherever they were available. We make no attempt to summarize fossil histories for those subterranean rodents without extant taxa (e.g., Anomalomyinae, Tachyoryctoidinae, Paleocastorinae, Mylagaulidae). Timelines (see figs. 9.2–9.8) were constructed using biostratigraphic chronologies (Marshall and Sempere 1993, Cione and Tonni 1995, Pascual, Jaureguizar, and Prado 1996 for South America; Fejfar and Heinrich 1986 and Steininger, Bernor, and Fahlbusch 1990 for Europe; Markova et al. 1995 for Eurasia; Berggren et al. 1985 for North America; and Flynn 1982 for Asia). In the case of ambiguous age assignments (i.e., late Miocene: 12 to 6 Ma), the earliest age was assigned (12 Ma). "Present" is used to designate extant, non-fossil taxa; "Holocene" refers to a fossil representative from the past 10,000 years.

Because of problems in sampling and taxonomic classification, fossil data rarely include species-level identifications for entire groups (Stucky 1990); exceptions for the subterranean rodents are the Geomyidae and Rhizomyinae. While there is an underlying evolutionary relationship between diversity at the species level and diversity at the genus level (Preston 1962), these two levels of diversity are not necessarily correlated. In fact, their relationship may reveal much about the macroevolutionary trajectories of the clades.

Geomyidae

This North American family presently occupies one of the largest geographic ranges of the eight subterranean clades (see Lacey, Patton, and Cameron, Introduction, this volume; Hall 1981; Nowak 1991). The geomyids also have the oldest fossil record and the greatest overall generic diversity (fifteen genera) of any subterranean clade (fig. 9.1). They originated in the

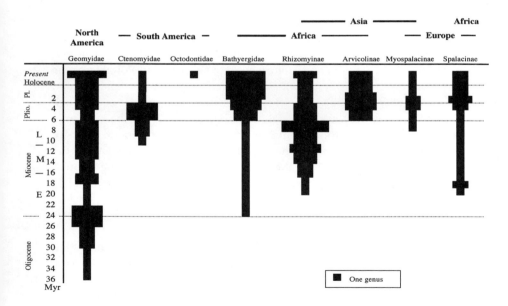

FIGURE 9.1. Worldwide comparison of the number of presumed subterranean genera present in subterranean rodent clades with extant members since the Oligocene. Scale is at lower right. "Present" denotes extant representatives; all others are fossil records.

early Oligocene (Korth 1994) and underwent two pulses of diversification: in the latest Oligocene and in the early Pliocene to the present (fig. 9.2). Four genera of geomyids in the late Oligocene were reduced to one genus by the early Miocene, and diversity then fluctuated between two and three genera until the early Pliocene, when it increased and held steady until the present peak of five genera (Korth 1994). The high generic diversity of this clade in the present is matched only by the Bathyergidae. The geomyid record of total number of genera is relatively stable through time (between two and three genera) and is characterized by relatively constant generic extinction and origination rates.

Arvicolinae

The subfamily Arvicolinae originated in the late Miocene of North America (Repenning 1987; Repenning, Fejfar, and Heinrich 1990); however, the first known subterranean genera appear in the early Pliocene of Asia (Repenning 1984; Repenning, Fejfar, and Heinrich 1990). Only the history of the subterranean and semi-subterranean lineages found in eastern central Asia and the Caucasus Mountains, presently represented by *Ellobius, Prometheomys,* and *Hyperacrius,* is summarized here. These clades, which probably are not sister taxa, diversified rapidly until the late Pliocene, with a peak diversity of four genera (fig. 9.3). Three of the Pliocene genera extend to the present. Fossil representatives are found in Israel, northern Africa, eastern Europe, and China (Kowalski 1966; Repenning 1984; Nadachowski 1990; Repenning, Fejfar, and Heinrich 1990; Gromov and Polyakov 1992; Markova 1992; Tchernov 1992). Subterranean Arvicolinae demonstrated relative stability in number of genera (either three or four genera) through the last 6 million years, and although they originated at about the same time as the Myospalacinae, they maintained relatively greater generic diversity. The subterranean members of the Arvicolinae presently occupy one of the largest geographic ranges of all the subterranean clades, extending through much of south and central Asia (Corbet 1978; Nowak 1991).

Myospalacinae

The fossil record of this northeastern Asian subfamily is sparse, and there are only two genera known (although see Lawrence 1991). This clade showed relative taxonomic stability in diversity from its origination during the late Miocene to the present, peaking at two genera during the late Pliocene and early Pleistocene (fig. 9.4) (Repenning, Fejfar, and Heinrich 1990; Lawrence 1991). The genus *Myospalax* presently occupies one of the most

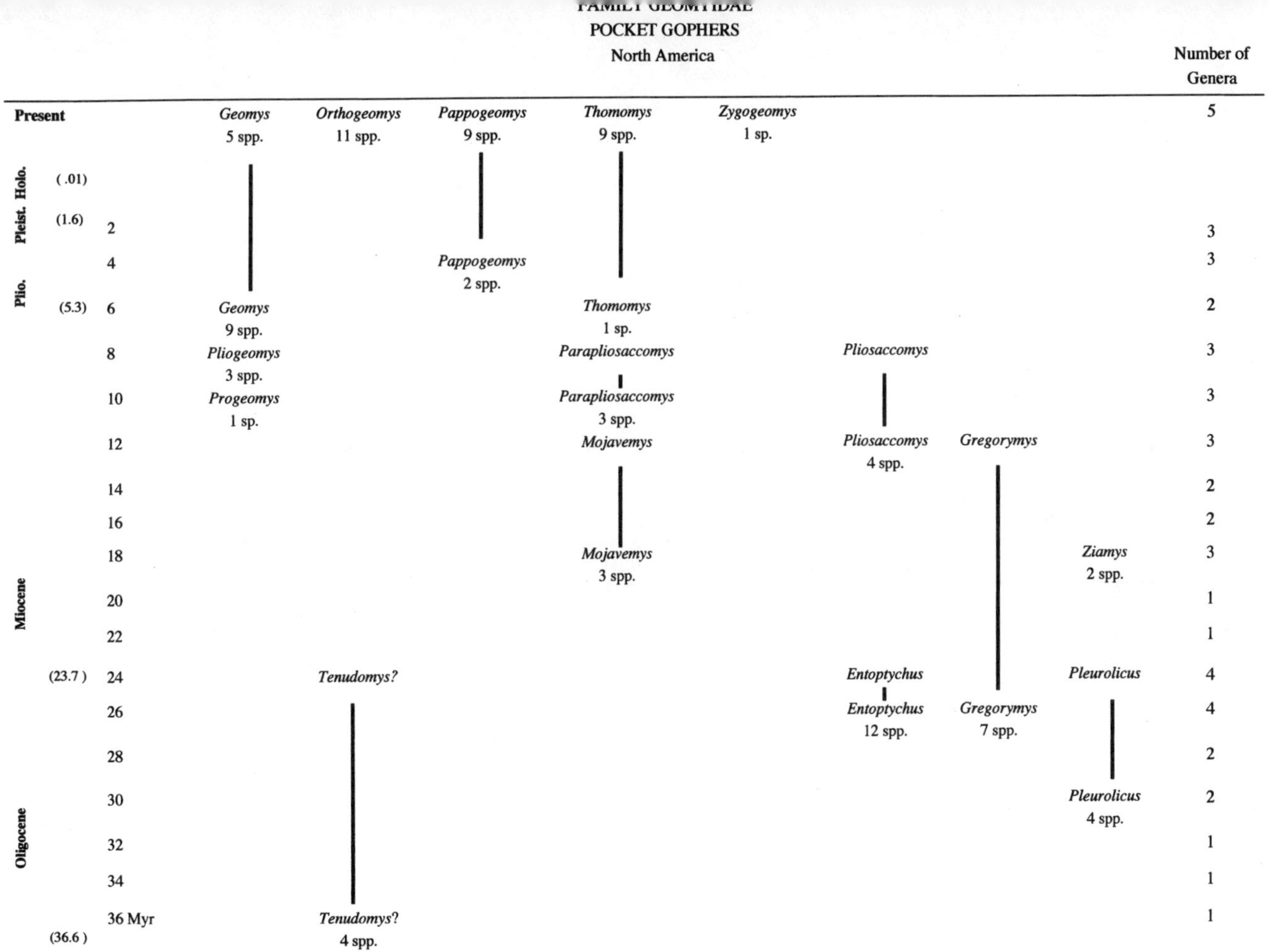

FAMILY GEOMYIDAE
POCKET GOPHERS
North America

FIGURE 9.2. Temporal duration of extinct and extant genera of subterranean rodent clades in the family Geomyidae. (Data from Berggren et al. 1985; Fahlbusch 1985; Korth 1993, 1994; Savage and Russell 1983; Wood 1955)

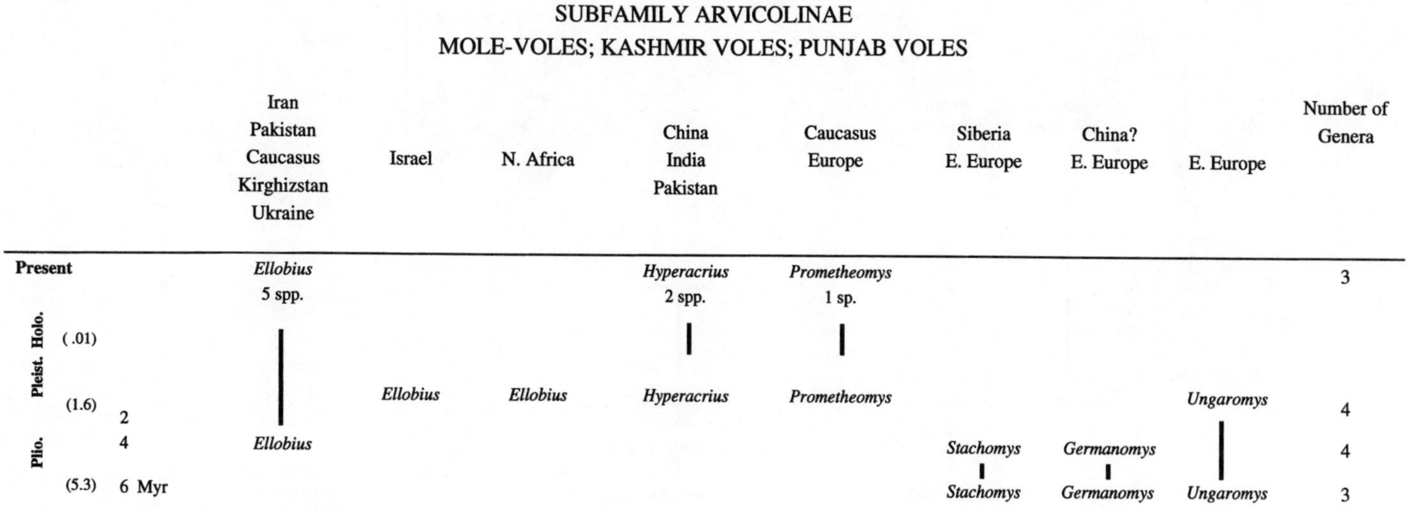

SUBFAMILY ARVICOLINAE
MOLE-VOLES; KASHMIR VOLES; PUNJAB VOLES

	Iran Pakistan Caucasus Kirghizstan Ukraine	Israel	N. Africa	China India Pakistan	Caucasus Europe	Siberia E. Europe	China? E. Europe	E. Europe	Number of Genera
Present	*Ellobius* 5 spp.			*Hyperacrius* 2 spp.	*Prometheomys* 1 sp.				3
Pleist. Holo. (.01)	—			—	—				
(1.6)		*Ellobius*	*Ellobius*	*Hyperacrius*	*Prometheomys*	*Stachomys*	*Germanomys*	*Ungaromys*	4
Plio.	*Ellobius*					—	—	—	4
(5.3) 6 Myr						*Stachomys*	*Germanomys*	*Ungaromys*	3

FIGURE 9.3. Temporal duration of extinct and extant genera of subterranean rodent clades in the subfamily Arvicolinae. Extinct forms from Europe and the Ural Mountains are as per Gromov and Polyakov 1992. This figure does not include *Goniodontomys* or *Microtoscoptes* of North America or Asia (Repenning, Fejfar, and Heinrich 1990). (Data from Kowalski 1966; Vereshchagin 1967; Corbet 1978; Savage and Russell 1983; Ünay and de Bruijn 1984; Fejfar and Heinrich 1986; Jánossy 1986; Repenning 1987; Jaeger 1988; Fejfar 1990; Fejfar and Heinrich 1990; Fejfar and Storch 1990; Kowalski 1990; Qiu 1990; Repenning, Fejfar, and Heinrich 1990; Storch and Fejfar 1990; Zheng and Li 1990; Baryshnikov 1991; Gromov and Polyakov 1992; Markova 1992; Tchernov 1992)

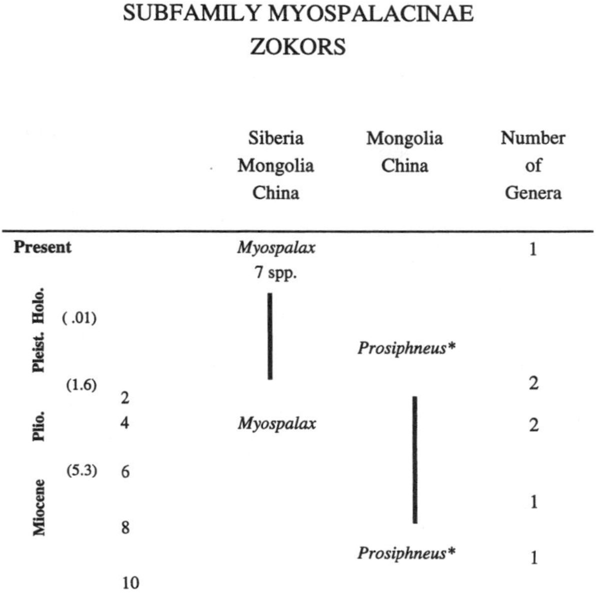

SUBFAMILY MYOSPALACINAE
ZOKORS

Epoch		Siberia Mongolia China	Mongolia China	Number of Genera
Present		*Myospalax* 7 spp.		1
Pleist. Holo. (.01)		|	*Prosiphneus**	
Plio. (1.6)	2	|	|	2
	4	*Myospalax*	|	2
Miocene (5.3)	6		|	1
	8			
	10		*Prosiphneus**	1

FIGURE 9.4. Temporal duration of extinct and extant genera of subterranean rodent clades in the subfamily Myospalacinae (*recognized as *Myospalax* in Lawrence 1991). (Data from Savage and Russell 1983; Carleton and Musser 1984; Jin and Xu 1984; Repenning, Fejfar, and Heinrich 1990; Lawrence 1991; Gromov and Polyakov 1992; Markova et al. 1995)

disjunct geographic ranges, extending throughout eastern and central Asia (Corbet 1978; Nowak 1991), with fossil occurrences in eastern Europe and Asia (Repenning, Fejfar, and Heinrich 1990; Lawrence 1991; Gromov and Polyakov 1992; Markova et al. 1995).

Rhizomyinae

Extinct genera of the subfamily Rhizomyinae occupied a somewhat larger, more continuous distribution than at present, which included eastern, central, and southern Asia, Turkey, and northeastern Africa (fig. 9.5) (Alpagut, Andrews, and Martin 1990; Flynn 1990; Nowak 1991; Kingdon 1997). The subfamily extends from the early Miocene to the present (Alpagut, Andrews, and Martin 1990; Flynn 1990). Diversification of the subfamily began in the middle Miocene, several million years after its first appearance in the fossil record. This diversification is coincident with the evolution of characters indicative of true fossoriality, about 8 to 10 million years ago (Flynn 1982). Generic diversity in the Rhizomyinae increased from two to a peak of six

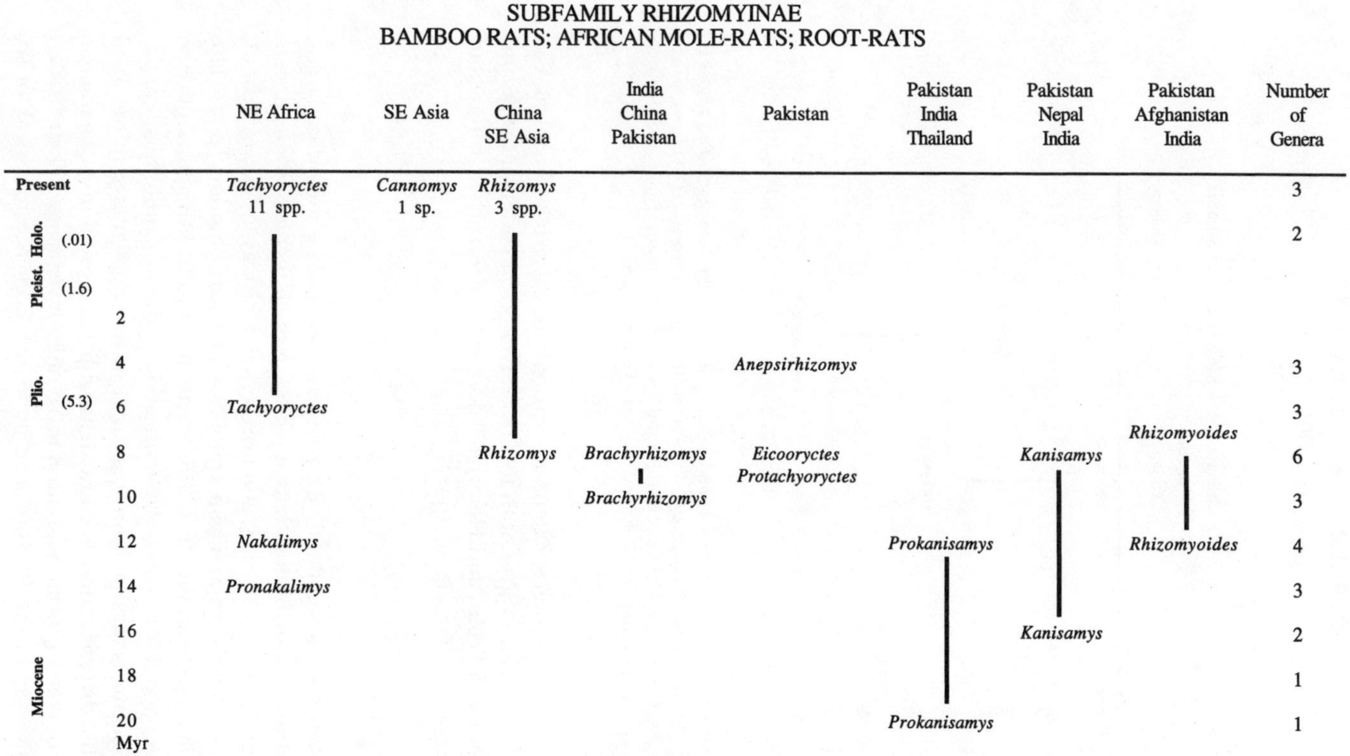

SUBFAMILY RHIZOMYINAE
BAMBOO RATS; AFRICAN MOLE-RATS; ROOT-RATS

	NE Africa	SE Asia	China SE Asia	India China Pakistan	Pakistan	Pakistan India Thailand	Pakistan Nepal India	Pakistan Afghanistan India	Number of Genera
Present	*Tachyoryctes* 11 spp.	*Cannomys* 1 sp.	*Rhizomys* 3 spp.						3
Pleist. Holo. (.01)									2
(1.6) 2									
4				*Anepsirhizomys*					3
Plio. (5.3) 6	*Tachyoryctes*		*Rhizomys*	*Brachyrhizomys*					3
8					*Eicooryctes* *Protachyoryctes*			*Rhizomyoides*	6
10				*Brachyrhizomys*					3
12	*Nakalimys*						*Kanisamys*	*Rhizomyoides*	4
14	*Pronakalimys*								3
16						*Prokanisamys*	*Kanisamys*		2
18									1
Miocene 20 Myr						*Prokanisamys*			1

FIGURE 9.5. Temporal duration of extinct and extant genera of subterranean rodent clades in the subfamily Rhizomyinae. (Data from Sabatier 1978; Flynn 1982; Savage and Russell 1983; Carleton and Musser 1984; Dawson and Krishtalka 1984; de Bruijn 1984; Flynn and Sabatier 1984; Flynn, Jacobs, and Lindsay 1985; Alpagut, Andrews, and Martin 1990; Barry and Flynn 1990; Flynn 1990; Jacobs, Flynn, and Downs 1990; Jacobs et al. 1990; Savic and Nevo 1990; Flynn 1993; Tong and Jaeger 1993)

genera during the late Miocene; it has since fluctuated between two and three genera. This subfamily is second only to the Geomyidae in overall generic diversity, although it has the fourth oldest fossil record (see fig. 9.1).

Spalacinae

The Spalacinae extend from the late Oligocene to the present in the eastern Mediterranean region of Europe, Israel, Turkey, and Africa (fig. 9.6) (Corbet 1978; Popov 1988; Alpagut, Andrews, and Martin 1990; Mein 1990; Sümengen et al. 1990; Nadachowski 1992; Tchernov 1992; Hugueney and Mein 1993; Wilson and Reeder 1993). The Spalacinae and the Bathyergidae have the second oldest fossil records of the subterranean groups (Savage 1990; Hugueney and Mein 1993). The subfamily exhibited relative stability throughout most of its history, diversifying from one to two genera briefly during the middle Miocene, and not again until the Plio-Pleistocene, with a peak diversity of three genera (see fig. 9.1). Diversity in the present has declined to two genera.

Bathyergidae

The family Bathyergidae extends from the earliest Miocene to the present in Africa (fig. 9.7) (Jarvis and Bennett 1990; Savage 1990; Nowak 1991; Kingdon 1997). There also is a record for the family from the Miocene of Israel (Tchernov 1992). In addition to the five extant genera of this group, there are at least four extinct genera. Generic diversity in this family did not increase until the early Pliocene, reaching its peak during the early Pleistocene to the present at five genera. This clade has the second oldest fossil history of the subterranean groups, and exhibits one of the longest records of stability in generic diversity during the first 14 million years of its history.

Ctenomyidae

The family Ctenomyidae extends from the late Miocene of South America to the present (fig. 9.8) (Wood 1974; Montalvo and Casadio 1988; Redford and Eisenberg 1992; Reig et al. 1990; Nowak 1991; Verzi, Montalvo, and Vucetich 1991; Reig and Quintana 1992; Quintana 1994; Pascual, Jaureguizar, and Prado 1996). There are at least six extinct genera of Ctenomyinae. Together with the Arvicolinae, the Ctenomyinae exhibited perhaps the most rapid diversification of all the subterranean groups. The peak diversity of four genera occurred by the late Miocene, shortly after the origination of this clade. The Plio-Pleistocene boundary resulted in extinction of all but one genus *(Ctenomys)* (Tonni et al. 1992; Vucetich and Verzi 1995; Verzi

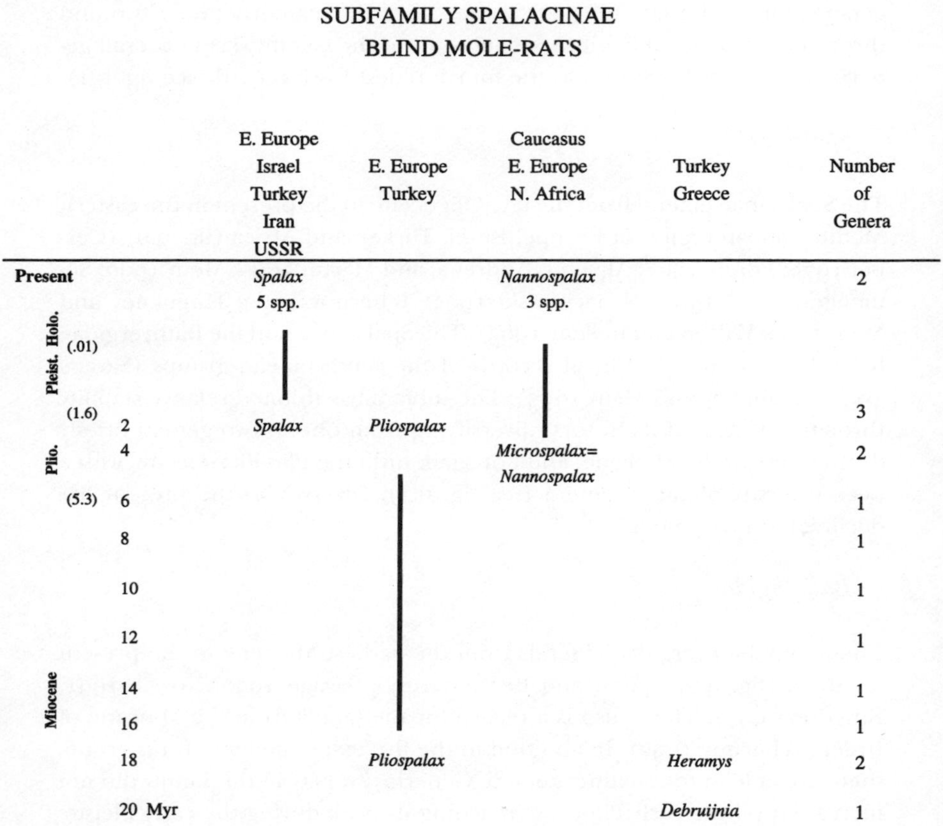

SUBFAMILY SPALACINAE
BLIND MOLE-RATS

		E. Europe Israel Turkey Iran USSR	E. Europe Turkey	Caucasus E. Europe N. Africa	Turkey Greece	Number of Genera
Present		*Spalax* 5 spp.		*Nannospalax* 3 spp.		2
Pleist. Holo. (.01)						
(1.6)	2	*Spalax*	*Pliospalax*			3
Plio.	4			*Microspalax=* *Nannospalax*		2
(5.3)	6					1
	8					1
	10					1
	12					1
Miocene	14					1
	16					1
	18		*Pliospalax*		*Heramys*	2
	20 Myr				*Debruijnia*	1

FIGURE 9.6. Temporal duration of extinct and extant genera of subterranean rodent clades in the subfamily Spalacinae. (*Allospalax, Prospalax, Pterospalax, Rhizospalax,* Anomalomyinae, and Tachyoryctoidinae are not included, as per Hugueney and Mein 1993, Savic and Nevo 1990, and de Bruijn 1984.) (Data from Wood 1955; Ognev 1963; Topachevskii 1976; Savage and Russell 1983; Carleton and Musser 1984; Dawson and Krishtalka 1984; de Bruijn 1984; Fahlbusch 1985; Flynn, Jacobs, and Lindsay 1985; Hofmeijer and de Bruijn 1985; Jánossy 1986; Popov 1988; Alpagut, Andrews, and Martin 1990; Fejfar 1990; Fejfar and Heinrich 1990; Kowalski 1990; Mein 1990; Nadachowski 1990, 1992; Savic and Nevo 1990; Storch and Fejfar 1990; Sümengen et al. 1990; Tchernov 1992; Hugueney and Mein 1993; Musser and Carleton 1993; Markova et al. 1995)

and Lezcano 1996). However, the present number of genera does not reflect subsequent diversification of this group, as most of the present diversity is at the species level. Ten species of *Ctenomys* known from the early Pleistocene diversified into approximately sixty species today (Reig et al. 1990).

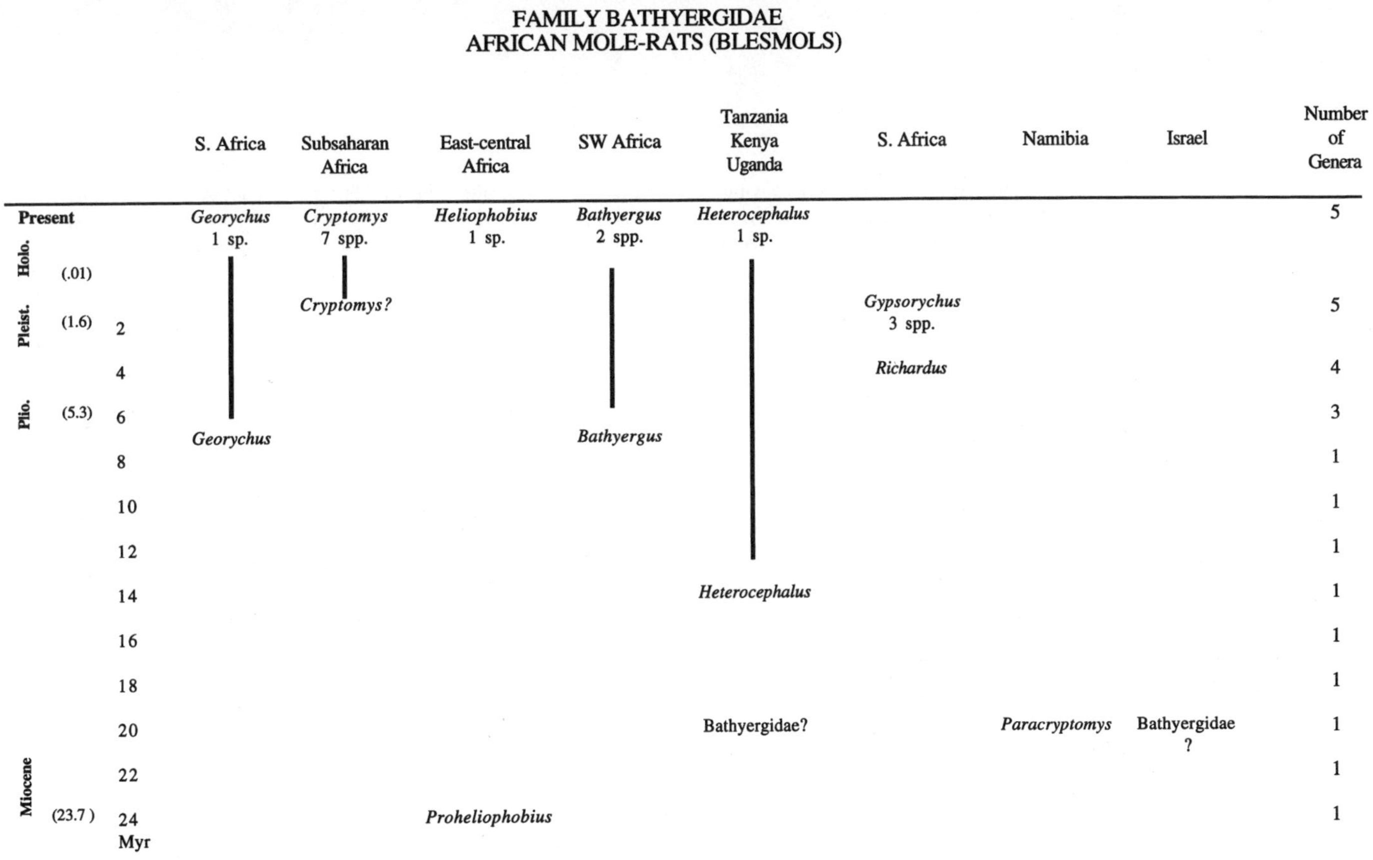

FAMILY BATHYERGIDAE
AFRICAN MOLE-RATS (BLESMOLS)

	S. Africa	Subsaharan Africa	East-central Africa	SW Africa	Tanzania Kenya Uganda	S. Africa	Namibia	Israel	Number of Genera
Present	*Georychus* 1 sp.	*Cryptomys* 7 spp.	*Heliophobius* 1 sp.	*Bathyergus* 2 spp.	*Heterocephalus* 1 sp.				5
Holo. (.01)									
Pleist. (1.6)		*Cryptomys?*							5
Plio. (5.3)	*Georychus*			*Bathyergus*		*Gypsorychus* 3 spp.			4
						Richardus			3
									1
									1
					Heterocephalus				1
									1
									1
Miocene					Bathyergidae?		*Paracryptomys*	Bathyergidae ?	1
									1
(23.7)			*Proheliophobius*						1

FIGURE 9.7. Temporal duration of extinct and extant genera of subterranean rodent clades in the family Bathyergidae. (*Bathyergoides* is not included, as per Wood 1985 and Jarvis and Bennett 1990.) (Data from Wood 1955, 1974, 1985; Hendey 1976; Lavocat 1978; Savage and Russell 1983; Dawson and Krishtalka 1984; Woods 1984; Denys 1987; Jarvis and Bennett 1990; Savage 1990; Tchernov 1992)

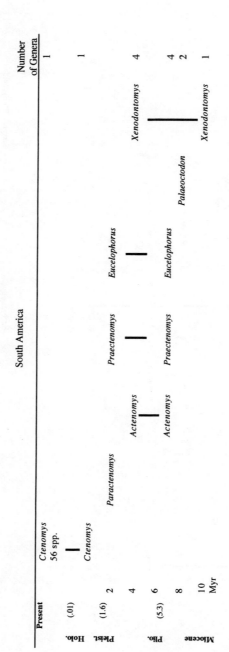

FIGURE 9.8. Temporal duration of extinct and extant genera of subterranean rodent clades in the subfamily Ctenomyinae. (Data from Pearson and Pearson 1982; Woods 1984; Tonni, Bargo, and Prado 1988; Alberdi et al. 1989; Genise 1989; Reig et al. 1990; Reig and Quintana 1992; Marshall and Sempere 1993; Quintana 1994; Pardiñas and Lezcano 1995; Vucetich and Verzi 1995; Pascual, Jaureguizar, and Prado 1996; Verzi and Lezcano 1996)

Octodontidae

The fossil record of the family Octodontidae extends from the Oligocene through the Holocene of southern South America (Wood 1974; Savage and Russell 1983; Reig and Quintana 1991). However, of this family, only one extant genus is known to be subterranean *(Spalacopus)*. There are no known fossils of this genus.

Synthesis of Macroevolutionary Patterns

The patterns of origination in most of these subterranean groups are not synchronous (see fig. 9.1). Only the Geomyidae extends to the Oligocene. Three groups, the Spalacinae, Bathyergidae, and Rhizomyinae originated in the early Miocene. Three groups (Myospalacinae, Ctenomyidae, and subterranean Arvicolinae) first appeared in the late Miocene and early Pliocene. The octodontid genus *Spalacopus* has no fossil record.

Nor is the initiation of diversification in these lineages globally synchronous. The Geomyidae and Spalacinae are alone in having an early diversification pulse (late Oligocene to early Miocene and early to middle Miocene, respectively) that is not subsequently followed by the peak diversity for the group. All other diversification pulses presage a peak diversity for the clade, which in some cases is followed by a rapid reduction in diversity (Spalacinae, Arvicolinae, Myospalacinae, Rhizomyinae, and Ctenomyidae). Only three of the lineages (Octodontidae, Geomyidae, and Bathyergidae) have their peak diversity in the present, suggesting that sampling bias (i.e., number of localities) alone does not explain diversity patterns. Diversity peaks for four groups (Ctenomyidae, Rhizomyinae, Arvicolinae, and Myospalacinae) occur in the Pliocene, and five of the groups reach or maintain their peak of diversity during the Pleistocene and present (Bathyergidae, Geomyidae, Spalacinae, Octodontidae, and Myospalacinae). Declines in generic diversity for some of these groups during the Plio-Pleistocene suggest possible extinction events that may be somewhat synchronous for at least five of the clades (Ctenomyidae, Rhizomyinae, Arvicolinae, Myospalacinae, and Spalacinae). This observation should be investigated more closely.

Despite over 20 million years of evolution for several of the subterranean clades (Bathyergidae, Rhizomyinae, and Spalacinae), their diversification is delayed until the late Miocene and Pliocene. Also during the late Miocene and early Pliocene, three other groups first appear and subsequently diversify (Ctenomyidae, Arvicolinae, and Myospalacinae). The Pliocene peak in generic diversity for four groups (Ctenomyidae, Rhizomyinae, Arvicolinae, and Myospalacinae) may indicate the only potentially contemporaneous series of events across many clades and would be consistent with a global, perhaps climatic, perturbation. The early Pliocene period (5 to 3 million years ago) was warmer than the present (Crowley 1991); however, global cooling and glaciation began in the later Pliocene (Raymo et al. 1989; Raymo 1992). Although the overall trend was toward cooling, oscillations between warm and cold conditions occurred between 3.1 and 2.6 million years ago. Cooling during the Pliocene is correlated with an increase in grasses and composites in the Northern Hemisphere and with renewed diversification of large, cursorial grazing herbivores (Potts and Behrensmeyer 1992). However, it is not clear why cooling would necessarily lead to an increase in the frequency of subterranean taxa. The changes in diversity for the four groups that exhibit peak diversity during the Pliocene and Pleistocene, together with careful age controls, would provide a fruitful arena for further study. The long period of relative stability in the diversity of many of these clades suggests that evolution of the subterranean lifestyle was not a key innovation leading to an adaptive radiation. The only potential exceptions are the Ctenomyidae and the Arvicolinae, which diversified soon after their first appearance.

For the subterranean clades, faunal turnover (the number of originations and extinctions) (MacFadden 1992; Raup 1978; Stanley 1979) is highest during the Pliocene. The total number of originations for all clades is highest during the early Pliocene, with eleven generic originations. Generic extinctions also peak during the Pliocene, with the loss of eight genera. Whether these are true extinctions or pseudoextinctions is not known. These results suggest that diversification in these clades is somehow enhanced during this interval, which is concordant with other studies of rodents, fissiped carnivores, and artiodactyls, all of which show high faunal turnover during the Pliocene relative to the middle to late Miocene and the Pleistocene (Barnosky 1989). Thus neither a dramatic cooling during the Oligocene (Briggs 1995; Crowley and North 1991; Potts and Behrensmeyer 1992; Stucky 1990) nor the expansion of open grassland habitats during the middle Miocene (i.e., Briggs 1995; Janis 1993; Potts and Behrensmeyer 1992) are correlated with increased origination and diversification of these clades.

Of the twenty-one extant genera of subterranean rodents, six (or 29%) have no fossil representatives. If the fossil record becomes less representative of total diversity as it gets older, then we may be drastically underrepresenting the total diversity of fossorial rodents. However, only three groups reach peak diversity in the present, suggesting that sampling bias alone does not explain the pattern of diversification during the Neogene.

Diversification Rates Based on Molecular Phylogenies

Diversification rate is the difference between speciation and extinction rates (Stanley 1979). We employed four methods (reviewed by Purvis 1996; Sanderson and Donoghue 1996; Mooers and Heard 1997) to identify possible shifts in diversification rates. These tests are drawn from Cook and Lessa (1998) and use explicit null models that incorporate different levels of information derived from an understanding of evolutionary relationships. Herein, we expand on these tests by highlighting the possibilities and shortcomings of these methods and reviewing the existing data for subterranean rodents. The first three tests require knowledge of tree shape, while the fourth test also includes information about branch lengths (fig. 9.9). Mooers and Heard (1997) note that adding estimates of branch length to a phylogeny provides opportunities to examine diversification rates across extant lineages at different times in the past (horizontal slices) as well as rate variation within clades through time (vertical slices).

Standing Diversity: Two-Taxon Tests

The two-taxon test compares the species richness of sister taxa (Slowinski and Guyer 1989) and relies on the null Markovian model (birth/death or Yule model) of homogeneous diversification rates. Our null hypothesis is that the acquisition of the subterranean lifestyle has not promoted diversification in rodents. The simple comparison of sister group diversities is of limited value, because the direction of rate change is not recovered. For example, if a significant difference in rate is detected, it may be due to an increase in diversification in one of the lineages, or a slowdown in the sister lineage, or some combination of the two. This test has limited power to discriminate among alternative hypotheses because it requires a nearly twentyfold difference in the number of species to detect a significant difference (one-tailed test, $p < .05$). Indeed, no significant difference in diversity between the subterranean clades and their non-subterranean sister clades (e.g., Geomyidae and Heteromyidae) was detected (table 9.1). Although subterranean lineages generally have more named species than their non-subterranean sister clades, the simple paired two-taxon test would require over a fourfold increase in the number of species of tuco-tucos, for example, before finding a significant increase in the rate of diversification in that lineage.

We also used three sets of cumulative tests. In one set, the cumulative probabilities were based on the highest published estimate of diversity for the subterranean and the lowest estimate for the corresponding sister taxa (e.g., *Ctenomys* = 56; bathyergid sister taxon = 3). This test was modified

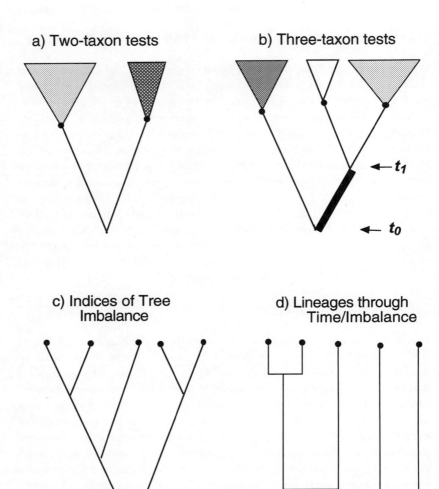

FIGURE 9.9. Tests used to identify possible shifts in diversification rates. Statistical power increases from the simplest two-taxon tests through three-taxon tests and indices of tree imbalance to the lineages-through-time and imbalance tests because an increasing amount of information is incorporated into the analyses.

TABLE 9.1 TWO-TAXON TESTS OF DIVERSIFICATION

Sister taxa (subterranean/ non-subterranean)	Number of species		P value	Source
	Subterranean	Non-subterranean		
Pairs of sister taxa				
1. Ctenomyidae/Octodontidae[a]	38	9	.196	Woods 1993
2. Ctenomyidae/Octodontidae	56	9	.141	Reig et al. 1990
3. Geomyidae/Heteromyidae	35	59	.624	Patton 1993
4. Bathyergidae/Hystricidae[b]	12	11	.500	Woods 1993
5. Bathyergidae/ Thryonomyidae	12	3	.214	Woods 1993
6. Rhizomyinae	15	1		Musser and Carleton 1993
7. Spalacinae	8	1		Musser and Carleton 1993
8. Myospalacinae	7	1		Musser and Carleton 1993
Cumulative tests				
Pairs 1, 3, and 4			.470	
Pairs 2, 3, and 5			.242	
Pairs 2, 3, 5, 6, 7 and 8 (Best case)			.011	

Note: Two-taxon tests of imbalance (Slowinski and Guyer 1989, 1993) between subterranean lineages and their sister taxa (where known), with results of cumulative tests across pairs of taxa (Slowinski and Guyer 1993).
[a] The monotypic octodontine genus *Spalacopus* independently converged on the subterranean niche.
[b] Two possible outgroup taxa are used for the Bathyergidae, following Nedbal, Allard, and Honeycutt 1994.

from Cook and Lessa (1998) to include the most conservative estimate of sister taxon diversity ($n = 1$) for the Spalacinae, Rhizomyinae, and Myospalacinae. In this best attempt to reject the hypothesis of no increase in diversification rate across these independent clades, the test was marginally significant. The higher-level relationships of these three taxa are not well known (Carleton and Musser 1984; Flynn 1990; Lawrence 1991). They are not now generally thought to have a recent common ancestor; however, additional work is needed to clarify their higher-level relationships.

Standing Diversity: Three-Taxon Tests

The three-taxon test is based on a Yule model of diversification and allows for one or more rate parameters in different parts of a phylogeny. This test addresses the problem of identifying the source of the rate change by using a directional maximum likelihood method that can identify shifts in diversification in three-taxon comparisons. We used the LRDIVERSE program (Sanderson and Wojciechowski 1996), which is more sensitive to shifts in diversification rates than the two-taxon tests because of its increased statistical power.

Three-taxon tests have been applied to tuco-tucos (Cook and Lessa 1998), in combination with their sister octodontines and the outgroup to both of these groups, which may be the Echimyidae, Abrocomidae, or (less likely) Myocastoridae. These tests (Sanderson and Donoghue 1994) were employed under four different conditions. Two of these tests set the common ancestor of ctenomyines and octodontines to half the age of the common ancestor of all three taxa based on paleontological assessments. One of the tests used these assumptions to set the diversification rate of the internal branch (i.e., the one linking the two nodes of the tree) equal to that of the outgroup. The other set the rate as equal to the average of the three external branches (i.e., those leading to the three terminal taxa). The remaining two tests were performed making no assumption about the relative timing of divergence of the three lineages and, again, setting the rate for the internal branch as either equal to that of the outgroup or as an average rate. Under all four conditions, the three-taxon tests failed to reject the null model that assumes equal rates of diversification for all three taxa (Cook and Lessa 1998).

Tree Balance Tests

More detailed tests can be carried out for complete phylogenies and may be particularly sensitive to detecting rate change when combined with estimates of time of cladogenic events based on either the fossil record or rate homogeneity. It is clear that phylogenies reconstructed from extant lineages will change form as lineage birth or death rates are altered. Nonparametric tests of tree balance (Shao and Sokal 1990; Kirkpatrick and Slatkin 1993) are more powerful than the sister clades tests because they use information from every branching point. Nonetheless, speciation rates must be substantially different to be detected. Statistical power increases with increasing numbers of taxa; however, the null model is not likely to be rejected for single phylogenies (Purvis 1996). Since phylogenies are incomplete for all of the subterranean clades, diversification rates based on molecular genetic data have been explored only for the Ctenomyidae and Octodontidae (Lessa and Cook 1998; Cook and Lessa 1998; D'Elía, Lessa, and Cook 1999). We suspect that similar tests can be applied to the Bathyergidae, Geomyidae, and Spalacinae in the near future. The Colless (1982) index of tree imbalance for the Ctenomyidae and Octodontidae ($Ic = 0.2356$) was not significant, indicating that we could not reject the null expectation that the tree was balanced. This result corroborated previous tests suggesting no increase in diversification in the ctenomyid lineage (Cook and Lessa 1998).

Lineages through Time

Under the assumption of a molecular clock, branch lengths are proportional to time. This additional information about a phylogeny can be used to produce plots of lineages through time with the Endemic-Epidemic program (Rambaut, Harvey, and Nee 1997). The advantage of this approach is that sample size is increased in proportion to the number of taxa included in the analysis, resulting in less biased estimates of rate and a more powerful set of tests. Cook and Lessa (1998) developed a maximum likelihood tree (DNAMLK procedure of PHYLIP, which assumes a molecular clock) for the ctenomyid lineages. This phylogeny was used in a lineages-through-time analysis (Harvey, May, and Nee 1994). The relative cladogenesis statistic calculates the probability under a constant-rates birth-death model that a particular lineage existing at time t (in the past) will have k extant tips (compared with the total number of tips). This statistic allows the identification of branches that have higher than expected rates of cladogenesis. The semilogarithmic plot indicated an early and significant increase in diversification rate in the *Ctenomys* lineage (fig. 9.10), which was consistent with the hard polytomy previously described (Lessa and Cook 1998) and the early diversification of species in the fossil record (see fig. 9.8). This was the single significant test statistic suggesting a higher diversification rate for subterranean rodents in our exploration of these taxa.

Overview of Diversification Test Results

Species Diversity

A preliminary examination of diversification rates at the species level in subterranean rodents suggests little support for a key innovation—namely, the subterranean lifestyle—having repeatedly caused increased diversification rates in these independently derived rodent clades. The application of a more powerful set of tests of diversification rate was possible only in the Ctenomyidae, the most speciose group. The lineage-through-time test suggested a higher rate of diversification in the branch leading to *Ctenomys*. This higher rate may be indicative of a burst of speciation early in the history of the Ctenomyidae and agrees with the previously identified "hard polytomy" in the clade (Lessa and Cook 1998). However, the high rate of speciation in the genus *Ctenomys* is supported by the fossil record, which suggests that all of the morphological diversity found in extant species occurred subsequent to the first appearance of the genus, within the last two million years.

The six independent lineages of subterranean rodents do not offer col-

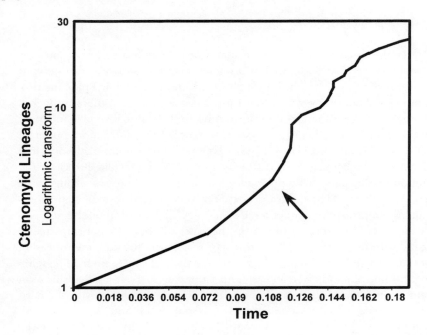

FIGURE 9.10. The lineages-through-time plot for a phylogeny of the Octodontidae and Ctenomyidae (Cook and Lessa 1998), illustrating the suspected increase in diversification rate (arrow) at the base of the radiation of the Ctenomyidae. This plot is based on DNAMLK estimates of relative time, assuming a molecular clock. Units of time on the plot are arbitrary, but should be proportional to actual time if the assumptions of this model are met.

lective evidence of increased diversification rates associated with the subterranean lifestyle, but this conclusion is based on paired and cumulative two-taxon tests, which are known to have low statistical power. Ample opportunities exist for further testing, but there does not appear to be a solid case for strikingly high diversification rates in subterranean rodents at this time. These conclusions are tentative and need to be explored further with additional information for each clade. In particular, data related to the timing of divergence, the duration of lineages (estimated from the fossil record and molecular data), phylogenetic relationships, and even species boundaries (Steinberg and Patton, chap. 8, this volume) will be necessary to enhance our ability to examine diversification rates.

We know little about the sensitivity of these statistical tests to violations of the underlying assumptions (Purvis 1996). For example, the tests of tree balance that we used are thought to require complete phylogenies (Kirkpatrick and Slatkin 1993). Incomplete trees are thought to be more imbalanced than full trees (Brown 1994; Mooers and Heard 1997). The effect of this bias appears to be relatively mild, however (Heard 1992; Mooers 1995; Mooers and Heard 1997), particularly if taxa have not been systematically omitted from the analysis (Slowinski and Guyer 1989). Other potential complications with respect to the assumptions of these statistical analyses include (1) possible deviations from random extinctions across lineages, (2) potential problems of simultaneous speciation in some lineages, (3) phylogenetic uncertainty, and (4) rejection of the molecular clock (Cook and Lessa 1998).

On the other hand, the results of our preliminary tests may suggest the need to rephrase and expand our questions. In this context, chromosomal variation (Reig 1989) and population structure (Patton 1985; Steinberg and Patton, chap. 8, this volume) with regard to speciation need to be further explored. Are these issues necessarily peculiar to subterranean rodents? In some cases, it appears that we can be relieved of the task of trying to account for higher diversification rates. The sister taxon of geomyids, for example, is fairly speciose. Chromosomal diversity should be viewed in a similar light. For example, the karyotypic variation of non-subterranean octodontines (Gallardo 1992) and echimyids (Reig 1989) may be as impressive as is that of tuco-tucos.

Generic Diversity

The fossil record of subterranean clades is inadequate for testing relative rates of speciation at this time because species-level taxonomy is unresolved for many of the lineages. Potential exceptions include the Geomyidae and the Rhizomyinae. However, genus-level analyses of diversity provide a means of understanding the timing of events that led to enough morphological divergence that a systematist would recognize a generic split, an event certainly correlated with speciation.

Although there is a strong relationship between generic and species diversity across modern groups (Preston 1962), in particular cases, high generic diversity does not necessarily correlate with high diversity at the species level. Generic diversity indicates the diversity of morphotypes within a clade, rather than the diversity of taxa with very similar morphotypes (e.g., Bogert et al. 1943; Nadachowski 1993). Therefore, changes in generic diversity may well be driven by processes that occur on a different scale than processes that drive speciation (Stanley 1979). For example, extinction may eliminate all "intermediate" forms in a clade, leaving the most diverse forms. The number of species thus may decline, leaving a proportionately greater number of genera. Conversely, a genus may have many species, yet there may be no coexisting genera of the same clade. For example, the Bathyergidae has five extant genera, yet three are monospecific; the Ctenomyidae has only one extant genus, yet that genus may include over sixty species. If similar processes drive generic diversification in all subterranean taxa, similar generic diversity patterns through time would be expected in all subterranean groups. The basic patterns of generic diversification found in this study of subterranean taxa are as follows.

1. The first appearances of subterranean rodents through the Cenozoic are variable. Subterranean rodents within certain lineages first appeared in the Oligocene, the Miocene, the Pliocene, and the Quaternary. Therefore, no single event apparently triggered entry into the rodent subterranean niche.
2. Three clades (Bathyergidae, Rhizomyinae, and Spalacinae) originated in the early Miocene and exhibited several millions of years of stability before diversification. This pattern suggests that the origination of the subterranean lifestyle itself was not a key innovation leading to adaptive radiation in at least these three groups. The Rhizomyinae are thought to have begun diversification before the acquisition of true fossoriality (Flynn 1982).
3. Two clades (Ctenomyidae and Myospalacinae) originated in the late Miocene, and two clades (Ctenomyidae and Rhizomyinae) diversified then. The synchrony of these events thus may be evidence of a global driver. However, three subterranean groups already present (Geomyidae, Bathyergidae and Spalacinae) exhibited no change in diversity during the late Miocene.
4. Three clades (Ctenomyidae, Arvicolinae, and Myospalacinae) reached their peak diversity in the Pliocene, and three others (Geomyidae, Bathyergidae, and Spalacinae) diversified then. The evolutionary patterns of the Pliocene may be the strongest evidence of synchrony in the evolution of subterranean existence in these rodent clades.
5. Most groups exhibited only short periods of high generic diversity (less than 2 million years).
6. The oldest subterranean clade does have the most genera (Geomyidae: 15), and the youngest clade the fewest (Octodontidae: 1); however, the rest of the groups are variable, with some shorter-lived clades producing many genera (Rhizomyinae: 12), and older groups producing only a few (Spalacinae: 4). The relationship between number of genera produced and the duration of the clade is positive, but not significant ($p = .16$, Spearman rank correlation).
7. A macroevolutionary trend that may be indicative of the diversification rate of a group is the relative duration of a genus. Mammalian generic duration varies from 2.4 million years (Plio-Pleistocene mammals) (Lundelius et al. 1987) to 5–7.5 million years (general) (Simpson 1953; Stan-

ley 1979) to 8.4 million years (Equidae) (McFadden 1992). The mean generic duration for all the subterranean rodent clades is 4.6 million years. The shortest mean generic durations are found in the Arvicolinae (2.8 myr) and the Ctenomyidae (3.1 myr). These generic durations are less than some species durations, such as those found in fossil Equidae (3.3 myr) (McFadden 1992). The longest generic durations are found in the Spalacinae (6.5 myr) and the Geomyidae (5.8 myr). It is unclear whether the durations found in these subterranean clades are comparable to those in other rodent taxa.

8. Five of the eight subterranean groups do not have a peak of generic diversity in the present. If the Ctenomyidae are excluded, there is a positive relationship ($p = .05$) between the number of genera and the number of species in these subterranean clades. This regression suggests that in some, but not all, cases, generic diversity is congruent with species diversity.

Integration of Fossil and Molecular Data

The continued development of fossil and molecular data will allow us to ascertain whether the patterns demonstrated by one data set are concordant with the other. A better understanding of how fossil and modern species concepts differ is essential. More complete summaries of the range of morphological variation (or lack thereof) in modern subterranean species should be used to standardize fossil species diversity. Sampling bias could be controlled by comparing total faunal diversity with subterranean rodent diversity, and both could be evaluated against the number of fossil localities per unit time. There is a need for summaries of the morphological features associated with a subterranean mode of life in order to determine whether specific members of these groups were fossorial. A chronology of all subterranean groups (including those without extant members) would provide additional evidence of potential global influences on the evolution of fossoriality (e.g., Lessa 1990; Reig and Quintana 1992). Finally, an increased resolution of the fossil macroevolutionary patterns presented herein would result from more detailed chronologies and more data on global biotic and abiotic conditions during the late Cenozoic.

In summary, we find little support for rapid specific or generic diversification in subterranean rodents. The long period of relative stability in many of these clades suggested by the fossil record may mean that the subterranean lifestyle was not a key innovation leading to adaptive radiation, or even increased diversification. Still, there is plenty of additional paleontological, phylogenetic, and taxonomic work to be completed (table 9.2). For all sub-

TABLE 9.2 FEASIBILITY AT THE PRESENT TIME OF SELECTED TESTS OF DIVERSIFICATION FOR EIGHT SUBTERRANEAN CLADES

Family/ subfamily	Test feasible?			
	Two-taxon, standing diversity	Three-taxon, standing diversity	Tree imbalance (branching pattern)	Lineages through time and imbalance
Ctenomyidae	Yes	Yes[a]	Yes[b]	Yes[c]
Octodontidae	?	?	?	Any tests for them?
Geomyidae	Yes	?	Probably[d]	Possibly soon[e]
Bathyergidae	Yes[f]	?	Yes?	Yes?
Spalacinae	No	No	No	No
Rhizomyinae	No	No	No	No
Arvicolinae	No	No	No	No
Myospalacinae	No	No	No	No

[a]The echimyids are the most widely accepted outgroup to octodontids + ctenomyids, but abrocomids and *Myocastor* remain at least possible.
[b]Subject to the limitation of an incomplete phylogeny.
[c]Subject to limitations stemming from an incomplete phylogeny and rejection of molecular clock.
[d]A working hypothesis of relationships for geomyids could be derived based on a composite of analyses of chromosomes, morphology, allozymes, and DNA sequences.
[e]Additional work on geomyines and the subgenus *Thomomys* is needed.
[f]Some or several of these tests may be possible for bathyergids.

terranean groups, this will require adding previously unstudied taxa to existing phylogenies. Poorly resolved sections of phylogenies should be reexamined with new data and new analytical techniques, but we should remember that some may be largely intractable due to rapid pulses of speciation, hybridization (Steinberg and Patton, chap. 8, this volume), or other events that may preclude our recovering a strictly dichotomous branching pattern. Minimally, all phylogenetic hypotheses could stand further testing with additional characters.

Other Prospects for Phylogenetic Studies in Subterranean Rodents

Knowledge of phylogenetic relationships is central to the construction of natural classifications of organisms (Wiley 1981). Recently, phylogenies have become central to comparative biology because they provide a historical framework for fields as diverse as ecology, ethology, physiology, and biogeography (Brooks and McLennan 1991; Harvey and Pagel 1991; Ricklefs and Schluter 1993; Riddle 1996). In this chapter, we have examined some of the potential uses of phylogenies to investigate macroevolutionary patterns in subterranean mammals. There are numerous other exciting opportunities to explore convergent evolution with this comparative approach.

Chromosomal Evolution and Speciation

A discussion of evolutionary relationships in subterranean rodents would be incomplete without mention of how knowledge of phylogenies might enhance our understanding of chromosomal evolution. Diploid variation in some of these clades is among the highest recorded for mammals (e.g., $2N = 10-70$ in the genus *Ctenomys:* Cook, Anderson, and Yates 1990). Early attempts to trace the history of chromosomal change in some of these groups were based on rather simple models. For example, George and Weir (1974) hypothesized that the general model of chromosome evolution in hystricomorph rodents was a decrease in diploid number due to tandem fusions. We now know that chromosomal change in this taxon was extensive and involved complex rearrangements in some instances. Recent phylogenetic treatments of the ctenomyids have suggested tremendous complexity of chromosomal rearrangements (e.g., $2N = 10$ *C. steinbachi* as sister to $2N = 44$ *C. boliviensis:* Cook and Yates 1994; Lessa and Cook 1998). Speciation models based on the most parsimonious treatment of standard (George and Weir 1974), C-banded (e.g., Gallardo 1992), or G-banded chromosomes (Ortells 1995) should be tested with phylogenies derived from independent character sets (e.g., morphological characters or DNA sequences). A phylogenetic framework would also help in attempts to examine the processes or mechanisms responsible for these extensive karyotypic rearrangements. For example, interpretation of the relationship between chromosomal evolution and a highly repetitive sequence that is apparently unique to the Ctenomyidae (Rossi et al. 1995) would benefit from a phylogenetic framework.

Comparative Biology

In addition to further investigations of macroevolutionary patterns across subterranean rodents, it is evident that most comparative studies of these taxa would benefit from the ability to frame questions within an evolutionary setting. Recent examples of this approach include studies of structural constraints (Lessa and Patton 1989), host/parasite cospeciation and comparisons of rates of molecular evolution (Hafner, Demastes, and Spradling, chap. 10, this volume), the evolution of sociality (Lacey, chap. 7, this volume), the evolution of features related to fossoriality (Lessa 1990; Stein, chap. 1, this volume), and the evolution of genome size (Ruedas et al. 1993). This abbreviated list of recent research suggests that there will be many additional exciting prospects once we have a reasonable understanding of the historical relationships of subterranean organisms.

Summary and Future Directions

The subterranean niche was invaded by eight extant independent lineages of subterranean rodents, the Geomyidae, Arvicolinae, Myospalacinae, Rhizomyinae, Spalacinae, Bathyergidae, Ctenomyidae, and Octodontidae. Numerous studies have attempted to elucidate the mechanisms that have produced high diversity in several of these taxa. Some investigators have suggested that these rodents are classic examples of adaptive radiations associated with the independent invasion of the subterranean niche.

The basic premise that the subterranean lifestyle triggered explosive speciation has not been tested previously across phylogenetically independent lineages. Further, the partitioning of variation, rate of evolution, and speciation pulses have not been adequately characterized. Some clades are highly speciose and previously were suggested to be actively speciating based on the presence of intraspecific (and intrapopulation) chromosomal variation and high levels of genetic differentiation as measured by allozymes (and, more recently, DNA sequences). The case for high rates of diversification in other clades, however, is equivocal or poorly demonstrated.

We explored the apparent increase in species diversification rate in these groups using four statistical tests, including two-taxon, three-taxon, tree balance, and lineages-through-time analyses. We also reviewed the occupation of the subterranean niche and the pattern of generic diversification of these lineages in the fossil record. Further, we examined patterns of variation, speciation, and extinction as demonstrated by the shape of phylogenetic trees and the timing and potential causes of generic diversification through the Cenozoic. We found little support for the assertion that subterranean rodents experienced rapid species or generic diversification. The long periods of relative stability seen in the fossil record for these clades suggests that adaptations related to the subterranean lifestyle were not key innovations that spurred diversification. Additional paleontological, phylogenetic, and taxonomic work needs to be completed to test these preliminary conclusions.

Acknowledgments

We appreciate the invitation by Eileen Lacey, Guy Cameron, and Jim Patton to participate in this synopsis of subterranean rodents. Chris Conroy assisted with the development of the figures. We appreciate comments from Chris Bell, Norma Roche, and three anonymous reviewers. We were supported partially by a Sloan Foundation postdoctoral fellowship and grants from CONICYT, CSIC, and PEDECIBA (E. P. L.), a Fulbright Fellowship and a

National Science Foundation visitation award (J. A. C.), and a grant from the National Science Foundation (EPS9640667 to E. A. H.).

Literature Cited

Akersten, W. A. 1973. Evolution of geomyoid rodents with rooted cheek teeth. Ph.D. dissertation, University of Michigan, Ann Arbor.

Alberdi, M. T., A. N. Menegaz, J. L. Prado, and E. P. Tonni. 1989. La fauna local Quequen Salado-Indio Rico (Pleistoceno Tardio) de la provincia de Buenos Aires, Argentina. Aspectos paleoambientales y biostratigraficos. *Ameghiniana* 25:225–36.

Alpagut, B., P. Andrews, and L. Martin. 1990. Miocene paleoecology of Pasalar, Turkey. In *European mammal chronology*, edited by E. H. Lindsay, 443–59. New York: Plenum Press.

Anderson, S. 1997. Mammals of Bolivia, taxonomy and distribution. *Bulletin of the American Museum of Natural History*, 231. New York: American Museum of Natural History.

Barnosky, A. D. 1989. The late Pleistocene event as a paradigm for widespread mammal extinction. In *Mass extinctions*, edited by S. Donovan, 235–54. New York: Columbia University Press.

Barraclough, T. G., P. Harvey, and S. Nee. 1995. Sexual selection and taxonomic diversity in passerine birds. *Proceedings of the Royal Society of London* B 259:211–15.

Barrett, S. C. H., and S. W. Graham. 1997. Adaptive radiation in the aquatic plant family Pontederiaceae. In *Molecular evolution and adaptive radiation*, edited by T. J. Givnish and K. J. Sytsma, 225–58 Cambridge: Cambridge University Press.

Barry, J. C., and L. J. Flynn. 1990. Key biostratigraphic events in the Siwalik sequence. In *European mammal chronology*, edited by E. H. Lindsay, 557–71. New York: Plenum Press.

Baryshnikov, G. F. 1991. Vertebrates of the Barakayevskaya Mousterian site in the northern Caucasus. *Proceedings of the Zoological Institute, USSR Academy of Science* 238:139–66.

Berggren, W. A., D. V. Kent, L. J. Flynn, and J. A. Van Couvering. 1985. Cenozoic geochronology. *Geological Society of America Bulletin* 96:1407–18.

Bogert, C. M., W. F. Blair, E. R. Dunn, E. R. Hall, C. L. Hubbs, E. Mayr, and G. G. Simpson. 1943. Criteria for vertebrate subspecies, species and genera. *Annals of the New York Academy of Sciences* 44:105–88.

Briggs, J. C. 1995. *Global biogeography*. New York: Elsevier.

Brooks, D. R., and D. A. McLennan. 1991. *Phylogeny, ecology, and behavior*. Chicago: University of Chicago Press.

Brown, J. K. M. 1994. Probabilities of evolutionary trees. *Systematic Biology* 43:78–91.

Bush, G. L. 1975. Modes of animal speciation. *Annual Review of Ecology and Systematics* 6:339–64.

Carleton, M. D., and G. G. Musser. 1984. Muroid rodents. In *Orders and families of Recent mammals of the world*, edited by S. Anderson and J. K. Jones, Jr., 289–380. New York: John Wiley & Sons.

Cione, A. L., and E. P. Tonni. 1995. Chronostratigraphy and "Land-Mammal Ages" in the Cenozoic of southern South America: Principles, practices, and the "Uquian" problem. *Journal of Paleontology* 69:135–59.

Colless, H. 1982. *Phylogenetics: The theory and practice of phylogenetic systematics* II. Book review. *Systematic Zoology* 31:100–104.

Contreras, L. C., J. C. Torres-Mura, and A. E. Spotorno. 1990. The largest known chromosome number for a mammal in a South American desert rodent. *Experientia* 46:506–9.

Contreras, L. C., J. C. Torres-Mura, A. Spotorno, and F. M. Catzeflies. 1993. Morphological variation of the glans penis of South American octodontid and abrocomid rodents. *Journal of Mammalogy* 74:926–35.

Cook, J. A., S. Anderson, and T. L. Yates. 1990. Notes on Bolivian mammals 6: The genus *Ctenomys* (Rodentia: Ctenomyidae) in the highlands. *American Museum Novitates* 2980: 1–27.

Cook, J. A., and E. P. Lessa. 1998. Are rates of diversification in subterranean South American tuco-tucos (genus *Ctenomys*, Rodentia: Octodontidae) unusually high? *Evolution* 52:1521–27.

Cook, J. A., and T. L. Yates. 1994. Systematic relationships of the Bolivian tuco-tucos, genus *Ctenomys* (Rodentia: Ctenomyidae). *Journal of Mammalogy* 75:583–99.

Corbet, G. B. 1978. *The mammals of the Palearctic region: A taxonomic review.* London: British Museum of Natural History.

Crowley, T. J. 1991. Modeling Pliocene warmth. *Quaternary Science Reviews* 10:275–82.

Crowley, T. J., and G. R. North. 1991. *Paleoclimatology.* Oxford Monographs on Geology and Geophysics, 18. Oxford: Oxford University Press.

Dawson, M., and L. Krishtalka. 1984. Fossil history of the families of Recent mammals. In *Orders and families of Recent mammals of the world,* edited by S. Anderson and J. K. Jones, Jr., 11–57. New York: John Wiley & Sons.

de Bruijn, H. 1984. Remains of the mole-rat *Microspalax odessanus* Topachevskii from Karaburun (Greece, Macedonia) and the family Spalacidae. *Proceedings Koninklijke Nederlandse Akademie van Wetenschappen* B 87:417–25.

D'Elía, G., E. P. Lessa, and J. A. Cook. 1999. Molecular phylogeny of tuco-tucos, genus *Ctenomys* (Rodentia: Octodontidae): Evaluation of the *mendocinus* group and the evolution of asymmetric sperm. *Journal of Mammalian Evolution* 6:19–38.

Denys, C. 1987. Fossil rodents (other than Pedetidae) from Laetoli. In *Laetoli, a Pliocene site in northern Tanzania,* edited by M. D. Leakey and J. M. Harris, 118–70. Oxford: Oxford University Press.

Fahlbusch, V. 1985. Origin and evolutionary relationships among geomyoids. In *Evolutionary relationships among rodents: A multidisciplinary analysis,* edited by W. P. Luckett and J. L. Hartenberger, 617–29. New York: Plenum Press.

Faulkes, C. G., D. H. Abbott, H. P. O'Brien, L. Lau, M. R. Roy, R. K. Wayne, and M. W. Bruford. 1997. Micro- and macrogeographical genetic structure of colonies of naked mole-rats *Heterocephalus glaber. Molecular Ecology* 6:615–28.

Fejfar, O. 1990. The Neogene VP sites of Czechoslovakia: A contribution to the Neogene terrestrial biostratigraphy of Europe based on rodents. In *European mammal chronology,* edited by E. H. Lindsay, 211–36. New York: Plenum Press.

Fejfar, O., and W.-D. Heinrich. 1986. Biostratigraphic subdivision of the European Cenozoic based on muroid rodents (Mammalia). *Memoirs Sociedad Geologie Italia* 31:185–90.

———. 1990. Muroid rodent biochronology of the Neogene and Quaternary in Europe. In *European mammal chronology,* edited by E. H. Lindsay, 91–117. New York: Plenum Press.

Fejfar, O., and G. Storch. 1990. A Pliocene (late Ruscinian) small mammal fauna from Gundersheim, Rheinhessen. *Senckenbergiana lethaea* 71:139–84.

Filippucci, M. G., H. Burda, E. Nevo, and J. Kocka. 1992. Allozyme divergence and systematics of common mole-rats (*Cryptomys,* Bathyergidae, Rodentia) from Zambia. *Zeitschrift für Säugetierkunde* 59:42–51.

Flynn, L. J. 1982. Systematic revision of Siwalik Rhizomyidae (Rodentia). *Geobios* 15: 328–89.

———. 1990. The natural history of rhizomyid rodents. In *Evolution of subterranean mammals at the organismal and molecular levels,* edited by E. Nevo and O. A. Reig, 155–83. Progress in Clinical and Biological Research, vol. 335. New York: Wiley-Liss.

———. 1993. A new bamboo rat from the late Miocene of Yushe Basin. *Vertebrata PalAsiatica* 31:97–101.

Flynn, L. J., L. L. Jacobs, and E. H. Lindsay. 1985. Problems in muroid phylogeny: Relationship to other rodents and origin of major groups. In *Evolutionary relationships among rodents: A multidisciplinary analysis*, edited by W. P. Luckett and J.-L. Hartenberger, 589–616. New York: Plenum Press.

Flynn, L. J., and M. Sabatier. 1984. A muroid rodent of Asian affinity from the Miocene of Kenya. *Journal of Vertebrate Paleontology* 3:160–65.

Gallardo, M. 1992. Karyotypic evolution in octodontid rodents based on C-band analysis. *Journal of Mammalogy* 73:89–98.

Genise, J. F. 1989. Las cuevas con *Actenomys* (Rodentia, Octodontidae) de la formación Chapadmalal (Plioceno Superior) de Mar del Plata y Miramar (Provincia de Buenos Aires). *Ameghiniana* 26:33–42.

George, W., and B. J. Weir. 1974. Hystricomorph chromosomes. In *The biology of hystricomorph rodents*, edited by I. W. Rowlands and B. J. Weir, 79–108. Symposia of the Zoological Society of London, 34. London: Academic Press.

Givnish, T. J. 1997. Adaptive radiation and molecular systematics: Issues and approaches. In *Molecular evolution and adaptive radiation*, edited by T. J. Givnish and K. J. Sytsma, 1–54. Cambridge: Cambridge University Press.

Givnish, T. J., and K. J. Sytsma. 1997. *Molecular evolution and adaptive radiation*. Cambridge: Cambridge University Press.

Gromov, I. M., and I. Y. Polyakov. 1992. *Fauna of the USSR*. Vol. 3. *Mammals*. No. 8. *Voles (Microtinae)*. English translation. Washington, DC: Smithsonian Institution Libraries and the National Science Foundation.

Hafner, M. S. 1982. A biochemical investigation of geomyoid systematics (Mammalia: Rodentia). *Zeitschrift für Zoologische Systematik und Evolutionsforschung* 20:118–30.

———. 1991. Evolutionary genetics and zoogeography of Middle American pocket gophers, genus *Orthogeomys*. *Journal of Mammalogy* 72:1–10.

Hafner, M. S., P. D. Sudman, F. X. Villablanca, T. A. Spradling, J. W. Demastes, and S. A. Nadler. 1994. Disparate rates of molecular evolution in cospeciating hosts and parasites. *Science* 265:1087–90.

Hall, E. R. 1981. *Mammals of North America*. New York: John Wiley & Sons.

Harvey, P. H., R. M. May, and S. Nee. 1994. Phylogenies without fossils: Estimating lineage birth and death rates. *Evolution* 48:523–29.

Harvey, P. H., and M. D. Pagel. 1991. *The comparative method in evolutionary biology*. Oxford: Oxford University Press.

Heard, S. B. 1992. Patterns in tree balance among cladistic, phenetic, and randomly generated phylogenetic trees. *Evolution* 46:1818–26.

Hendey, Q. B. 1976. The Pliocene fossil occurrences in "E" Quarry Langabaanweg, South Africa. *Annals of the South African Museum, Cape Town* 69:215–47.

Hodges, S. A., and M. L. Arnold. 1995. Spurring plant diversification: Are floral nectar spurs a key innovation? *Proceedings of the Royal Society of London* B 262:343–48.

Hofmeijer, G. K., and H. de Bruijn. 1985. The mammals from the lower Miocene of Aliveri (Island of Evia, Greece), 4: The Spalacidae and Anomalomyidae. *Proceedings Koninklijke Nederlandse Akademie van Wetenschappen* B 88:185–98.

Honeycutt, R. L., M. W. Allard, S. V. Edwards, and D. A. Schlitter. 1991. Systematics and evolution of the Bathyergidae. In *The biology of the naked mole-rat*, edited by P. W. Sherman, J. U. M. Jarvis, and R. D. Alexander, 45–65. Princeton, NJ: Princeton University Press.

Honeycutt, R. L., S. V. Edwards, K. Nelson, and E. Nevo. 1987. Mitochondrial DNA variation and the phylogeny of the African mole rats (Rodentia: Bathyergidae). *Systematic Zoology* 36:280–92.

Honeycutt, R. L., and S. L. Williams. 1982. Genic differentiation in pocket gophers of the genus *Pappogeomys*, with comments on intergeneric relationships in the subfamily Geomyinae. *Journal of Mammalogy* 63:208–17.

Hugueney, M., and P. Mein. 1993. A comment on the earliest Spalacinae (Rodentia, Muroidea). *Journal of Mammalian Evolution* 1:215–23.

Jacobs, L. L., L. J. Flynn, and W. R. Downs. 1990. Neogene rodents of southern Asia. In *Papers on fossil rodents in honor of Albert Elmer Wood*, edited by C. C. Black and M. R. Dawson, 157–77. Science Series, no. 33. Los Angeles: Natural History Museum of Los Angeles County.

Jacobs, L. L., L. J. Flynn, Downs, W. R., and J. C. Barry. 1990. Quo vadis, *Antemus?* The Siwalik muroid record. In *European mammal chronology*, edited by E. H. Lindsay, 573–86. New York: Plenum Press.

Jaeger, J. J. 1988. Rodent phylogeny: New data and old problems. In *The phylogeny and classification of the tetrapods*, vol. 2, *Mammals*, edited by M. J. Benton, 177–99. Systematics Association Special Volume no. 35B. Oxford: Clarendon Press.

Janis, C. M. 1993. Tertiary mammal evolution in the context of changing climates, vegetation and tectonic events. *Annual Review of Ecology and Systematics* 24:467–500.

Jánossy, D. 1986. *Pleistocene vertebrate faunas of Hungary*. Budapest: Akadémiai Kiadó.

Jarvis, J. U. M., and N. C. Bennett. 1990. The evolutionary history, population biology, and social structure of African mole-rats: Family Bathyergidae. In *Evolution of subterranean mammals at the organismal and molecular levels*, edited by E. Nevo and O. A. Reig, 97–128. Progress in Clinical and Biological Research, vol. 335. New York: Wiley-Liss.

Jin, C., and Q. Xu. 1984. The Quaternary mammalian faunas from Qingshantou site, Jilin Province. *Vertebrata PalAsiatica* 22:314–23.

Just, W., W. Rau, W. Vogel, M. Akhverdian, K. Fredga, J. A. M. Graves, and E. Lyapunova. 1995. Absence of *SRY* in species of the vole *Ellobius*. *Nature Genetics* 11:117–18.

Kingdon, J. 1997. *The Kingdon guide to African mammals*. New York: Academic Press.

Kirkpatrick, M., and M. Slatkin. 1993. Searching for evolutionary patterns in the shape of a phylogenetic tree. *Evolution* 47:1171–81.

Korth, W. W. 1993. Review of the Oligocene (Orellan and Arikareean) genus *Tenudomys* Rensberger (Geomyoidea: Rodentia). *Journal of Vertebrate Paleontology* (Supplement): 41A.

———. 1994. *The Tertiary record of rodents in North America*. New York: Plenum Press.

Kowalski, K. 1966. The stratigraphic importance of rodents in the studies on the European Quaternary. *Folia Quaternaria* 22:1–16.

———. 1990. Stratigraphy of Neogene mammals of Poland. In *European mammal chronology*, edited by E. H. Lindsay, 193–210. New York: Plenum Press.

Lavocat, R. 1978. Rodentia and Lagomorpha. In *Evolution of African mammals*, edited by V. J. Maglio and H. B. S. Cooke, 69–89. Cambridge, MA: Harvard University Press.

Lawrence, M. A. 1991. A fossil *Myospalax* cranium (Rodentia: Muridae) from Shanxi, China, with observations on Zokor relationships. In *Contributions to mammalogy in honor of Karl F. Koopman*, edited by T. A. Griffiths and D. Klingener, 261–86. *Bulletin of the American Museum of Natural History*, 206. New York: American Museum of Natural History.

Lessa, E. P. 1990. Morphological evolution of subterranean mammals: Integrating structural, functional, and ecological perspectives. In *Evolution of subterranean mammals at the organismal and molecular levels*, edited by E. Nevo and O. A. Reig, 211–30. Progress in Clinical and Biological Research, vol. 335. New York: Wiley-Liss.

Lessa, E. P., and J. A. Cook. 1998. The molecular phylogenetics of tuco-tucos (genus *Ctenomys*, Rodentia: Octodontidae) suggests an early burst of speciation. *Molecular Phylogenetics and Evolution* 9:88–99.

Lessa, E. P., and J. L. Patton. 1989. Structural constraints, recurrent shapes, and allometry in pocket gophers (genus *Thomomys*). *Biological Journal of the Linnean Society* 36:349–63.

Lundelius, E. L., Jr., T. Downs, E. H. Lindsay, H. A. Semken, R. J. Zakrewski, C. S. Churcher, C. R. Harrington, G. E. Schultz, S. D. Webb. 1987. The North American

Quaternary sequence. In *Cenozoic mammals of North America*, edited by M. O. Woodburne, 211–35. Berkeley: University of California Press.

MacFadden, B. J. 1992. Interpreting extinctions from the fossil record: Methods, assumptions, and case examples using fossil horses (Family Equidae). In *Extinction and phylogeny*, edited by M. J. Novacek and Q. D. Wheeler, 17–45. New York: Columbia University Press.

Markova, A. 1992. Fossil rodents (Rodentia, Mammalia) from the Sel'Ungur Achelulian cave site (Kirghizstan). *Acta Zoologica Cracoviensia* 35:217–39.

Markova, A. K., N. G. Smirnov, A. V. Kozharinov, N. E. Kazantseva, A. N. Simakova, and L. M. Kitaev. 1995. Late Pleistocene distribution and diversity of mammals in northern Eurasia. *Paleontologia i Evolució* 28–29:5–143.

Marshall, L. G., and T. Sempere. 1993. Evolution of the Neotropical land mammal fauna in its geochronologic, stratigraphic, and tectonic context. In *Biological relationships between Africa and South America*, edited by P. Goldblatt, 329–92. New Haven, CT: Yale University Press,

Mayr, E. 1963. *Animal species and evolution*. Cambridge: Cambridge University Press.

Mein, P. 1990. Updating of MN zones. In *European mammal chronology*, edited by E. H. Lindsay, 73–90. New York: Plenum Press.

Mitter, C., B. Farrell, and B. Wiegmann. 1988. The phylogenetic study of adaptive zones: Has phytophagy promoted insect diversification? *American Naturalist* 132:107–28.

Montalvo, C. I., and S. Casadio. 1988. Presencia del genero *Palaeoctodon* (Rodentia, Octodontindae) en el Huayqueriense (Mioceno Tardio) de la Provincia de la Pampa. *Ameghiniana* 25:111–14.

Mooers, A. O. 1995. Tree balance and tree completeness. *Evolution* 49:379–84.

Mooers, A., and S. B. Heard. 1997. Inferring evolutionary process from phylogenetic tree shape. *Quarterly Review of Biology* 72:31–54.

Musser, G. G., and M. D. Carleton. 1993. Family Muridae. In *Mammal species of the world: A taxonomic and geographic reference*, 2d ed., edited by D. E. Wilson and D. M. Reeder, 501–756. Washington, DC: Smithsonian Institution Press.

Nadachowski, A. 1990. Review of fossil Rodentia from Poland (Mammalia). *Senckenbergiana Biologica* 70:229–50.

———. 1992. Early Pleistocene *Predicrostonyx* (Rodentia, Mammalia) from Poland. *Acta Zoologica Cracoviensia* 35:203–16.

———. 1993. The species concept and Quaternary mammals. *Quaternary International* 19:9–11.

Nedbal, M. A., M. A. Allard, and R. L. Honeycutt. 1994. Molecular systematics of hystricognath rodents: Evidence from the mitochondrial 12S rRNA gene. *Molecular Phylogenetics and Evolution* 3:206–20.

Nevo, E. 1979. Adaptive convergence and divergence of subterranean mammals. *Annual Review of Ecology and Systematics* 10:269–308.

Nevo, E., R. Ben Shlomo, A. Beiles, J. U. M. Jarvis, and G. C. Hickman. 1987. Allozyme differentiation and systematics of the endemic subterranean mole-rats of South Africa (Rodentia: Bathyergidae). *Biochemical Systematics and Ecology* 15:489–502.

Nowak, R. M. 1991. *Walker's mammals of the world*. 5th edition. Baltimore: Johns Hopkins University Press.

Ognev, S. I. 1963. *Mammals of the USSR and adjacent countries*. Vol. 5. *Rodents*. Moscow: Academy of Sciences USSR.

Ortells, M. O. 1995. Phylogenetic analysis of G-banded karyotypes among the South American subterranean rodents of the genus *Ctenomys* (Caviomorpha: Octodontidae), with special reference to chromosomal evolution and speciation. *Biological Journal of the Linnean Society* 54:43–70.

Pardiñas, U. F. J., and M. J. Lezcano. 1995. Cricetidos (Mammalia: Rodentia) del Pleistoceno Tardio del nordeste de la provincia de Buenos Aires (Argentina). Aspectos sistematicos y paleoambientales. *Ameghiniana* 32:249–65.

Pascual, R., E. O. Jaureguizar and J. L. Prado. 1996. Land mammals: Paradigm for Cenozoic South American geobiotic evolution. *Münchner Geowissenschaftliche Abhandlungen* 30:265–319.

Patton, J. L. 1985. Population structure and the genetics of speciation in pocket gophers, genus *Thomomys*. *Acta Zoologica Fennica* 170:109–14.

———. 1993. Family Geomyidae. In *Mammal species of the world: A taxonomic and geographic reference*, 2d ed., edited by D. E. Wilson and D. M. Reeder, 469–76. Washington, DC: Smithsonian Institution Press.

Patton, J. L., and M. F. Smith. 1990. The evolutionary dynamics of the pocket gopher *Thomomys bottae*, with emphasis on California populations. *University of California Publications in Zoology* 123:1–161.

Pearson, O. P., and A. K. Pearson. 1982. Ecology and biogeography of the southern rainforests of Argentina. In *Mammalian biology in South America*, edited by M. A. Mares and H. H. Genoways, 129–42. Special Publications Series, vol. 6. Linesville, PA: Pymatuning Laboratory of Ecology, University of Pittsburgh.

Popov, V. V. 1988. Middle Pleistocene small mammals (Mammalia: Insectivora, Lagomorpha, Rodentia) from Varbeshnitsa (Bulgaria). *Acta Zoologica Cracoviensia* 31: 193–234.

Potts, R., and A. K. Behrensmeyer. 1992. Late Cenozoic terrestrial ecosystems. In *Terrestrial ecosystems through time*, edited by A. K. Behrensmeyer, J. D. Damuth, W. A. DiMichele, R. Potts, H.-D Sues, and S. L. Wing, 419–541. Chicago: University of Chicago Press.

Preston, F. W. 1962. The canonical distribution of commonness and rarity. *Ecology* 43:185–215; 410–32.

Purvis, A. 1996. Using interspecies phylogenies to test macroevolutionary hypotheses. In *New uses for new phylogenies*, edited by P. H. Harvey, A. J. Leigh Brown, J. Maynard Smith, and S. Nee, 153–68. Oxford: Oxford University Press.

Qiu, Z. 1990. The Chinese Neogene mammalian biochronology: Its correlation with the European Neogene mammalian zonation. In *European mammal chronology*, edited by E. H. Lindsay, 527–56. New York: Plenum Press.

Quintana, C. A. 1994. Ctenominos primitivos (Rodentia, Octodontidae) del Mioceno de la Provincia de Buenos Aires, Argentina. *Boletín de la Real Sociedad Española de Historia Natural* (Sección Geológica) 89:19–23.

Rambaut, A., P. H. Harvey, and S. Nee. 1997. End-Epi: An application for reconstructing phylogenetic sequence and population processes from molecular sequences. *Computer Applications in the Biological Sciences* 13:303–6.

Raymo, M. E. 1992. Global climate change: A three million year perspective. In *Start of a glacial*, edited by G. J. Kukla and E. Went, 207–23. NATO ASI Series, series 1, Global Environmental Change, vol. 3. Berlin: Springer-Verlag.

Raymo, M. E., W. F. Ruddiman, J. Backman, B. M. Clement, and D. G. Martinson. 1989. Late Pliocene variation in Northern Hemisphere ice sheets and North Atlantic deep water circulation. *Paleoceanography* 4:413–46.

Raup, D. M. 1978. Cohort analysis of generic survivorship. *Paleobiology* 4:1–15.

Redford, K. H., and J. F. Eisenberg. 1992. *Mammals of the Neotropics*. Vol. 2. *The southern cone*. Chicago: University of Chicago Press.

Reig, O. A. 1989. Karyotypic repatterning as the triggering factor in cases of explosive speciation. In *Evolutionary biology of transient, unstable populations*, edited by A. Fontedevila, 246–89. New York: Springer-Verlag.

Reig, O. A., C. Busch, M. O. Ortells, and J. Contreras. 1990. An overview of evolution, systematics, population biology, cytogenetics, molecular biology and speciation in *Ctenomys*. In *Evolution of subterranean mammals at the organismal and molecular levels*, edited by

E. Nevo and O. A. Reig, 71–96. Progress in Clinical and Biological Research, vol. 335. New York: Wiley-Liss.
Reig, O. A., and C. A. Quintana. 1991. A new genus of fossil Octodontine rodent from the early Pleistocene of Argentina. *Journal of Mammalogy* 72:292–99.
———. 1992. Fossil Ctenomyine rodents of the genus *Eucelophorus* (Caviomorpha: Octodontidae) from the Pliocene and early Pleistocene of Argentina. *Ameghiniana* 29: 363–80.
Repenning, C. A. 1984. Quaternary rodent biochronology and its correlation with climatic and magnetic stratigraphies. In *Correlation of Quaternary chronologies*, edited by W. C. Mahaney, 105–19. Norwich: Geo Books.
———. 1987. Biochronology of the microtine rodents of the United States. In *Cenozoic mammals of North America*, edited by M. O. Woodburne, 236–68. Berkeley: University of California Press.
Repenning, C. A., O. Fejfar, and W.-D. Heinrich. 1990. Arvicolid biochronology of the Northern Hemisphere. In *International symposium: Evolution, phylogeny, and biostratigraphy of arvicolids (Rodentia, Mammalia)*, edited by O. Fejfar and W.-D. Heinrich, 385–417. Prague: Geological Survey.
Ricklefs, R. E., and D. Schluter, eds. 1993. *Species diversity in ecological communities: Historical and geographical perspectives*. Chicago: University of Chicago Press.
Riddle, B. R. 1996. The molecular phylogeographic bridge between deep and shallow history in continental biotas. *Trends in Ecology and Evolution* 11:207–12.
Roig, V. G., and O. A. Reig. 1969. Precipitin test relationships among Argentinian species of the genus *Ctenomys* (Rodentia: Octodontidae). *Comparative Biochemistry and Physiology* 30:665–72.
Rossi, M. S., C. A. Redi, G. Viale, A. I. Massarini, and E. Capanna. 1995. Chromosomal distribution of the major satellite DNA of South American rodents of the genus *Ctenomys*. *Cytogenetics and Cell Genetics* 69:179–84.
Ruedas, L., J. A. Cook, T. Yates, and J. Bickham. 1993. Conservative genome size evolution in tuco-tucos (Rodentia: Ctenomyidae). *Genome* 36:449–58.
Russell, R. J. 1968. Evolution and classification of the pocket gophers of the subfamily Geomyinae. *Miscellaneous Publications, Museum of Natural History, University of Kansas* 16:473–579.
Sabatier, M. 1978. Un nouveau *Tachyoryctes* (Mammalia, Rodentia) du bassin Pliocene de hader (Ethiopie). *Geobios* 11:95–99.
Sanderson, M. J., and M. J. Donoghue. 1994. Shifts in diversification rate with the origin of angiosperms. *Science* 264:1590–93.
———. 1996. Reconstructing shifts in diversification on phylogenetic trees. *Trends in Ecology and Evolution* 11:15–20.
Sanderson, M. J., and M. F. Wojciechowski. 1996. Diversification rates in a temperate legume clade: Are there "so many species" of *Astragalus* (Fabaceae)? *American Journal of Botany* 83:1488–1502.
Sage, R. D., J. R. Contreras, V. G. Roig, and J. L. Patton. 1986. Genetic variation in the South American burrowing rodents of the genus *Ctenomys* (Rodentia: Ctenomyidae). *Zeitschrift für Säugetierkunde* 51:158–72.
Savage, D. E., and D. E. Russell. 1983. *Mammalian paleofaunas of the world*. London: Addison-Wesley.
Savage, R. J. G. 1990. The African dimension in European early Miocene mammal faunas. In *European mammal chronology*, edited by E. H. Lindsay, 587–99. New York: Plenum Press.
Savic, I. R., and E. Nevo. 1990. The Spalacidae: Evolutionary history, speciation and population biology. In *Evolution of subterranean mammals at the organismal and molecular levels*, edited by E. Nevo and O. A. Reig, 129–53. Progress in Clinical and Biological Research, vol. 335. New York: Wiley-Liss.

Shao, K.-T., and R. R. Sokal. 1990. Tree balance. *Systematic Zoology* 39:266–76.
Simpson, G. G. 1953. *The major features of evolution*. New York: Columbia University Press.
———. 1961. *Principles of animal taxonomy*. New York: Columbia University Press.
Slowinski, J. B., and C. B. Guyer. 1989. Testing the stochasticity of patterns of organismal diversity: An improved null model. *American Naturalist* 134:907–21.
Slowinski, J. B., and C. B. Guyer. 1993. Testing whether certain traits have caused amplified diversification; an improved method based on a model of random speciation and extinction. *American Naturalist* 142:1019–24.
Smith, M. F. 1998. Phylogenetic relationships and geographic structure in pocket gophers in the genus *Thomomys*. *Molecular Phylogenetics and Evolution* 9:1–14.
Spotorno, A. E. 1979. Contrastación de la macrosistematica de roedores caviomorfos por analisis de la morfologia reproductiva masculina. *Archivos de Biología y Medicina Experimentales* 12:97–106.
Stanley, S. M. 1979. *Macroevolution*. San Francisco: W. H. Freeman.
Steininger, F., R. L. Bernor, and V. Fahlbusch. 1990. European Neogene marine/continental chronologic correlations. In *European mammal chronology*, edited by E. H. Lindsay, 15–46. New York: Plenum Press.
Storch, G., and O. Fejfar. 1990. Gundersheim-Findling, a Ruscinian rodent fauna of Asian affinities from Germany. In *European mammal chronology*, edited by E. H. Lindsay, 405–12. New York: Plenum Press.
Stucky, R. K. 1990. Evolution of land mammal diversity in North America during the Cenozoic. *Current Mammalogy* 3:375–432.
Sümengen, M., E. Ünay, G. Sarac, H. de Bruijn, I. Terlemez, and M. Gürbüz. 1990. New Neogene rodent assemblages from Anatolia (Turkey). In *European mammal chronology*, edited by E. H. Lindsay, 61–72. New York: Plenum Press.
Tchernov, E. 1992. The Afro-Arabian component in the Levantine mammalian fauna: A short biogeographical review. *Israel Journal of Zoology* 38:155–92.
Thaeler, C. S., Jr. 1980. Chromosome number and systematic relations in the genus *Thomomys* (Rodentia: Geomyidae). *Journal of Mammalogy* 61:414–22.
Tong, H., and J.-J. Jaeger. 1993. Muroid rodents from the middle Miocene Fort Ternan locality (Kenya) and their contribution to the phylogeny of muroids. *Palaeontographica* A 229:51–73.
Tonni, E. P., M. T. Alberdi, J. L. Prado, M. S. Bargo, and A. L Cione. 1992. Changes of mammal assemblages in the pampean region (Argentina) and their relation with the Plio-Pleistocene boundary. *Palaeogeography, Palaeoclimatology, Palaeoecology* 95:179–94.
Tonni, E. P., M. S. Bargo, and J. L Prado. 1988. Los cambios ambientales en el Pleistoceno Tardio y Holoceno del sudeste de la provincia de Buenos Aires a traves de una secuencia de mamiferos. *Ameghiniana* 25:99–110.
Topachevskii, V. A. 1976. *Fauna of the USSR*. Vol. 3. *Mammals*. No. 3. *Mole-rats, Spalacidae*. Washington, DC: Smithsonian Institution.
Ünay, E. 1978. *Pliospalax primitivus* n. sp. and *Anomalomys gaudryi* Gaillard from the *Anchitherium* fauna of Sricay (Turkey). *Bulletin of the Geological Society of Turkey* 21:121–28.
Ünay, E., and H. de Bruijn. 1984. On some Neogene rodent assemblages from both sides of the Dardanelles. *Newsletters on Stratigraphy* 13:119–32.
Vereshchagin, N. K. 1967. The mammals of the Caucasus: A history of the evolution of the fauna. English translation. Jerusalem: Israel Program for Scientific Translations.
Verzi, D. H., and M. Lezcano. 1996. Status sistematico y antigüedad de *Megactenomys kraglievichi* Rusconi, 1930 (Rodentia, Octodontidae). *Revista Museo La Plata, Paleontologica* 9:239–46.
Verzi, D. H., C. I. Montalvo, and M. G. Vucetich. 1991. Nuevos restos de *Xenodontomys simpsoni* Kraglievich y la sistematica de los mas antiguos ctenomyinae (Rodentia, Octodontidae). *Ameghiniana* 28:325–31.

Vucetich, M. G., and D. H. Verzi. 1995. Los roedores caviomorfos. In *Evolución biológica y climática de la región pampeana durante los últimos cinco millones de años,* edited by M. T. Alberdi, G. Leone, and E. Tonni, 211–26. Museo Nacional de Ciencias Naturales Monografías, 12. Madrid: CSIC,
White, M. J. D. 1978. *Modes of speciation.* San Francisco: W. H. Freeman.
Wiley, E. O. 1981. *Phylogenetics: The theory and practice of phylogenetic systematics.* New York: John Wiley & Sons.
Wilson, D. E., and D. M. Reeder, eds. 1993. *Mammal species of the world: A taxonomic and geographic reference.* 2d edition. Washington, DC: Smithsonian Institution Press.
Wood, A. E. 1955. A revised classification of rodents. *Journal of Mammalogy* 36:165–86.
———. 1974. The evolution of the Old World and New World hystricomorphs. In *The biology of hystricomorph rodents,* edited by I. W. Rowlands and B. J. Weir, 21–60. Symposia of the Zoological Society of London, 34. London: Academic Press.
———. 1985. The relationships, origin, and dispersal of the hystricognath rodents. In *Evolutionary relationships among rodents: A multidisciplinary analysis,* edited by W. P. Luckett and J.-L. Hartenberger, 475–513. New York: Plenum Press.
Woods, C. A. 1982. The history and classification of South American rodents: Reflections on the far away and long ago. In *Mammalian biology in South America,* edited by M. A. Mares and H. H. Genoways, 377–92. Special Publications Series, vol. 6. Linesville, PA: Pymatuning Laboratory of Ecology, University of Pittsburgh.
———. 1984. Hystricognath rodents. In *Orders and families of Recent mammals of the world,* edited by S. Anderson and J. K. Jones, Jr., 389–446. New York: John Wiley & Sons.
———. 1993. Suborder Hystricognathi. In *Mammal species of the world: A taxonomic and geographic reference,* 2d ed., edited by D. E. Wilson and D. M. Reeder, 771–806. Washington, DC: Smithsonian Institution Press.
Zheng, S., and C. Li. 1990. Comments on fossil arvicolids of China. In *International symposium on evolution, phylogeny, and biostratigraphy of arvicolids (Rodentia, Mammalia),* edited by O. Fejfar and W.-D. Heinrich, 431–42. Prague: Geological Survey.

CHAPTER TEN

Coevolution and Subterranean Rodents

Mark S. Hafner, James W. Demastes, and Theresa A. Spradling

The word "coevolution" is a broad term that often means different things to different users. Accordingly, we begin our review of coevolution in subterranean rodents with a brief history and definition of the term. It was coined by Ehrlich and Raven (1964) in reference to a variety of ecological interactions between plants and the insects that feed on them. The fact that Ehrlich and Raven did not provide an explicit definition of their new term led immediately to its widespread application to describe almost any kind of interaction between or among species. As early as the late 1970s, the word "coevolution" was losing its usefulness because it was beginning to mean all things to all people. For example, "coevolution" was used in the literature almost interchangeably with ecological terms such as "symbiosis" and "mutualism." In response to this growing terminological confusion, Janzen (1980) argued that there was need for a precisely defined term "coevolution," and he proposed a restrictive definition of the term. Janzen's definition of coevolution required reciprocal evolution in coexisting species, meaning that *a trait in one species must have evolved in response to a trait in another species, which trait itself has evolved in response to the trait in the first* (Janzen 1980; Futuyma and Slatkin 1983). Despite almost universal acceptance of Janzen's (1980) restrictive definition of coevolution, well-documented and widely accepted examples of this phenomenon (which we term "strict coevolution") are essentially nonexistent in the literature.

In this chapter, we use a more relaxed definition of coevolution because we believe that strict application of Janzen's (1980) definition places an almost insurmountable obstacle in the path of researchers attempting to discover and understand coevolution in natural systems. For example, to document strict coevolution between symbiotic species A and B, one must demonstrate adaptation in species A, adaptation in species B, and—most importantly—that adaptation in species A is a direct response to the presence of species B, and vice versa. Any one of these requirements is formidable enough, but the combined set is effectively impossible to demonstrate. For example, many purported cases of coevolution in the literature show fairly strong evidence of adaptation on the part of species A, but not species B. This leaves most of the criteria for strict coevolution unmet. Also, it is no trivial matter to distinguish between a feature or character of an organism that evolved to serve its present-day function (hence is an adaptation) and a feature or character that evolved to serve another function, but has been co-opted to serve its present-day function (hence is an "exaptation": Gould and Vrba 1982). Thus, in many supposed instances of coevolution, it is likely that the so-called adaptations we see in one or both species are actually exaptations from an earlier biological association involving a third (or fourth) species that no longer is associated with the present-day system. These associations, although intimately symbiotic today, may represent recent and fortuitous "marriages of convenience" between organisms that evolved in separate evolutionary contexts. Clearly, such associations do not represent coevolution, but they can mimic coevolution amazingly well if viewed solely in an ecological (present-day) context. We refer to this phenomenon as "pseudo-coevolution."

The Role of Phylogenetics in the Study of Coevolution

Because documentation of causality in the study of adaptations and exaptations requires knowledge of history, it is impossible to distinguish between strict coevolution and pseudo-coevolution simply by studying the present-day ecology, morphology, or behavior of symbiotic organisms. Accordingly, a study of coevolution should begin with a historical examination of the symbiotic relationship in question. To document strict coevolution, we believe it is first necessary to demonstrate a long history of association between the symbiotic species in question. This requires a good fossil record for both species (which is almost never available for either, much less both) or a comparative phylogenetic analysis of the symbiotic lineages that documents a significant pattern of cospeciation. Importantly, evidence of a long-term association between two symbiotic lineages provides the *opportunity* for reciprocal adaptation in both lineages and, at the same time, eliminates the possibility of a recent, fortuitous association between the lineages (pseudo-coevolution). Although one cannot rule out rapid (and reciprocal) adaptive change in a relatively young symbiotic association (such as the well-known case of the rabbit and myxoma virus in Australia: Fenner 1965), the possibility always lingers that supposed reciprocal adaptations in a young symbiotic association actually evolved in another, older, evolutionary context and have been co-opted for use in the present-day symbiotic relationship. In contrast, it is difficult to imagine two lineages of organisms living symbiotically for millennia that have not adapted to each other's presence in some, perhaps imperceptible, way. To quote Pirozynski and Hawksworth (1988: 3): "The more intimate and stable a symbiosis, the greater is the probability of it being the outcome of coevolution."

Preoccupation with the elusive concept of "reciprocal adaptation" in the study of coevolution may cause us to mistake pseudo-coevolution for coevolution because we fail to consider the evolutionary origin of the supposed reciprocal adaptations. This same focus may cause us to conclude that an ancient symbiotic association is not coevolution simply because we lack evidence of reciprocal, or bidirectional, adaptation. Clearly, it is a mistake to assume that all intimate ecological interactions we see in nature are the result of coevolution (Janzen 1980), but it also is a mistake to assume that a demonstrably long-term ecological association between two species is not coevolution simply because we lack evidence of an adaptive response on the part of one of the two species. We agree with Futuyma and Slatkin (1983: 3) that "the study of coevolution is the analysis of reciprocal genetic changes *that might be expected to occur* in two or more ecologically interacting species and the analysis of whether the expected changes are actually realized" [emphasis ours]. Our point is simply this: "reciprocal genetic changes" are much more likely to occur in a symbiotic association that is demonstrably old. Accordingly, our operational definition of coevolution requires only that the association between two species be *intimate* (to exclude casual ecological associations) and *demonstrably old* (to reduce the likelihood of mistaking pseudo-coevolution for strict coevolution). Any association that meets both of these criteria almost certainly will have involved some level of reciprocal adaptation, which is the essential feature of strict coevolution.

Subterranean Rodents and the Study of Coevolution

Although subterranean rodents account for only about 6% of all living species of rodents (Hafner and Hafner 1988), their unusual lifestyle makes them ideal candidates for the study of coevolution. For a number of reasons, the fossorial mode of existence usually entails a solitary lifestyle, with individuals often living in single-occupant burrow systems within small, isolated populations (Pearson 1959; Patton 1972; Nevo 1979). This patchy distribution of populations usually reflects the patchy distribution of friable and flood-free soils, compounded by the often dispersed nature of suitable food resources, including the need for year-round availability of edible roots and tubers (Busch et al., chap. 5, this volume). Within these isolated populations, individual subterranean rodents often live alone in burrow systems that are defended vigorously against intruders, including conspecifics. Presumably, the enormous energy invested in the excavation of a complex system of tunnels argues against sharing of burrows in most subterranean rodents.

In the act of foraging through the soil, subterranean rodents create a cave environment that is ecologically very different from the typical surface environment. Through evolutionary time, many creatures have invaded this new environment created by subterranean rodents (Cameron, chap. 6, this volume), and some of these burrow occupants show adaptations that are specific to the unique biotic and abiotic regime in which they now live. Often, cave-specific adaptations, such as reduction of eyes and limbs, render the organism unfit for life on the surface, meaning that subterranean rodents and many of the organisms that coexist with them may be evolutionarily "locked" into a long-term ecological association. If so, then the likelihood of eventual biological interaction between these ecological partners is high—and biological interaction is the necessary precursor to reciprocal adaptation, or coevolution. Minimally, the distribution of a subterranean rodent species will influence (if not determine) the distribution of organisms that live in its burrow, which means that the latter creatures often will show the patchy distribution characteristic of so many species of subterranean rodents.

In most species of subterranean rodents, these factors work in concert to isolate individual rodents and their burrow commensals from conspecific individuals, from other subterranean organisms, and from surface-dwelling creatures in general. This geographic isolation, coupled with characteristically small population sizes in subterranean organisms, is a powerful catalyst for evolutionary change and provides ample opportunity for the evolution of long-term symbiosis between subterranean rodents and the many organisms that coexist with them.

Coevolution of Subterranean Rodents and Their Burrow Associates

For the purposes of discussion, we will recognize three classes of organisms that coexist with subterranean rodents. We will use the term "inquiline" to refer to an organism that is an obligate burrow associate of a subterranean rodent. Most inquilines both feed and reproduce in the burrow system, with the result that they often show evolutionary adaptations for life in caves, including reduction of eyes and limbs. We will use the term "transient" to refer to organisms that are casual or opportunistic users of burrow systems constructed by subterranean rodents. A transient may use a burrow system occasionally for shelter, or it may go so far as to forage and reproduce opportunistically in a burrow system. Regardless, the transient is only a facultative user of the burrow system, and rarely will a transient show adaptations for life in caves. Finally, we will use the term "parasite" to refer to organisms that gain most or all of their energy resources by living in or on a subterranean rodent. Parasites often show adaptations for their parasitic lifestyle, and they may also show secondary adaptations for life in caves. Although the terms inquiline, transient, and parasite are not mutually exclusive (e.g., an obligate parasite of a subterranean rodent also could be considered an inquiline), they represent three different lifestyles with different likelihoods for coevolution with their subterranean rodent host.

Subterranean Rodents and Their Burrow Inquilines

Coevolutionary studies of nonparasitic inquilines are exceedingly rare in the literature. In large part, this reflects the difficulty of studying small creatures that live in dark, narrow tunnels that rarely exceed 10 cm in diameter. One noteworthy exception is the thorough study of the burrows of the Florida pocket gopher, *Geomys pinetis*, conducted by Hubbell and Goff (1940). Hubbell and Goff showed that the burrow system of *G. pinetis* is a complex subterranean ecosystem (fig. 10.1) comprising the gopher and large numbers of burrow associates, including scavengers, phytophages, predators, coprophiles, and endo- and ectoparasites. Of the sixty-one species of burrow associates recorded by Hubbell and Goff (1940), thirteen species were determined to be inquilines (table 10.1).

Most of the inquilines listed in table 10.1 show adaptations for life in caves (e.g., the blind cricket, *Typhloceuthophilus*). This observation suggests a long history of subterranean existence on the part of these species, and it may reflect a long history of coexistence with this particular host, *G. pinetis* (this latter possibility could be explored by comparative phylogenetic analysis, which is discussed below). If so, then each of these inquilines represents a potential case study in coevolution, if it can be shown that the cave adaptations in the inquilines are direct responses to life with *G. pinetis* and if reciprocal adaptations can be demonstrated in *G. pinetis*. At the very least, each of these inquilines represents a potential case study in cospeciation (also discussed below).

COEVOLUTION 375

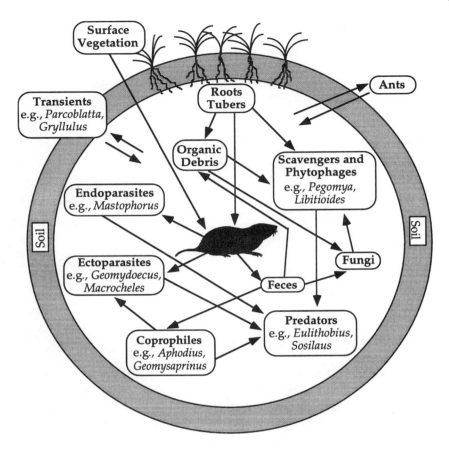

FIGURE 10.1. A diagrammatic depiction of the burrow ecosystem of a geomyid rodent and its burrow associates. Arrows indicate trophic relationships. (After Hubbell and Goff 1940)

Subterranean Rodents and Their Burrow Transients

The study of creatures that make only occasional or opportunistic use of burrow systems has little potential for revealing a pattern of coevolution between the transient and a subterranean rodent. As stated earlier, coevolution almost always requires a long history of symbiosis between two evolutionary partners, a requirement that clearly is not met in the case of subterranean rodents and burrow transients. Nevertheless, any study of a burrow ecosystem, including a study of coevolutionary partnerships within that ecosystem, must consider the ecological impact of the many transients that participate in the system (see fig. 10.1).

TABLE 10.1 INQUILINES (OBLIGATE BURROW ASSOCIATES) OF *GEOMYS PINETIS*

Order	Family	Species
Chilopoda	Lithobiidae	*Eulithobius hypogeus*
		Pholobius goffi
Arachnida	Aranae	*Sosilaus spiniger*
Thysanura	Lepismatidae	*Nicoletia* sp.
Orthoptera	Gryllacrididae	*Typhloceuthophilus floridanus*
Coleoptera	Histeridae	*Spilodiscus floridanus*
		Atholus minutus
		Geomysaprinus goffi
		Geomysaprinus tibialis
	Scarabaeidae	*Aphodius goffi*
		Aphodius geomysi
Lepidoptera	Tineidae	*Amydria* sp.
Diptera	Anthomyiidae	*Pegomya* sp.

Source: Data from Hubbell and Goff 1940.

Subterranean Rodents and Their Parasites

Perhaps the most conclusive evidence for coevolution comes from the study of host-parasite relationships (Stone and Hawksworth 1986). The parasite's dependence on the host for some or all of its required resources and the resultant energetic drain on the host combine to generate an evolutionary "arms race" as the host attempts to defend its energy resources and the parasite responds by countering the host's defenses. Although there are many well-studied host-parasite assemblages, few studies have examined the association from a historical perspective, and fewer yet are unequivocal examples of strict coevolution. Most studies document an adaptive response on the part of the parasite but fail to show a reciprocal response on the part of the host. Those that show apparent evolutionary responses in both the host and parasite tend to lack evidence of a long-term relationship between the two, which leaves open the possibility of a recent association between the symbionts (i.e., pseudo-coevolution).

Endoparasites have been studied in several groups of subterranean rodents (table 10.2). Both cestodes and nematodes have been reported from ctenomyids (Voge 1954; Gardner 1991) and geomyids (Ubelaker and Downhower 1965; Todd, Lepp, and Tryon 1971; Gardner 1983, 1985). Nematodes also have been reported from octodontids (Hall 1916). Finally, coccidian parasites are known from both geomyids (Todd, Lepp, and Tryon

TABLE 10.2   ENDOPARASITES OF SUBTERRANEAN RODENTS

Host	Protozoa	Cestoda	Nematoda
Bathyergidae			*Trichuris contorta*
			Trichuris muris
Ctenomyidae		*Taenia* sp.	*Paraspidodera* sp.
Geomyidae	*Eimeria thomomysis*	*Catenotaenia linsdalei*	*Heligmosomoides thomomyos*
	E. fitzgeraldi	*Paranoplocephala infrequens*	*Ransomus rodentorium*
		P. variabilis	*Capillaria hepatica*
		Hymenolepis horrida	*Litomosoides thomomydis*
		H. tualatinensis	*Vexillata vexillata*
		H. diminuta	
		H. geomydis	
		H. weldensis	
Octodontidae			*Oxyuris hilgerti*
			O. hamata
			Seuratum tacapense
			Nematodirus spathiger
			Filaria diacantha
			F. bifida
			Trichuris muris
			Strongylus isotrichus
Spalacinae	*Eimeria spalacensis*		
	E. anzanensis		
	E. carmelensis		
	Isospora spalacensis		

Source: Data are compiled from sources cited in text.

1971; Gardner 1983) and spalacines (Couch, Duszinski, and Nevo 1993). Because of their intimate symbiosis with their hosts, endoparasites would seem to be ideal candidates for coevolutionary studies (e.g., see the ctenomyid-nematode example discussed below). However, the solitary existence of individuals of many subterranean rodent species makes them less than ideal vectors for the spread of parasites. As a result, individuals of many species of subterranean rodents, particularly geomyids (T. A. Spradling, personal observation), harbor few, if any, endoparasites.

Ectoparasite faunas have been described for many species of subterranean rodents (table 10.3). Ectoparasitic arthropods provide outstanding opportunities for the study of coevolution in subterranean rodents because many of them are restricted to a single host taxon, are geographically widespread, and show high prevalence and abundance on their hosts. Taken together, these factors facilitate sampling of parasite taxa across the entire geographic range of their host. For example, Costa and Nevo (1969) sampled the arthropods associated with forty-seven breeding nests of *Nannospalax ehrenbergi* from throughout its geographic range. Of the fifty-three species of

TABLE 10.3 SUBTERRANEAN RODENTS AND THEIR ASSOCIATED ECTOPARASITIC ARTHROPODS

Host	Parasitic arthropod	Common name
Bathyergidae	Polyplacidae: *Eulinognathus*	Sucking louse
	Chimaeropsyllidae: *Cryptopsylla*	Flea
	Hystrichopsyllidae: *Dinopsyllus*	Flea
	Pulicidae: *Xenopsylla*	Flea
	Ascidae: *Myonyssoides*	Mite
	Atopomelidae: *Bathyergolichus*	Mite
	Ixodidae: *Haemaphysalis*	Tick
Ctenomyidae	Polyplacidae: *Eulinognathus*	Sucking louse
	Gyropidae: *Gyropus, Phtheiropoios*	Chewing louse
	Staphylinidae: *Edrabius, Megamblyopinus*	Rove beetle
	Stephanocircidae: *Tiarapsylla*	Flea
	Rhopalopsyllidae: *Ectinorus, Tiamastus*	Flea
	Ixodidae: *Ixodes*	Tick
Geomyidae	Trichodectidae: *Geomydoecus, Thomomydoecus*	Chewing louse
	Ceratophyllidae: *Dactylopsylla, Foxella*	Flea
	Pulicidae: *Pulex*	Flea
	Haemogomasinae: *Ischyropoda*	Mite
	Pygmephoridae: *Pygmephorus*	Mite
	Listrophoridae: *Geomylichus*	Mite
	Ixodidae: *Amblyomma, Haemaphysalis, Ixodes*	Tick
Ellobius	Pulicidae: *Amphipsylla, Xenopsylla*	Flea
Octodontidae	Hoplopleuridae: *Hoplopleura*	Sucking louse
	Rhopalopsyllidae: *Delostichus, Ectinorus*	Flea
Rhizomyinae	Polyplacidae: *Polyplax*	Sucking louse
	Xiphiopsyllidae: *Xiphiopsyllus*	Flea
	Hystrichopsyllidae: *Ctenopthalmus*	Flea
	Laelapidae: *Rhyzolaelaps, Tylolaelaps*	Mite
	Ixodidae: *Haemaphysalis, Ixodes*	Tick
Spalacinae	Hystrichopsyllidae: *Ctenopthalmus*	Flea
	Ixodidae: *Haemaphysalis, Ixodes, Dermacentor*	Tick
Prometheomys	Data unavalable	

Source: Adapted from Kim 1985a.

mites and five species of fleas recorded by Costa and Nevo (1969), three of the mite species *(Hyrstionyssus ellobii spalacis, Androlaelaps hirstionnyssoides,* and *Ctenoglyphus evansi)* were determined to be host-specific and widespread throughout the range of *N. ehrenbergi*. These species of mites are particularly good candidates for the study of coevolution between *N. ehrenbergi* and its ectoparasitic arthropods.

Both chewing lice (Mallophaga [Phthiraptera]) and sucking lice (Anoplura) have limited dispersal ability and cannot survive long periods off their host. As a result, lice generally rely on host-to-host contact for dispersal, unlike other more vagile arthropods. Because host-to-host contact is almost exclusively intraspecific, there are few opportunities for lice to colonize new host taxa; hence we see a high degree of host specificity among lice (Hopkins 1957; Marshall 1981). High host specificity, in turn, makes lice ideal candidates for the study of host-parasite coevolution. Although current taxonomy suggests that chewing lice are generally species-specific, whereas sucking lice are only genus-specific, this may be more an artifact of taxonomic opinion than a biological reality (Hopkins 1957). The louse genus *Hoplopleura*, which parasitizes octodontids, is a particularly promising candidate for the study of coevolution because it shows a distributional pattern that is highly correlated with that of its host (Kim 1985b).

Parasitic arthropods that are generally considered vagile, such as fleas, may evolve reduced dispersal abilities when they parasitize subterranean rodents. For example, fleas of the genus *Pulex* that parasitize terrestrial mammals have well-developed eyes (Kim 1985a). In contrast, *Pulex sinoculus*, which parasitizes geomyid rodents, lacks eyes altogether. The same is true of fleas of the genus *Dinopsyllus*, which parasitize bathyergids. Other adaptations in fleas for life in the subterranean environment include reduction of spines and combs used for attachment to the host (presumably, the host is always nearby) and reduction in the pleural arch, which is well developed in fleas with good jumping abilities (Kim 1985a). At least one of these adaptations for life in the subterranean environment is evident in all fleas reported thus far from subterranean rodents. These morphological modifications suggest that subterranean fleas are not equipped for surface dispersal or life on a terrestrial host. This, in turn, increases the probability of long-term symbiosis between subterranean fleas and their rodent hosts.

The Study of Cospeciation in Host-Parasite Assemblages

Host-parasite systems are intrinsically interesting to evolutionary biologists because they signal a long and intimate association between two or more groups of organisms that are distantly related and quite dissimilar biologically. This long history of association may lead to reciprocal adaptations in the hosts and the parasites (strict coevolution) as well as contemporaneous cladogenic events in the two lineages (termed "parallel cladogenesis" or "cospeciation"). The phenomenon of cospeciation is of particular interest to comparative phylogeneticists because cospeciation events identify temporal links between the host and parasite phylogenies, and thus provide an internal time calibration for comparative studies of rates of evolution in the two groups. Evidence of cospeciation also can be used to test hypotheses of coadaptation in the hosts and parasites.

Despite the exceptional population biology and behavioral characteristics that make many subterranean rodents ideal candidates for studies of coevolution, there are few studies in the literature that examine historical associations of subterranean rodents and their parasites. Although many examples of parasites and inquilines of subterranean rodents are known (see tables 10.1–10.3), more thorough knowledge of the basic biology of these organisms would do much to facilitate further coevolutionary studies. Given the unusual natural history of many species of subterranean rodents—including their asocial behavior, relative isolation from parasites of terrestrial animals, and patchy geographic distributions—coevolutionary studies involving these rodents should prove unusually rewarding, especially if lineages of host-specific parasites can be identified.

Cospeciation in Tuco-Tucos and Their Endosymbionts

One of the few coevolutionary studies focusing on subterranean rodents and their parasites is an examination by Gardner (1991) of six species of tuco-tucos *(Ctenomys)* and their nematode parasites of the genus *Paraspidodera*. Relationships among the *Ctenomys* taxa involved in Gardner's study were estimated primarily from morphological data, and relationships among the parasites were estimated from a combination of morphological and allozyme data. Gardner suggested, based on these data, that there is a general history of parallel cladogenesis between these subterranean rodents and their parasites, including limited instances of host switching. Our reanalysis of the host and parasite trees compared by Gardner suggests that the host and parasite trees are not more similar than would be expected by chance ($P = .29$, based on 1,000 randomizations of the host tree by TreeMap: Page 1994). Nevertheless, the lack of congruence between the host and parasite trees may be largely the result of poorly resolved relationships among closely related host taxa and among closely related parasite taxa. Given the large number of *Ctenomys* species (>40 species: Nowak 1991) and the apparent host specificity of *Paraspidodera* species, this host-parasite assemblage clearly represents a system that could be studied profitably using comparable, high-resolution data for both groups (e.g., nucleotide sequence data from orthologous genes in the hosts and parasites).

Although it is a considerable challenge to document parallel phylogenesis between hosts and their parasites (Hafner and Page 1995), it is even more difficult to determine the *cause* of such parallel phylogenesis. The tuco-tucos and their nematode parasites investigated by Gardner (1991) may, in fact, be coevolving in the strict sense of the term. However, it is also possible that instances of congruence in the phylogenies of these rodents and their parasites are the result of common vicariance and dispersal events. Like many other subterranean rodents, *Ctenomys boliviensis* shows high levels of population subdivision (Gallardo, Köhler, and Araneda 1995). If a highly subdivided population structure is characteristic of most species of *Ctenomys*, then populations of their host-specific parasites probably are subdivided as well. This kind of population structure may foster parallel cladogenesis between populations of these hosts and parasites by reducing opportunities for host switching. On a larger scale, Gardner (1991) suggested that the generally nonoverlapping geographic distributions of species of *Ctenomys* also reduce opportunities for host switching and, therefore, favor parallel cladogenesis in *Ctenomys* and *Paraspidodera*.

Cospeciation in Pocket Gophers and Chewing Lice

Cospeciation in pocket gophers and their chewing lice has been investigated using a variety of methods, including morphology (Timm 1983), allozymes (e.g., Hafner and Nadler 1988; Demastes and Hafner 1993; Highland 1996), and nucleotide sequences (Hafner et al. 1994). In each case, evidence of cospeciation in this assemblage has been so dramatic that the gopher-louse system has become literally a "textbook example" of cospeciation (e.g., Noble et al. 1989; Esch and Fernández 1993; Ridley 1993). In the first DNA-based study of the gopher-louse system, Hafner et al. (1994) obtained nucleotide sequences from a 379-base pair region of the cytochrome *c* oxidase subunit I (COI) gene from the mitochondria of fifteen taxa of pocket gophers and seventeen taxa of lice that parasitize these gophers. Hafner et al. (1994) used four tree-building methods to reconstruct gopher and louse relationships and showed that major portions of the phylogenies (fig. 10.2) were insensitive to method of analysis.

A simple test of the hypothesis of cospeciation is to ask whether the structure of the parasite tree is independent of that of the host tree. If it is, then one would expect the number of cospeciation events observed between the two phylogenies to be no greater than that expected between the host tree and random parasite trees (Page 1994). Applying this test to the phylogenies in figure 10.2 (using the program COMPONENT: Page 1993a), Hafner et al. (1994) rejected the hypothesis that the louse phylogeny is independent of the gopher phylogeny ($p = .004$, computed using 1,000 random trees). It is possible, of course, that recent host switching could produce spurious congruence between the host and parasite trees, especially if the parasites preferentially colonize hosts that are closely related. Similarly, incongruence between the host and parasite phylogenies could result from

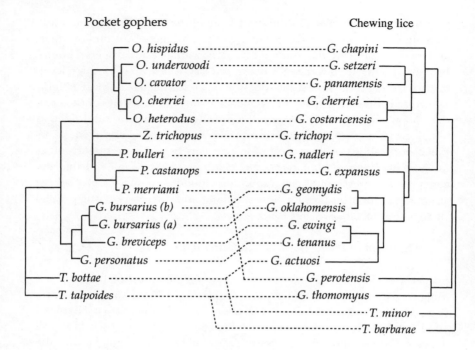

FIGURE 10.2. Phylogenies for pocket gophers and their chewing lice based on nucleotide sequence data analyzed by Hafner et al. (1994). Shown are consensus trees based on multiple methods of phylogenetic analysis. Branch lengths are proportional to expected numbers of substitutions at the third codon position in the COI gene, estimated using Felsenstein's (1989) maximum likelihood algorithm (DNAML, with transition/transversion ratio of 4.0 for both clades). Coexisting hosts and parasites are connected by dashed lines. Pocket gopher genera are *Orthogeomys, Zygogeomys, Pappogeomys, Geomys,* and *Thomomys. Geomys bursarius* is represented by two subspecies (a, *G. b. halli;* b, *G. b. majusculus*). Chewing louse genera are *Geomydoecus* and *Thomomydoecus*.

differential survival of parasite lineages, rather than host switching (Page 1993b). If genetic data such as those used in Hafner et al.'s (1994) study are available for hosts and parasites, information on amounts of genetic divergence (or relative coalescence times) can assist our efforts to discriminate between these possibilities (Page 1993b).

Component analysis (Page 1993a) can identify pairs of equivalent nodes in the host and parasite trees that reflect the same historical event. Hypotheses of coadaptation can be tested using these nodes. For example, Harvey and Keymer (1991) used simplified phylogenies of gophers and lice taken from Hafner and Nadler (1988) to show that the evolution of body size in lice and in their hosts is highly correlated. Numerous other morphological, physiological, and ecological attributes of the hosts and parasites can be compared using the cospeciation framework.

Perhaps the most exciting aspect of cospeciation analysis has been its recent application to comparative studies of evolutionary rates (e.g., Hafner and Page 1995; Page and Hafner 1996). Again, the gopher-louse system has figured prominently in this research area. There are many ways to convert molecular data (including data from allozymes, restriction fragment patterns, and protein and DNA sequences) into estimates of genetic divergence (Swofford and Olsen 1990). Each method has inherent advantages and limitations, and each involves assumptions about the nature of evolutionary change at the molecular level. Recent comparative studies of genetic differentiation in hosts and parasites have used either pairwise estimates of genetic distance (e.g., Hafner and Nadler 1990; Page 1990) or estimates of length of homologous branches in the host and parasite trees (e.g., Hafner et al. 1994; Page 1996).

Hafner and Nadler (1990) proposed a theoretical framework for comparing host and parasite genetic divergence, whether measured as genetic distance or as relative length of branches on a phylogenetic tree. Fitting a line to a plot of parasite divergence against host divergence (fig. 10.3) allows us to describe simultaneously two aspects of host-parasite divergence. The slope of the line (fig. 10.3A) is an estimate of the relative rate of genetic change in the two groups. The y-intercept of the line (fig. 10.3B) measures genetic divergence in the parasites at the time of host speciation. For example, an intercept of zero indicates synchronous cospeciation, wherein hosts and parasites diverge simultaneously. A negative intercept suggests delayed cospeciation, in which case the parasites tend to diverge consistently after their hosts. Finally, a positive intercept signals preemptive cospeciation, in which case the parasites diverge prior to their hosts.

Although it is widely acknowledged that estimates of DNA sequence divergence should be adjusted for the effects of saturation, there is no general consensus as to how this should be done. For example, Hafner et al. (1994) attempted to correct for transition bias in the gopher and louse COI data by using the largest observed pairwise transition bias in a maximum likelihood phylogeny reconstruction. They reasoned that this value, which is usually measured between the most recently diverged taxa, is least likely to be affected by saturation and, therefore, is the most reasonable estimate of the actual transition bias for this gene region. In contrast, Page (1996) recommends use of the transition bias estimate that maximizes the likelihood of the phylogeny. The use of these different correction factors can have a profound influence on estimates of branch length. For example, Hafner et al.'s analysis suggests lice are evolving ten to eleven times more rapidly than pocket gophers at selectively neutral sites. In contrast, Page's analysis sug-

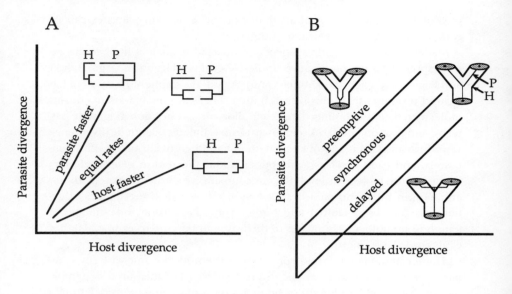

FIGURE 10.3. Bivariate plots of the relationship between parasite divergence and host divergence. The slope of the relationship (A) indicates relative rates of evolution in the two clades. The trees (insets in A) are drawn with branch lengths proportional to amount of genetic change in the hosts (H) and parasites (P). The y-intercept (B) indicates the relative timing of speciation events. The insets in B illustrate the relative timing of speciation events in the hosts (outer portion of each figure) and their parasites (thin line within each figure). (Adapted from Hafner and Nadler 1990)

gests that lice are evolving only two to three times as fast as gophers. Research into the effects of transitional saturation (and evolutionary models in general) is now moving at a rapid pace (Kelly and Rice 1996; Yang 1996; Neilsen 1997), and we expect that some degree of consensus will be reached in the near future.

Future studies comparing the population structure of hosts and their parasites will reveal whether the structuring of a parasite population on an individual host (and founder events as new hosts are colonized) tends to accelerate parasite evolution relative to that of the host (Nadler et al. 1990). To be convincing, such a test would have to demonstrate that short-term population-level phenomena (such as decreased heterozygosity and polymorphism in the parasites) have persistent and long-term phylogenetic consequences. Similarly, studies of the molecular genetics of parasites at zones of hybridization between host taxa can yield important information about the history of those zones (e.g., Patton et al. 1984; Nadler et al. 1990; Hafner et al. 1998) or about modes of parasite transmission (Demastes 1996).

If genetic introgression is present in both the hosts and parasites, then rates and patterns of introgression can be compared to reveal common demographic patterns. In other cases, parasites can be treated as "genes" of their hosts to serve as an independent measure of the extent of host introgression (Bohlin and Zimmerman 1982; Patton et al. 1984).

Although, at present, there are few published studies of cospeciation explored from a molecular perspective, we anticipate rapid growth in this research area as molecular tools become more widely available and the advantages of this approach better known. We believe that subterranean rodents and their parasites will continue to play a prominent role in cospeciation research, largely because of the unusual life history of subterranean rodents. Unfortunately, many host-parasite systems will show little or no evidence of cospeciation (e.g., Baverstock, Adams, and Beveridge 1985), which will preclude comparative studies of higher-order phenomena such as evolutionary rates. However, in those systems with appreciable cospeciation, researchers will have the unparalleled opportunity to compare evolution in the same genes, and over the same period of time, in distantly related organisms. Within this framework, the potential for discovery of large-scale evolutionary patterns that apply to diverse groups of organisms is great.

Summary and Prospectus for Future Research

We have argued that studies of coevolution should begin with a comparative phylogenetic analysis to elucidate the evolutionary history of a symbiotic association. This research protocol, which seeks to document cospeciation between symbiotic lineages, is elaborated elsewhere (Hafner and Nadler 1990; Hafner and Page 1995; Page and Hafner 1996). Given statistically significant evidence of cospeciation, which signals a long history of association between two lineages of symbiotic organisms, the researcher next is faced with the daunting challenge of demonstrating reciprocal adaptation in the two groups. Because it is difficult, if not impossible, to document causality in the study of adaptation, the researcher often must resort to "just-so stories" to explain the evolutionary origin of supposed reciprocal adaptations. Some of these stories will be more convincing than others. Importantly, however, an absence of evidence for reciprocal adaptation between two cospeciating lineages in no way undermines the prior evidence for cospeciation, which is based on a statistical test of similarity between the phylogenies of the two symbiotic groups. Thus, in practice, statistically significant evidence of cospeciation may turn out to be the most rigorous component in the overall analysis of coevolution for most symbiotic associations.

We have identified no fewer than 238 possible research projects that have the potential for revealing cospeciation—and, possibly, coevolution—between subterranean rodents and their symbionts. Among the most promising candidates for study are the thirty families of parasitic arthropods and their rodent hosts listed in table 10.3. Also included in the above total are at least eight potential studies of the five subterranean rodent families and their endoparasitic protozoans, cestodes, and nematodes listed in table 10.2. Finally, if we make the conservative assumption that each subterranean rodent genus has about ten inquilines associated with its burrow system (there are thirteen for *Geomys pinetis* alone: see table 10.1), then there are at least 200 potential studies involving twenty subterranean rodent genera and their inquilines. Of these 238 potential studies of coevolution, only two have been carried out in detail to date (on ctenomyids and their nematode parasites [Gardner 1991] and on geomyids and their chewing lice [Hafner et al. 1994]). We anticipate that a wealth of new studies involving subterranean rodents and their symbionts will appear in the near future as increasingly sophisticated methods for the analysis of cospeciation and coevolution are developed.

Acknowledgments

We thank David J. Hafner for his assistance in all phases of this research. Mark Hafner's research on pocket gophers and chewing lice is funded by the National Science Foundation.

Literature Cited

Baverstock, P. R., M. Adams, and I. Beveridge. 1985. Biochemical differentiation in bile duct cestodes and their marsupial hosts. *Molecular Biology and Evolution* 2:321–37.

Bohlin, R. G., and E. G. Zimmerman. 1982. Genic differentiation of two chromosomal races of the *Geomys bursarius* complex. *Journal of Mammalogy* 63:218–28.

Costa, M. and E. Nevo. 1969. Nidicolous arthropods associated with different chromosomal types of *Spalax ehrenbergi* Nehring. *Zoological Journal of the Linnean Society* 48:199–215.

Couch, L., D. W. Duszinski, and E. Nevo. 1993. Coccidea (Apicomplexa), genetic diversity, and environmental unpredictability in four chromosomal species of the subterranean superspecies *Spalax ehrenbergi* (mole-rat) in Israel. *Journal of Parasitology* 79:181–89.

Demastes, J. W. 1996. Analysis of host-parasite cospeciation: Effects of spatial and temporal scale. Ph.D. dissertation, Louisiana State University, Baton Rouge.

Demastes, J. W., and M. S. Hafner. 1993. Cospeciation of pocket gophers *(Geomys)* and their chewing lice *(Geomydoecus)*. *Journal of Mammalogy* 74:521–30.

Ehrlich, P. R., and P. H. Raven. 1964. Butterflies and plants: A study in coevolution. *Evolution* 18:586–608.

Esch, G. W., and J. C. Fernández. 1993. *A functional biology of parasitism: Ecological and evolutionary implications.* London: Chapman and Hall.

Felsenstein, J. 1989. PHYLIP-Phylogenetic inference package, version 3. 2. *Cladistics* 5: 164–66.
Fenner, F. 1965. Myxoma virus and *Oryctolagus cuniculus:* Two colonizing species. In *Genetics of colonizing species*, edited by H. G. Baker and G. L. Stebbins, 485–99. New York: Academic Press.
Futuyma, D. J., and M. Slatkin, eds. 1983. *Coevolution*. Sunderland, MA: Sinauer Associates.
Gallardo, M. H., N. Köhler, and C. Araneda. 1995. Bottleneck effects in local populations of fossorial *Ctenomys* (Rodentia, Ctenomyidae) affected by vulcanism. *Heredity* 74:638–46.
Gardner, S. L. 1983. Endoparasites of North American pocket gophers. M. S. thesis, University of Northern Colorado, Greeley.
———. 1985. Helminth parasites of *Thomomys bulbivorus* (Richardson) (Rodentia: Geomyidae), with the description of a new species of *Hymenolepis* (Cestoda). *Canadian Journal of Zoology* 63:1463–69.
———. 1991. Phyletic coevolution between subterranean rodents of the genus *Ctenomys* (Rodentia: Hystricognathi) and nematodes of the genus *Paraspidodera* (Heterakoidea: Aspidoderidae) in the Neotropics: Temporal and evolutionary implications. *Zoological Journal of the Linnean Society* 102:169–201.
Gould, S. J., and E. S. Vrba. 1982. Exaptation: A missing term in the science of form. *Paleobiology* 8: 4–15.
Hafner, J. C., and M. S. Hafner. 1988. Heterochrony in rodents. In *Heterochrony in evolution: A multidisciplinary approach*, edited by M. L. McKinney, 217–35. New York: Plenum.
Hafner, M. S., J. W. Demastes, D. J. Hafner, T. A. Spradling, P. D. Sudman, and S. A. Nadler. 1998. Age and movement of a hybrid zone: Implications for dispersal distance in pocket gophers and their chewing lice. *Evolution* 52:278–82.
Hafner, M. S., and S. A. Nadler. 1988. Phylogenetic trees support the coevolution of parasites and their hosts. *Nature* 332:258–59.
———. 1990. Cospeciation in host-parasite assemblages: Comparative analysis of rates of evolution and timing of cospeciation events. *Systematic Zoology* 39:192–204.
Hafner, M. S., and R. D. M. Page. 1995. Molecular phylogenies and host-parasite cospeciation: Gophers and lice as a model system. *Philosophical Transactions of the Royal Society of London* B 349:77–83.
Hafner, M. S., P. D. Sudman, F. X. Villablanca, T. A. Spradling, J. W. Demastes, and S. A. Nadler. 1994. Disparate rates of molecular evolution in cospeciating hosts and parasites. *Science* 265:1087–90.
Hall, M. C. 1916. Nematode parasites of mammals of the orders Rodentia, Lagomorpha and Hyracoidea. *Proceedings of the United States National Museum* 50:1–258.
Harvey, P. H., and A. E. Keymer. 1991. Comparing life history using phylogenies. *Philosophical Transactions of the Royal Society of London* B 332:31–39.
Highland, R. 1996. Cospeciation of Midwestern pocket gophers and their ectoparasitic chewing lice. M. S. thesis, Northern Illinois University, DeKalb.
Hopkins, G. H. E. 1957. The distribution of Pthiraptera on mammals. In *First symposium on host specificity amongst parasites of vertebrates*, edited by J. G. Baer, 88–119. Institute of Zoology, University Neuchatel, Switzerland.
Hubbell, T. H., and C. C. Goff. 1940. Florida pocket-gopher burrows and their arthropod inhabitants. *Proceedings of the Florida Academy of Sciences* 4:127–66.
Janzen, D. H. 1980. When is it coevolution? *Evolution* 34:611–12.
Kelly, C., and J. Rice. 1996. Modeling nucleotide evolution: A heterogeneous rate analysis. *Mathematical Biosciences* 133:85–109.
Kim, K. C. 1985a. Coevolution of fleas and mammals. In *Coevolution of parasitic arthropods and mammals*, edited by K. C. Kim, 295–437. New York: John Wiley & Sons.
———. 1985b. Evolution and host associations of Anoplura. In *Coevolution of parasitic arthropods and mammals*, edited by K. C. Kim, 197–231. New York: John Wiley & Sons.

Marshall, A. G. 1981. *The ecology of ectoparasitic insects.* London: Academic Press.
Nadler, S. A., M. S. Hafner, J. C. Hafner, and D. J. Hafner. 1990. Genetic differentiation among chewing louse populations (Mallophaga: Trichodectidae) in a pocket gopher contact zone (Rodentia: Geomyidae). *Evolution* 44:942–51.
Neilsen, R. 1997. Site-by-site estimation of the rate of substitution and the correlation of rates in mitochondrial DNA. *Systematic Biology* 46:346–53.
Nevo, E. 1979. Adaptive convergence and divergence of subterranean mammals. *Annual Review of Ecology and Systematics* 10:269–308.
Noble, E. R., G. A. Noble, G. A. Schad, and A. J. MacInnes. 1989. *Parasitology: The biology of animal parasites.* 6th edition. Philadelphia: Lea and Febiger.
Nowak, R. M. 1991. *Walker's mammals of the world.* 5th edition. Baltimore: Johns Hopkins University Press.
Page, R. D. M. 1990. Component analysis: A valiant failure? *Cladistics* 6:119–36.
———. 1993a. COMPONENT 2. 0. Tree comparison software for use with Microsoft Windows. London: The Natural History Museum
———. 1993b. Genes, organisms, and areas: The problem of multiple lineages. *Systematic Biology* 42:77–84.
———. 1994. Parallel phylogenies: Reconstructing the history of host-parasite assemblages. *Cladistics* 10:155–73.
———. 1996. Temporal congruence revisited: Comparison of mitochondrial DNA sequence divergence in cospeciating pocket gophers and their chewing lice. *Systematic Biology* 45:151–67.
Page, R. D. M., and M. S. Hafner. 1996. Molecular phylogenies and host-parasite cospeciation: Gophers and lice as a model system. In *New uses for new phylogenies,* edited by P. H. Harvey, A. J. L. Brown, J. Maynard Smith, and S. Nee, 255–70. Oxford: Oxford University Press.
Patton, J. L. 1972. Patterns of geographic variation in karyotype in the pocket gopher, *Thomomys bottae* (Eydoux and Gervais). *Evolution* 26:574–86.
Patton, J. L., M. F. Smith, R. D. Price, and R. A. Hellenthal. 1984. Genetics of hybridization between the pocket gophers *Thomomys bottae* and *Thomomys townsendii* in northeastern California. *Great Basin Naturalist* 44:431–40.
Pearson, O. P. 1959. Biology of the subterranean rodent, *Ctenomys,* in Peru. *Memorias del Museo de Historia Natural "Javier Prado"* 9:1–56.
Pirozynski, K. A., and D. L. Hawksworth, eds. 1988. *Coevolution of fungi with plants and animals.* London: Academic Press.
Ridley, M. 1993. *Evolution.* Boston: Blackwell Scientific Publications.
Stone, A. R., and D. L. Hawksworth, eds. 1986. *Coevolution and systematics.* Oxford: Clarendon Press.
Swofford, D. L., and G. J. Olsen. 1990. Phylogeny reconstruction. In *Molecular systematics,* edited by D. M. Hillis and C. Moritz, 411–501. Sunderland, MA: Sinauer Associates.
Timm, R. M. 1983. Farenholz's rule and resource tracking: A study of host-parasite coEvolution. In *Coevolution,* edited by M. H. Nitecki, 225–66. Chicago: University of Chicago Press.
Todd, K. S., D. L. Lepp, and C. A. Tryon. 1971. Endoparasites of the northern pocket gopher from Wyoming. *Journal of Wildlife Diseases* 7:100–104.
Ubelaker, J. E., and J. F. Downhower. 1965. Parasites recovered from *Geomys bursarius* in Douglas County. *Transactions of the Kansas Academy of Science* 68:206–8.
Voge, M. 1954. Exogenous proliferation in a larval taeniid (Cestoda: Cyclophyllidea) obtained from the body cavity of Peruvian rodents. *Journal of Parasitology* 40:411–13.
Yang, Z. 1996. Among-site rate variation and its impact on phylogenetic analyses. *Trends in Ecology and Evolution* 11:367–72.

CHAPTER ELEVEN

The Evolution of Subterranean Rodents: A Synthesis

Enrique P. Lessa

My aims in this chapter are, first, to provide a conceptual framework for synthesizing the evolutionary biology of subterranean rodents, second, to provide a historical framework for understanding the development of ideas about subterranean rodents, and third, to highlight present and future directions of research on these taxa. In the conceptual synthesis presented in the first section of this chapter, I attempt to show how the diverse research programs reflected in the preceding chapters are framed by general lines of thought in evolutionary biology. In doing so, I touch on many issues that are highly controversial, and all I can offer is to express my own conclusions and biases as candidly as possible. The historical section of the chapter emphasizes areas in which subterranean rodents have for decades been important in crafting and testing ideas in evolutionary biology. This is the legacy of intellectual work upon which current developments are built. I hope this succinct historical summary stimulates, rather than replaces, the direct study of the original literature, especially among students who are new to this field of work. The third and final section of the chapter emphasizes a few present and future directions of research, although suggestions in this respect are made throughout the chapter.

I provide references to the preceding chapters at critical points throughout my discussion, but I cannot possibly do justice to the wealth of information they cover. I urge readers to examine those chapters first-hand, with full assurance that it will be a rewarding exercise.

Subterranean Rodents in Evolutionary Biology: An Interpretive Framework

In this section, I briefly outline the core of the adaptationist view that constitutes the modern neo-Darwinian synthesis (Gould 1982) as it pertains to subterranean rodents. In particular, I consider what, if anything, may be contributed by an examination of constraints on adaptation. My emphasis here is on individual-and population-level processes. Another important set of challenges to the modern synthesis—namely, those set to produce a hierarchical expansion of the theory (e.g., Eldredge 1996)—is not considered here, but will be touched upon in relation to phylogenetics and macroevolution.

The Adaptationist Program

Evolutionary explanations of biological diversity may be seen as drawing from three main sources of interpretive thought. In the functionalist tradition, the diversity of life is thought to reflect primarily the particulars of adaptation to organismal survival and reproduction. In contrast, the formal, structuralist viewpoint seeks general rules of construction and modification of structure. Finally, phylogenetics and the temporally protracted process of change of life on earth constitutes the focus of historicism. These three viewpoints may be visualized as corners of a triangle of interpretive thought (fig. 11.1; Raup 1972).

Obviously, these are not mutually exclusive viewpoints, but research programs have varied in the relative emphasis placed upon each of these components. The functionalist viewpoint, for instance, does not deny that biological adaptation can only be built upon available variation, and that such variation is both contingent upon history and subject to the rules of struc-

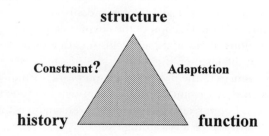

FIGURE 11.1. Seilacher's triangle depicting lines of interpretive thought in evolutionary biology. The study of adaptation has been viewed as emphasizing the line connecting structure and function, whereas constraints presumably arise from structure and history. (Adapted from Raup 1972)

tural construction. If, however, variation is widely available, the rules of construction are rather plastic and therefore reflect the dictates of past adaptation, and selection is highly efficient, then explanations will naturally lean toward the functionalist vertex of the triangle in figure 11.1.

Much of the research on subterranean rodents discussed in this volume is framed in a primarily functionalist perspective, with an emphasis on the organismal level as the focus of analysis and on adaptation as the outcome of evolution, driven by natural selection. This emphasis is to be expected for several reasons. First, the adaptationist program in general has been a most successful endeavor. True, adaptation itself is a difficult concept (see Burian 1992 and references therein): demonstrating adaptation at any level is not an easy task (it is, after all, the product of a process we can seldom observe), and the wholesale assumption of adaptation is not uncommon. It seems, however, that the adaptationist program has grown in sophistication (see West-Eberhardt 1992) in response to criticism by Gould and Lewontin (1979) and others. Second, subterranean rodents (as well as insectivores) have long been the focus of a specific version of the adaptationist program, championed by Nevo and summarized in his influential 1979 review. No other modern biologist has carried the flag of panselectionism as enthusiastically (e.g., Nevo 1988).

My personal view is that the notion that a single process, natural selection, has been equally effective in bringing nearly every single aspect of an animal's biology to an adaptive optimum is untenable. After all, even adaptive optima may lead to suitable conditions for substantial genetic drift. For example, the optimal utilization of patchy resources may produce highly subdivided populations composed of demes of small effective size, in which drift will inevitably contribute to the fate of alleles even in the presence of selection (see discussion below on the niche width hypothesis).

In spite of these problems, the niche occupied by subterranean rodents offers interesting possibilities for carrying out the study of adaptation. First, as emphasized below and described in detail by Busch et al. (chap. 5, this volume), this niche is relatively simple and poses distinct challenges to organisms. It has been relatively easy to propose adaptive hypotheses, at least at the most general level, about requirements for coping with the particulars of subterranean life. For example, long-distance communication through a dense medium such as the soil is possible only at certain wavelengths (Francescoli, chap. 3, this volume), and a serious commitment to life underground poses substantial physiological challenges (Buffenstein, chap. 2, this volume). Since reproduction is key to individual success, functionalist perspectives are extremely useful in the examination of social systems (Lacey, chap. 7, this volume) and reproductive behavior and physiology (Bennett, Faulkes, and Molteno, chap. 4, this volume).

In sum, most, if not all, of the research at the organismal level has leaned, in the classic triangle of interpretive trends, toward the adaptive viewpoint, thus combining structure and function. No one denies that this dyad develops through history, but much of the emphasis of the adaptationist program has relied upon the notion that an adaptive equilibrium driven by selection quickly brings structure in line with function. Nearly instantaneous responses to selection relieve us of the need to bring history into the picture, except in the most general way. I will return to this issue below, but will now turn my attention to the actual and potential contributions of the structuralist viewpoint, still at the organismal level.

Constraints

The term "constraint" has been used more or less erratically in the biological literature, leading Antonovics and van Tienderen (1991) to dismiss it as a vague notion lacking any heuristic value, often used to indicate any directional change. Thus, soil properties "constrain" the propagation of sound and dictate the use of low frequencies in communication. This is not wrong, but it is a mere rephrasing of the classic notion of environmentally mediated selective pressures. Indeed, as recognized by Gould (1989), a more restrictive meaning of the term is needed if constraints are to offer anything new. Gould (1989) resorts to a definition by Stearns (1986), who suggested that "we can preserve it in a relative sense if we recognize that it only has meaning in a local context where one concentrates on the possibilities latent in certain processes and views the limitations on those possibilities as arising from outside that context." Gould (1989) proposed that the historical and structural (which he called formal) corners of the triangle in figure 11.1 bring constraints to the functionalist corner represented by the adaptationist program.

I see three widespread problems with the literature about constraints that must be overcome if the term is to have any relevance. One is the inconclusive rephrasing of already established patterns. We could say that the long gestation periods of hystricognaths represent a "phylogenetic constraint." Nothing is added to the known facts by this statement, and the gestation periods of these rodents remain just as puzzling. I am not suggesting here that there is no use in the notion of a constraint, but rather pointing out that these types of vague statements are not useful. The second problem lies in the notion, rightly dispelled by Stearns's definition (above), that constraints must be absolute, and therefore insurmountable, to have any relevance. In fact, one of the most interesting and far-reaching analyses of constraints has been Carrier's (1987) examination of the conflict between lateral undulation and respiration in tetrapods. Cowen (1996) and Carrier himself have shown how many features of tetrapod evolution may have run along pathways allowing them to overcome that constraint or mitigate its effects. A third problem is the somewhat naive setting of viewpoints as mutually exclusive (as in "are these features due to natural selection or to phylogenetic history?").

It seems clear that in the study of subterranean rodents, as well as in the rest of biology, the structuralist viewpoint epitomized by constraints has played a rather minor role, at least since the forging of the modern synthesis. This may be due to the bias introduced by our own neo-Darwinian training (a sort of historical constraint on our intellectual perspective), to the great heuristic appeal of the adaptationist program (if applied with due caution), to the intrinsic limitations of the structuralist viewpoint, or to any combination thereof. Whatever the cause, I will resort (as I did in the Rome symposium: Lessa 1990) to my own studies to illustrate the potential of the structuralist perspective and its interactions with the adaptationist program in the study of subterranean rodents.

When confronted with a handful of species exhibiting, for example, morphological variation, most biologists attempt to identify likely adaptive reasons for such differences. Gould and Lewontin (1979) pointed to allometry as one potential alternative, nonadaptive explanation for such morphological differences. Variation in incisor procumbency, for instance, may reflect differential adaptation for chisel-tooth digging (Hildebrand 1985; Lessa and Thaeler 1989). However, Lessa and Patton (1989) have shown that procumbency may also develop as a by-product of allometric skull growth. It is inevitable as long as (1) the allometry of the rostrum is such that its length increases faster than its depth and (2) the incisors are contained within the rostrum, have their roots at its proximal base, emerge at its distal end, and are ever growing, being therefore forced to follow a helical growth trajectory. These rules set lower limits on procumbency in relation to size. The very same geometric property (procumbency) may arise along two very different evolutionary pathways, one as a passive outcome of allometry and the other entailing a significant modification of growth patterns (fig. 11.2).

The allometric relationship that passively produces increased procumbency is but an example of a very widespread feature of mammalian skull growth, which entails the relative elongation of the rostrum with respect to the braincase (Radinsky 1985). It certainly applies to all pocket gophers, irrespective of their digging habits. Thus, Stearns's definition applies here. If we consider the potential direction of evolutionary change of a species, this general allometric trend acts as a local constraint. Certain directions are simply more likely than others, regardless of the use of incisors for chisel-tooth digging.

Importantly, this explanation says nothing about the basis for the allomet-

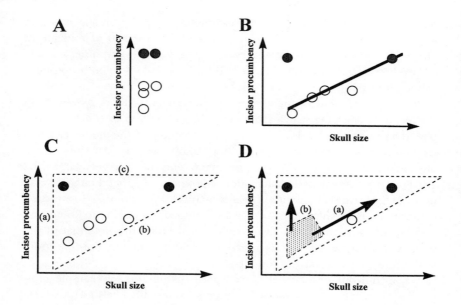

FIGURE 11.2. Simplified diagram of the relationship between skull size and incisor procumbency in pocket gophers. Each data point represents the average morphology of a hypothetical species or subspecies. (A) Range of values for incisor procumbency, in which two taxa (cross-hatched circles) stand out as highly procumbent. (B) Allometric analysis, which suggests that only one taxon departs substantially from a general allometric trend. (C) Overall range of variation represented as a triangle, with each of its sides caused by a different factor: (a) minimal skull size, (b) size-related constraint on minimal procumbency, and (c) Landry's mechanical constraint. (D) Range of variation represented as in C, but with a polygon representing the range that encompasses the vast majority of the taxa, and arrows representing directions of departure from such standard morphology in two directions: (a) size-related, allometric change, and (b) increase in procumbency independent from size. (Adapted from Lessa and Patton 1989; Lessa 1990)

ric trend itself. In principle, allometric trends such as this one result from genetic covariances of polygenic traits. Quantitative geneticists know well that genetic covariances can be readily changed (by directional selection, for instance) and, furthermore, are not likely to persist unless actively maintained by purifying selection. It may well be that the allometry itself is shaped and maintained by selection. Indeed, one well-established approach in morphology is to predict the form of allometric relationships on the basis of functional considerations and test the predictions with observations (e.g., Alexander 1983; West, Brown, and Enquist 1997). At least in some cases, allometric relationships are readily modified by selection (e.g., Wilkinson 1993), suggesting that stabilizing selection, not lack of variation, explains their maintenance.

In sum, allometry does not go against adaptation in general. Rather, it directs our attention to the fact that explanations of diversity may not lie in the particulars of a species' adaptations (for incisor use, in our example), but rather at a more general level that encompasses an entire group of species. We are taken outside the local context in search of more economical explanations. Although this is purely speculative in the case of rostral allometry, it is possible that both changes in body size and the persistence of allometric trends are dictated by selective pressures of one sort or another. Yet I would argue that allometry represents a local constraint that sets biases for potential evolution, and that recognition of this constraint is crucial for understanding this particular case of morphological variation.

The constraint reflected in the allometry of the rostrum, whatever its basis, is not absolute, and some of the exceptionally procumbent pocket gophers represent not a passive result of large body size, but a repatterning of growth that may well be adaptive. There are constraints on this type of repatterning that have to do with the conflicting mechanical consequences of procumbency. These constraints were examined by Landry (1957) in a pioneering article. Again, as in many interesting cases of constraints, Landry's mechanical one can be overcome by repositioning the incisor roots. This has taken place in a few lineages of subterranean rodents (see Lessa 1990; Stein, chap. 1, this volume).

In this brief analysis of constraints, we have moved gradually from the particulars of differences between a few species to an overall allometry representing a relative constraint; species departing from the overall allometry are rare, but suggest the potential to overcome the constraint, presumably under directional selection. These departing morphologies, in turn, take us to another constraint, related to mechanical conflicts, which, again, may be overcome in a few lineages by means of structural changes. At each level, we see interactions between relative constraints setting biases that shape the form and the potential for selection. Furthermore, selection at one level of analysis may be the basis of constraints at another. I have represented the morphospace relating incisor procumbency to body size as an unevenly populated triangle (see fig. 11.2). The overall result of this analysis is, first, the realization that morphological variation is highly clumped, so that only a small fraction of available morphospace is in fact occupied, and unevenly so. Second, the several edges of such morphospace, although likely "soft," relative barriers, are dictated by rather different factors. Whereas the mean trend, as picked up by classic interspecific allometry, is relevant, perhaps the edges themselves are more revealing of constraints.

Edges such as those in figure 11.2 are, of course, particularly interesting if they have biological meaning, as I think they have in this case. In his book *Macroecology*, Brown (1995) shows convincingly that edges in large-scale ecological patterns can indeed be very revealing of underlying biological fac-

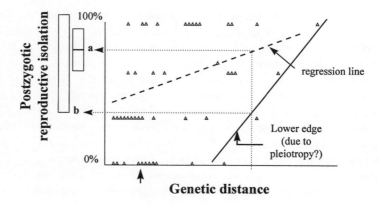

FIGURE 11.3. Diagrammatic representation of relationships between postzygotic reproductive isolation and genetic distance in *Drosophila*. Each data point represents a pairwise comparison between two hypothetical species. For any given genetic distance between two species, a regression line predicts a mean level of postzygotic reproductive isolation (a) and some variation about it. Alternatively, the notion that genetic distance (as a reflection of overall differentiation) simply imposes a lower limit to reproductive isolation offers a different prediction (b). (Adapted from Coyne and Orr 1989)

tors. Our intellectual eye is poorly trained to detect and interpret edges, as we have been taught that mean trends are important and variation about them represents noise. We may be missing interesting opportunities because of this bias. Consider, for instance, the pattern of relationship between overall genetic differentiation (as estimated by allozyme distances) and reproductive isolation in *Drosophila* (fig. 11.3; see also Coyne and Orr 1989). A mean relationship can be estimated with a suitable regression model and shown to be statistically significant even after corrections for phylogenetic effects. This finding suggests a linear relationship between reproductive isolation and genetic differentiation, and noise about that relationship can be estimated as well. Such an exercise, however, may be misleading. First, reproductive isolation may be reached quite independently from overall differentiation, as demonstrated by the fact that the full range of isolation is found at extremely low genetic distances. Second, substantial differentiation must take place before any reproductive isolation arises as an inevitable—presumably pleiotropic—by-product. I would argue that the mean regression line is irrelevant, as anything goes above the lower edge of the relationship. The pleiotropic effects of overall differentiation upon reproductive isolation are best reflected by that lower edge.

The use of allometry and of the edges of biological variation are but two aspects of a structuralist perspective. In an analysis of structural complexity and phylogenetic mapping, Lessa and Stein (1992) examined the consequences of structural and mechanical characteristics of jaws, forelimb bones, and associated muscles for the potential for variation in the phylogenetic history of pocket gophers. That study showed a relationship between structural "degrees of freedom" (i.e., numbers of articulating bones and muscles) and phylogenetic flexibility that is in line with a general hypothesis put forth by Vermeij (1974).

Gould (1989) suggested that much of the analysis of constraints could be pursued with little reference to adaptation. My limited experience in this field suggests otherwise, and it seems to me that the structuralist perspective, to be fruitful, must place itself in permanent interaction with the other corners of the interpretive triangle of figure 11.1. In this context, the structuralist approach promises to provide refreshing insights that, although not necessarily in direct conflict with the adaptationist program, are not easily derived within its restricted traditions. This is well illustrated by what may be a breakthrough in our understanding of allometric scaling laws (West, Brown, and Enquist 1997).

Subterranean Rodents in Evolutionary Biology: A Historical Overview

Subterranean rodents have been focal subjects of study and controversy in several areas of evolutionary biology, including at least the following: (1) niche concepts in behavioral and evolutionary ecology, (2) evolutionary genetics and the selectionist-neutralist debate, (3) species concepts, and (4) the mechanisms of speciation. This section provides an overview of these issues, not merely to satisfy the reader's historical curiosity, but also to illustrate how new conceptual and empirical tools offer numerous opportunities to revisit these long-standing problems and produce fresh perspectives.

The Subterranean Niche: Behavioral and Evolutionary Ecology

Subterranean rodents have a sizable effect on their environment. They often represent a substantial fraction of mammalian herbivores (in terms of biomass) and have a large effect on the landscape, soil structure, and energy flows (Cameron, chap. 6, this volume). Perhaps the most dramatic potential example is mima mounds and comparable structures, thought by some to be the long-term products of the animals' working and reworking of the soil. Naturally, these relationships are reciprocal, and the texture of the landscape has profound influences upon the animals (Cameron, chap. 6, this volume).

Subterranean rodents have inspired or served as examples of key, and heatedly disputed, concepts in ecology, particularly the concept of the biological niche and the idea of interspecific competition. One of the most widely accepted historical definitions of the biological niche concept came from Grinnell (1914), a prominent pocket gopher systematist. The Grinnellian niche emphasized the intricate interactions of a species with its environment (including both biotic and abiotic components) and the roles it plays in the ecosystem (reviewed in James et al. 1984). As it turns out, unlike their non-subterranean counterparts, subterranean rodents are rather monotonous in their observable interactions with their ecotope. Drawing from comparisons of pocket gophers and tuco-tucos, Pearson (1959) indicated that rodents, unlike moles, are unable to subdivide their subterranean niche. Save for a few exceptional situations of marginally overlapping ranges, there usually is but one species of subterranean rodent in any given area of habitable soil. Species of the many lineages of subterranean rodents throughout the world tend, therefore, to show contiguous allopatric or parapatric distributions. This distribution pattern is very much unlike that among rodents above ground, where any single area is inhabited by several species, and resources are partitioned along several axes of variation.

One important consequence of the indivisibility of the niche occupied by subterranean rodents is that species and subspecies appear to be mutually exclusive. Any patch of suitable habitat is in the possession of a single species. What simpler system could one aspire to utilize to examine interspecific competition? In a classic paper, Miller (1964) took advantage of this opportunity to examine interspecific competition in pocket gophers. To this end, he resorted to the Hutchinsonian niche concept, which portrays the niche as that fraction of a conceptual multidimensional space occupied by a species. The niche dimensions are the several environmental parameters to which species respond. Miller thought that pocket gophers could be ranked according to their specialization for subterranean life. Thus, he considered large-clawed *Geomys* as highly specialized, small-clawed *Thomomys* as less modified, and *Pappogeomys (Cratogeomys)* as somewhat intermediate. Ecological specialization followed the same pattern in Miller's view; he proposed that the most specialized taxa kept privileged control of the best habitat patches of deep, friable, sandy soils. Successively less specialized taxa were displaced by more specialized ones from good patches, in an ecological pecking order dictated by specialization. Thus, ecological separation of pocket gophers reflected competitive displacement and a hierarchy of competitive abilities. Miller's ideas can be traced back to Merriam (1895), who provided comparable assessments of pocket gopher morphology. Others have qualified or disputed this line of thought on several grounds. Best (1973), for instance, emphasized differential ecological specialization, leading to reciprocal, rather than unidirectional, exclusion in pocket gopher interactions. A related line of argument has suggested that subterranean rodents have explored two different directions of morphological specialization to different degrees: one of them (scratch digging) does lead to narrow niches of friable soils, whereas the other does not (Thaeler 1968; Lessa and Thaeler 1989). Finally, history may greatly affect subterranean rodent distributions by giving early colonizers an advantage over latecomers (Thaeler 1968; Patton and Smith 1990).

By and large, the debate over niche dimensions and competition in subterranean rodents has centered on pocket gophers, although there have been some suggestions as to how those same issues are reflected in other taxa (Lessa and Thaeler 1989; Lessa 1990). Importantly, there are gaps in our understanding of these issues even in pocket gophers. For instance, discussions have focused on morphology and its relationship to ecology, with relatively little attention paid to the obviously intervening issues of digging behavior and energetics (but see Vleck 1979, 1981; Lovegrove 1989). Also, the morphological and behavioral diversity of subterranean rodents clearly is much broader than that found among pocket gophers (Stein, chap. 1, this volume), and this limits our power to understand other cases by analogy (Lessa 1990; Wake 1993).

Despite some unresolved problems and controversies, the discussions of interspecific competition outlined above fit well into the neo-Darwinian tradition of this field. Interspecific competition is seen as an extension of competition between individuals, examined for convenience at a higher level (Pianka 1988). At the basis of this viewpoint is the notion that typical subterranean rodents are largely solitary animals that maintain individual territories (Nevo 1979): the exclusivity of individual territories is here seen as the basic competitive mechanism behind the exclusive distributions of species and subspecies. To put it in the terminology developed by macroevolutionists, the processes and resulting patterns at and above the species level examined by classic models of interspecific competition are effects of classic properties of individuals (Vrba and Gould 1986).

Evolutionary Genetics and the Selectionist-Neutralist Debate

The core of the synthetic, neo-Darwinian theory of evolution lies in three of its major ingredients: (1) its favoring of the population level of organization as the primary evolutionary playing field ("population thinking," as Mayr [1991] put it), (2) its view of natural selection as the primary guiding process in the evolution of populations, and (3) the notion that evolution is gradual and proceeds by the slow accumulation of small changes. Just as theories of stasis and punctuation challenged the primacy of population-level processes and of gradualism (Gould 1982; Gould and Eldredge 1993), new molecular data and the neutral mutation model have challenged the predominance of positive natural selection at the population level (Gillespie 1991).

Specifically, the neutral mutation theory of molecular evolution has sought support in, and provided explanations for, a growing wealth of information constituting two lines of evidence: first, the seemingly clocklike nature of amino acid (and more recently, nucleotide) substitution through time, as suggested by interspecific comparisons of sequence variation, and second, the abundance and pattern of protein (primarily enzyme) polymorphisms within species, as evidenced by starch gel electrophoresis (see Gillespie 1991; Nei 1987 for reviews). Subterranean rodents have figured prominently in the accumulation of this evidence, primarily of the latter kind, by providing a focal case in which to examine the driving factors accounting for allelic variation in natural populations.

The role of subterranean rodents in this debate stems directly from the nature of the subterranean niche. In his influential review, Nevo (1979) depicted this niche as rather sheltered from aboveground variation in temperature and humidity, and therefore less variable both locally and geographically. Resorting to the Hutchinsonian concept of niche dimensions, the subterranean niche was perceived as being much narrower and more predictable than aboveground niches. On the basis of these notions, Nevo and co-workers saw subterranean rodents as occupying narrow niches that should sustain low levels of genetic variation as a result of homogenizing selection (a version of the niche width hypothesis: see Nevo, Filippucci, and Beiles 1990). The cumulative result of a substantial number of allozyme studies has been to show that, on average, subterranean rodents have reduced levels of heterozygosity compared with their non-subterranean counterparts (e.g., Nevo, Filippucci, and Beiles 1990). Although these analyses are not without problems (e.g., non-independence of species as data points [Felsenstein 1985] and the difference between correlation and causation), the basic observation of reduced average heterozygosity among subterranean rodents is well documented and finds parallels among subterranean insectivores (Yates and Moore 1990).

Yet alternative explanations are not so easily dismissed. Very often habitats suitable for subterranean rodents are very patchily distributed. Census and, more importantly, effective deme sizes may be extremely low, and genetic drift may consequently play a significant role in reducing heterozygosity (Patton and Feder 1981). Capitalizing on their sampling of a tuco-tuco population before its disruption by a volcanic eruption, Gallardo, Köhler, and Araneda (1995) resampled the area afterward and found dramatic losses of genetic variation. Low heterozygosity in this area may well remain as a footprint of this historical accident for a long time. Without the fortuitous fact that current levels of variability could be compared with those found in samples collected before the eruption, it would have been all too easy to see this as one more case of low variation in accordance with the niche width hypothesis!

More generally, Patton and co-workers (Steinberg and Patton, chap. 8, this volume) have shown that populations of subterranean rodents show dramatic variation in their levels of "connectedness" (by gene flow) across geography, by virtue of the variation in the distribution of suitable habitats. Accordingly, substantial differences in the levels and partitioning of genetic variation are to be expected. Whereas low heterozygosity is in accordance with both a neutralist and a selectionist perspective, Patton (1990) emphasizes that cases of high heterozygosity, and therefore the broad range of heterozygosities found among subterranean rodents (e.g., Apfelbaum et al. 1991; Patton and Yang 1977; Patton and Smith 1990), are in conflict with the niche width hypothesis, but are easily accounted for on the basis of historical processes such as gene flow and genetic drift.

This historically important debate has lost much of its fuel for several reasons. First, a trend toward low variation is broadly predicted by both the selectionist niche width hypothesis and the neutralist derivations from demography and history in the subterranean habitat. Second, broad-scale comparisons and correlational approaches in general are marred by interpretive problems and inherently charged with substantial noise. Even if there is a general selective signal, it may not be possible to detect it from these types of data (Gillespie 1991). Third, whereas the buffered nature of the subterranean niche is well supported for certain environmental variables (e.g., temperature, humidity), it probably does not apply at all to other, equally important variables such as oxygen and carbon dioxide content, which are known to vary radically after rains (Arieli 1979; Buffenstein, chap. 2; Busch et al., chap. 5, this volume). Strictly adaptationist perspectives along the lines of the niche width hypothesis might suggest increased (rather than reduced) variation in, for example, molecules involved in respiration and gas transport. More generally, the types of positive selection, if any, acting upon different genes may be very different, and therefore hard to subsume under a single, overriding hypothesis.

Given these difficulties, the search for evidence of positive selection (or lack thereof) has shifted to detailed analyses of specific loci (Kreitman and Akashi 1995) or gene families of functionally and phylogenetically related loci (Nei and Hughes 1991). The realization that different types of selection, including purifying, directional, and balancing selection, leave distinctive footprints in patterns of variation at the DNA level (Kreitman 1991) has opened new avenues for examining the neutralist-selectionist issue. Furthermore, molecular phylogenies of the loci of interest may help us detect episodes of accelerated evolution in relation, for example, to new selective regimes (Messier and Stewart 1997; Zhang, Kumar, and Nei 1997). Notice also that, as with many other important issues in evolutionary biology, the neutralist-selectionist debate has turned into one about relative importance, rather than about all-encompassing explanation.

Studies of subterranean rodents have rarely taken advantage of these growing opportunities to examine the factors driving variation at the molecular level, despite plenty of fascinating possibilities. For instance, can we detect bouts of accelerated evolution driven by directional selection in specific loci in relation to the invasion of the subterranean niche? Think of molecules such as hemoglobins and myoglobins, whose physiological properties must change in predictable directions if they are to adapt to hypoxic, hypercapnic conditions (Arieli 1979). How important are those possible episodic selection events (Gillespie 1991) with respect to sustained adaptation or neutral evolution driven by mutation, drift, and purifying selection? Subterranean rodent taxa of different ages and degrees of commitment to underground life provide independent evolutionary experiments that should allow us to examine these issues.

To highlight the potential rewards of these types of analyses at this level, I wish to conclude this section by referring to two examples. First, de Jong et al. (1990) have demonstrated accelerated rates of amino acid substitution in the α-crystallin of Mediterranean mole-rats *(Nannospalax)*, whose eyes are regressed and covered by skin and, though still functional in photoreception, are under relaxed purifying selection. Second, an elegant study by Xia, Hafner, and Sudman (1996) has dissected apart the relative roles of mutational bias and purifying selection in third codon positions of pocket gopher mitochondrial DNA. Purifying selection, as illustrated in these two cases, is widely acknowledged by both neutralists and selectionists to play a significant role in molecular evolution, and thus the controversy concerns the relative role of "positive" types of selection.

Species Concepts

Subterranean rodents typically exhibit substantial, geographically structured morphological and genetic variation. The often localized diversity of subterranean rodents triggered descriptive frenzies that generated a plethora of specific and subspecific names for pocket gophers (summarized in Hall and Kelson 1959; Hall 1981) and tuco-tucos (see Reig et al. 1990). Surely there were excesses in the descriptive zeal of some of this early work. At the same time, variation in morphology and genetics is truly remarkable in many of these subterranean lineages. Diploid numbers in *Ctenomys*, for example, range from 10 to 70, thus encompassing much of the variation found among mammals (Reig et al. 1990), and allozyme variation between geographic units within *Thomomys bottae* is as large as that between many species, or even genera, of mammals (Patton and Smith 1990).

Observations and assertions such as the ones just made beg for clarification on two related fronts. First, just what exactly do we mean by "species," and how do alternative species concepts apply to the puzzling variation of subterranean rodents? Second, what are the processes of speciation in these highly variable organisms? Subterranean rodents have been prominent in discussions of these two issues, and Steinberg and Patton (chap. 8, this volume) provides a concise summary of both. Here, I focus primarily on the historical development of ideas about the mechanisms of speciation in subterranean rodents, but some comments about species concepts are needed for clarity.

Recall that, by and large, the distributions of subterranean taxa are allopatric or parapatric, and that zones of contact or overlap between taxa are typically very narrow, if present at all. This fact alone poses great difficulty for the practical application of the biological species concept because tests of reproductive isolation are rarely possible. Furthermore, it has long been recognized that it is difficult to identify levels of reproductive isolation on the basis of the degree of differentiation of two taxa. Thaeler (1974) showed that pairs of taxa characterized by grossly similar karyotypic divergences behaved very differently in contact, displaying rather limited interbreeding in some cases and extensive crossing in others. Hafner et al. (1983) obtained a roughly comparable range of results using allozyme markers. When contacts can be examined, operational definitions of "species" as entities with a reasonable degree of reproductive isolation are possible (Steinberg and Patton, chap. 8, this volume).

The mutual exclusion so typical of subterranean taxa leads to parapatric or entirely allopatric distributions, and therefore geographic structuring, differentiation, and ultimately, speciation appear likely. Yet the inescapable conclusion is that reproductive isolation between even highly differentiated geographic units may be lacking or may remain largely incomplete for extensive periods of time. It may well be that, at least within genera or comparably divergent taxa, the only units that are completely devoid of some degree of interbreeding are those that are simply precluded from engaging in breeding by total geographic isolation! Strict adherents to the biological species concept who demand nothing less than complete reproductive isolation are therefore forced to lump extremely divergent taxa within a single species. In the case of pocket gophers, for example, Hall and Kelson (1959), while retaining hundreds of subspecies to reflect their geographic distinctness, subsumed *Thomomys townsendii, T. bottae,* and *T. umbrinus* under a single specific name. Yet interbreeding among these taxonomic units is largely limited to F_1 individuals, posing little or no threat to their distinctness.

Relaxing the requirement of zero interbreeding to adopt the aforementioned operational rules leads to further complications because reasonable "species" defined by degree of reproductive isolation may be paraphyletic (Patton and Smith 1994). This is to be expected if geographic fragmentation is the norm and reproductive isolation appears erratically, bearing little correspondence to the overall degree of genetic differentiation or time of isolation. Paraphyly, of course, is not a problem for the classic biological species concept, since it makes no reference to cladistic identity. Furthermore, paraphyly is the expected pattern if one of Mayr's (1970) favored modes of speciation—peripatric speciation—takes place. Under that scenario, a geographically structured "central" species should be paraphyletic with respect to incipient species derived from peripheral isolates.

But paraphyly is a problem from a strictly cladistic perspective. If the basic unit of systematics and evolution—the species—is so careless of the essential dictum of phylogenetic systematics—monophyly—then why should higher taxa be any different? The phylogenetic species concept attempts to resolve the challenges posed by paraphyletic species through redefinition of terms (Cracraft 1989). Consider one biological species consisting of three geographic units (A, B, and C) that are paraphyletic with respect to a distinct biological species (D). Suppose D and C are closer (phylo)genetically to each other than either is to the remaining units A and B. If all these units are diagnosable as distinct from one another in a cladistic sense, then we may think of them as four distinct monophyletic units, which Cracraft (1989) and others would call phylogenetic species.

Unfortunately, this alternative species concept is equally problematic in its application to subterranean rodents. Some well-differentiated units are characterized more by overall shifts in allele frequencies and resulting genetic distances than by fixed, diagnostic characters, as illustrated by the "genetic units" of *T. bottae* (Patton and Smith 1990). Nevertheless, these units have rather fuzzy edges, exhibiting genetic exchanges that vary dramatically in degree in contact zones and in levels of introgression away from such zones (Ruedi, Smith, and Patton 1997). Finally, reticulate evolution results in complex, discordant patterns across loci. Nuclear polymorphisms appear to persist for extended periods of time, with the result that units are defined more by degrees of differentiation than by neatly fixed differences concordant across loci and geography. Mitochondrial variants, while reaching fixation much faster, may nonetheless be notoriously discordant with nuclear differentiation, not only as a result of lineage sorting but also as a consequence of reticulation and transfers across rather distant taxa (Patton and Smith 1993; Ruedi, Smith, and Patton 1997).

I believe that the patterns we observe among subterranean rodents (e.g., strong geographic differentiation, a persistent ability to interbreed despite widespread allopatry and only rare, limited contacts) naturally lead to a patchwork of hierarchical differentiation and reticulation that will inevitably challenge the neat application of any species concept. Differentiation is too dirty a business to allow for such simplicity. We can rest assured that however troubling these complicated relationships may be for our learned species concepts, genetic units (be they termed species or otherwise) appear not to be disturbed by acquisitions of genes from other distant units. Some, in fact, have suggested that these reticulate events may stimulate adaptive evolution (Grant and Grant 1997).

Mechanisms of Speciation

No matter how complicated the definition of species may be, differentiation is the inevitable result of spatial subdivision and limited genetic exchange through time. In line with the biological species concept, reproductive isolation is the traditionally recognized point of no return that seals the process of differentiation (Mayr 1970). According to this notion, speciation may simply be defined as the evolution of reproductive isolating mechanisms. As Mayr (e.g., 1963: 431) wrote, "the problem of the multiplication of species . . . is to explain how a natural population is divided into several that are reproductively isolated." Under the "recognition species concept," the emphasis is placed on the evolution of mate recognition systems to the point at which they become incompatible, a strategy that may be viewed as a shift in emphasis about the same issue (Templeton 1989). If less or no attention is given to reproductive isolation and more to differentiation itself (in order to recognize monophyletic, or less ambitiously, simply distinct units), then speciation is nothing other than extensive differentiation.

Notice that none of these definitions implies a sharp difference between speciation and phyletic divergence unless special circumstances and/or types of differentiation are needed to produce the changes required by a particular concept to make two lineages distinct. Subterranean rodents have been a focus of sustained interest in the historical literature on speciation precisely because, by virtue of their extensive karyotypic variation (e.g., Reig and Kiblisky 1969), they were thought to be eloquent examples of "chromosomal speciation." For White (1978), these cases met two particularly exciting conditions: first, chromosomal rearrangements were the cause of speciation, by virtue of their resulting in total or partial sterility in hybrids (negative heterosis), and second, fixed chromosomal differences could be acquired in parapatry, without the mediation of geographic barriers characteristic of traditional allopatric speciation.

It is virtually impossible to review the vast literature on this subject here, but a few comments are in order. First, chromosomal speciation is necessarily based on the claim of a causal relationship between chromosomal differentiation and reproductive incompatibility. A simple association of chromosomal differences with the establishment of distinct species is insufficient if the genetic bases of reproductive incompatibility are not the chromosomal rearrangements themselves. It has been extremely difficult to dissect the direct effects of chromosomal differences from the correlated effects of differentiation (reviewed by Patton and Sherwood 1983; King 1993), except in model organisms such as the house mouse and *Drosophila*. Overall, two conclusions may be drawn from such studies: first, that the net effect of chromosomal differences on reproductive isolation varies dramatically for similar classes of gross rearrangements (e.g., centric fusions), and second, that the net effect is mediated, at least in part, by the capacity of different mechanisms (including those of hybrids) to handle chromosomal polymorphisms in meiosis.

In terms of the actual dynamics of speciation, models of chromosomal speciation have been cornered by their conflicting, nearly incompatible postulates. On the one hand, it is easy to demonstrate that strongly negatively heterotic rearrangements require rather stringent conditions for their fixation. These conditions include small local effective population sizes, on the order of $1/s$ if s is the selection coefficient against heterozygotes, assuming no gene flow. Postulating very small values of s makes the fixation process more likely, but lessens its importance in relation to speciation. On the other hand, the appeal of the early propositions of chromosomal speciation resided in the claim of its being a parapatric process, namely, one that takes place in spite of gene flow. Adding even modest levels of gene flow to the picture while retaining a respectable value for s makes the process even more stringent, if not entirely impossible.

All of these problems can be alleviated by departing from classic chromosomal speciation models in one or several respects. Relax the persistence of gene flow and you are back in a strictly allopatric model, only with chromosomes as a particular class of interesting characters. Lessen the value of s and the fixation of chromosomal rearrangements becomes substantially more likely (and less significant with regard to speciation!). Special cases of departures from strongly negative heterosis include strictly neutral rearrangements (no heterosis) and balanced chromosomal polymorphisms (positive heterosis à la fruit fly).

One particular scenario that overcomes many of these difficulties is that of "Robertsonian fans" (Baker and Bickham 1986; Capanna 1982), in which the fixation of centric fusions from all acrocentric parental forms takes place with little or no negative heterosis, but their accumulation leads to forms with incompatible sets of metacentric chromosomes. Interestingly, this is a special case of Dobzhansky's (1937: see Gavrilets 1997) classic proposition of populations that end up at different "adaptive peaks" without ever having to cross through a valley—that is, without having to overcome the continuous scrutiny of selection against transitional forms.

My impression is that no major conceptual advances can be expected on this topic on the sole basis of information on gross or banded karyotypes, their geographic distributions, or their behavior in polymorphic or hybrid populations. These types of data, of course, will continue to be useful in delineating units of evolution among subterranean rodents. But perhaps we need to complete the ongoing round of investigation of the genomic processes that underlie chromosomal changes and of the genetic control of meiosis (for examples of recent propositions about the role of certain repeated sequences in this respect, see Wichman et al. 1991; Bradley and Wichman 1994; Rossi et al. 1995). It is perhaps too difficult to understand the processes that affect chromosomal variation at the population level while treating these basic underlying processes as black boxes. Furthermore, hypotheses about the effects of specific types of DNA repeats and of their copy numbers and locations on the likelihood of chromosomal rearrangements can be addressed, at least initially, quite independently from the role of rearrangements in speciation.

Once fundamental mechanisms of change are better understood, chromosomal or any other proposed mechanisms of speciation will be best scrutinized in the framework of population genetics. Such process-based approaches (Steinberg and Patton, chap. 8, this volume) will provide a solid basis for the more careful rephrasing (or outright dismissal) of many colorful—but vague at best—propositions that have plagued the field of speciation (e.g., genetic revolutions and chromosomal orthoselection). It is somewhat ironic that both Mayr (1970) and White (1978), while supporting very different views of the process of speciation, expressed their opinions in rather qualitative terms, lacking detailed reference to population genetics ("beanbag genetics," as Mayr [1963: 263] called it). Speciation remains, of course, a process of which we still know relatively little. The realization that it seems to entail regular types of genes (Coyne 1992) subjected to regular population genetic processes (Smith et al. 1997; Steinberg and Patton, chap. 8, this volume), however, has stripped the subject only of its aura of unassailable mystery, not of its excitement or importance.

Present and Future Directions of Research

Organismal Perspectives

Perhaps one of the most striking realizations that comes from reading the chapters in the first part of this volume, which focuses on the organismal biology of subterranean rodents, is just how limited and skewed our knowledge of these animals is. For example, biomechanical analyses of most of the subterranean rodent families would have to be preceded by morphological and osteological descriptions, which at present are available for only a few taxa (Stein, chap. 1, this volume). Similarly, although detailed descriptions of life history and reproductive patterns have increased substantially for bathyergid mole-rats, they are considerably more sketchy for most other subterranean taxa (Bennett, Faulkes, and Molteno, chap. 4; Busch et al., chap. 5, this volume).

The organismal level of analysis is of fundamental importance, and should continue to be the focus of sustained research efforts. Simply to highlight a few of the most obvious possibilities, here is a very selective list of problems that require immediate attention.

MORPHOLOGY. As discussed above, perhaps one of the most bitterly debated problems in current evolutionary biology is the relative importance of adaptation and constraint in the evolution of morphology. The few proposed examples of constraints have not been adequately examined across diverse lineages of subterranean rodents or related taxa. Functional morphological assessments placed in an adaptationist framework have generally emphasized what makes subterranean lineages different from non-subterranean rodents. But what about variation among subterranean rodents? How does this variation map onto phylogenetic trees and relate to ecological diversity? With the advent of detailed phylogenies that include outgroup taxa, can we draw any generalizations regarding the sequence of morphological innovations associated with the occupation of and specialization within the subterranean niche?

PHYSIOLOGY AND REPRODUCTION. Enough research has been conducted to demonstrate that subterranean rodents do indeed differ significantly from their surface-dwelling relatives in many physiological attributes. This aboveground/belowground approach will continue to be fruitful and, as has happened recently with the analysis of vitamin D deficiency, should continue to produce unforeseen contrasts between the two lifestyles. As with morphology, however, one has the impression that the emphasis on these broad-scale contrasts may have obstructed the way for much-needed comparative analyses of physiological variation within lineages of subterranean rodents in relation to their phylogenetic position, commitment to fossoriality, and other ecological characteristics. Reproductive studies are rife with intriguing questions along the same lines. The drastically different gestation lengths found among subterranean rodents and their phylogenetic allies have yet to be accounted for. More generally, we have yet to complete sufficiently detailed analyses of energetic budgets to understand how subterranean rodents allocate energy to their diverse activities, including the very demanding activities of digging and reproduction.

COMMUNICATION. The issue of communication in subterranean rodents has understandably arisen in studies of social, burrow-sharing species such as the naked mole-rat and some octodontids and ctenomyids, the last of which are notorious for their vocalizations underground. In chapter 3, Francescoli sets forth a general hypothesis to account for variation in communication systems across subterranean rodents. The testing of this and alternative hypotheses will require substantial refinement and broadening of our data in this field of knowledge. As noted in chapter 3, information about sensory capabilities, the necessary counterpart of signal emission, is all but lacking save for studies of blind mole-rats *(Nannospalax)* and a few other species.

These directions for future work, diverse as they are, have significant points in common. The most obvious of these is that they will all require us to go beyond the current focus on taxa of special interest and on broad comparisons between surface-dwelling and subterranean species. As illustrated by the items just outlined, comparative analyses must (1) uncover variation across diverse families of subterranean rodents, (2) consider the diversity of forms within each one of these families in terms of phylogenetic lineages and morphological, ecological, and behavioral characteristics, and (3) incorporate related non-subterranean taxa into these analyses. Explicit phylogenies (e.g., Hafner, Demastes, and Spradling, chap. 10, this volume) will be required as frameworks for these types of analyses. Finally, I suspect that one of the rewards of this research agenda will be the ability to address with fresh data one of the most pressing issues of current evolutionary biology, namely, the interaction between constraint and adaptation in the evolution of form and function.

Behavioral and Evolutionary Ecology

The problems of organismal biology discussed in the preceding section are not independent of the concerns of behavioral and evolutionary ecology, although the latter disciplines encompass additional issues. Some of them will be briefly examined here.

Unquestionably, it was Jarvis's (1981) report of the unusual reproductive biology of the naked mole-rat that placed subterranean rodents in the mainstream of behavioral evolutionary ecology. Against the backdrop of widespread solitary lifestyles, the finding of eusociality in *Heterocephalus* (Jarvis 1981; Sherman, Jarvis, and Alexander 1991) and, more recently, in other bathyergids (Lacey, chap. 7, this volume) was highly unexpected. Hamilton's (1964) classic argument for a direct link between haplodiploidy and eusociality is relevant for hymenopterans, but hardly satisfactory for naked mole-rats or other subterranean rodents. Although the more general notion of inclusive fitness is still relevant in this context, naked mole-rats have prompted a search for ecological factors triggering, or at least facilitating, the evolution of sociality.

This line of analysis has been facilitated by the relative simplicity of the subterranean rodent niche. Could it be that the particular structure (e.g., food distribution) of this niche in the drylands of Africa has been a key factor in triggering, or at least facilitating, eusociality? This issue is analyzed in detail by Lacey in chapter 7. One important contribution of this chapter is the placing of the problem in the broader context of evolving mating and social systems. Hints of a diversity of social systems in some species of *Ctenomys* and *Spalacopus*, including burrow sharing by adults (Pearson 1959; Reig 1970), have culminated in the description of a social (in the broad sense) species of ctenomyid (*C. sociabilis:* Pearson and Christie 1985; Lacey, Braude, and Wieczorek 1997). Although very little is known about these social systems, they are best viewed as representing particular states along a continuum going from solitary, territorial species to eusocial systems (Lacey, chap. 7, this volume). Along the same lines, Lacey and Sherman (1997) view the reproductive skew found in naked mole-rats as part of a continuum potentially ranging from fairly even individual reproductive success to the very extreme cases of reproductive skew known among certain ants and termites.

These attempts to account for sociality on the basis of ecological considerations can be tested by taking advantage of the diverse, phylogenetically independent lineages of subterranean rodents and the various environments that they currently occupy. Yet it is possible that social and reproductive systems are subject to considerable phylogenetic inertia that keeps them out of equilibrium with habitat characteristics. Two factors may lead to this discordance: (1) the inertia caused by penalties imposed on individuals that deviate from established reproductive systems and (2) the fact that a species' niche is not merely passively received from its habitat.

In the chapters in this volume about population and community ecology (Busch et al., chap. 5; Cameron, chap. 6), it is apparent that the ecological data needed to test ideas about the evolution of social and reproductive systems are still insufficient. One particularly important task will be to direct efforts in these two subdisciplines of ecology to species that are prima facie candidates to span the range from solitary to eusocial species.

Metapopulation Dynamics: Demography and Genetics

The population level has been the classic focus of evolutionary theory, yet in practice we know much more about the results of evolution (in our understanding of individual-level adaptation, and to a lesser extent, constraint) than about the underlying processes. The discipline of population dynamics has two related facets, namely, demography and genetics. The coverage of these two issues is highly uneven for subterranean rodents. Whereas there have been extensive analyses of the population genetics of subterranean rodents (Steinberg and Patton, chap. 8, this volume), demographic analyses lag far behind, and are available only for some geomyids, spalacines, and bathyergids (Bennett, Faulkes, and Molteno, chap. 4; Busch et al., chap. 5, this volume). Even the most basic combinations of demographic and genetic data, required for developing and testing individual-based models of population genetic structure (Steinberg and Patton, chap. 8, this volume), are available for only a single species, the pocket gopher *Thomomys bottae*.

I want to reiterate that, in spite of some intricate relationships, determining the environmental parameters that dictate the distributions of subterranean rodents may be unusually simple. Thus, we have yet to explore fully the potential of our study organisms for testing ideas about metapopulation dynamics. For example, what is the role of extinction-recolonization processes in shaping the demography and genetics of subterranean rodents? Geneticists often invoke such processes to account for apparent departures from equilibrium observed at the genetic level (Steinberg and Patton, chap. 8, this volume), but these ideas have not been adequately tested at the strictly demographic level. Widely available aerial photographs, coupled with soil and vegetation maps, can take us a long way in predicting what the landscape should look like for these animals. Field records of a few environmental variables may suffice to supplement these maps. At a minimum, the areas heavily utilized by subterranean rodents are easily spotted in aerial photographs and satellite records. What is the distribution of suitable patch sizes and the corresponding distribution of population sizes? What is the frequency of occupancy of such patches? In other words, metapopulation dynamics can be easily examined at the strictly demographic level, and should give us an idea of whether there are substantial deviations from equilibrium at this level of analysis.

It is against this backdrop that population genetic models and empirical analyses must be examined. Notice, for instance, that the expected effects of metapopulation structure on genetic structure depend upon rather subtle aspects of demography, such as the frequency of extinction and sources of recolonization (Harrison and Hastings 1996). Clearly, demography must have a substantial effect on population genetic structure, but distinguishing between different genetic scenarios requires direct demographic analyses. For instance, if current populations deviate from static equilibrium at the demographic level, that observation may suffice to explain departure from equilibrium at the genetic level. If, however, demography is by and large in equilibrium with current conditions, then genetic disequilibrium must reflect historical demographic events. Slatkin (1993) has detected departures from simple isolation by distance expectations in pocket gopher population genetics. The temporal scale at which these phenomena are to be explained, however, is best determined with reference to demographic analysis of metapopulations.

Temporal lags between demographic changes and the expression of those changes at the genetic level are well known to analytical population geneticists, and also appear in individual-based modeling efforts (Steinberg and Patton, chap. 8, this volume). It remains to be determined, however, to what extent individual-based expectations depart from those derived from analytical (or simulated) studies of much more general models, such as those in vogue in population genetics. Individual-based models are in their infancy, and thus it is too early to tell how much closer they will bring us to the "truth" of population dynamics than general models have.

Regardless of the specific answers to these questions, it seems clear that both demographic and genetic analyses of metapopulations can share one significant factor: history. Indeed, history appears in these analyses almost as an uninvited guest, leaving its footprint in current patterns that are more reflective of past conditions or events than of current status. In other words, any time that we note a departure from static equilibrium, history may be required to account for our observations. This is certainly the case for population genetic studies of pocket gophers, in which nonequilibrium conditions may be the norm, rather than the exception, although at this point the issue remains open for investigation.

Phylogenetics and Macroevolution

Because of their patchy, usually parapatric distributions, subterranean rodents offer dramatic examples of hierarchical structuring of genetic variation, from local populations to species and higher phylogenetic units. But, as argued by Steinberg and Patton (chap. 8, this volume), it is perhaps precisely for the same reason that the ability to hybridize appears, at least in pocket gophers, to be retained for long periods of time, and, consequently, that reticulation in the form of horizontal gene transfer is also likely. In spite of the difficulties that the processes of hierarchical differentiation and reticulation pose for applications of species concepts, discrete units of evolution can be identified, although they may not be as neat or as monophyletically correct as some might like them to be.

No species concept can readily accommodate the reality of subterranean rodents, but, once species or other units of evolution are evident, lineage-level properties emerge and macroevolutionary processes arise. I submit that, just as constraints offer useful perspectives that do not necessarily conflict with adaptation, macroevolutionary theory brings in fresh viewpoints even if it is not complete when decoupled from microevolution.

For assessing evolutionary patterns and processes above the species level, a phylogenetic framework is essential. Phylogenetic perspectives have bearing on problems of macroevolution for several reasons. First, as suggested by Cook, Lessa, and Hadly (chap. 9, this volume), long-held views about the unusually high rates of speciation among subterranean rodents are macroevolutionary hypotheses (see also Vrba and Gould 1986) whose testing is possible only with phylogenetic information. It is all too evident that our limited understanding of subterranean rodent phylogeny is a major hurdle for these purposes. Even with well-supported phylogenies, tests of these hypotheses will require unambiguous definitions of species or other evolutionary units whose differential proliferation can be tested, as in comparisons of subterranean versus surface-dwelling rodents. Although it is currently impossible to pursue this question in adequate detail, one has to wonder whether such tests will be sensitive to the use of different operational units.

Second, and along a different line of thought, phylogenies are the necessary starting point for the study of cospeciation. Hafner, Demastes, and Spradling (chap. 10, this volume) argue that subterranean rodents and their parasites offer unique opportunities to develop a research program in coevolution. The subterranean niche appears to provide ideal conditions for strict cospeciation, although horizontal transfers of parasitic lineages, comparable to those of genes, are known to have occurred. These transfers are interesting on their own right, yet it is clear that this system is far simpler to track, and therefore to understand, than others in which host switches are much more likely. If host switches take place frequently, then the evolutionary interactions between host and parasite lineages, although just as important, will be far more difficult to track. Such is the case with classic examples of coevolution of broadly interacting groups, such as flowering plants and their pollinators (Darwin 1877) or armored prey and their predators (Vermeij 1987).

The examples included in this volume illustrate a more general point, namely, that phylogeny and, more generally, history (including both hierar-

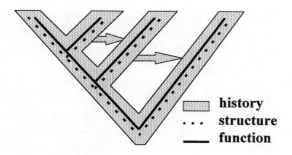

FIGURE 11.4. Historical (including hierarchical and reticulate patterns) framework for thinking about the evolution of structure and function, and, consequently, analyzing adaptation and constraint (compare with figure 11.1). The diagram shows structure and function as embedded within history, but no presumption is made as to their mode of evolution.

chy and reticulation) are the proper contexts in which to think about structure and function, constraint and adaptation. Although the recognition of history is as old as evolutionary biology itself, recent years have witnessed substantial growth in the explicit use of historical, especially phylogenetic, frameworks for the analysis of evolutionary questions (e.g., Huelsenbeck and Ranalla 1997 and references therein). The viewpoints reflected in the classic triangle of figure 11.1 should perhaps be reorganized as in figure 11.4 to reflect this growing historical awareness.

In this section, we have arrived at phylogenetic and macroevolutionary patterns on a pathway originating at lower organizational levels, but it is now evident that phylogenies and history reflect back upon our entire field of endeavor. Before we are carried away in an all-embracing cladistic sweep, however, I wish to mention that there are good biological reasons why history may show its footprint to a different extent at different levels of organization, or even in different aspects of the same problem. Is it possible that we need not worry as much about history when studying physiological traits, which may show nearly instantaneous adjustment to current requirements? The physiological tradition, an example of adaptationism at its best, would perhaps answer positively. Those of us closer to issues of historical remnants and constraints would tend to think differently, and would insist upon using a phylogenetic framework to test alternative hypotheses. There is currently no resolution to this debate, but it is entirely possible that different issues require thought and analyses that encompass different time scales. In other words, it may be possible to view the same phenomenon from different perspectives. For instance, balancing selection may bring a polymorphism to near equilibrium frequencies in a fraction of N generations. Yet it may take multiples of that time for us to see the typical excess of linked silent substitutions that constitutes the historical footprint of a balanced polymorphism at the DNA level.

Conclusions

This chapter is intended to have two take-home messages. First, subterranean rodents have a long and productive history as suitable subjects for developing and testing ideas in evolutionary biology. Second, the diversity and significance of the evolutionary issues that these animals allow us to address, coupled with the current availability of conceptual and empirical analytical tools to get the job done, should generate even more fertile research in this area. The work and ideas summarized in this volume, as well as in that edited by Nevo and Reig (1990), offer plenty of guidance for future work.

Acknowledgments

The organizers and volunteers of ITC-7 are to be thanked for providing a stimulating atmosphere for the symposium from which this volume emerged. I am grateful to Eileen Lacey, James Patton, and Guy Cameron for their organizational and editorial efforts and for their kind invitation to produce this synthesis. Finally, over the years I have been profoundly influenced by the work and ideas of many individuals. Among them, Carlos Altuna, Joe Cook, María Paz Echeverriarza, Thales de Freitas, Jim Patton, Peg Smith, Barbara Stein, Charlie Thaeler, and Marvalee Wake deserve special thanks for long hours of work and discussion.

Literature Cited

Alexander, R. M. 1983. *Animal mechanics*. London: Blackwell Scientific.
Antonovics, J., and P. H. van Tienderen. 1991. Ontoecogenophyloconstraints? The chaos of constraint terminology. *Trends in Ecology and Evolution* 6:166–68.
Apfelbaum, L. I., A. I. Massarini, L. E. Daleffe, and O. A. Reig. 1991. Genetic variability in the subterranean rodents *Ctenomys australis* and *Ctenomys porteousi* (Rodentia: Octodontidae). *Biochemical Systematics and Ecology* 19:467–76.
Arieli, R. 1979. The atmospheric environment of the fossorial mole rat *(Spalax ehrenbergi)*: Effects of season, soil texture, rain, temperature and activity. *Comparative Biochemistry and Physiology* 63A:569–75.
Baker, R. J., and J. W. Bickham. 1986. Speciation by monobrachial centric fusions. *Proceedings of the National Academy of Sciences, USA* 83:8245–48.
Best, T. L. 1973. Ecological separation of three genera of pocket gophers (Geomyidae). *Ecology* 54:1311–19.

Bradley, R. D., and H. A. Wichman. 1994. Rapidly evolving repetitive DNAs in a conservative genome: A test of factors that affect chromosomal evolution. *Chromosome Research* 2:354–60.
Brown, J. H. 1995. *Macroecology*. Chicago: University of Chicago Press.
Burian, R. M. 1992. Adaptation: Historical perspectives. In *Keywords in evolutionary biology*, edited by E. F. Keller and E. A. Lloyd, 7–12. Cambridge, MA: Harvard University Press.
Capanna, E. 1982. Robertsonian numerical variation in animal speciation: *Mus musculus*, an emblematic model. In *Mechanisms of speciation*, edited by C. Barigozzi, 155–74. New York: Alan R. Liss.
Carrier, D. R. 1987. The evolution of locomotor stamina in tetrapods: Circumventing a mechanical constraint. *Paleobiology* 13:326–41.
Cowen, R. 1996. Locomotion and respiration in aquatic air-breathing vertebrates. In *Evolutionary paleobiology*, edited by D. Jablonski, D. H. Erwin, and J. H. Lipps, 337–52. Chicago: University of Chicago Press.
Coyne, J. A. 1992. The genetics of speciation. *Nature* 355:511–15.
Coyne, J. A., and H. A. Orr. 1989. Two rules of speciation. In *Speciation and its consequences*, edited by D. Otte and J. A. Endler, 180–207. Sunderland, MA: Sinauer Associates.
Cracraft, J. 1989. Speciation and its ontology: The empirical consequences of alternative species concepts for understanding patterns and processes of differentiation. In *Speciation and its consequences*, edited by D. Otte and J. A. Endler, 28–59. Sunderland, MA: Sinauer Associates.
Darwin, C. 1877. *The different forms of flowers on plants of the same species*. London: John Murray.
de Jong, W., W. Hendriks, S. Sanyal, and E. Nevo. 1990. The eye of the blind mole rat *(Spalax ehrenbergi)*: Regressive evolution at the molecular level. In *Evolution of subterranean mammals at the organismal and molecular levels*, edited by E. Nevo and O. A. Reig, 383–95. Progress in Clinical and Biological Research, vol. 335. New York: Wiley-Liss.
Dobzhansky, T. H. 1937. *Genetics and the origin of species*. New York: Columbia University Press.
Eldredge, N. 1996. Hierarchies in macroevolution. In *Evolutionary paleobiology*, edited by D. Jablonski, D. H. Erwin, and J. H. Lipps, 42–61. Chicago: University of Chicago Press.
Felsenstein, J. 1985. Phylogenies and the comparative method. *American Naturalist* 125:1–15.
Gallardo, M. H., N. Köhler, and C. Araneda. 1995. Bottleneck effects in local populations of fossorial *Ctenomys* (Rodentia, Ctenomyidae) affected by vulcanism. *Heredity* 74:89–98.
Gavrilets, S. 1997. Evolution and speciation on holey adaptive landscapes. *Trends in Ecology and Evolution* 12:307–12.
Gillespie, J. H. 1991. *The causes of molecular evolution*. New York: Oxford University Press.
Gould, S. J. 1982. Darwinism and the expansion of evolutionary theory. *Science* 216:380–87.
———. 1989. A developmental constraint in *Cerion*, with comments on the definition and interpretation of constraint in evolution. *Evolution* 43:516–39.
Gould, S. J., and N. Eldredge. 1993. Punctuated equilibrium comes of age. *Nature* 366:223–27.
Gould, S. J., and R. C. Lewontin. 1979. The spandrels of San Marco and the Panglossian paradigm: A critique of the adaptationist programme. *Proceedings of the Royal Society of London* B 205:581–98.
Grant, P. R., and B. R. Grant. 1997. Genetics and the origin of bird species. *Proceedings of the National Academy of Science, USA* 94:7768–75.
Grinnell, J. 1914. The niche-relationships of the California thrasher. *Auk* 34:427–33.
Hafner, J. C., D. J. Hafner, J. L. Patton, and M. F. Smith. 1983. Contact zones and the ge-

netics of differentiation in the pocket gopher *Thomomys bottae* (Rodentia: Geomyidae). *Systematic Zoology* 32:1–20.
Hall, E. R. 1981. *The mammals of North America.* 2d edition. New York: John Wiley & Sons.
Hall, E. R., and K. R. Kelson. 1959. *The mammals of North America.* New York: Ronald Press.
Hamilton, W. D. 1964. The genetical evolution of social behavior. I, II. *Journal of Theoretical Biology* 7:1–52.
Harrison, S., and A. Hastings. 1996. Genetic and evolutionary consequences of metapopulation structure. *Trends in Ecology and Evolution* 11:180–83.
Hildebrand, M. 1985. Digging in quadrupeds. In *Functional vertebrate morphology,* edited by M. Hildebrand, D. M. Bramble, K. F. Liem, and D. B. Wake, 89–109. Cambridge, MA: Harvard University Press.
Huelsenbeck, J. P., and B. Ranalla. 1997. Phylogenetic methods come of age: Testing hypotheses in an evolutionary context. *Science* 276:227–32.
James, F. C., R. F. Johnston, N. O. Wamer, G. J. Neimi, and W. J. Boecklen. 1984. The Grinnellian niche of the wood thrush. *American Naturalist* 124:17–30.
Jarvis, J. U. M. 1981. Eusociality in a mammal: Cooperative breeding in naked mole-rat colonies. *Science* 212:571–73.
King, M. 1993. *Species evolution: The role of chromosome evolution.* Cambridge: Cambridge University Press.
Kreitman, M. E. 1991. Detecting selection at the level of DNA. In *Evolution at the molecular level,* edited by R. K. Selander, A. G. Clark, and T. S. Whittam, 204–21. Sunderland, MA: Sinauer Associates.
Kreitman, M., and H. Akashi. 1995. Molecular evidence of natural selection. *Annual Review of Ecology and Systematics* 26:403–22.
Lacey, E. A., S. H. Braude, and J. R. Wieczorek. 1997. Burrow sharing by colonial tucotucos *(Ctenomys sociabilis). Journal of Mammalogy* 78:556–62.
Lacey, E. A., and P. W. Sherman. 1997. Cooperative breeding in naked mole-rats: Implications for vertebrate and invertebrate sociality. In *Cooperative breeding in mammals,* edited by N. G. Solomon and J. A. French, 267–301. Cambridge: Cambridge University Press.
Landry, S. O., Jr. 1957. Factors affecting the procumbency of upper incisors. *Journal of Mammalogy* 38:223–34.
Lessa, E. P. 1990. Morphological evolution of subterranean mammals: Integrating structural, functional, and ecological perspectives. In *Evolution of subterranean mammals at the organismal and molecular levels,* edited by E. Nevo and O. A. Reig, 211–30. Progress in Clinical and Biological Research, vol. 335. New York: Wiley-Liss.
Lessa, E. P., and J. L. Patton. 1989. Structural constraints, recurrent shapes and allometry in pocket gophers (genus *Thomomys*). *Biological Journal of the Linnean Society* 36:349–63.
Lessa, E. P., and B. R. Stein. 1992. Morphological constraints in the digging apparatus of pocket gophers (Mammalia: Geomyidae). *Biological Journal of the Linnean Society* 47:439–53.
Lessa, E. P., and C. S. Thaeler, Jr. 1989. A reassessment of morphological specializations for digging in pocket gophers. *Journal of Mammalogy* 70:689–700.
Lovegrove, B. G. 1989. The cost of burrowing by the social mole rats (Bathyergidae) *Cryptomys damarensis* and *Heterocephalus glaber:* The role of soil moisture. *Physiological Zoology* 62:449–69.
Mayr, E. 1963. *Animal species and evolution.* Cambridge, MA: Harvard University Press.
———. 1970. *Populations, species, and evolution: An abridgment of Animal species and evolution.* Cambridge, MA: Harvard University Press.
———. 1991. *One long argument: Charles Darwin and the genesis of modern evolutionary thought.* Cambridge, MA: Harvard University Press.
Merriam, C. H. 1895. Monographic revision of the pocket gopher, family Geomyidae, exclusive of the species of *Thomomys. North American Fauna* 8:1–213.

Messier, W., and C.-B. Stewart. 1997. Episodic adaptive evolution of primate lysozymes. *Nature* 385:151–54.
Miller, R. S. 1964. Ecology and distribution of pocket gophers (Geomyidae) in Colorado. *Ecology* 45:256–72.
Nei, M. 1987. *Molecular evolutionary genetics*. New York: Columbia University Press.
Nei, M., and A. L. Hughes. 1991. Polymorphism and evolution of the major histocompatibility complex loci in mammals. In *Evolution at the molecular level*, edited by R. K. Selander, A. G. Clark, and T. S. Whittam, 222–47. Sunderland, MA: Sinauer Associates.
Nevo, E. 1979. Adaptive convergence and divergence of subterranean mammals. *Annual Review of Ecology and Systematics* 10:269–308.
———. 1988. Natural selection in action: The interface of ecology and genetics in adaptation and speciation at the molecular and organismal levels. In *The zoogeography of Israel*, edited by Y. Yom-Tov and E. Tchernov, 411–38. Dordrecht: D. W. Junk.
Nevo, E., M. G. Filippucci, and A. Beiles. 1990. Genetic diversity and its ecological correlates in nature: Comparisons between subterranean, fossorial, and aboveground small mammals. In *Evolution of subterranean mammals at the organismal and molecular levels*, edited by E. Nevo and O. A. Reig, 347–66. Progress in Clinical and Biological Research, vol. 335. New York: Wiley-Liss.
Nevo, E., and O. A. Reig, eds. 1990. *Evolution of subterranean mammals at the organismal and molecular levels*. Progress in Clinical and Biological Research, vol. 335. New York: Wiley-Liss. A. R. Liss, New York.
Patton, J. L. 1990. Geomyid evolution: The historical, selective, and random basis for divergence patterns within and among species. In *Evolution of subterranean mammals at the organismal and molecular levels*, edited by E. Nevo and O. A. Reig, 49–70. Progress in Clinical and Biological Research, vol. 335. New York: Wiley-Liss.
Patton, J. L., and J. H. Feder. 1981. Microspatial genetic heterogeneity in pocket gophers: Non-random breeding and drift. *Evolution* 35:912–20.
Patton, J. L., and S. W. Sherwood. 1983. Chromosome evolution and speciation in rodents. *Annual Review of Ecology and Systematics* 14:139–58.
Patton, J. L., and M. F. Smith. 1990. The evolutionary dynamics of the pocket gopher, *Thomomys bottae*, with emphasis on California populations. *University of California Publications in Zoology* 123:1–161.
———. 1993. Molecular evidence for mating asymmetry and female choice in a pocket gopher *(Thomomys)* hybrid zone. *Molecular Ecology* 2:3–8.
———. 1994. Paraphyly, polyphyly, and the nature of species boundaries in pocket gophers (genus *Thomomys*). *Systematic Biology* 43:11–26.
Patton, J. L., and S. Y. Yang. 1977. Genetic variation in *Thomomys bottae* pocket gophers: Macrogeographic patterns. *Evolution* 31:697–720.
Pearson, O. P. 1959. Biology of the subterranean rodents, *Ctenomys*, in Peru. *Memorias de Museo de Historial Natural "Javier Prado"* 9:1–56.
Pearson, O. P., and M. I. Christie. 1985. Los tuco-tucos (género *Ctenomys*) de los parques nacionales Lanin y Nahuel Huapi, Argentina. *Historia Natural* 5:337–43.
Pianka, E. R. 1988. *Evolutionary ecology*. 4th edition. New York: Harper & Row.
Radinsky, L. B. 1985. Approaches in evolutionary morphology: A search for patterns. *Annual Review of Ecology and Systematics* 16:1–14.
Raup, D. M. 1972. Approaches to morphological analysis. In *Models in paleobiology*, edited by T. J. M. Schopf, 28–44. San Francisco: Freeman, Cooper & Co.
Reig, O. A. 1970. Ecological notes on the fossorial octodontid rodent *Spalacopus cyanus* (Molina). *Journal of Mammalogy* 51:592–601.
Reig, O. A., C. Busch, M. O. Ortells, and J. R. Contreras. 1990. An overview of evolution, systematics, population biology, cytogenetics, molecular biology and speciation in *Cten-*

omys. In *Evolution of subterranean mammals at the organismal and molecular levels*, edited by E. Nevo and O. A. Reig, 71–96. Progress in Clinical and Biological Research, vol. 335. New York: Wiley-Liss.

Reig, O. A., and P. Kiblisky. 1969. Chromosome multiformity in the genus *Ctenomys* (Rodentia, Octodontidae). *Chromosoma* 28:211–44.

Rossi, M. S., C. A. Redi, G. Viale, A. I. Massarini, and E. Capanna. 1995. Chromosomal distribution of the major satellite DNA of South American rodents of the genus *Ctenomys*. *Cytogenetics and Cell Genetics* 69:179–84.

Ruedi, M., M. F. Smith, and J. L. Patton. 1997. Phylogenetic evidence of mitochondrial DNA introgression among pocket gophers in New Mexico (family Geomyidae). *Molecular Ecology* 6:453–62.

Sherman, P. W., J. U. M. Jarvis, and R. D. Alexander, eds. 1991. *The biology of the naked mole-rat*. Princeton, NJ: Princeton University Press.

Slatkin, M. 1993. Isolation by distance in equilibrium and nonequilibrium populations. *Evolution* 47:264–79.

Smith, T. B., R. K. Wayne, D. J. Girman, and M. W. Bruford. 1997. A role for ecotones in generating rainforest biodiversity. *Science* 276:1855–57.

Stearns, S. C. 1986. Natural selection and fitness, adaptation and constraint. In *Patterns and processes in the history of life*, edited by D. M. Raup and D. Jablonski, 23–44. New York: Springer-Verlag.

Templeton, A. R. 1989. The meaning of species and speciation: A genetic perspective. In *Speciation and its consequences*, edited by D. Otte and J. A. Endler, 3–27. Sunderland, MA: Sinauer Associates.

Thaeler, C. S., Jr. 1968. An analysis of the distribution of pocket gopher species in northeastern California (genus *Thomomys*). *University of California Publications in Zoology* 86: 1–46.

———. 1974. A study of four contacts between the ranges of different chromosome forms of the *Thomomys talpoides* complex. *Systematic Zoology* 23:343–54.

Vermeij, G. J. 1974. Adaptation, versatility, and evolution. *Systematic Zoology* 22:466–77.

———. 1987. *Evolution and escalation*. Princeton, NJ: Princeton University Press.

Vleck, D. 1979. The energy cost of burrowing by the pocket gopher *Thomomys bottae*. *Physiological Zoology* 52:122–36.

———. 1981. Burrow structure and foraging costs in the fossorial rodent, *Thomomys bottae*. *Oecologia* 49:391–96.

Vrba, E. S., and S. J. Gould. 1986. The hierarchical expansion of sorting and selection: Sorting and selection cannot be equated. *Paleobiology* 12:217–28.

Wake, M. H. 1993. The skull as a locomotor organ. In *The skull*, vol. 3, *Functional and evolutionary mechanisms*, edited by J. Hanken and B. K. Hall, 197–240. Chicago: University of Chicago Press.

West, G. B., J. H. Brown, and B. J. Enquist. 1997. A general model for the origin of allometric scaling laws in biology. *Science* 276:122–26.

West-Eberhardt, M. J. 1992. Adaptation: Current usages. In *Keywords in evolutionary biology*, edited by E. F. Keller and E. A. Lloyd, 13–18. Cambridge, MA: Harvard University Press.

White, M. J. D. 1978. *Modes of speciation*. San Francisco: W. H. Freeman.

Wichman, H. A., C. T. Payne, O. A. Ryder, M. J. Hamilton, M. Maltbie, and R. J. Baker. 1991. Genomic distribution of heterochromatin sequences in equids: Implications to rapid chromosomal evolution. *Journal of Heredity* 82:369–77.

Wilkinson, G. S. 1993. Artificial sexual selection alters allometry in the stalk-eyed fly *Cyrtodiopsis dalmanni* (Diptera: Diopsidae). *Genetical Research* 62:213–22.

Xia, X., M. S. Hafner, and P. D. Sudman. 1996. On transition bias in mitochondrial genes of pocket gophers. *Journal of Molecular Evolution* 43:32–40.

Yates, T. L., and D. W. Moore. 1990. Speciation and evolution on the family Talpidae (Mammalia: Insectivora). In *Evolution of subterranean mammals at the organismal and molecular levels,* edited by E. Nevo and O. A. Reig, 1–22. Progress in Clinical and Biological Research, vol. 335. New York: Wiley-Liss.

Zhang, J., S. Kumar, and M. Nei. 1997. Small-sample tests of episodic adaptive evolution: A case study of primate lysozymes. *Molecular Biology and Evolution* 14:1335–38.

Contributors

C. Daniel Antinuchi
Departamento de Biología
Facultad de Ciencias Exactas y Naturales
Universidad Nacional de Mar del Plata
Mar del Plata
Argentina
antinuch@mdp.edu.ar

Nigel C. Bennett
Department of Zoology and Entomology
University of Pretoria
Pretoria 0002
Republic of South Africa
ncbennett@zoology.up.ac.za

Rochelle Buffenstein
Department of Biology
City College
City University of New York
138th Street at Convent Avenue
New York, NY 10031
USA
rochelle@harold.sci.ccny.cuny.edu

Cristina Busch
Departamento de Biología
Facultad de Ciencias Exactas y Naturales
Universidad Nacional de Mar del Plata
Mar del Plata
Argentina
cbusch@mdp.edu.ar

Guy N. Cameron
Department of Biological Sciences
University of Cincinnati
Cincinnati, OH 54211
USA
g.cameron@uc.edu

Joseph A. Cook
University of Alaska Museum
907 Yukon Drive
Fairbanks, AK 99775
USA
ffjac@aurora.alaska.edu

James W. Demastes
Department of Biology
University of Northern Iowa
Cedar Falls, IA 50614
USA
jim.demastes@uni.edu

Christopher G. Faulkes
Department of Biological Sciences
Queen Mary & Westfield College
University of London
London E1 4NS
United Kingdom
c.g.faulkes@qmw.ac.uk

Contributors

Gabriel Francescoli
Sección Etología, Instituto de Biología
Facultad de Ciencias
Universidad de la República Oriental del Uruguay
Montevideo 11400
Uruguay
gabo@fcien.edu.uy

Elizabeth A. Hadly
Department of Biological Sciences
Stanford University
Stanford, CA 94305
USA
hadly@stanford.edu

Mark S. Hafner
Museum of Natural Science
Department of Biological Sciences
Louisiana State University
Baton Rouge, LA 70803
USA
namark@lsu.edu

Marcello J. Kittlein
Departamento de Biología
Facultad de Ciencias Exactas y Naturales
Universidad Nacional de Mar del Plata
Mar del Plata
Argentina
kittlein@mdp.edu.ar

Eileen A. Lacey
Museum of Vertebrate Zoology
Department of Integrative Biology
University of California
Berkeley, CA 94720
USA
ealacey@socrates.berkeley.edu

Enrique P. Lessa
Laboratorio de Evolución
Facultad de Ciencias
Universidad de la República Oriental de Uruguay
Montevideo 11400
Uruguay
lessa@fcien.edu.uy

Ana I. Malizia
Departamento de Biología
Facultad de Ciencias Exactas y Naturales
Universidad Nacional de Mar del Plata
Mar del Plata
Argentina

Andrew J. Molteno
Department of Zoology and Entomology
University of Pretoria
Pretoria 0002
Republic of South Africa
ajmolteno@zoology.up.ac.za

James L. Patton
Museum of Vertebrate Zoology
Department of Integrative Biology
University of California
Berkeley, CA 94720
USA
patton@uclink.berkeley.edu

Theresa A. Spradling
Department of Biology
University of Northern Iowa
Cedar Falls, IA 50614
USA
theresa.spradling@uni.edu

Barbara R. Stein
Museum of Vertebrate Zoology
University of California
Berkeley, CA 94720
USA
bstein@socrates.berkeley.edu

Eleanor K. Steinberg
Department of Zoology
University of Washington
Seattle, WA 98195
USA
steinbek@zoology.washington.edu

J. Cristina del Valle
Departamento de Biología
Facultad de Ciencias Exactas y Naturales
Universidad Nacional de Mar del Plata
Mar del Plata
Argentina
delvalle@mdp.edu.ar

Aldo I. Vassallo
Departamento de Biología
Facultad de Ciencias Exactas y Naturales
Universidad Nacional de Mar del Plata
Mar del Plata
Argentina
avassall@mdp.edu.ar

Roxana R. Zenuto
Departamento de Biología
Facultad de Ciencias Exactas y Naturales
Universidad Nacional de Mar del Plata
Mar del Plata
Argentina
rzenuto@mdp.edu.ar

Taxonomic Index

Acanthosicyos naudinianus, 96
Acomys subspinosus, 90t
Aconaemys sp., 233, 235
acouchi *(Myoprocta pratti)*, 164
African mole-rat. *See* Bathyergidae
African skunk predator activity, 215
Amblysomus (golden mole), 235
Ambystoma tigrinum (tiger salamander), 237
amphibians in burrows, 238
Antilocapra americana (pronghorn), 236
Aplodontia (mountain beaver), 2
Arizona elegans, 238
Arvicola terrestris (water vole), 3
Arvicolinae. *See also Ellobius; Prometheomys*
  diversification test feasibility, 358
  evolutionary history, 335
  geographic distribution, 3, 337f, 338, 340f
  macroevolutionary patterns, 337f, 338, 340f, 347–48
Asio sp. (owl), 213–14
Aspidelaps scutatus (snakes), 215

bamboo rats. *See Cannomys; Rhizomys*
bank voles, 163
Bathyergidae. *See also Bathyergus; Cryptomys; Georychus; Heliophobius; Heterocephalus*
  auditory ability, 130
  BMR adaptations, 73
  bone size relation to force exertion, 42
  breeding life span, 161
  burrow configuration, 188, 189
  burrow sharing, 263t
  calcium deposition in teeth, 96
  captive animal studies resources, 260
  chisel-tooth digging, 30
  digging apparatus elements, 29t
  digging seasonality, 190
  dispersal behaviors, 162, 211
  diversification test feasibility, 358
  ecological attributes, 217t
  ectoparasitic arthropods of, 378t
  endoparasites of, 377t
  energy-conserving adaptations, 75–76
  energy costs of foraging, 69
  evolutionary history, 335–36
  eye size implications, 123
  food locating ability, 197
  food resources, 191
  food selectivity and feeding, 192t
  foraging activities, 46
  geographic distribution, 8–9, 337f, 343, 345f
  gestation period, 151, 152
  gut fermentation temperature, 71–72, 72f
  hearing sensitivity, 65
  juvenile dispersal, 162
  litter sizes, 154, 207t
  macroevolutionary patterns, 337f, 343, 345f, 347–48
  mate detection using seismic signals, 148
  mineral balance, 95–96
  nest sharing, 264
  neuroendocrine studies, 66
  non-shivering thermogenesis, 80
  paraphyletic/polyphyletic relationships study resources, 321t
  pelage and skin, 22
  pelvic bone, 47

426 TAXONOMIC INDEX

Bathyergidae (*cont.*)
  phylogenetic influences on sociality, 281
  predation data lack, 215
  pup appearance at birth, 155–56
  pup development, 156
  radiotelemetry study resources, 260
  reproductive data, 153*t*
  reproductive division of labor, 120
  seismic signal use, 134
  sexual maturity, 160, 204
  sociality trends, 276–77, 277*f*
  social organization, 146–47
  social suppression of reproduction: breeding season, 167, 169; familiarity avoidance, 167; females vs. males, 169; induced infertility, 163, 168*f*; inhibition of ovulation, 163–64; nonreproductive female characteristics, 165–66, 166*f*; pheromone effects, 163; reproductive male characteristics, 164–65
  soil removal techniques, 31
  spatial food distributions and sociality, 278–79
  spatial system overlap, 261
  spontaneous ovulators, 151
  tails, 24
  trapping methods, 259
  two-taxon tests of diversification, 351*t*
  visual abilities, 122
  visual system, 26
  vitamin D_3 deficiency tolerance, 96
  vitamin D_3 metabolism, 92–94
Bathyergus (blesmols). *See also* Bathyergidae
  acoustic signal production, 126
  body size, 21
  chisel-tooth digging, 30
  digging behavior, 28
  digging mode and habitat soil relationship, 186
  food selectivity and feeding, 192*t*
  foraging behavior, 191
  foreclaws, 43
  forefeet and hind feet, 24
  genetic studies resources, 305*t*
  geographic distribution, 8–9
  incisor groove function, 36
  incisor procumbency, 33, 34
  janetta: basal metabolic rate, 74*t*; digestive efficiencies, 70*t*; growth rates, 159*t*; litter sizes, 155; morphological and behavioral development, 158*t*; pup growth rates, 156, 159, 159*t*; reproductive data, 153*t*; sexual dimorphism, 147; sparring among pups, 156; temperature regulation, 79*t*; vertical stratification and competition, 234
  osteology and myology study resources, 32
  pinnae, 23
  scratch digging, 30
  secondary digging apparatus studies, 46
  seismic signal production, 131
  sensory and communicative characteristics, 115*t*
  suillus: basal metabolic rate, 74*t*; competitive exclusion, 233; digestive efficiencies, 70*t*; gestation period, 152; growth rates, 159*t*; juvenile dispersal, 161–62; litter sizes, 155; mode of dispersal, 212; morphological and behavioral development, 158*t*; pup growth rates, 159; reproductive data, 153*t*; seasonality of breeding, 152; sexual dimorphism, 147; sparring among pups, 156; sympatry and habitat selection, 232; tactile communication during mating, 125; temperature regulation, 79*t*; vertical stratification and competition, 234
Bertoeroa incana, 243
bighorn sheep *(Ovis canadensis)*, 236
bird predator activity, 215
bison *(Bos bison)*, 215, 236, 245
black-tailed prairie dogs *(Cynomys ludovicianus)*, 65
blesmols. *See* Bathyergus; Cryptomys; Georychus; Heliophobius; Heterocephalus
Bos bison (bison), 215, 236, 245
Bothrops neuwwidii (pit vipers), 215
Bubo virginianus (owls), 213–14
Bufo sp., 238
Buteo sp. (hawks), 213–14

Canis latrans (coyotes), 214
Cannomys (bamboo rats). *See also* Rhizomyinae
  auditory system, 27
  claw morphology, 43
  enamel color, 35
  geographic distribution, 7–8
  hind limb structure, 48
  incisor procumbency, 33, 35
  pelage and skin, 22
  skull shape and size, 38

TAXONOMIC INDEX 427

Cavia porcellus (guinea pigs)
  hearing frequency, 131*f*
  luteal phase length, 164
Cervus canadensis (elk), 236
cestode parasites, 376, 377*t*
Chaetodipus (pocket mice), 311
Chelemys macronyx, 233
chewing lice, 377–78
Chinchilla laniger (chinchilla), 164
Clethrionomys glareolus (bank voles), 163
Cnemidophorus sexlineatus (six-lined racerunners), 237
coccidian parasites, 376
Conepatus chinga (skunks), 215
coruro. See *Spalacopus*
coyotes predator activity, 214
Crotalus viridis (prairie rattlesnakes), 237–38
Crotaphopeltis hotamboeia (snakes), 215
Cryptomys (blesmols). See also Bathyergidae
  acoustic signal production, 126
  anselli: familiarity avoidance, 167; gestation period, 152; litter sizes, 155; morphological and behavioral development, 158*t*; pup growth rates, 159; reproductive data, 153*t*
  auditory system, 27
  behavioral specialization, 276
  bocagei, 79*t*
  body size, 21
  claw morphology, 43
  courtship and copulation, 150
  damarensis: alloparental care, 274; basal metabolic rate, 74*t*; behavioral specialization, 276; breeding life span, 161; burrow sharing, 262; calcium deposition in teeth, 96; courtship and copulation, 150; courtship sounds, 127; digestive efficiencies, 70*t*; group sizes, 268*t*; growth rates, 159*t*; Harderian gland size, 68; induced infertility, 163; lack of seismic communication, 134–35; lifetime reproductive success, 271; litter sizes, 155; loss of breeding animal consequences, 273; morphological and behavioral development, 158*t*; nonreproductive female characteristics, 165; non-shivering thermogenesis, 80*t*; pup development, 156; pup growth rates, 159; reaction to foreign conspecifics, 273; reproductive behavior inhibition, 167; reproductive data, 153*t*; reproductive male characteristics, 164; seasonality of breeding, 152; sensory and communicative characteristics, 115*t*; socially induced infertility, 168*f*; tactile communication during mating, 125; temperature regulation, 79*t*; temporal patterns of activity, 67; thermoneutral zones, 78; trapping studies resources, 259; vocalizations in courtship, 149
  darlingi: behavioral specialization, 276; gestation period, 152, 205*t*; group sizes, 268*t*; growth rates, 159*t*; induced infertility, 163; litter sizes, 155; morphological and behavioral development, 158*t*; reproductive behavior inhibition, 167; reproductive data, 153*t*; seasonality of breeding, 152; socially induced infertility, 168*f*
  digging mode and habitat soil relationship, 186
  eye size implications, 123
  food locating ability, 197
  food selectivity and feeding, 192*t*
  foraging activities, 46
  genetic studies resources, 305*t*
  geographic distribution, 8–9
  group sizes, 268*t*
  hind limbs structure, 48
  hottentotus: acid-base balance and hypercapnia, 86; alloparental care, 274; anti-predator behavior, 237; auditory ability, 130; behavioral specialization, 276; blood oxygen transport properties, 84, 85*f*, 86*t*; burrow interactions with other mammals, 235; burrow sharing, 262; competitive exclusion, 233; courtship sounds, 127; digestive efficiencies, 70*t*; gestation period, 152; group sizes, 268*t*; hearing frequency, 131*f*; hypercapnic response, 82; lack of seismic communication, 134–35; non-shivering thermogenesis, 80*t*; pup development, 156; reaction to foreign conspecifics, 273; sensory and communicative characteristics, 115*t*; sympatry and habitat selection, 232; tactile communication during mating, 125; taste capabilities, 117; temperature regulation, 79*t*; temporal patterns of activity, 67; 2,3-DPG levels, 86
  hottentotus amatus: basal metabolic rate, 74*t*;

Cryptomys (blesmols) (*cont.*)
   gestation period, 205*t*; temperature regulation, 79*t*
   hottentotus darlingi: basal metabolic rate, 74*t*; temperature regulation, 79*t*; thermoneutral zones, 78
   hottentotus hottentotus: breeding life span, 161; breeding season, 167; courtship and copulation, 149, 150; effects of acclimation on BMR, 73; growth rates, 159*t*; litter sizes, 155; morphological and behavioral development, 158*t*; pup growth rates, 159; reproductive data, 153*t*; seasonality of breeding, 152; temperature regulation, 79*t*; vertical stratification and competition, 234
   hottentotus natalensis, 79*t*
   hottentotus nimrodi, 79*t*
   incisor procumbency, 34, 35
   juveniles dispersal, 162
   mechowi: acoustic signal production, 126; basal metabolic rate, 74*t*; behavioral specialization, 276; gestation period, 152, 205*t*; group sizes, 268*t*; growth rates, 159*t*; lack of seismic communication, 135; litter sizes, 155; morphological and behavioral development, 158*t*; reproductive data, 153*t*; seasonality of breeding, 152; sensory and communicative characteristics, 115*t*; temperature regulation, 79*t*; trapping studies resources, 259; vocalizations, 127
   natality, 206
   olfaction, 27
   olfactory center size, 117
   population density, 198
   sexual dimorphism in burrows, 265
   sexual maturity, 160, 204
   skull shape and size, 39
   soil quantities moved, 229
   temperature regulation, 79*t*
   thermal conductance, 77*t*
   urogenital system study, 51
   visual system, 26
   vocalizations, 127
Ctenomyidae. See also *Ctenomys*
   burrow configuration, 188
   burrow sharing, 263*t*
   captive animal studies resources, 260
   coprophagy practice, 71
   diet content and foraging behavior, 191
   digging apparatus elements, 29*t*
   diversification test feasibility, 358
   ecological attributes, 217*t*
   ectoparasitic arthropods of, 378*t*
   endoparasites of, 377*t*
   evolutionary history, 336
   food caching, 196
   food locating ability, 197
   geographic distribution, 10, 337*f*, 343–44, 346*f*
   gestation period, 152
   incisor microstructure relationship to function, 36
   life span, 161
   lineages through time tests, 353
   macroevolutionary patterns, 337*f*, 343–44, 346*f*, 347–48
   paraphyletic/polyphyletic relationships study resources, 321*t*
   population density factors, 199
   radiotelemetry study resources, 260
   sensory physiology, 65
   sexual maturity, 160, 204
   sociality trends, 277–78, 277*f*
   spatial system overlap, 261
   species diversity, 353
   three-taxon tests results, 352
   trapping methods, 259
   tree balance tests results, 352
   two-taxon tests of diversification, 351*t*
   visual incisor displays, 121
Ctenomys (tuco-tucos). *See also* Ctenomyidae
   absence of incisor grooves, 36
   auditory bullae, 129
   australis: aggressive behavior, 234; body size and habitat distribution, 187; diet content and foraging behavior, 191, 194; dispersal behaviors, 211; dispersal rates, 212; food selectivity and feeding, 193*t*; influences on distribution, 185–86; litter sizes, 154, 207*t*; population age structure, 202; population density, 199; predation of, 214; sex ratio, 200*t*; sympatry and habitat selection, 232
   chasiquensis, 214
   claw morphology, 43
   cospeciation with nematode parasites, 380–81
   digestive system study sources, 51
   digging mode and habitat soil relationship, 186
   enamel color, 35

TAXONOMIC INDEX   429

eye size implications, 124
forefeet and hind feet, 24, 42–43
genetic studies resources, 303
geographic distribution, 10
haigi: burrow interactions with other mammals, 235; communication between solitary species, 284; diet content and foraging behavior, 194; male/female burrow distribution, 265; predation of, 214
hind limb structure, 48
incisor procumbency, 35
lack of seismic communication, 134, 135
maulinus: competitive exclusion, 233; genetic studies resources, 305*t;* geographic partitioning levels, 310; population age structure, 202; sex ratio, 200*t*
mendocinus: food selectivity and feeding, 193*t;* foraging activities, 46; gestation period, 151, 205*t;* litter sizes, 154; longevity, 203; olfactory center size, 117; sex ratio, 200*t*, 201; sexual maturity, 204
minutus, 265
molar structure intergeneric differences, 37
opimus: foraging activities, 46; gestation period, 151, 205*t;* litter sizes, 154, 207*t;* sex ratio, 200*t*
osteology and myology study resources, 32
pearsoni: acoustic signal production, 126; courtship and copulation, 149; digging mode and habitat soil relationship, 186; scent marking, 119–20; tactile communication during mating, 125; vocalizations, 127, 128*f*
pelage and skin, 22
pelvis shape, 48*f*
peruanus: burrow sharing, 262; litter sizes, 154, 207*t;* male/female burrow distribution, 265; sex ratio, 200*t*, 201; social organization, 147
pheromonal communication, 118
pinnae, 23
porteousi, 262
quantity of muscle, 44
rionegrensis: genetic studies resources, 305*t;* scent marking, 119–20
scratch digging, 30
seismic and vocal signals, 137*f*
seismic signal production, 131
sensory and communicative characteristics, 115*t*
similarity to *Spalacopus,* 10

sociabilis: alloparental care, 274–75; behavioral specialization, 276; burrow sharing, 262; diet content and foraging behavior, 194; group sizes, 268*t;* group structure, 269; lifetime reproductive success, 271–72; nest sharing, 264; reproductive specialization, 270; social organization, 147; trapping studies resources, 259
soil quantities moved, 229
soil removal techniques, 31
tails, 24, 125
talarum: aggressive behavior, 234; body size and habitat distribution, 187; burrow design, 189*f;* courtship and copulation, 150; diet content and foraging behavior, 191, 194; dispersal behaviors, 162, 211; dispersal rates, 212; food selectivity and feeding, 193*t;* gestation period, 151, 205*t;* induced ovulation, 151; influences on distribution, 185; juvenile dispersal, 162; life span, 161; litter sizes, 154, 207*t;* longevity, 203; male/female burrow distribution, 265; mark-recapture studies resources, 259; mode of dispersal, 212; morphological and behavioral development, 157*t;* mortality from aggression, 209; population age structure, 202; population density, 198, 199; predation of, 214; pup appearance at birth, 155; sex ratio, 200*t*, 201; sexual maturity, 204; sympatry and habitat selection, 232
ulna modifications, 42
urogenital structure study, 52
visual acuity, 122
visual system, 26
cuis *(Galea masteloides),* 164
Cynomys eudovicianus (black-tailed prairie dogs)
behavior studies, 258
hearing sensitivity, 65
pup growth rates, 159

Dasypus novemcinctus, 86
desert cottontails *(Sylvilagus auduboni),* 238
Desmodillus auricularis, 90*t*
Didelphis albiventris (opossums), 215
Dinopsyllus (fleas), 379
Dipodomys ordii (kangaroo rat)
burrow occupation, 238
geographic partitioning levels, 311
hearing frequency capability, 131*f*

earless lizards *(Holbrookia maculata)*, 237
eastern moles *(Scalopus aquaticus)*, 238
elk *(Cervus canadensis)*, 236
Ellobius (mole-vole). *See also* Arvicolinae
  body size, 21
  chisel-tooth digging, 30
  claw morphology, 43
  digestive system study sources, 51
  digging behavior, 28
  ectoparasitic arthropods of, 378*t*
  enamel color, 35
  foraging activities, 46
  forefeet and hind feet, 24
  geographic distribution, 3
  glandular system studies, 52
  head and neck muscles, 40
  head-lift digging, 30–31
  incisor procumbency, 33, 34
  incisor root placement, 34
  lutescens, 123
  molar structure intergeneric differences, 37
  pelage and skin, 23
  pinnae, 23
  sensory and communicative characteristics, 115*t*
  skull shape and size, 38, 39
  soil removal techniques, 31
  tails, 24
  talpinus, 123
  urogenital structure study, 52
  visual system, 26
European mole *(Talpa europaea)*, 86

fleas, 379
forbs *(Lomatium)*, 241
Fouquieri splendens (ocotillo), 241
foxes predator activity, 213–14

Galea masteloides (cuis), 164
Galictis cuja (grison), 214–15
Geomyidae. *See also Geomys; Orthogeomys; Pappogeomys; Thomomys; Zygogeomys*
  BMR adaptations, 73
  burrow configuration, 188
  burrow sharing, 263*t*
  claws, 43
  coprophagy practice, 71
  diet content and foraging behavior, 191
  digging apparatus elements, 29*t*
  diversification test feasibility, 358
  ecological attributes, 217*t*
  ectoparasitic arthropods of, 378*t*
  endoparasites of, 377*t*
  energy-conserving adaptations, 75
  energy costs of foraging, 69
  evolutionary history, 334
  food locating ability, 197
  geographic distribution, 9–10, 337–38, 337*f*, 339*f*
  gestation period, 151
  hearing sensitivity, 65
  hind limbs structure, 49
  incisor microstructure relationship to function, 36
  life span, 161
  litter sizes, 154
  macroevolutionary patterns, 337–38, 337*f*, 339*f*, 347–48
  mate detection using seismic signals, 148
  paraphyletic/polyphyletic relationships study resources, 321*t*
  pelage and skin, 22
  pinnae, 23
  quantity of muscle, 44–45
  scratch digging, 29–30
  secondary digging apparatus studies, 46
  sexual maturity, 160, 204
  spatial system overlap, 261
  trapping methods, 259
  two-taxon tests of diversification, 351*t*
  visual system, 26
Geomys (pocket gophers). *See also* Geomyidae
  attwateri: dietary preferences impact on plant diversity, 241; diet content and foraging behavior, 191; dispersal behaviors, 210; dispersal mode, 212; dispersal rates, 212; food selectivity and feeding, 193*t*; litter sizes, 154, 207*t*; predation of, 214; sex ratio, 200*t*; soil quantities moved, 228
  breeding season length, 206
  breviceps: floral abundance and, 242; predation of, 214
  bursarius: aggressive behavior, 234; allopatric distributions due to soil preferences, 232; auditory ability, 130; basal metabolic rate, 74*t*; burrow occupation by other species, 237, 238; dietary overlap with surface herbivores, 236; dietary preferences impact on plant diversity, 241; diet content and foraging behavior, 191; eye size implications, 123; food locating ability, 197; genetic studies re-

TAXONOMIC INDEX 431

sources, 305*t*; gestation period, 151, 205*t*; hearing frequency, 131*f*; litter sizes, 154, 207*t*; mound effects on plants, 243, 244; osteology and myology study resources, 32; population age structure, 202; seasonality of breeding, 152; sex ratio, 200*t*; soil chemistry changes from, 229, 230; temperature regulation, 79*t*; thermal conductance, 77*t*
dietary preferences impact on plant diversity, 241
digging mode and habitat soil relationship, 186
enamel color, 35
foreclaws, 43
forefeet and hind feet, 24
genetic drift and natural selection observations, 308–9
geographic distribution, 9–10
glandular system studies, 52–53
hind limbs structure, 48
incisor groove function, 36
incisor procumbency, 35
lower incisor growth rate, 35
pelvic bone, 47
personatus, 232
pinetis: burrow associate studies, 374; soil quantities moved, 228
predation of, 214
scratch digging, 30
seismic signal detection, 132–33
sensory and communicative characteristics, 115*t*
soil quantities moved, 229
tail as navigational tool, 125
ulna modifications, 42
upper incisor growth rate, 34
visual acuity, 122
Georychus (blesmols). *See also* Bathyergidae
acoustic signal production, 126
capensis: anti-predator behavior, 237; breeding life span, 161; calcium deposition in teeth, 96; communication between solitary species, 284; competitive exclusion, 233; courtship and copulation, 149; courtship sounds, 127; digestive efficiencies, 70*t*; dispersal mode, 212; food caching, 196; gestation period, 152, 205*t*; growth rates, 159*t*; juvenile dispersal, 161–62; litter sizes, 154–55; morphological and behavioral development,

158*t*; pup appearance at birth, 156; pup growth rates, 156, 159, 159*t*; reproductive data, 153*t*; seasonality of breeding, 152; seismic signal production, 131, 132; sexual dimorphism, 147; sexual maturity, 160; sparring among pups, 156; sympatry and habitat selection, 232; tactile communication during mating, 125; temporal patterns of activity, 67
claw morphology, 43
food selectivity and feeding, 192*t*
foraging behavior, 191
genetic studies resources, 305*t*
geographic distribution, 8–9
hind limb structure, 48
incisor procumbency, 34, 35
litter sizes, 155
osteology and myology study resources, 32
secondary digging apparatus studies, 46
sensory and communicative characteristics, 115*t*
sparring among pups, 156
visual system, 26
Geoxus valdivanus
burrow interactions with other mammals, 235
competitive exclusion, 233
gerbil *(Meriones unguiculatus)*
hearing frequency, 131*f*
urine concentrating abilities, 90*t*
golden mole *(Amblysomus)*, 235
gopher snakes *(Pituophis catenifer)*, 237
grasshoppers, 236
grison predator activity, 214–15
ground squirrels *(Spermophilus)*, 2
burrow occupation, 238
studies of, 258
guinea pigs *(Cavia porcellus)*
hearing frequency, 131*f*
luteal phase length, 164

hawks predator activity, 213–14
Heliophobius (blesmols). *See also* Bathyergidae
argenteocinereus: basal metabolic rate, 74*t*; morphological and behavioral development, 158*t*; reproductive data, 153*t*; seasonality of breeding, 152; temperature regulation, 79*t*; temporal patterns of activity, 67; thermal conductance, 77*t*
claw morphology, 43
flexibility of pelvis, 47
food selectivity and feeding, 192*t*

Heliophobius (blesmols) (*cont.*)
  foraging activities, 46
  foraging behavior, 191
  geographic distribution, 8–9
  incisor procumbency, 34, 35
  kapeti, 79*t*
  molar structure intergeneric differences, 37
  olfaction, 27
  sensory and communicative characteristics, 115*t*
  temperature regulation, 79*t*
Heterocephalus (blesmols). *See also* Bathyergidae
  anti-predator behavior, 237
  body size, 21
  claw morphology, 43
  comparative lung study, 50
  digging mode and habitat soil relationship, 186
  flexibility of pelvis, 47
  food selectivity and feeding, 192*t*
  forefeet and hind feet, 24
  geographic distribution, 8–9
  glaber: acoustic signal production, 126; alloparental care, 274–75; auditory ability, 130; basal metabolic rate, 74*t;* blood oxygen transport properties, 84, 85*f,* 86*t;* breeder replacement, 272–73; breeding life span, 161; burrow design, 189*f;* burrow sharing, 262; colony cohesion, 120; coprophagy practice, 71; dietary preferences impact on plant diversity, 241; digestive efficiencies, 70*t;* dispersal behaviors, 211; dispersal mode, 212; division of labor in colonies, 275–76; food caching, 196; food locating ability, 197; food resources, 191; genetic studies resources, 303, 305*t;* geographic partitioning levels, 310; gestation period, 205*t;* glandular system studies, 52; group sizes, 268*t;* Harderian gland size, 68; hearing frequency, 131*f;* induced infertility, 163; juvenile dispersal, 162; lack of seismic communication, 134, 134–35; lifetime reproductive success, 271; litter sizes, 155, 207*t;* longevity, 203; lung morphology, 82; mark-recapture studies resources, 259; metabolic changes due to pregnancy, 97; molecular genetic studies, 260; morphological and behavioral development, 158*t;* nonreproductive female characteristics, 165; non-shivering thermogenesis, 80*t;* pheromonal communication, 118; reaction to foreign conspecifics, 273; reproductive behavior inhibition, 167; reproductive data, 153*t;* reproductive male characteristics, 164; reproductive specialization, 270; scent trail use, 120–21; seasonality of breeding, 152; sex ratio, 200*t,* 201; sexual maturity, 160, 204; socially induced infertility, 168*f;* social organization, 147; soil quantities moved, 228; tactile communication during mating, 125; tail as navigational tool, 125; temperature regulation, 78, 79*t;* temporal patterns of activity, 67; thermal conductance, 77*t;* thermoneutral zones, 78; 2,3-DPG levels, 86; urine concentrating abilities, 90*t;* vitamin D_3 tolerance, 96; vocalization range, 126–27; vocalizations in courtship, 149; water balance and, 91
  incisor groove function, 36
  incisor procumbency, 34, 35
  natality, 206
  olfaction, 27
  pelage and skin, 22
  pinnae, 23
  population density, 198
  sensory and communicative characteristics, 115*t*
  soil removal techniques, 31
  sparring among pups, 156
  tails, 24, 27
Heterodon platyrhinos, 238
Heteromyidae, 351*t*
Holbrookia maculata (earless lizards), 237
Hoplopleura (lice), 379
Hyperacrius wynnei (mole-voles), 3. *See also* Arvicolinae
Hystricidae, 351*t*

Ictonyx striatus (African skunk), 215
insects in burrows, 238

kangaroo rat *(Dipodomys ordii)*
  burrow occupation, 238
  geographic partitioning levels, 311
  hearing frequency capability, 131*f*

lice *(Hoplopleura)*
  gopher-louse cospeciation, 381–85, 382*f,* 384*f*

host-parasite coevolution study and, 378–79
Lomatium (forbs), 241
long-clawed mole-voles. *See Prometheomys*

Marmota sp.
  behavior studies, 258
  hypercapnic response, 82
meadow voles. *See Microtus* sp.
Megascaphus (pocket gophers), 33
Mehelya capensis (snakes), 215
Meriones unguiculatus (gerbils)
  hearing frequency, 131*f*
  urine concentrating abilities, 90*t*
Micranthus, 242
Microtus sp. (meadow voles)
  burrow occupation, 238
  dietary preferences impact on plant diversity, 241
  dispersal rates, 212
  exclusion from burrows, 236
mites, 378
mole-rats. *See Nannospalax;* Spalacinae; *Spalax*
mole-voles. *See Ellobius; Hyperacrius wynnei*
mountain beaver *(Aplodontia),* 2
mouse hearing frequency, 131*f*
mule deer *(Cervus canadensis),* 236
Muridae. *See also* Arvicolinae; Myospalacinae; Rhizomyinae; Spalacinae
  digging apparatus elements, 29*t*
  geographic distribution, 3, 6–8
Mustela frenata (long-tailed weasels), 238
Myoprocta pratti (acouchi), 164
Myospalacinae. *See also Myospalax*
  diversification test feasibility, 358
  evolutionary history, 335
  geographic distribution, 3, 6, 337*f,* 338, 341, 341*f*
  macroevolutionary patterns, 337*f,* 338, 341, 341*f,* 347–48
  radiotelemetry study resources, 260
  spatial system overlap, 261
  two-taxon tests of diversification, 351*t*
Myospalax (zokors). *See also* Myospalacinae
  absence of incisor grooves, 36
  aspalax, 3
  baileyi, 186–87
  enamel color, 35
  foraging activities, 46
  force facilitating modifications to muscles, 44
  foreclaws, 43

geographic distribution, 3, 6
pelvic bone, 47
pinnae, 23
scratch digging, 30
soil removal techniques, 31
tails, 24
visual system, 26

naked mole-rat. *See Heterocephalus*
Nannospalax (mole-rats). *See also* Spalacinae
  auditory system, 27
  claw morphology, 43
  digestive system study sources, 51
  ehrenbergi: acid-base balance and hypercapnia, 86; acoustic signal production, 126; basal metabolic rate, 74*t;* blood oxygen transport properties, 84, 85*f,* 86*t;* burrow design, 188–89; burrowing activities, 247; cardiac response to hypoxia, 83–84; chemical signal identification ability, 118; courtship and copulation, 149, 150; dietary preferences impact on plant diversity, 241; digestive efficiencies, 70*t;* dispersal mode, 211–12; ear structure and function, 129–30; ectoparasites of, 377–78; enetic studies resources, 305*t;* food caching, 196; food selectivity and feeding, 192*t;* foraging behavior, 191; geographic distribution, 6–7; geographic partitioning levels, 310; gestation period, 152, 205*t;* Harderian gland function, 67–68; hearing frequency, 131*f;* hypercapnic response, 82; incisor microstructure relationship to function, 36; induced ovulation, 151; litter sizes, 207*t;* mandible size and shape, 39; mate detection using seismic signals, 148; morphological and behavioral development, 157*t;* non-shivering thermogenesis, 80*t;* olfactory center size, 117; pheromonal communication, 118; population density factors, 199; predation upon, 215; pup appearance at birth, 156; retinal development, 122; seasonality of breeding, 152; seismic signal detection, 132–33; seismic signal production, 131; seismic and vocal signals, 137*f;* sensory physiology, 65; sex ratio, 200*t;* soil quantities moved, 228; somatosensory organs, 66; temperature regulation, 78; temporal patterns of activity, 67; thermoneutral zones, 78; urine concentrat-

Nannospalax (mole-rats) (*cont.*)
    ing abilities, 90*t;* urine odor use, 118–19; ventilation rates, 82; water balance and, 91
  enamel color, 35
  foraging activities, 46
  force facilitating modifications to muscles, 44
  forefeet and hind feet, 24
  geographic distribution, 6–7
  head and neck muscles, 40
  head-lift digging, 30
  hind limbs structure, 48
  incisor procumbency, 33, 35
  incisor root placement, 34
  leucodon, 6–7
  molar structure intergeneric differences, 37
  olfaction, 27
  osteology and myology study resources, 32
  pelage and skin, 22
  pelvic bone, 47*f*
  pelvis shape, 48*f*
  pinnae, 23
  reproductive isolation, 315, 316
  scratch digging, 30
  sensory and communicative characteristics, 115*t*
  skull shape and size, 38, 39
  soil removal techniques, 31
  speciation mechanisms study, 313–14
  tails, 24
  urogenital structure study, 52
  urogenital system study, 51
  visual abilities, 122
  visual cortex size, 122–23
  visual system, 26
nematode parasites, 376, 377*t*
Notiomys edwardsii, 235
Notomys alexis, 90*t*

ocotillo *(Fouquieri splendens),* 241
Octodontidae. See also *Spalacopus*
  auditory bullae, 39
  burrow sharing, 263*t*
  digging apparatus elements, 29*t*
  diversification test feasibility, 358
  ectoparasitic arthropods of, 378*t*
  endoparasites of, 377*t*
  evolutionary history, 336
  food selectivity and feeding, 193*t*
  geographic distribution, 10, 337*f*
  incisor microstructure relationship to function, 36
  macroevolutionary patterns, 337*f,* 347–48
  three-taxon tests results, 352
  tree balance tests results, 352
  two-taxon tests of diversification, 351*t*
Odocoileus hemionus (mule deer), 236
Onychomys leucogaster (grasshopper mouse), 238
opossums *(Didelphis albiventris),* 215
ornate box turtles *(Terrapene ornata),* 237
Orthogeomys (pocket gophers). *See also* Geomyidae
  cherriei: litter sizes, 154, 207*t;* sex ratio, 200*t*
  geographic distribution, 9–10
  hispidus hispidus, 152
  incisor groove function, 36
  pelage and skin, 22
  seasonality of breeding, 152
Ovis canadensis (bighorn sheep), 236
owl predator activity, 213–14

Pappogeomys (pocket gophers). *See also* Geomyidae
  castanops: adult mortality, 209; aggressive behavior, 234; allopatric distributions due to soil preferences, 232; breeding season length, 206; competitive exclusion, 233; food selectivity and feeding, 195; geographic distribution, 9; gestation period, 151, 205*t;* life span, 161; litter sizes, 154, 207*t;* longevity, 203; seasonality of breeding, 152; sex ratio, 200*t;* vertical stratification and competition, 234
  digging mode and habitat soil relationship, 187
  geographic distribution, 9–10
  gymnurus, 233, 234
  incisor groove function, 36
  merriami merriami, 152
  seasonality of breeding, 152
  skull shape and size, 39
  tylorhinus, 214
Paraspidodera (nematode), 380
Penstemon grandiflorus, 243
Perognathus sp. (pocket mouse), 238
Peromyscus maniculatus (deer mice), 238
Pituophis catenifer (gopher snakes), 237
pit vipers predator activity, 215

TAXONOMIC INDEX 435

Pitymys pinetorum, 74*t*
pocket gophers. See also *Geomys; Megascaphus; Orthogeomys; Pappogeomys; Thomomys; Zygogeomys*
  bone size relations to force exertion, 42
  cospeciation with chewing lice: genetic divergence estimates, 383–84, 384*f*; parasite evolution acceleration hypothesis, 384–85; phylogeny studies, 381–83, 382*f*
  force facilitating modifications to muscles, 44
  geographic distribution, 9–10
  hind limbs structure, 48–49
  incisor groove function, 36
  pelvic bone, 47
  soil removal techniques, 31
Poecilogale albinucha (striped weasel), 215
prairie dog. See *Cynomys eudovicianus*
prairie rattlesnakes *(Crotalus viridis)*, 237–38
Prometheomys (mole-voles). See also Arvicolinae
  body size, 21
  claw morphology, 43
  digestive system study sources, 51
  digging behavior, 28
  ectoparasitic arthropods of, 378*t*
  enamel color, 35
  foraging activities, 46
  forefeet and hind feet, 24
  geographic distribution, 3
  glandular system studies, 52
  incisor groove function, 36
  incisor procumbency, 33, 34
  pelage and skin, 22
  pinnae, 23
  schaposchrikowi, 262
  scratch digging, 30
  secondary digging apparatus studies, 46
  soil removal techniques, 31
  tails, 24
  urogenital structure study, 52
  visual system, 26
pronghorn *(Antilocapra americana)*, 236
protozoa parasites, 377*t*
Pseudalopex gymnocercus (foxes), 214
Pseudapsis cana (snakes), 215
Pulex sinoculus (fleas), 379

Rattus
  auditory system, 27
  urine concentrating abilities, 90*t*
  reptiles in burrows, 238

Rhamphiophis oxyrhynchus rostratus (snakes), 215
Rhizomyinae. See also *Rhizomys; Tachyoryctes*
  chisel-tooth digging, 30
  diversification test feasibility, 358
  ectoparasitic arthropods of, 378*t*
  evolutionary history, 335
  eye size implications, 123
  foraging activities, 46
  geographic distribution, 7–8, 337*f*, 341, 342*f*, 343, 347
  macroevolutionary patterns, 337*f*, 341, 342*f*, 343, 347
  pelvic bone, 47
  pinnae, 23
  predation data lack, 215
  skull shape and size, 38
  soil removal techniques, 31
  two-taxon tests of diversification, 351*t*
  visual incisor displays, 121
  visual system, 26
Rhizomys (bamboo rats). See also Rhizomyinae
  auditory system, 27
  body size, 21
  claw morphology, 43
  enamel color, 35
  geographic distribution, 7–8
  glandular system studies, 52–53
  incisor procumbency, 35
  osteology and myology study resources, 32
  pelage and skin, 22
  secondary digging apparatus studies, 46
  skull shape and size, 38
root rats. See *Tachyoryctes*

Scalopus aquaticus (eastern moles), 238
Scaphiopus sp. (spadefoot toads), 237, 238
six-lined racerunners *(Cnemidophorus sexlineatus)*, 237
skunk predator activity, 215
snakes predator activity, 215, 237
spadefoot toads *(Scaphiopus* sp.), 237
Spalacinae. See also *Nannospalax; Spalax*
  BMR adaptations, 73
  bone size relations to force exertion, 42
  burrow configuration, 188
  captive animals studies resources, 260
  digging seasonality, 190
  diversification test feasibility, 358
  ecological attributes, 217*t*
  ectoparasitic arthropods of, 378*t*
  endoparasites of, 377*t*

Spalacinae (*cont.*)
  energy-conserving adaptations, 75
  evolutionary history, 335
  food locating ability, 197
  food resources, 191
  foraging activities, 46
  geographic distribution, 6–7, 334*f*, 337*f*, 343
  gestation period, 151–52
  hearing sensitivity, 65
  macroevolutionary patterns, 334*f*, 337*f*, 343, 347–48
  neuroendocrine studies, 66
  non-shivering thermogenesis, 80
  paraphyletic/polyphyletic relationships study resources, 321*t*
  pup appearance at birth, 155–56
  radiotelemetry study resources, 260
  seismic signal use, 134
  spatial system overlap, 261
  two-taxon tests of diversification, 351*t*
Spalacopus (coruro). *See also* Octodontidae
  auditory bullae, 39
  chisel-tooth digging, 30
  claw morphology, 43
  cyanus: acoustic signal production, 126; burrow design, 189*f;* dietary preferences impact on plant diversity, 241; energy budgets, 245; food locating ability, 197; food selectivity and feeding, 193*t;* geographic distribution, 10; group sizes, 268*t;* lack of seismic communication, 134; mound effects on plants, 243; social organization, 146–47; trapping methods, 259; trapping studies resources, 259; vocalizations, 127
  dispersal behaviors, 210
  enamel color, 35
  eye size implications, 124
  foraging activities, 46
  forefeet and hind feet, 24
  incisor procumbency, 33, 35
  molar structure intergeneric differences, 37
  pelage and skin, 22
  pinnae, 23
  sensory and communicative characteristics, 115*t*
  soil removal techniques, 31
  tails, 24
  visual acuity, 122
  visual system, 26

Spalax (mole-rats). *See also* Spalacinae
  geographic distribution, 6–7
  microphthalmus: burrow sharing, 262; dispersae, 210; longevity, 203
Spermophilus (ground squirrels), 2
  burrow occupation, 238
  studies of, 258
striped weasel predator activity, 215
sucking lice, 377–78
Sylvilagus auduboni (desert cottontails), 238

Tachyglossus aculeatus, 82
Tachyoryctes (root rats). *See also* Rhizomyinae
  acoustic signal production, 126
  chisel-tooth digging, 30
  claw morphology, 43
  comparative lung study, 50
  digging seasonality, 190
  enamel color, 35
  eye size implications, 123
  food selectivity and feeding, 192*t*
  foraging behavior, 191
  forefeet and hind feet, 24
  geographic distribution, 7–8
  head and neck muscles, 40
  hind limb structure, 48
  incisor procumbency, 35
  macrocephalus, 32
  mandible size and shape, 39
  osteology and myology study resources, 32
  pelage and skin, 22
  pelvic bone, 47
  pelvis shape, 48*f*
  population age structure, 202
  pup growth rates, 159
  ruandae, 159
  sensory and communicative characteristics, 115*t*
  soil removal techniques, 31
  splendens: courtship and copulation, 149; litter sizes, 207*t;* lung morphology, 82; mate detection using seismic signals, 148; morphological and behavioral development, 157*t;* mortality rates, 208; osteology and myology study resources, 32; population age structure, 202; population density, 198; predation upon, 215; pup appearance at birth, 156; seasonality of breeding, 152; seismic communication lack, 134; sex ratio, 200*t;* temperature regulation, 78, 79*t;* temporal

TAXONOMIC INDEX 437

patterns of activity, 67; thermal conductance, 77t
ulna modifications, 42
visual acuity, 122
visual system, 26
Talpa europaea (European mole), 86
Tamias striatus, 82
Tatera brantsii, 159
Terrapene ornata (ornate box turtles), 237
Thomomys (pocket gophers). *See also* Geomyidae
  absence of incisor grooves, 36
  acoustic signal production, 126
  body size, 21
  bottae: acid-base balance and hypercapnia, 88; aggressive behavior, 234; allopatric distributions due to soil preferences, 232; blood oxygen transport properties, 84, 85f, 86t; burrow occupation by other species, 237; competition avoidance, 232–33; courtship and copulation, 150; degree of sociality, 284; dietary overlap with surface herbivores, 236; enamel color, 35–36; energy budgets, 245; food locating ability, 197; genetic studies resources, 305t; gene tree inaccuracy, 318–19, 319f; geographic partitioning levels, 310–11; gestation period, 151, 205t; hybridization, 315–16; hypercapnic response, 82; isolation by distance genetic evidence, 306–7; juvenile dispersal, 162; life span, 161; litter sizes, 154, 207t; longevity, 203; male/female burrow distribution, 265; mark-recapture studies resources, 259; molecular genetic studies, 260; mortality from aggression, 209; mound effects on plants, 243; natal philopatry, 266; population density factors, 199; seasonality of breeding, 154; sex ratio, 200t, 201; sexual dimorphism, 147, 265; speciation mechanisms study, 314; temporal patterns of activity, 67; 2,3- DPG levels, 86; ventilation rates, 82; vertical stratification and competition, 234
  breeding season length, 206
  bulbivorus: genetic studies resources, 305t; litter sizes, 154; seasonality of breeding, 152
  chisel-tooth digging, 30
  claw morphology, 43
  courtship and copulation, 149

  courtship initiation, 146
  dietary preferences impact on plant diversity, 241
  digging mode and habitat soil relationship, 186, 187
  enamel color, 35
  forefeet and hind feet, 24
  genetic drift and natural selection observations, 308–9
  genetic studies resources, 303, 305t
  geographic distribution, 9–10
  geographic partitioning levels, 310
  head and neck muscles, 40
  hind limb structure, 48
  incisor procumbency, 33, 35
  lower incisor growth rate, 35
  monticola: food selectivity and feeding, 195; litter sizes, 154, 207t; longevity, 203; sex ratio, 202
  osteology and myology study resources, 32
  pelvic bone, 47
  pelvis shape, 48f
  predation of, 213–14
  secondary digging apparatus studies, 46
  sensory and communicative characteristics, 115t
  sexual dimorphism, 147, 265
  soil chemistry changes from, 229, 230
  soil quantities moved, 229
  talpoides: aggressive behavior, 234; allopatric distributions due to soil preferences, 232; basal metabolic rate, 74t; burrow occupation by other species, 237; competitive exclusion, 232–33; courtship and copulation, 150; courtship sounds, 127; dietary overlap with surface herbivores, 235–36; diet content and foraging behavior, 191; energy budgets, 245; food caching, 196; food selectivity and feeding, 193t; genetic studies resources, 305t; gestation period, 151, 205t; innate responses, 119; litter sizes, 154, 207t; longevity, 203; morphological and behavioral development, 157t; mound effects on plants, 243; pup appearance at birth, 156; pup growth rates, 159; seasonality of breeding, 152; sex ratio, 200t; thermal conductance, 77t
  townsendii, 305t
  ulna modifications, 42
  umbrinus: basal metabolic rate, 74t; genetic

Thomomys (pocket gophers) (*cont.*)
  studies resources, 305*t;* hybridization, 315–16; thermal conductance, 77*t;* vertical stratification and competition, 234
  upper incisor growth rate, 34
Thryonomyidae, 351*t*
tiger salamander *(Ambystoma tigrinum)*, 237
Tragopogon dubius, 241
tuco-tucos. *See Ctenomys*
Tyto alba (owl), 213–14

Urocyon cinereus (foxes), 213–14

voles, 3, 163. See also *Ellobius; Microtus* sp.; *Prometheomys*

water vole *(Arvicola terrestris)*, 3

zokors. *See Myospalax*
Zygogeomys (pocket gophers). *See also* Geomyidae
  competitive exclusion, 233
  geographic distribution, 10
  incisor groove function, 36
  trichopus, 233
  vertical stratification and competition, 234

Subject Index

acid-base balance, 87–88
acoustic receptors
   auditory range of taxa, 129–30, 130f
   inner ear structure, 128–29
   inner/outer hair cells, 129
   pinnae size and function, 127–28
acoustic signals
   prevalence in taxa, 126
   reception, 133
   vocalization ranges, 126–27, 128f
adaptations. *See also* coevolution; morphology
   allometric relationships, 393–95, 394f
   basal metabolic rate, 73
   digestive efficiencies, 70
   evolutionary biology: functionalist perspective, 391; triangle of interpretive thought, 390–91, 390f
   hypoxia and blood oxygen transport, 84
   hypoxia and hypercapnia, 81–82
   of individuals, 16–18
   metabolic scope, 75
   physiological, 88–89, 100–101
   reciprocal, 372
   thermoneutral zones, 78, 80
aerial plant parts in diet, 194
African geographic distributions of rodents, 8–9
allometry
   adaptations, 393–95, 394f
   of the rostrum, as a constraint, 393
alloparental care, 274–75, 280
anatomy, internal
   circulatory system, 50–51
   digestive system, 51; coprophagy, 71; efficiencies in, 69–71, 70t; fermentation temperature, 71–72, 72f; microbial fermentation, 71
   glandular system, 52–53
   respiratory system, 50–51
aridity-food distribution hypothesis, 279–80
auditory system
   acoustic receptors: auditory range of taxa, 129–30, 130f; inner ear structure, 128–29; inner/outer hair cells, 129; pinnae size and function, 127–28
   hearing physiology, 65
   morphology, 26–27
   signals for communication, 113

basal metabolic rate (BMR)
   advantages of low, 73, 75–76
   burrow location and, 187
   definition of, 72–73
   effects of acclimation, 73
   energy-conserving adaptations, 75
   reduction relative to body size, 73
   table of, 74t
behavioral ecology, 409–11
blood pH. *See* acid-base balance
blood transport
   hemoglobin affinity for oxygen, 84, 86–87
   hypoxia adaptations, 84
   oxygen transport properties, 85f
   taxa responses to hypoxia/hypercapnia, 84
BMR. *See* basal metabolic rate
body form and size, 21, 187
body temperature, 77–78, 79t
Bohr effect, 85f, 87f

boundary populations model, 313, 313f
breeding. *See* reproduction
burrow systems. *See also* habitat selection
  architecture, 188
  characteristics of, 11–12
  coevolution and: inquilines, 374, 376t; parasites, 376–79, 377t, 378t; term definitions, 373–74; transients, 375, 375f
  energetic costs of, 68, 279
  environment in, 20–21
  foraging activities and, 188–89
  habitat plant distribution and, 246–47
  location determinants: digging behavior, 186–87; metabolic rate and body size, 187; soil and vegetation, 185–86
  occupation by other species, 237–39
  population ecology and, 183
  seasonality of excavations, 190
  spatial relationships among individuals, 262, 262f, 263t, 264
  temperature variations in, 247
  temporal variation in, 247

calcium homeostasis
  absorption, 95–96, 95f
  deposition in teeth, 96
  function of, 91–92
carbon dioxide. *See* hypoxia and hypercapnia
cardiac responses to hypoxia and hypercapnia, 83–84
cheek pouches, 23
chemical communication
  chemical reception, 113, 117–18
  functional significance: colony cohesion, 120; Harderian gland and, 119; perineal gland and, 119–20; reproductive division of labor, 120; scent trails, 120–21; urine odor use, 118–19; VNO role in interactions, 119
  non-dominance of, 121
  pheromones, 116
  receptors system, 27–28
chisel-tooth digging
  calcium requirements, 96
  claw morphology, 43
  elements of apparatus, 33f
  morphology, 30
  skull shape and size, 38–39
  soil types in habitats, 186
chromosomal evolution and speciation study, 359

circadian rhythms
  absence of photoperiodic entrainment, 67–68
  mechanisms for, 66–67
  melatonin secretion and, 68
  patterns of activity, 67
  visual system development and, 123
circulatory system, 50–51
claws, 30, 43
coevolution. *See also* adaptations; morphology
  burrow associates and: inquilines, 373–74, 376t; parasites, 376–79, 377t, 378t; term definitions, 373–74; transients, 375, 375f
  cospeciation: benefits of study, 379–80; pocket gophers and chewing lice, 381–85, 382t; tuco-tucos and nematode parasites, 380–81
  criteria for, 370–71
  history and definition, 370
  operational definition, 372
  phylogenetics role in study, 371–72
  subterranean lifestyle influences on, 372–73
  summary and future directions, 385–86
communication, 114, 115t, 116f
  chemical: chemical reception, 117–18; functional signficance, 118–21; non-dominance of, 121; pheromones, 116
  development under constraints, 113–14
  functions of, 111–12
  need for further study, 409
  overview, 136–37
  research challenges, 112
  scope of discussion, 112–13
  summary and future directions, 137–38
  tactile: courtship and mating behavior, 125; input related to burrow life, 124–25; tail function, 125
  vibrational: acoustic receptors, 127–31, 130f; acoustic signals, 126–27, 128f; described, 125–26; seismic receptors, 132–33; seismic signals, 131–32, 284; signal differences between taxa, 133–36
  visual: acuity relationship to foraging habits, 122; eye size implications, 123–24; retinal development, 122; visual cortex size, 122–23; visual display use, 121

SUBJECT INDEX 441

community ecology
    competitive exclusion mechanisms: aggressive behavior, 233–34; allopatric distributions due to, 232; avoidance of, 232–33; spatial separation, 233; sympatry, 232
    conclusion and future directions, 248–49
    indirect effects and impact: energy flow and nutrient cycling, 244–45; floral and rodent abundance correlation, 242; plant production, 239–40; plant species diversity, 240–41; soil improvements, 239; vegetation biomass changes, 240; vegetation succession, 243–44
    interspecies interactions: with organisms in burrows, 237–39; with predators, 236–37; with subterranean mammals, 235; with surface herbivores, 235–36
    intraspecific aggression considerations, 231
    keystone species concept, 227
    scope of discussion, 227–28
    soil characteristics due to rodent effects: chemistry changes, 229–30; impact of, 230–31; nutrient enrichment, 230; quantity moved, 228–29
    spatial and temporal variations, 246–48
    vertical stratification and, 234
competitive exclusion mechanisms
    aggressive behavior, 233–34
    allopatric distributions due to, 232
    avoidance of, 232–33
    spatial separation, 233
    sympatry, 232
conduction, 77
cooperative breeding, 278
coprophagy, 71
cospeciation. *See also* adaptations; coevolution
    benefits of study, 379–80
    gopher-louse system: genetic divergence estimates, 383–84, 384*f;* parasite evolution acceleration hypothesis, 384–85; phylogeny studies, 381–83, 382*f*
    phylogenetic perspective of, 413
    tuco-tucos and nematode parasites, 380–81
courtship. *See also* reproduction
    copulation and, 148–50
    mating behavior: described, 286–88; location detection, 148; multiple matings, 150; social organization, 146–47; tactile communication during, 125; vocalizations, 126

demography study needs, 411–12
diet
    digestion: coprophagy, 71; efficiencies in, 69–71, 70*t;* fermentation temperature, 71–72, 72*f;* microbial fermentation, 71; system, 51
    foraging activities and: caching, 196–97; energy acquisition, 69, 279; food locating ability, 197–98; food resources and behavior, 191, 192*t*–93*t*, 194; generalists vs. specialists, 190, 195; selectivity, 194–95
    preferences, impact on plant diversity, 240–41
    pregnant/lactating female needs, 97
    water balance and, 91
differentiation, 404–5
digging activities and apparatus
    breaking up soil methods. *See* soil breaking up methods
    burrow systems and, 186–87
    elements of apparatus, 29*t*
    energetic costs and sociality, 279
    habitat selection and, 186
    primary components: element comparison, 29*t;* study resources, 32; teeth (*see main entry*)
    process/functional units, 28
    secondary components: foraging activities, 46; hind limbs, 48–49; pelvis and axial skeleton, 47–48, 48*f;* study resources, 46
    soil removal methods, 31
dispersal
    age and sex of individuals, 210–11
    modes of, 211–12
    mortality and, 208
    rates of, 212–13
    reproduction and, 161–62
diversification
    rates tests: lineages through time, 350*f*, 353, 354*t;* three-taxon, 350*f*, 351–52; tree balance, 350*f*, 352; two-taxon, 349, 350*f*, 351, 351*t*
    test results: fossil and molecular data integration, 357; generic diversity patterns, 355–57; species diversity, 353–55; summary, 357–58, 358*t*

ears. *See* auditory system
ecological correlates of sociality, 278–80
ecological effects, 246–48
ectoparasite faunas, 377–79, 378*t*

endoparasites, 376–77
energy flux
  basal metabolic rate: advantages of low, 73, 75–76; definition of, 72–73; effects of acclimation, 73; energy-conserving adaptations, 75; reduction relative to body size, 73; table of, 74t
  body temperature, 77–78, 79t
  burrowing influence on, 64
  digestion: coprophagy, 71; efficiencies in, 69–71, 70t; fermentation temperature, 71–72, 72f; microbial fermentation, 71; system, 51
  energetic costs of burrow life, 68, 279
  food caching and energetic conservation, 196–97
  foraging and energy acquisition, 69, 279
  non-shivering thermogenesis, 80
  nutrient cycling and ecosystem function, 244–45
  physiological challenges of habitat, 64
  reproduction and, 97–98, 98f
  thermal flux, 76–77, 76f, 77t
  thermoneutral zones, 78, 80
environment in burrows, 20–21
Eurasian and African geographic distributions of rodents, 3, 6, 7–8
evolutionary biology
  adaptationist program: functionalist perspective, 391; triangle of interpretive thought, 390–91, 390f
  behavioral and evolutionary study needs, 409–11
  communication study needs, 409
  constraints and structuralist perspective: allometric relationships and, 393–95, 394f; definitions of, 392–93; edges of a biological variation, 395–96
  diversification rates tests: lineages through time, 350f, 353, 354t; three-taxon, 350f, 351–52; tree balance, 350f, 352; two-taxon, 349, 350f, 351, 351t
  diversification test results: fossil and molecular data integration, 357; generic diversity patterns, 355–57; species diversity, 353–55; summary, 357–58, 358t
  genetics and natural selection: molecular evolution and, 400–401; molecular studies, 401–2
  history of taxa, 334–36
  macroevolutionary patterns from fossil records: Arvicolinae, 337f, 338, 340f; Bathyergidae, 337f, 343, 345f; Ctenomyidae, 337f, 343–44, 346f; Geomyidae, 337–38, 337f, 339f; methods used, 336–37, 337f; Myospalacinae, 337f, 338, 341, 341f; Octodontidae, 337f, 347; Rhizomyinae, 337f, 341, 342f, 343; Spalacinae, 334f, 337f, 343; taxa overview, 347–48
  metapopulation data, 411–12
  morphology study needs, 408
  niche concepts, 11–12, 397–99
  phylogenetic framework, 413–14, 414f
  phylogenetic studies prospects, 358–59
  physiology and reproduction study needs, 408–9
  scope of discussion, 298–99, 333–34
  speciation mechanisms, 405–7
  species concepts: differentiation, 404–5; paraphyly, 404; reproductive isolation, 403–4
  summary and future direction, 360
evolutionary ecology study needs, 409–11
exaptation, 371
eyes. See visual sensory system

fixation index, 303
food sources. See also diet
  caching, 196–97
  habitat selection and, 185
foraging
  apparatus for, 46
  burrow systems and, 188–89
  caching, 196–97
  energy acquisition and, 69, 279
  food locating ability, 197–98
  food resources and behavior, 191, 192t–93t, 194
  generalists vs. specialists, 190, 195
  selectivity, 194–95
  visual acuity and, 122
forbs in diet, 241, 246
forefeet and hind feet
  characteristics of, 24
  scratch digging using forelimbs, 29–30, 42–43
forelimbs
  myology: fiber structure, 45; force facilitating modifications, 43–44; quantity of muscles, 44–45
  osteology: claws, 43; fore and rear feet, 42–43; muscle and ligament attach-

ments, 41–42; size and shape, 40–41, 41f; ulna modifications, 42
scratch digging using, 29–30, 42–43
fossil records macroevolutionary patterns
Arvicolinae, 337f, 338, 340f
Bathyergidae, 337f, 343, 345f
Ctenomyidae, 337f, 343–44, 346f
Geomyidae, 337–38, 337f, 339f
methods used, 336–37, 337f
Myospalacinae, 337f, 338, 341, 341f
Octodontidae, 337f, 347
Rhizomyinae, 337f, 341, 342f, 343
Spalacinae, 334f, 337f, 343
taxa overview, 347–48
fossorial definition, 1–2
F statistics use, 303

gas exchange
characteristics in burrows, 63–64
ventilatory responses, 82, 83f
genetic structure. *See* population genetic structure
geographic distributions
in Africa, 8–9
Arvicolinae, 3, 337f, 338, 340f
Bathyergidae, 8–9, 337f, 343, 345f
Ctenomyidae, 10, 337f, 343–44, 346f
in Eurasia and Africa, 3, 6, 7–8
Geomyidae, 9–10, 337–38, 337f, 339f
map, 6f
Myospalacinae, 3, 6, 337f, 338, 341, 341f
in North America, 9–10
Octodontidae, 10, 337f
Rhizomyinae, 337f, 341, 342f, 343, 347
in South America, 10
Spalacinae, 334f, 337f, 343
table of, 4f–5f
gestation, 151–52, 205, 205t
glandular system, 52–53
gopher-louse cospeciation system
genetic divergence estimates, 383–84, 384f
parasite evolution acceleration hypothesis, 384–85
phylogeny studies, 381–83, 382f

habitat selection. *See also* burrow systems
burrow location determinants: digging behavior, 186–87; metabolic rate and body size, 187; soil and vegetation, 185–86
climatic influences, 184
energetic feasibility and, 185

microclimatic conditions, 63
physiological challenges of: absence of light, 63; energy flux, 64; gas exchange, 63–64; intertaxa study of, 64
Harderian glands, 67–68, 119
head and neck
force required for digging, 37
myology, 39–40
osteology: adaptations, 37–38; mandible size and shape, 39; skull shape and size, 38–39
head-lift digging
calcium requirements, 96
claw morphology, 43
elements of apparatus, 33f
morphology, 30–31
skull shape and size, 38–39
hearing physiology, 65. *See also* auditory system
heat exchange, 76–77
hemoglobin affinity for oxygen, 84, 86–87
hind limbs, 48–49
hybridization, 315–17
hypoxia and hypercapnia
acid-base balance, 87–88
blood transport: differences between taxa, 84; hemoglobin affinity for oxygen, 84, 86–87; hypoxia adaptations, 84; oxygen transport properties, 85f
cardiac, 83–84
conditions, 83
morphological adaptations, 81–82
ventilatory responses, 82, 83f

incisors. *See also* teeth
chisel-tooth digging and, 30
curvature, 34
enamel color, 35–36
groove function, 36
growth rates, 35
head-lift digging and, 30
microstructure relationship to function, 36
procumbency, 33, 34–35
root placement and length, 33–34
size and shape functionality, 32–33
inquilines, 373–74, 376t
interspecific competition, 398–99
isolation by distance model, 306–7

keystone species concept, 239
K-selection, 202

LH concentrations, 165–66, 166f
life history traits
  breeding season length, 206
  gestation period, 151–52, 205, 205t
  longevity, 161, 203
  mortality: aggressive interactions and, 209; dispersal and, 208; reproductive contact and, 208; types of, 208
  natality, 206–7, 207t
  sexual maturity, 204–5
  sociality and, 280
  survival vs. production, 202
life span, 161, 203
light absence in burrows. *See also* visual sensory system
  physiology influenced by, 63
  vitamin D_3 synthesis and, 92
lineages through time tests, 350f, 353, 354t
litter sizes, 154–55, 206–7, 207t
luteinized unruptured follicles (LUFs), 165–66

mammary glands, 52
mandible, 39
mating systems. *See also* reproduction
  described, 286–88
  location detection, 148
  multiple matings, 150
  social organization, 146–47
  tactile communication during, 125
  vocalizations, 126
melatonin and circadian activities, 68
metabolism. *See* basal metabolic rate (BMR)
metapopulations model, 313, 313f, 411–12
mineral homeostasis
  calcium absorption, 95–96, 95f
  calcium deposition in teeth, 96
  calcium function, 91–92
  vitamin D_3: deficiency in, 94, 96; functions of, 92; hormone regulation, 93f, 94; hydroxylation process, 92; metabolism of, 92–94, 93f
Miocene period, 356
modes of communication. *See* communication
molars, 37. *See also* teeth
morphology. *See also* adaptations; coevolution
  digging activities (*see also* soil breaking up methods): process/functional units, 28; soil removal methods, 31
  digging apparatus, primary components: element comparison, 29t; study resources, 32; teeth *(see main entry)*
  digging apparatus, secondary components: foraging activities, 46; hind limbs, 48–49; pelvis and axial skeleton, 47–48, 48f; study resources, 46
  external: body form and size, 21; burrow environment, 20–21; forefeet and hind feet, 24; pelage and skin, 22–23; pinnae, 23, 26; tails, 23–24
  forelimbs: myology, 43–45; osteology, 40–43, 41f; scratch digging using, 29–30
  head and neck: force required for digging, 37; myology, 39–40; osteology, 37–39
  hypoxia and hypercapnia adaptations, 81–82
  internal anatomy: circulatory system, 50–51; digestive system, 51; glandular system, 52–53; respiratory system, 50–51; urogenital system, 51–52
  modifications due to burrow system, 12
  scope of discussion, 19–20
  sensory systems: auditory, 26–27; chemomechano receptors, 27–28; enhanced aspects of, 25; visual, 25–26
  study needs, 408
  summary and future direction, 49, 53–54
mortality
  aggressive interactions and, 209
  dispersal and, 208
  reproductive contact and, 208
  types of, 208
myology and osteology study resources, 32

natality, 206–7, 207t
natal philopatry, 266, 269
natural selection and genetic drift, 308–9
neuroendocrinology physiology, 66
neutral mutation theory of molecular evolution, 400
niche, subterranean
  behavioral and evolutionary ecology, 11–12, 397–99
  burrow system characteristics, 11–12
  coevolution influenced by, 372–73
  molecular evolution and, 400–401
  sociality role, 280
nitrogen intake, 244–45
non-shivering thermogenesis, 80
North American geographic distributions of rodents, 9–10

olfaction
  chemical reception and, 117
  learned responses to mediation, 119
  as mode of communication, 113
  morphology, 27
  physiology, 65–66
  vomeronasal organ, 117–18
Oligocene period, 356
osteology and myology study resources, 32
ovulation, 150–51, 163–64. See also reproduction
oxygen transport. See blood transport

paleontology. See evolutionary biology
parallel cladogenesis, 379
paraphyletic/polyphyletic relationships study resources, 321t
paraphyly, 404
parasites, 374, 376–79, 377t, 378t
paternity, 150
pelage and skin
  functional differences in, 22–23
  pelage presence at birth, 155–56
pelvis and axial skeleton, 47–48, 48f
perineal gland, 119–20
peripheral isolation, 319
pheromones
  chemical signaling with, 116, 121
  Harderian glands and, 119
  reproductive axis inhibition, 163
  VNO role in detection, 117–18
photosensitivity. See visual sensory system
phylogenetic framework
  coevolution and, 371–72
  evolutionary biology, 413–14, 414f
  studies prospects, 358–59
physiology
  adaptations, 88–89, 100–101
  circadian rhythms: absence of photoperiodic entrainment, 67–68; mechanisms for, 66–67; melatonin secretion and, 68; patterns of activity, 67
  energy flux: basal metabolic rate, 72–76, 74t; body temperature, 77–78, 79t; digestion, 69–72, 70t; energetic costs of burrow life, 68; foraging and energy acquisition, 69; non-shivering thermogenesis, 80; thermal flux, 76–77, 76f, 77t; thermoneutral zones, 78, 80
  habitat challenges: absence of light, 63; energy flux, 64; gas exchange, 63–64; intertaxa study of, 64

  hypoxia and hypercapnia responses: acid-base balance, 87–89; blood transport, 84–87; cardiac, 83–84; morphological adaptations, 81–82; ventilatory, 82, 83f
  mineral homeostasis: calcium absorption, 95–96, 95f; calcium deposition in teeth, 96; calcium function, 91–92; vitamin D_3 deficiency tolerance, 96; vitamin D_3 metabolism, 92–94, 93f
  need for further study, 408–9
  neuroendocrinology, 66
  reproduction: energetic costs of, 97; heat flux changes, 97–98, 98f
  scope of discussion, 62
  sensory: hearing, 65; somatosensory organs, 65–66; vision, 63, 65
  study needs, 99–100
  water flux: loss mechanisms, 89–90, 89f; shifts in water balance, 91; urine concentrating abilities, 90
PI (progression index), 117
pinnae
  acoustic receptor role, 127–28
  auditory function, 26
  size and shape, 23
plants. See also diet
  aerial plant parts in diet, 194
  gopher effects on composition, 246–47
Pliocene period, 348, 356
population biology. See also population genetic structure
  emergent properties, 180
  interrelated features of, 301
  study needs, 180–81
population ecology
  dispersal: age and sex of individuals, 210–11; modes of, 211–12; rates of, 212–13
  foraging: caching, 196–97; food locating ability, 197–98; food resources and behavior, 191, 192t–93t, 194; generalists vs. specialists, 190, 195; selectivity, 194–95
  habitat selection: burrow location determinants, 185–88; burrow systems, 188–90; climatic influences, 184; energetic feasibility and, 185
  influences on, 183–84
  life history traits: breeding season length, 206; gestation period, 205, 205t; longevity, 203; mortality, 207–9; natality,

SUBJECT INDEX

population ecology (*cont.*)
  206–7, 207*t*; sexual maturity, 204–5; survival vs. production, 202
  need for metapopulation study, 411–12
  population structure: adult sex ratio, 200–201, 200*t*; age distributions, 201–2; density and distribution, 198–99
  predation: degree of surface activity and, 215; differences between taxa, 215–16; lack of data on frequency of, 213; predator types, 213–15
  summary, 217–18
  variations in responses to habitat, 216
population genetic structure. *See also* population biology
  extinction-recolonization dynamics, 307–8
  factors effecting, 302–3
  F statistics use, 303
  genetic differentiation: isolation by distance model, 306–7; need for further study, 411–12
  genetic divergence: chromosome races, 314; patterns of, 355–57; population models, 313, 313*f*; speciation favoring conditions, 312–13; speciation mechanisms study, 314
  genetic drift and natural selection, 308–9
  gene trees: horizontal gene transfer complications, 319; new species formation geographic scenarios, 319–20, 320*f*, 321*t*; phyletic relationships inaccuracies, 317–19, 318*f*, 319*f*
  hybridization effects, 316–17
  natural selection and: molecular evolution, 400–401; molecular studies, 401–2
  reproductive isolation, 315–16
  scope of discussion, 301–2
  spatial array of variations: geographic partitioning levels, 310–11; study resources, 309
  spatial configuration influences, 306–7
  species boundaries, 320–21
  species formation and geography: allopatric speciation effects, 312; sympatric and parapatric models, 311–12
  study resources, 303–4, 305*t*
  summary and future direction, 322–24
predation
  degree of surface activity and, 215
  differences between taxa, 215–16
  lack of data on frequency of, 213
  predator avoidance, 119

  predator types, 213–15
  social behavior influenced by, 281
preputial gland, 52–53
procumbency
  allometry of the rostrum, 393
  morphology, 33, 34–35
progression index (PI), 117
pseudo-coevolution, 371
pup ontogeny
  appearance at birth, 155–56
  development, 156, 157*f*–58*f*
  growth rates, 156, 159–60, 159*t*
  sparring among pups, 156

Quaternary period, 356

reciprocal adaptation, 372
recognition species concept, 405
reproduction
  breeding: seasonality of, 152, 154; season length, 206
  courtship and copulation, 148–50
  dispersal, 161–62
  division of labor and chemical communication, 120
  gestation, 151–52, 205, 205*t*
  isolation reinforcement hypothesis, 315–16
  lifetime reproductive success, 161, 203
  litter sizes, 154–55, 206–7, 207*t*
  longevity of breeding, 161
  mating systems: described, 286–88; location detection, 148; multiple matings, 150; social organization, 146–47; tactile communication during, 125; vocalizations, 126
  mortality and, 208
  need for further study, 408–9
  ovulation, 150–51, 163–64
  photoperiod signal lack, 146
  physiology, 97–98, 98*f*; energetic costs of, 97
  pup ontogeny: appearance at birth, 155–56; development, 156, 157*f*–58*f*; growth rates, 156, 159–60, 159*t*; sparring among pups, 156
  reproductive isolation and species concepts, 403–4
  scope of discussion, 145
  sexual dimorphism, 147, 265
  sexual maturity, 160, 204–5

social suppression of: breeding season, 167, 169; familiarity avoidance, 167; females vs. males, 169; induced infertility, 163, 168*f*; inhibition of ovulation, 163–64; nonreproductive female characteristics, 165–66, 166*f*; pheromone effects, 163; reproductive male characteristics, 164–65
social systems and, 146–47
specialization and sociality: lifetime reproductive success, 271–72; multi-breeding females, 270; reproductive skew, 163, 272; single-breeding females, 269–70
study needs, 408–9
summary and future direction, 169–70
reproductive skew, 163, 272
respiration and ventilatory responses, 82, 83*f*. *See also* hypoxia and hypercapnia
respiratory system, 50–51. *See also* hypoxia and hypercapnia
Robertsonian fans, 406–7
r-selection, 202

salt loading and water balance, 91
scent marking
perineal gland and, 119–20
trails, 120–21
urine odor use, 118–19
scratch digging
calcium requirements, 96
claw morphology, 43
elements of apparatus, 33*f*
morphology, 29–30
soil types in habitats, 186
seismic signals
communication between species, 131–32, 284
differences between taxa: biological parameters and signal types, 134; predictions for taxa, 135; social classification and signal use, 134–35
mate location detection, 148
receptors, 132–33
sensory systems
morphology: auditory, 26–27; chemomechano receptors, 27–28; enhanced aspects of, 25; visual, 25–26
physiology: hearing, 65; somatosensory organs, 65–66; vision, 65
sex ratio, 200–201, 200*t*
sexual dimorphism, 147, 265
skull shape and size, 38–39

snouts and digging activities, 30
social systems
logistic challenges to study: captive animals studies, 260; lack of direct observation opportunity, 258–59; mark-recapture studies, 259–60; molecular genetic studies, 260; trapping methods, 259
mating systems, 286–88
scope of discussion, 258
sociality: alloparental care, 274–75, 280; behavioral specialization, 275–76; ecological factors, 278–80; group size influences, 267–68, 268*t*; group structure, 269; inbreeding and breeder replacement, 272–73; life history traits, 280; phylogentic influences, 281–82; predation influences on, 281; reproductive specialization, 269–72; social complexity elements, 266–67; solitary vs. social taxa, 266; subterranean niche role, 282–83; trends among bathyergids, 276–77, 277*f*; trends among ctenomyids, 277–78, 277*f*
solitary taxa, 283–85
solitary vs. social taxa, 285
spatial relationships among individuals: burrow sharing, 262, 262*f*, 263*t*, 264; degree of overlap, 261–62, 262*f*; lack of intermediate spatial systems, 264–65; natal philopatry, 266; nest sharing, 264; sexual dimorphism influences, 265; social system dynamics and, 260
summary and future direction, 288–89
suppression of reproduction: breeding season, 167, 169; familiarity avoidance, 167; females vs. males, 169; induced infertility, 163, 168*f*; inhibition of ovulation, 163–64; nonreproductive female characteristics, 165–66, 166*f*; pheromone effects, 163; reproductive male characteristics, 164–65
soil breaking up methods. *See also* digging activities
chisel-tooth digging: calcium requirements, 96; claw morphology, 43; elements of apparatus, 33*f*; morphology, 30; skull shape and size, 38–39; soil types in habitats, 186
head-lift digging: calcium requirements, 96; claw morphology, 43; elements of apparatus, 33*f*; morphology, 30–31; skull shape and size, 38–39

448 SUBJECT INDEX

soil breaking up methods (*cont.*)
   scratch digging: calcium requirements, 96; claw morphology, 43; elements of apparatus, 33*f*; morphology, 29–30; soil types in habitats, 186
soil characteristics due to rodent effects
   chemistry changes, 229–30
   impact of, 230–31
   nutrient enrichment, 230
   quantity of soil moved, 228–29
soil removal, 31
somatosensory organs
   physiology, 65–66
   seismic signal reception, 133
South American geographic distributions of rodents, 10
spatial relationships among individuals
   burrow sharing, 262, 262*f*, 263*t*, 264
   degree of overlap, 261–62, 262*f*
   lack of intermediate spatial systems, 264–65
   natal philopatry, 266
   nest sharing, 264
   sexual dimorphism influences, 265
   social system dynamics and, 260
species concepts
   in evolutionary biology: differentiation, 404–5; paraphyly, 404; reproductive isolation, 403–4
   mechanisms of speciation, 405–7
sweat glands, 52

tactile sensory system
   communication: courtship and mating behavior, 125; input related to burrow life, 124–25; tail function, 125
   senses: physiology, 65–66; uses of, 27
   signals as mode of communication, 114
tails, 23–24, 27, 125
taxa table, 4*t*–5*t*
teeth
   allometry of the rostrum, 393
   calcium deposition in, 96
   incisors: chisel-tooth digging and, 30; curvature, 34; enamel color, 35–36; groove function, 36; growth rates, 35; head-lift digging and, 30; microstructure relationship to function, 36; procumbency, 33, 34–35; root placement and length, 33–34; size and shape functionality, 32–33
   molars, 37

temperature regulation
   body, 77–78, 79*t*
   fermentation, in digestion, 71–72, 72*f*
   thermal flux, 76–77, 76*f*, 77*t*
   thermoneutral zones, 78, 80
   variations in burrows, 247
three-taxon tests, 350*f*, 351–52
torpor, 80
transients in burrows, 374, 375, 375*f*
tree balance tests, 350*f*, 352
triangle of interpretive thought, 390–91, 390*f*
tunnels. *See* burrow systems; digging activities
two-taxon tests, 349, 350*f*, 351, 351*t*
2,3-diphosphoglycerate, 84, 86–87

ulna modifications for digging, 42
Umwelt, 113
urine odor use, 118–19
urogenital system, 51–52, 90

vegetational succession, 243–44
ventilation in burrows, 63–64
ventilatory responses to hypoxia and hypercapnia, 82, 83*f*
vibrational communication
   acoustic receptors: auditory range of taxa, 129–30, 130*f*; inner ear structure, 128–29; inner/outer hair cells, 129; pinnae size and function, 127–28
   acoustic signals: prevalence in taxa, 126; reception, 133; vocalization ranges, 126–27, 128*f*
   described, 125–26
   effectiveness of, 114
   seismic receptors, 132–33
   seismic signals, 131–32, 284
   signal differences between taxa: biological parameters and signal types, 134; predictions for taxa, 135; social classification and signal use, 134–35
vibrissae, 27, 124
vicariant allopatry, 319
visual sensory system
   communication: acuity relationship to foraging habits, 122; eye size implications, 123–24; retinal development, 122; visual cortex size, 122–23; visual display use, 121
   light absence in burrows: physiology influenced by, 63; vitamin D_3 synthesis and, 92

as mode of communication, 113
  morphology, 25–26
  physiology, 63, 65
vitamin D_3
  deficiency in, 94
  functions of, 92
  hormone regulation, 93*f*, 94
  hydroxylation process, 92
  metabolism of, 92–94, 93*f*
vocal signals
  communication between solitary species, 284
  in courtship and copulation, 126, 149

vomeronasal organ (VNO)
  chemical reception and, 117–18
  innate responses to mediation, 119
  role in interactions, 119

water flux
  loss mechanisms, 89–90, 89*f*
  shifts in water balance, 91
  urine concentrating abilities, 90
Wright, Sewall, 303